PIMLICO

247

TIME TO KILL

Paul Addison teaches history at the University of Edinburgh where he is Director of the Centre for Second World War Studies. He is the author of *The Road to 1945*, acclaimed in *The Times Literary Supplement* as 'a landmark in the writing of contemporary history', of *Churchill on the Home Front*, described by David Cannadine in the *Observer* as 'the best one-volume study of Churchill yet available', and of *Now The War is Over*, a study of the ideals and realities that went into the making of post-war Britain. All three books are available in Pimlico.

Angus Calder was formerly Reader in Cultural Studies and Staff Tutor in Arts with the Open University in Scotland and is now a freelance writer. He read English at Cambridge and received his D.Phil from the School of Social Studies at the University of Sussex. He was Convener of the Scottish Poetry Library when it was founded in 1984. His other books include *The People's War, The Myth of the Blitz* (both available in Pimlico) and *Revolutionary Empire*. He lives in Edinburgh.

TIME TO KILL

The Soldier's Experience of War in the West
1939–1945

Edited by Paul Addison and Angus Calder
with a Foreword by Len Deighton

PIMLICO

PIMLICO

An imprint of Random House
20 Vauxhall Bridge Road, London SWiV 2SA

Random House Australia (Pty) Ltd
20 Alfred Street, Milsons Point, Sydney
New South Wales 2061, Australia

Random House New Zealand Ltd
18 Poland Road, Glenfield
Auckland 10, New Zealand

Random House South Africa (Pty) Ltd
Endulini, 5A Jubilee Road, Parktown 2193, South Africa

Random House UK Ltd Reg. No. 954009

First published by Pimlico 1997

1 3 5 7 9 10 8 6 4 2

Selection, Preface and Editorial Matter © Paul Addison and Angus Calder 1997
Foreword © Len Deighton 1997
For copyright of contributors see page xiv

Papers used by Random House UK Limited are natural,
recyclable products made from wood grown in sustainable
forests. The manufacturing processes conform to the
environmental regulations of the country of origin

Printed and bound in Great Britain by
Mackays of Chatham PLC

ISBN 0-7126-7376-8

Contents

FOUR: SEVEN ARMIES IN EUROPE

FIVE: TWO MEMOIRS

SIX: REFLECTIONS

Foreword

Len Deighton

Time to Kill is the most stimulating collection of military history that I have yet encountered. Here is John Keegan on the subject of the demon drink:

> the Sherwood Forester Brigade on 1 July 1916, [attracted disapproval only when] the men in the front trench got the second wave's rum as well and were roaring drunk at zero hour. With the First World War evidence in mind, I was less surprised to discover that full barrels of spirits were rolled into squares at Waterloo or to read testimony that the French before Agincourt had spent the night sozzling all the wine they could get.

Time to Kill is a collection of talks given at Edinburgh University during the conference: 'The Soldier's Experience of War 1939–1945'. The speakers included many of the most respected authorities on military history and it was my pleasure and privilege to attend.

One recurring theme was that the Second World War was – in parts and places – little different from the First World War. G. D. Sheffield persuasively argues that this was so, and the idea is echoed by John Ellis. It was also made by Richard Holmes – a splendid writer and lively speaker too – in his account of 'five armies' fighting in Italy.

In contrast to the rigours of front-line action S. P. MacKenzie's talk dealt with service closer to home. His research shows us that the TV programme *Dad's Army* closely resembled the real thing. But MacKenzie reminds us that most Home Guards assembled in towns, rather than bucolic places like Walmington-on-Sea. About 112,000 Home Guards were assigned to the anti-aircraft guns and searchlights. Furthermore there was an officially unrecognised 'Mum's Army' of clerks, typists and cooks.

The contribution women made to the fighting forces of the Soviet Union remains equally unrecognised. No less than three-quarters of the Red Army's drivers were women. All nurses were women and so were over 40 per cent of the doctors, surgeons, paramedics and medical orderlies. Front-line medics were not non-combatant. Many of them were armed, and were expected to fight when not attending casualties. Female medics suffered a casualty rate comparable with that of the Red Army infantry. One of them explained:

> We carried men twice or three times our own weight. On top of that, we carried their weapons, and the men themselves were wearing greatcoats and

boots . . . we would throw a man off and go and get another one . . . And we did this five or six times during an attack.

Women medics with tank units were expected to accompany the tanks on foot, or ride on them, exposed to the shells and bullets. Some women managed to wangle assignments to tank units as drivers or mechanics. One woman became a lieutenant-colonel in a T-34 tank unit, and was awarded the Hero of the Soviet Union medal. A married couple donated their savings on condition they be crewed together in a tank. The wife became the tank commander and her husband the mechanic. There were women combat pilots, while other women became snipers; one claiming 309 kills. Reina Pennington's meticulously researched chapter about 'offensive women' is a startling revelation to those of us who previously supposed that war was a man's profession.

Jeremy Crang points out that the majority of British soldiers spent the greater part of the war in 'camps and depots across the country polishing their brasses and wondering why they were there'. In the British army squalor and discomfort were endemic. The money paid to families was inadequate, and many soldiers, unused to being treated so badly, suffered a crippling lack of self-respect. Army morale reports eavesdropped on a post-Dunkirk army cold, hungry and paid less than one pound a week. I found these innermost thoughts of Britain's wartime servicemen heartbreaking. 'The bell tent in which we sleep leaks . . . Haven't had clean clothes for 3 weeks or bath . . . If we want a wash or shave we have to go half a mile to wash in a *ditch*.' The reports reveal a strained relationship between officers and rankers during the war, providing an interesting insight on the post-war election of a socialist government.

I came to the end of each contribution with the earnest hope that it would soon be expanded into a book. As well as solid research and basic history, there are insightful anecdotes and asides that bring revelations and questions too. Ian S. Wood's piece, about Ireland and the war, shows that potent political decisions and intriguing religious dimensions overlap, as when a chaplain hears confessions aboard the ship taking His Majesty's Irish Guards to rescue the Dutch royal family.

If you didn't know that black African troops fought in Europe in the Second World War you are not alone. The bureaucrats in Whitehall denied all knowledge of them and didn't include them in the celebrations that marked the fiftieth anniversary of VE Day. And while we need no reminders about the extensive commitment and valour of Indian army units, it is as well to remember that, by accident or design, the weapons and equipment they were given were very much below standard.

Angus Calder's account of the invasion of Crete cleverly reconciles Evelyn Waugh's crotchety perceptions with the conflicting accounts of the Allied defender's courage and steadiness as recounted in the words of another eye-witness; the New Zealand subaltern Dan Davin. Calder wonders whether these contrasting impressions represent the difference between a decaying imperial ruling class and a rising nation. Perhaps the whole war was a spasm that defined old and new on each side, making post-war realignments inevitable.

Readers looking for hard facts will find plenty in Terry Copp's chapter about battle exhaustion. In passing he mentions that during the fighting in Normandy, 60 per cent of Allied tank losses resulted from a single shot from a 7.7-cm or 8.8-cm gun, and two-thirds of all tanks hit 'brewed up'. I stopped reading at that point while I tried to grasp this terrible conjunction of statistics. Copp's account of the psychological cost of the Rhineland battle shows that the 1945 fighting was not a final gasp of a defeated enemy. The Anglo-American armies fought a desperate and bloody campaign against a well-armed, ruthless army which did not stop fighting, even after all hope had gone. And then only when its generals gave the order. Hew Strachan in his 'The Soldier's Experience in Two World Wars: Some Historiographical Comparisons' suggests that the execution of 15,000 German soldiers may have played a part in this motivation.

The motivation of the German armed forces is the subject of Jürgen Förster's chapter. I strongly agree with his theory that the German military tradition was strengthened during the time of Ebert and the 'Weimar Republic'. Thus there was a continuing ideology which seamlessly connected Wilhelm's regimented Germany to that of Hitler's Third Reich. Germany's preparation for the war of 1939, such as building the warships *Deutschland, Admiral Graf Spee* and *Admiral Scheer*, started before Hitler took power. And it was Article 48 of the Weimar Constitution that made it possible to commit citizens to concentration camps without trial. Jürgen Förster points out that Hitler's army did not need political commissars. Colonel-General Walther von Brauchitsch told his officers that they must provide not only a tactical but a (Nazi) political leadership for their men. And the men of Hitler's armed forces all wore the badge of his political party on their uniform, as did those of Stalin.

But one question about Germany's army still puzzles me. I know the men of the Anglo-American armies regarded their military service as a temporary and inconvenient interruption to their real civilian life. The German army seemed mostly made up from men who had put aside the prospect of becoming a civilian again. Letters home revealed love and affection and even home-sickness, but the German soldier was not impatient to get home in the desperate way that we see in the attitudes of the British and the Americans. Was it pride that gave the German his willingness to soldier on? Was it the low status of the peacetime soldier in Britain, the United States, Canada, New Zealand and Australia, and so on, that made these men so keen to get the war over, get out of uniform and go home? 'Never again. I won't even join hands once I get out of this lot,' I heard men saying. But they stayed, they served, they fought and they won.

Which was the fourth largest army fighting on the Allied side? This question would have baffled me until I heard Mark Axworthy talking about the

Romanian army. For when, in August 1944, Romania defected to the Allies, their army qualified for that position. Axworthy is mostly concerned with the siege of Odessa in August and September 1941. The Axis forces pushed the Red Army back, leaving the Russian naval base at Odessa surrounded but supplied by sea. During the fierce and lengthy battle the Romanians suffered 98,000 casualties. Odessa was proclaimed a 'hero city' along with Leningrad and Stalingrad. The widespread illiteracy of the survivors, and post-war communist censorship, have contrived to make the Odessa battle one of the least-known great engagements of the war.

The collapse of the Soviet Union has provided interesting documentation for historians. John Erickson told us: 'reports on all levels, morale, crime and punishment, executions, vodka deliveries have recently seen the light of day'. What kind of army was it? How did the crude Red Army of 1941 transform itself into the flexible force of the amazing 1944 battles, and the comparatively sophisticated fighting machine that closed around Berlin? No one knows better than John Erickson, and this chapter – 'Red Army Battlefield Performance: the System and the Soldier' – is a fascinating and revealing one.

I'm running out of space, but so far I have mentioned only a small part of the contents of this valuable book and only a few of its truly remarkable contributors. I would like to emphasise that the chapters I've mentioned have been a random sampling. But let me plead space for John Ellis, perhaps the most provocative of all our modern military historians. Our impressions of the Second World War are dominated by the memories of men who never knew its horrors: 'for when the talk in the factory and the saloon bar is almost all of binges in Brussels or the travails of Lincolnshire ground crew and Bari store clerks, it becomes more and more difficult to try and describe trench-foot, battle exhaustion, soiled trousers and eviscerated mates.'

Readers will enjoy an advantage over those who attended the Edinburgh University conference. Many of the talks were presented simultaneously and we had the distressing dilemma of choosing between them. Another advantage that comes from having the talks in book form is the valuable notes and references that accompany each chapter. One thing you can't have is the question-and-answer session that rounded off each of the talks. Sometimes the exchanges were sharp. We learned to be less surprised at members of the audience getting to their feet to say: 'I may be able to add something to your research. I was with the regiment when it went into action that day...'.

Edinburgh was an amazing gathering. I wish you could have been there. But this book is the next best thing.

Len Deighton,
Portugal, 1996

Preface

This book arose out of a conference on 'The Soldier's Experience of War, 1939–45' organised by the History Department at the University of Edinburgh on 22–4 September 1995. We were surprised to learn, as organisation proceeded, that no one had ever set up an academic conference on this subject before. We were 'at the sharp end' of the New Military History and distinguished university scholars were enticed by that. At the same time, we attracted the early and enthusiastic support of Len Deighton, an excellent scholar but not a university man, and we knew that our subject would be of acute interest to veteran soldiers (some of whom did indeed attend) and other members of the general public.

Some of the contributions here stand much as they were delivered at the conference – the first three are of this type, and may convey some of the friendly atmosphere in which people exchanged doubts and self-criticism as well as unusual information. Other papers have been altered or expanded. Two have been 'added'. Dr Hamish Henderson and Dr Steve Weiss were important contributors to discussion. We are grateful to *Cencrastus* magazine (Edinburgh) for permission to reproduce a piece by Dr Henderson which first appeared in its pages (Issue 53, Winter 1995/6). Dr Weiss's memoir is the full version of an article which first appeared in a shorter form in *War Studies Intelligencer* (vol. 1, no. 3, 1992), a production of the King's College London War Studies Department.

Time to Kill knowingly and willingly reflects the wide range of different perspectives brought to bear by contributors to the conference. Their remit was simply to help us appreciate the many factors which determined what soldiering was like for the private soldiers, NCOs and junior officers of the major combatant armies in Europe and North Africa. Such military factors as army organisation, strategy and tactics, weaponry and operational conditions are legitimately here. But so too are national traditions, whether the ancient one of the 'fighting Irish', the disciplinary legacy of Frederick the Great to the Wehrmacht, or a New Zealand ethos still emergent. So too are the effects of fascism and communism on the troops who fought beneath their banners. *Time to Kill* of course cannot touch on all the national armies, let alone all the units within those armies, which took part in the war in the west. But we think that the general area is opened up by our different contributors in their different ways.

The agenda is set, as of right, by John Keegan and John Ellis. Keegan's *The Face of Battle* (1976) transformed the way historians, ourselves included, thought about military matters, an effect reinforced by John Ellis's *The Sharp End* (1980). Our organisation thereafter is geographical – Britain, Commonwealth (incidentally, Eire remained within the Commonwealth till 1949), then Europe, with its American and Russian invaders. After the memoirs of Dr Henderson and Dr Weiss, with their vivid strands of individual experience, we end with essays reflecting on the historiography of the Second World War, and the ways in which it both affects and is affected by broad social and cultural forces.

Though each contributor has a separate tale to tell, a number of themes run through the book as a whole. First, the extraordinary diversity of the factors – from the quality of NAAFI tea to the inspirational presence of 'the big man' – which affected the morale and fighting capacity of the ordinary soldier. Second, the importance of doing justice to the 'Poor Bloody Infantry' by looking at the history of armies from below, instead of viewing events exclusively through the eyes of war lords and commanders. Third, the extent to which the front-line soldiers of North Africa and Western Europe faced conditions remarkably similar to those of the Western Front during the First World War. Fourth, the contrast between the relatively conventional warfare in Western Europe, and the war of almost unlimited barbarism conducted by Nazi Germany and Stalinist Russia in the east. Fifth, the great potential of the common ground which military and social historians have begun to establish. Most of the contributors to this book are military historians by profession. But in reconstructing the 'soldier's experience' they have moved half-way to social history. The editors (like some of the contributors) come from the opposite direction. They are social historians fascinated by the ways in which histories of nationality, religion, class and gender, converge with the history of mass militarisation in the twentieth century.

We must thank, first, Len Deighton for supplying a generous Foreword which enables us to dispense with the often long and ponderous introduction made by editors to books of this kind; then fellow members of the organising committee of the conference, Terry Cole and Jill Stephenson, who mightily helped Paul Addison and Jeremy Crang sustain the main burden of organisation.

But the whole project would never have started without a munificent offer of sponsorship from David McWinnie of Lamancha Productions, who could well be called its 'onelie begetter'. For additional financial support we also gratefully acknowledge the assistance of Vickers plc, the Arts Faculty Group Research Fund of the University of Edinburgh, and the Development Office of the University of Edinburgh.

To Professor John Erickson, whose encouragement and advice at every stage of this venture were crucial to its success, we owe a very special debt of gratitude. For the individual support they so generously gave, many thanks are due to Professor H. T. Dickinson, Dr F. D. Dow, Dr Diana Henderson,

Professor Arthur Marwick, Professor Maurice Larkin, Professor J. S. Richardson, Professor George Shepperson and Mr George O. Sutherland. Nor would the conference have succeeded without the organisational skills of Mr Ronnie Galloway and Ms Marnie Simpson of Unived Technologies Ltd, to whom we express our very great appreciation and thanks.

Paul Addison
Angus Calder

Centre for Second World War Studies,
University of Edinburgh

To every thing there is a season, and a time to every purpose under the heaven;

A time to be born, and a time to die; a time to plant, and a time to pluck up that which is planted;

A time to kill, and a time to heal; a time to break down, and a time to build up;

A time to weep, and a time to laugh; a time to mourn, and a time to dance;

A time to cast away stones, and a time to gather stones together; a time to embrace, and a time to refrain from embracing;

A time to get, and a time to lose; a time to keep, and a time to cast away;

A time to rend, and a time to sew; a time to keep silence, and a time to speak;

A time to love, and a time to hate; a time of war, and a time of peace.

Ecclesiastes 3:1–8

Part One
Setting the Agenda

Towards a Theory of Combat Motivation

John Keegan

'A rational army would run away.' Let me repeat that quotation. 'A rational army would run away.' Of all sources of quotations I have lost in my life that is the one whose loss I regret most. Who wrote it? Montesquieu, I sometimes confidently say. The curtness and intellectuality are Montesquieuan enough; but war was not really Montesquieu's subject. Who else then could it have been? Diderot? I suspect we are looking for a Frenchman. Might it have been Voltaire, that supreme self-proclaimed rationalist? But Voltaire's real targets were kings and priests, not soldiers. He may well have thought them not worth his shafts of wit, since they were unlikely to read what he wrote. Selection and maintenance of the aim was a principle of war of which Voltaire would have approved.

Perhaps someone can put me out of my misery. No matter if not. I am confident we all endorse the point the unnamed author is making, which is that, while war may be rational, combat is not. It defies one of the strongest of all human instincts, that of self-preservation. Animals may be reactively impelled to combat when violation of distance mobilises their fight rather than flight reflexes; but animals do not reason. Man also is subject to the fight/flight reflex: but in him reason moderates its operation. Why?

The correct question at this point, it might be objected, is not why but how? If human biology could lay bare the workings of the limbic system – the seat of aggression in the human brain – we would understand not only the motivation to combat but also to war, crime, imperialism, politics, the battle of the sexes and, indeed, the struggle of life itself. The truth is, however, that the human biologists cannot. They have succeeded in demonstrating, under laboratory conditions, that electrical stimulation of the limbic system will produce aggressive response; but warriors do not exist in laboratories. More important, human biologists recognise and demonstrate that the limbic system, located in the lower brain, communicates with and is influenced by the higher brain; two media of communication – there may be others – are hormones and chemicals. While the process of communication is not yet fully clarified, it seems clear that calculation in the higher brain – certainly of risk but perhaps of more complex self-interest or even of values such as honour and compassion – works to moderate the instinctual response of the lower brain to fear and anger. Man, in short, is not an animal, nor a machine. He is something else altogether.

What is he? The simplest thing to say about *homo sapiens sapiens* – you and me, who have existed in the form we personify for only about 40,000 years in the 5 million in which some identifiable ancestor of ours has existed, or about two to three weeks in a nominal humanoid year – is that he is a social being. He lives in a group and he relates to others, sexually, emotionally, intellectually, economically and politically. It is because he is a social being that, having abandoned, I think rightly, the search for an answer to the question *how* human reason influences human instinct in the awful business of combat, we are entitled to ask *why*?

Twenty years ago, or a little more, I set out to find answers to that question. It was not a deliberate enquiry. The proposal I made to the publisher, in a letter recently unearthed by my literary agent, was to write a book on the theme 'What happens in a battle?' I genuinely wanted to know. I was then a young lecturer at the Royal Military Academy Sandhurst, surrounded by veterans of the Second World War still in the prime of life, and teaching teenagers who sought to follow in their footsteps. I was supposed to tell them about battles. My difficulty – apart from the almost insuperable one that I had not been in battle myself – was that the military history I read – and, even more, the military history I had been taught at Oxford, Clausewitz and the rest of it – simply did not coincide with the talk I overheard among the veterans in the officers' mess. 'Poor little Gurkhas. They just weren't up to the Hermann Goering Division.' 'I never smoked before Alamein – got through a tin of fifty Players in my first day in the OP [Observation Post].' 'Dennis was just too brave at Vaagso – no wonder he got shot through the head' – Dennis, I might add, was alive and well and commanding a Sandhurst company. 'I told the soldiers that fifty feet of standing wheat stopped a bullet, they believed me. I'd made it up.' 'We shot Japanese trying to swim the Irrawaddy like fish in a barrel, horrible, really.' 'The battalion commander wouldn't give the order to assault – when I touched him he'd gone to sleep, pure nerves.' 'The anti-tank platoon commander ran away – he was a Canadian replacement officer, nothing against Canadians, of course.' 'I took one of my subalterns by the lapels and shook him like a naughty child. He was killed in the next minefield.' 'Just the sort of officer you want in a battle. Made witty remarks. Turned it into a sort of social occasion.' 'One of my parachutists said he wanted to go sick on the startline. I said, "Let's both go sick." He forgot about it. I recommended him for the Military Medal later.' 'Lost all my officers in the Bois de Bavent – liked it really, meant I had to do everything myself.' 'I'd kill anybody in a battle if they're in the wrong uniform – stretcher-bearers, cooks, bugger the Geneva Convention.' 'I gave my Jocks rum after the battle, not before. They liked the reward. I didn't want them drunk.'

There was nothing about rum in Clausewitz, or about commanding officers having nervous breakdowns, or about one sort of warrior being better than another, or about officers bullying their subordinates. The warrior in Clausewitz was a sort of cypher – a being subject to fear and fatigue and capable of bravery – but faceless, unindividualistic and asocial, for all that Clausewitz had to say about the enthusiasm of popular armies. Conventionally narrative military history was little better: episodes and anecdotes might single out the occasional individual; but the range of feeling allowed to the man in the ranks was unrealistically narrow. The military historian's man in uniform bore no resemblance to the man in the street. He was a being without family or friends, without future or past, without values, good or bad, except for the incidental flash of courage or self-sacrifice. The traditional military historian's *homo pugnans* seemed a being different and separate from *homo oeconimicus, amans, ludens* or even just plain *sentiens*.

Friendship with soldiers, immersion in their company, told me that someone must be wrong: either Clausewitz and the conventional military historians must be misleading me, or else I was being misinformed by the evidence of testimony, my own ears. As I liked and trusted the soldiers I knew, I decided that it was not *they* who were misinforming me. When I sat down to write the book which eventually became *The Face of Battle* I therefore decided to examine such evidence as there was of why men behaved in battle as they did – what their motivation to combat was – not in the light of what other writers had to say, or more usually did not have to say, on the matter, but through the hints and allusions I had picked up from people whose exposure to combat was part of their life experience.

Under that pressure, the evidence began to yield a volume of revelation far larger than I had expected. To take but one example: drink. We all know about the rum ration in the trenches. I had carelessly assumed that it was a specific against cold and wet feet, the military equivalent of Nightnurse or Lemsip. Repetitive references to the rum issue before going over the top alerted me to the idea that it was, on the contrary, doled out as a specific against fear, literally Dutch courage, and that the practice attracted disapproval only when, as with the Sherwood Forester Brigade on 1 July, 1916, the men in the front trench got the second wave's rum as well and were roaring drunk at zero hour. With the First World War evidence in mind, I was less surprised to discover that full barrels of spirits were rolled into squares at Waterloo or to read testimony that the French before Agincourt had spent the night sozzling all the wine they could get. It was with some trepidation that I hinted on the association between drink and the quelling of fear to military audiences later. To my great surprise, I evoked nothing but endorsement. An infantry company commander who had survived the ordeal of the Mareth Line stood up to say that he and his company sergeant-major had been able to manage nothing for several days but a little tinned fruit, washed down by heavy intakes of rum. An air marshal survivor of Bomber Command's campaign over Germany told an audience to whom I had said I was sure that drink was not as freely used by fliers as it was by soldiers that the fourteen-pint-a-night rear gunner was a well-known Bomber Command figure. He got back from a mission to drink himself insensible and took off on the next still anaesthetised by his hangover against the terror with which his lonely station in a Lancaster afflicted him.

This raised questions about the effectiveness of the second counterweight I identified: coercion. Clearly the authorities were turning a blind eye to an indulgence in drink which must have threatened operational efficiency. The authorities are not usually so indulgent. 'Command', we are always being lectured about, and 'leadership', and we are often invited to reflect on the difference between the two; I will have something to say about leadership later. What we used never to be told about, and still rarely are, is enforcement. 'You can lead a horse to water but you can't make him drink' is a neat proverbial summary of the shortcomings of both leadership and command in battle. When the killing starts, when men are confronted by the realisation that to go further forward, to expose more of their bodies than is exposed already, entails a likelihood of death, both command and leadership risk failing. It is at that point – a mentally defined but physically identifiable time within the killing zone – that other measures become necessary. What are those measures? Historically we can identify a number: the sergeant with his half-pike pushing against the backs of the rear rank to hold the line in place in gunpowder warfare; the calling forward of cavalry to block the line of the infantry's potential flight, again in gunpowder warfare; the policing of communication trenches in trench warfare, where flight above ground was intrinsically danger-ous; the action of strong-minded officers, weapon in hand, in all warfare; and the threat of execution for desertion in the face of the enemy in a majority of organised armies in the modern age. 'Kill or be killed' is the logic of battle – to which military law adds the rider, 'Risk being killed by the enemy or else risk being killed by your own provost-marshal', an unenviable but ominous choice.

Coercion, even if softened by tolerance of chemical palliatives, is an arid means for authority to achieve its ends. Could there be some other emollient, I asked myself? It was clear that there was. If self-preservation is the strongest of human impulses, self-interest and its satisfaction follow close behind. Self-enrichment through combat has a long history. Women were a warrior's reward in the most primitive societies of which we have knowledge and they remained a valued prize among the nomad predators who terrorised settled civilisation far down into recorded history; the Mongols and the Turks aroused the horror they did because of their tribal habits of accumulating trains of women seized from the defeated quite as much as for their cruelty and destructiveness. Slaves, who were not necessarily women, were also an impor-tant capital asset yielded by success in combat, to be kept for personal or corporate use but, with the rise of commerce, increasingly to be sold on for profit; Western European outrage over the massacre of Chios in 1821 at the outbreak of the Greek War of Independence was largely caused by slave-taking by freelance interlopers from the mainland, out to enrich themselves while winning the Sultan's approval by putting down rebellion.

The capture of human loot was the cruellest outcome of the urge to self-enrichment through combat; animal stealing, probably an even older habit than enslavement, looks positively benign by comparison, as does looting – the appropriation of portable wealth – in its more conventional sense. The opportunity to loot depends, of course, on availability, which increased sharply as the range over which armies campaigned grew; baggage trains and war-chests offered bonanzas to the winning side, some directly to the individual, later indirectly through the distribution of prize money. Naval warfare, which long overlapped with the practice of piracy, promised particularly high returns, because of the large and highly concentrated value of a captured ship and its contents. We should remember, however, that the distribution of prize money, or the official administration of looting, as at Peking in 1860, remained a practice even of civilised armies well into the nineteenth century; while the ransom system may be seen in retrospect as a mediated form of the ancient habit of enslavement of captives.

By the end of the nineteenth century, however, looting had effectively lost its motivating force. The wealth was simply not there on the battlefield for the taking any longer. I often think that we should regard the institution of awards for bravery – the Legion of Honour, the Victoria Cross, the Medal of Honour – which originated at the beginning of that century as officialdom's effort to find a token substitute for booty's increasing rarity and eventual disappearance. Such awards, after all, sometimes attracted a pension, initially of some value. The success of the substitution was so marked, however, that attached pensions came to lose their point. We have only to recall that the current Victoria Cross pension of £100 a year, which the government, not without opposition from holders, plans to raise to £1,000 a year, was initially fixed at £10 a hundred and fifty years ago, evidence that the immaterial value of the award detached itself from the monetary inflation rate many years ago. Symbolic as opposed to real inducement has come to supply one of the most powerful motivations of combat performance ever known.

Inducement: I introduce the word deliberately. Reflecting on *The Face of Battle* some years after it was written, I began to see what it was about. I am no friend of deconstruction theory; I do not believe that authors need critics to decode their texts for the reader, or for themselves. Authors are not simple-tons, however simple-minded they may often feel themselves to be. Never-theless, I think we have all experienced a fuller realisation of what it was that we were trying to say only after a difficult piece of work has been completed; the work would, indeed, not have been worth doing were we not wiser after the effort than before. So it was with *The Face of Battle*: all my discoveries about drink and loot and punishment suddenly conceptualised themselves into three words: inducement, coercion and narcosis. I was delighted by the simplicity of the revelation. Having started merely with the desire to investigate and describe the reality of combat, I had stumbled on a total theory of combat motivation. Men fought partly because they were forced to do so, and partly because there were rewards for fighting – which might be material but could be symbolic or emotional, the emotional including self-esteem or the esteem of close comrades, friends, family, state and nation – and they suppressed the reflexive antipathy to the risks involved, when necessary, with alcohol or drugs. The theory, I say, needed a little refinement: the concept of coercion needed to embrace the idea of moral coercion – the fear of shame through show of cowardice foremost; but, with refinement, it seemed valid and universal.

Everyone enjoys the construction of a theory; nowadays it is about the only achievement which brings prestige in academic life. So, for a while, I was very pleased with myself indeed, and even more pleased with the triad of coercion, inducement and narcosis. Then, one day at the University of North Carolina, I heard someone at the other end of the seminar table observe, in the most good-natured way, that no doubt coercion, inducement and narcosis were important and perhaps principal ingredients of motivation to combat, but that they seemed a pretty good explanation also of how most people get through life itself. He knew a good few university departments that ran on coercion, inducement and narcosis and come to think of it, not a few marriages that worked by those mechanisms also.

I joined in the laugh; but, in time, I came to see the point. Recently, I have even come to welcome its having been made. The latest line of defence established by the military Leninists seeking to protect their claim that Clausewitz defined the nature of war once and for all is that he demonstrates how it is determined by the operation of reason, unreason and chance – or, as a variety of writers, including Michael Howard and Henry Summers, have reified it, the trinity of government, people and army. Now viewing war as an interplay between reason, unreason and chance sounds impressively conceptual; it has the ring of academic correctness. With only the briefest reflection, however, we can see that, as with the Keegan triad, the analysis is too general. It is not wrong: reason, unreason and chance do determine the nature of war; but so do they the workings of the market, the process of politics, the act of artistic creation, the path of intellectual discovery, love, marriage and life itself, both as experienced by the multitude and the solitary individual.

That reflection is not something against which to rail or protest. We all know perfectly well that, outside the hard sciences, there are no universally valid theories of exact explicative value – not in economics, not in political science, not in sociology, not in anthropology. Even in science, as Karl Popper has argued so persuasively, there are no indestructible theories, only hypotheses. Why should strategy be any different? It, of course, is not. The Clausewitzian trinity ought to be recognised for the hypothesis, vague and overgeneral, that it is; I have abandoned the Keegan triad as nothing more than a starting point and am now looking for new directions in which to go.

What suggests itself? I am increasingly interested by the anthropological concept of the 'big man', noted in Polynesian warrior society, who transcends the hierarchy and settled order of island society to acquire wealth, wives and followers by force of personality. The 'big man' seems to me a key figure in the way battles work. I feel confident that there are some in the audience, who know armies, navies or air forces from the inside, who will recognise the sort of person I mean. He is not necessarily a man of high rank or of any rank at all. Authority may disapprove of him and his comrades may even dislike him. He is nevertheless, at the risk of an oxymoron, the person who brings combat alive. He has counterparts in many other activities – the star without whom the film is a flop, the diva without whom an opera is only a recital, the trader who can take a business from breaking even to making a killing – forgive the analogy. Brigadier John Durnford-Slater, in his memoir of commando warfare, describes the type:

> Corporal Lofty King was a hard fellow and very hard with his men; he didn't give a damn if he knocked a man down. Sometimes I told him he was being too rough. Lofty would say, 'It's good for them, Colonel.' He would mean it and believe it. He genuinely enjoyed fighting and looked happiest, indeed inspired, in battle. In the field he was kinder to his men, as if it were a sort of release to him.

Sandhurst colleagues in the audience will recall other Lofty Kings, notably Desmond Lynch, D.C.M., Irish Guards, who had also unmistakably enjoyed fighting. He had strangled a German hospital orderly with his bare hands – 'I hadn't got anything against the poor little fellow' – in his urge to get back to his unit and to combat after having been wounded and captured, and he frightened his comrades quite as much as he did the enemy. I once asked a senior Irish Guards officer if it was true that Desmond had put a guardsman on a charge for being blown up by a shell. 'Quite true,' he answered approvingly, 'the guardsmen were very impressed.' Desmond was only a company sergeant-major when he made his name at the Battle of the Bou in Tunisia. Years later he was heard complaining for several weeks at Sandhurst that an official account, recently published, had not done him justice. 'I was the star of the battle,' he insisted, 'the star.' I will not labour the analogy with Hollywood or grand opera, but combat, which takes place in a theatre of war, is indeed a sort of performance and, however elaborate the *mise en scène*, however expensive the props, however distinguished the producer and director, it will flop without its stars.

Why do big men – Robert Maxwell was one in the business world, General Ratko Mladic is patently another – exert the influence they do? They are usually not nice (though Desmond Lynch, a wild beast on the battlefield, is a social lamb, who is now a devoted prison visitor with a particular concern for the welfare of wife-killers), while they quite often instigate disaster, as Maxwell did and Ratko Mladic may prove to have done as well. They have that power over other men, none the less, that great sporting stars in major team games exert. Let us introduce another concept: mimicry. I toyed with substituting 'conformism', and believe conformism is a major force in the human response to warfare; we probably all remember Lytton Strachey's description of his fellow conscripts at the recruiting station in 1916 who were – I quote from memory – 'brave enough to risk their own lives but not brave enough to protest against having to do so' – as he, a conscientious objector, saw himself to be. Conformism, however, is a neutral behavioural attitude. Mimicry is an active one. I believe it is the active aspect at which we should be looking. Mimicry explains a great deal.

In any competitive or dangerous activity, those who are seen best to meet its challenge set the standard which others, not necessarily voluntarily or consciously, emulate. They copy. They copy in all sorts of ways. They copy their opponents, if their opponents are strong and dangerous, wearing their skins if they are hunters, almost always adopting part or the whole of their uniforms if they are victorious in war: who will write the monograph on the transmission of the Sam Browne belt to almost every army in the world after 1916, including the Red Army and the Reichswehr? They copy, and this is the relevant point, their comrades. The Lofty Kings and Desmond Lynches, far more than the Lees or Grants or Wellingtons or Zhukovs, are those who energise the human raw material with which combat is conducted. They neither lead nor command; leadership and command, those set texts of staff college curricula, may be actual dead ends in the study of war. The Lofty Kings and Desmond Lynches *are*. They exist. They function. Their existence and their functioning explain a great deal of what we need to know about motivation to combat.

Big men, however, have their limitations. Combat mobilises profound emotions, which my triad – coercion, inducement, narcosis – goes some way to controlling and directing. Fear is a principal among such emotions, and the triad helps to explain how fear is contained. Combat, however, also plumbs deeper, into the realms of cruelty, frenzy and fantasy, which feed and are fed by each other. Men can behave disgustingly in combat, killing everything that moves, animal as well as human, when the frenzy is aroused and it is no accident that an undergrowth of semi-obscene literature of combat is found and bought in the sort of shop which purveys the real literature of pornography also. It can be found in a more respectable context: the most recent translation of the *Iliad* describes what the heroes of Troy did to each other in the sort of details which would occupy the attention of a serial killer – and that material invaded the Western consciousness nearly 3,000 years ago.

Raw killing revolts. It is not surprising, therefore, that – finally – I introduce another motivation to combat which I would add to coercion, inducement, narcosis, big man stardom, mimicry and the mechanistic impulses of cruelty, frenzy and fantasy. It is that of honour. Colin Lucas, Master of Balliol, the distinguished historian of the French Revolution, told me recently that his current interest is in revolutions' search for a means to 'good violence': the painless purge, the passionless executioner. The search is fruitless, of course: violence always taints. From as early as we have records, nevertheless, we can identify the figure of the honourable warrior. He is, though he may consciously or unconsciously use the 'big man', his antithesis. He seeks to practise good violence on the battlefield, through the ethic of the fair fight against his own like on the other side. His values are internal, they stand in moral opposition to inducement, coercion and narcosis, basing themselves on the principles of respect for the weak and of risk-giving only for equal risk-taking. The exemplar of the honourable warrior finds his personification today in the product of the military academy, though the knight of chivalry, the mameluke, the Chinese chariot warrior of the Spring and Autumn period and the officer of *ancien régime* Europe would be at home in his company.

Honour, though, is only an influence in combat, however much it is to be applauded. It is not a determinant, less still a motivation. What these motivations are has still to be discovered. Science may tell us one day, though I doubt it. In the meantime scholars must plug away. They will make little headway in university departments, or libraries, or even in the most ample archives. They must go elsewhere, nearer the action. Strategic and war studies have failed as university subjects because they have indeed been taught *in* universities by people who have been nowhere but universities. They know nothing of the realities of their subject, nor can they; in consequence, their students learn nothing either. Those of us who, by contrast, have taught in military academies or been members of the armed forces or have perhaps even been to a war know how lucky we are: some present qualify under several of those headings. Their students are lucky also. Still, there is no need to feel downcast, under-privileged or excluded if your world hitherto has been bounded solely by a university perimeter. There is much you can do to repair the deficiencies; join the Territorial Army, take a sabbatical at Sandhurst or even go to war. There are plenty of them and the trip is easy – a few hundred pounds for the airfare, a few hundred deutschmarks for living expenses. Your local newspaper – the *Western Daily Press*, the *Liverpool Echo*, the *Oban Times* – will be glad to give you an accreditation. The trip will be worth every minute of the time spent away from your books. You will learn some invaluable truths. You may even come back equipped to push us a shade further towards a convincing theory of combat motivation.

Reflections on the 'Sharp End' of War

John Ellis

I begin with two shards of autobiographical speculation. My first book on front-line life was *Eye-Deep in Hell*, a study of the (mainly British) infantry experience on the Western Front in the First World War. Its central thesis, that life in the trenches was ghastly, was of course monumentally unoriginal, being in fact part of the very bedrock of British social history between the wars and since. Yet for almost sixty years no one had sought to provide a considered *overview* of the front-line experience, a fact which I still find remarkable. (Though, of course, there had been some good anthologies of front-line life and the famous unit history, *The War the Infantry Knew* [1938], had explicit ambitions to record more than mere battalion peregrinations.)

In fact, my original notion with *Eye-Deep in Hell* had been to write a comparative book, in which the sufferings of the infantryman on the Western Front would be contrasted with what I assumed to have been the much less onerous conditions experienced by fighting troops during the Second World War. In almost every respect, I felt sure, the lot of the soldier of 1939–45 would prove to have been far less demanding than that of the Tommy, *poilu*, *Landser* or doughboy. Actual research, of course, soon disabused me, though interestingly it took my literary agent and myself two to three years to convince a publisher that a book about the front line in the Second World War could include much material with dramatic, popular impact.

As personal anecdote all this is of limited interest but it does, I suggest, throw up a couple of important general questions. First, why had the trench experience of 1914–18 not been subjected to comparative analysis and, second, why was the popular perception of the combat soldier's life in the Second World War so wide of the mark?

For an answer to the first question does seem pertinent to the study of *both* world wars, in that it concerns the status of oral and memoir accounts within military history. Up until recently that status was fairly low. Military history was written very much like the sports pages of the history syllabus as a whole, being largely concerned with *ad hoc* decision-making, personalities, technique and zero-sum outcomes. The perceptions and instincts of *commanders* were always the focus of attention, especially when much of the history was written by these commanders themselves or their peers and intimates, and the experience of the ordinary soldier, as a sentient human being, was deemed pretty much irrelevant to the overall contest. Reflect, for example, that until quite recently the only illustrations of importance in battle and campaign studies were the maps, in which thousands of individuals are compacted into compliant red and blue rectangles. (Now the reverse is true of course, which prompts the ultra-determinist notion that the new penchant for front-line studies might merely reflect the relative costs of maps and much cheaper photographs.)

But for a great many years most military accounts saw it as pretty much a cerebral exercise in which victory went to the commander who was best able to devise superior stratagems (rather like Baldrick's 'cunning plan', a late flowering of the Sun Tzu approach to warfare) and best able to surmount the various Clausewitzian 'frictions' of war. Yet the exact nature of most of these frictions remained of only limited interest. And not just the travails and obduracy of the rank and file, for even such a crucial constraint as logistics was only properly 'outed' in 1977, in Martin van Creveld's *Supplying War*. With all due deference to Mr van Creveld, it seems a faintly ridiculous state of affairs that a book on the problems of supply should be regarded as ground-breaking over thirty years after the end of the *Second* World War. But so it had been right through the 1950s, 1960s and beyond. If anything, the cult of personality had become still more marked and book after book, especially in the non-academic market, lauded the exploits of the likes of Montgomery, Patton and Rommel.

And this even though the very three mentioned are amongst the most overrated commanders in history, all in their own ways more or less oblivious to the frictions of war. If I might be permitted to expand upon my earlier sports pages analogy, the example of American football seems pertinent here, a sport upon which Marshal Foch commented, on seeing an Army–Navy game in the 1920s, 'Mais, c'est la guerre!' If so, then Rommel was the coach who thought he could play the whole game with only the original twelve men on the field and not the full squad of forty-four; Montgomery was the coach who spent three hours orchestrating a ten-yard drive against a high-school team (implacably ignoring all advice from his aides in the stands); and Patton the coach with the brilliant play-book that only worked during the opposing team's time-out.

But the serious point I wish to make here is that until very recently military history has been 'The Commander's Tale' and the role of the PBI (Poor Bloody Infantry) has been marginalised to a quite remarkable extent. To such an extent, in fact, that books like *The Face of Battle* (1976) and *The Sharp End* (1980) could be seen as being genuinely innovative. Heretofore, the sort of demotic perspective that had for decades been part of the warp and weft of economic and much political history was still seen as too reminiscent of a vulgar brawl for military studies. These remained something of a gentlemanly preserve in which only the generals used the front door. Victuallers and quartermasters slunk round to the tradesmen's entrance and the other ranks remained billeted below stairs.

Yet even servants are allowed out sometimes. During the inter-war years military histories 'proper' of the First World War remained immersed in the problems of the senior generals, but there did also emerge a popular demand for front-line reminiscences of the war, especially on the Western Front. After the censorship of the war years, but also after an ostrich-like lack of curiosity on much of the home front, there developed a sharp appetite for frank accounts of mud and blood, machine-guns and barbed wire, sacrifice and disillusionment. With the appearance of overtly literary memoirs like those of Blunden, Graves, Read or Williamson, and the poetry of Owen or Sassoon, these personal accounts achieved intellectual respectability and helped shape the attitudes of a whole generation. Though they never made much impression upon an Edmonds, a Charteris or a Rawlinson, and have only served to irritate a later generation of 'revisionist' historians who remain committed to the generals' view of history, they did at least make significant numbers of people aware that day-to-day existence in the trenches was a foul, terrifying and hideously perilous experience.

So why, and here I come belatedly to my second question, have the experiences of front-line troops in the Second World War, as often as not equally grim, had nothing like the same impact on the popular consciousness? The Kokoda Trail in New Guinea, flooded Dutch polders, the Hürtgen Forest and the Reichswald, an Arakan monsoon, frozen foxholes in the Ardennes and the Apennines, the beaches of Tarawa and the putrid slime of Okinawa, have all been graphically documented and yet do not seem to have taken hold in the popular imagination. And this despite an enormous number of widely disseminated war titles. These were a dominant genre in mass-market paperbacks throughout the 1950s and early 1960s and included frank accounts by the likes of Martin Lindsay, R. M. Wingfield, Alexander Baron, Neil McCallum and Fred Majdalany.[1]

There are, I think, two main reasons for this muffled impact. The first is the low proportion of men in the Second World War who actually served at the front. On the one hand this meant much lower aggregate casualties than in the First World War and thus much less of a shock to community and national consciousness. (My remarks here are, of course, limited to the United Kingdom, the Commonwealth and the United States.) On the other hand it meant that the memories recounted by servicemen (and women) back home were overwhelmingly those of rear echelon life, mainly featuring male boisterousness and relatively minor discomforts and ennui. The stories of Gazala and Cassino, Kohima and Caen, were simply swamped. Even if they were told at all: for when the talk in the factory and the saloon bar is almost all of binges in Brussels or the travails of Lincolnshire ground crew and Bari store clerks, it becomes more and more difficult to try and describe trench-foot, battle exhaustion, soiled trousers and eviscerated mates. The sheer volume of fond memories of a 'good war' simply swamped the recollections of those who had endured the worst of times and made them wary of disputing the more benign folk memory.

Here, perhaps we have a variation upon the notion of 'The Fog of War', where the experiences of the actual combatants are swallowed up in a miasma of mundane recollections about the rear areas and lines of communication. Not that I have any particular desire to disparage these recollections. If I'd been a working-class lad in 1943 arriving in, say, a supply base in Tripoli, I'm sure I'd be bubbling with my subsequent experiences to this day. All I wish to attempt to explain is why the experiences of the front-line soldier have been submerged. And the fog of recollection becomes still thicker when one considers the overall character of the Second World War. In 1914–18, the abiding image of war was the infantryman, either in his trench or going over the top. To be sure, fighter aces, dreadnoughts and U-boats also commanded some attention, but just a glance at the illustrated magazines of the time shows that it was the Tommy in Flanders who loomed largest in the popular consciousness. And this remained the case after the war, especially as the uniform ghastliness of the infantryman's lot became apparent. For the public the First World War was the war in the trenches.

But the Second World War offers no such unitary iconography. Then and since, especially in Great Britain, the perception of the actuality of the war has become very diffuse. It began with nine months of *sitzkrieg* and then a sudden crescendo of military disaster. Defeat and retreat also punctuated such ground action as followed during the next thirty months, in North Africa and the Far East, and up until 1943 the Battle of Britain, the bomber offensive, and the Battles of the Atlantic and the Mediterranean, versus both submarine and surface ship, were regarded as equally, if not more important than the land campaigns.

Of course, especially during the war itself, the Tommy, with regulation fag, thumbs up and mug of tea, was still a central figure, but in the subsequent march-pasts through popular memory he became somewhat shadowy, dwarfed by the rumbling columns of tanks and drowned out by the fighter screens and bomber streams passing over. Even post-1918 the machinery of war still had something of a Fred Karno air and could still be regarded as a rather ungainly and unreliable accessory to the man (or should that be the man and the horse?). And it has been the man, trapped in his muddy hole in the ground, who has continued to dominate the popular imagination to this day.

During the Second World War, however, the machines came of age. Caterpillar tracks became a reasonably reliable form of advance, aircraft achieved ever-better performance figures, submarines posed a still more critical threat, missiles were guided, the atom was split to horrific military effect, and the humble machine tool now turned out this weaponry in quite unprecedented quantities. *Arma virumque cano*, announced Virgil. But in the popular epics fashioned out of the two world wars, the man dominates the First and the arms the Second.

Given this techno-centrism in much Second World War historiography, I do take some pride in having played an early part in what, nevertheless, will doubtless come to be known as the 'Keeganisation' of Second World War studies. As this conference emphatically demonstrates, the ordinary soldiers' perspective on the war has now at least some historical credibility and I hope I might be forgiven for highlighting a few basic themes from *The Sharp End* that still seem to have some relevance for modern research:

> that the front-line soldier in the Second World War spent almost as much time in holes in the ground as his First World War predecessor. Important advances in fire and movement were made, but the proliferation of mortars, machine-guns and self-propelled artillery meant that movement was still extremely hazardous;

> that it was not just attackers who went to ground. The dominant German and Japanese mode of warfare, perforce, was to establish static defensive positions and hang on to them for as long as possible;

> that movement, in fact, was so hazardous that casualties amongst *front-line* troops were of the same order as the First World War;

> that whilst Flanders 1914–18 was for much of the time one of the most God-forsaken theatres in the history of warfare, it was no more arduous than, say, New Guinea, Burma, Luzon, Okinawa, the Vosges, northern Italy or much of north Africa;

> that the rigours of combat in such conditions were an enormous psychological strain that would, eventually, grind down any man. The 'shell shock' that once seemed a special feature of the Western Front is in fact an ineluctable consequence of all modern combat;

> that the Second World War on the ground remained an infantryman's war. So-called *blitzkrieg* represented an abrupt breakdown of the opposing command system rather than a lasting vindication of massed panzers. Once the initial shock effect of such weapons had been dissipated, the tank had to revert to being a very circumspect creature, or risk being lured into precisely plotted killing fields;

> that despite the geographic sweep of the war and despite the stupendous deployment of men and materiel, the war at the sharp end remained a remarkably intimate affair. Possibly the only effective check on the inevitable breakdown of individual resilience was the support of the men in one's immediate ambit. Whatever the delusions of a commander dispensing cigarettes like some latter-day Caesar throwing coins to the plebs, armies held together, as fighting machines, at squad and platoon level. At that level, comradeship was no fleeting, boozy sentimentality but a fierce and enduring commitment of men one to another.

I sometimes wonder just how far this seemingly conventional wisdom has actually percolated. To be sure, personal memoirs of the Second World War now appear regularly and some do very good business. Most Second World War battle books now also include a ration of pertinent anecdotes from those who did the actual fighting. Yet one still feels that front-line accounts of the war are still something of a sub-genre. What continue to dominate the popular literature are the Great Captain biographies (usually about the trio cited at the beginning of this contribution) and the technology books. Weaponry and commanders, guns and bombast, are still the subjects that really sell books about the Second World War.

Perhaps more worrying is the lack of impact that front-line studies have had on military establishments. I hasten to add that when I wrote my book I had no thought of influencing any brass hat or mandarin. Indeed, I have increasingly come to realise how banal were some of my motives for writing *The Sharp End*, reflecting a definite predilection for minor detail. I have, for example, always loved movies, especially westerns, but the cinéaste who ponders John Ford's pictorialism and montage has always been balanced by the student of the prosaic, who queries just when and where cowboys went to the lavatory. Giving some impression of the sufferings of the ordinary soldier has always, of course, been a basic impulse in my work but I have to confess that another has been a very mundane curiosity about the day-to-day routine that keeps huge armies in the field for years on end. The du Picq or S. L. A. Marshall in me has often had to give way to the Mrs Dale.

The Sharp End, then, was never intended to attract the attention of the powers that be, nor even to have any prescriptive content. Nevertheless, as this book is now but one in a fast-expanding library of front-line studies, I cannot but feel some disappointment that combat soldiers still seem to be accorded a fairly low priority. A recent British army reorganisation was dubbed 'Front Line First', but one has to assume that this was little more than an airy counsel of logistical perfection when one learns that the army still has no adequate counselling procedures for disturbed Falklands veterans, or that it has yet to devise a decent army boot. At the time of writing, this latter problem had been handed over to Timberland, though one has some reservations about the relevance of military chic to trench-foot. Or, given the last ten years' obeisance to the omniscience of the market, do we detect here the beginnings of privatisation, or at least sponsorship in the British army? Should we expect the imminent formation of the Timberland Light Infantry? And if so, can the raising of the Trust House Fortieth Foot and the re-equipment of the Royal Armoured Securicorps be far behind?

A final reflection on *The Sharp End* concerns this conference directly in that it has been most heartening to discover such a vibrant consensus about the importance of an aspect of military history that has for me, hitherto, been something of a solitary preoccupation. And this sense of isolation has sometimes been increased rather than diminished by certain reviews which the book has received. I cite a few examples just to emphasise how valuable such a conference as this can be in reassuring a writer working on his own. Confusion can arise at the most basic level. Thus a British reviewer of *The Sharp End* stated that 'the source material is preponderantly American', whilst an American soldier regretted that 'Mr Ellis relied mainly on British sources'. A special concern of mine, of course, was that I have never been at the sharp end, nor even in uniform, so had my attempts at empathy through wide reading and introspection got anywhere near the truth of the matter? The *Army Quarterly* was encouraging. A general wrote: 'Mr Ellis has got it exactly right as all those of us who were at the sharp end will recognise.' For an anonymous reviewer in the *British Army Review*, however, 'Generally speaking, conditions were nothing like as bad as this book might lead the student to suppose...The book is frankly a misfire and must be read with extreme caution.' It has been most gratifying, therefore, to find that most of the research presented at this conference tends to favour the general over Anon.

But it would be most presumptuous of me to end on a merely personal note. So I turn with gratitude to a great writer. Having last night had the privilege to hear Hamish Henderson reciting some of his own superb poetry, my coda inevitably loses a little of its impact. Nevertheless, the short poem below does, I think, perfectly encapsulate the sympathy, admiration and humility that underpin most of our collective research and our conviction that the valour of the common man is the bedrock of all military endeavour. I offer you, therefore, 'A Lullaby' by Randall Jarrell:

> For wars his life and half a world away
> The soldier sells his family and days.
> He learns to fight for freedom and the State;
> He sleeps with seven men within six feet.
>
> He picks up matches and he cleans out plates;
> Is lied to like a child, cursed like a beast.
> They crop his head, his dog tags ring like sheep
> As his stiff limbs shift wearily to sleep.
>
> Recalled in dreams or letters, else forgot,
> His life is smothered like a grave, with dirt;
> And his dull torment mottles like a fly's
> The lying amber of the histories.

Part Two
The British

The Scottish Soldier: Reality and the Armchair Experience

Diana M. Henderson

On 26 August 1914 the 1st Battalion Gordon Highlanders were in action at Le Cateau. The war was some three weeks old. In the confusion of the retreat from Mons, in darkness and in a state of exhaustion, the 1st Battalion the Gordon Highlanders, a regular regiment of the British army, lost their way, found themselves surrounded and were taken prisoner.[1] The circumstances were virtually unprecedented. It led to strong feelings of disgrace and dishonour in the north-east and after the war to bitter recriminations and even a court case.[2] These were pre-First World War professional soldiers, extremely well trained and very competent and I began to do some work on what it was like to be there, what happened and how, and of course the inevitable question in modern journalism, and a trap which I fell into, who was to blame?

At some point I came across a private letter, written in 1978 by an officer, Second Lieutenant Fraser, who was present at the action, to an old and trusted friend. Fraser wrote in this letter of the action at Le Cateau: 'Lumsden the best of us all was killed, Malcolm Hay who would have been a tower of strength was shot in the head and was kicking poor chap like a rabbit. Lyon if he had been at hand would have been really good but it was pitch dark and no one was very sure where anyone was. But, you know, it's very easy to be critical from an armchair.'[3]

This last single sentence has made me think more than anything else about drawing too hasty conclusions, about taking time to think through not what it was like to be there, but what it might have been like to be there, and to remember that every soldier's experience is a deeply personal one.

I would like first of all to take a pace back and look at the whole question of 'experience' in the sense of the actual observation of facts or events. From the soldier's point of view this was very much more than observation and the facts and events were often absolutely unique and physically and emotionally demanding. We as historians are asking in a sense, what was it like to be there? I believe that we should be asking what might it have been like to be there? Those of us who are under fifty were not there and we have to recognise that. Many of the remainder have never 'worn a red coat' and therefore I advocate that we take great care when expounding on 'the soldier's experience'. I certainly would like to think that I have become a little more understanding, a little more forgiving and much less judgemental, particularly when interpreting events and the remembrances of those who were there.[4]

How many of us, I wonder, have not been able to have a bath or a shower for three months; have had the clothes we are wearing reduced to rags in the space of three or four days, or even three or four hours; been covered from head to foot in mud and blood, both our own and other people's; been cold and wet and tired and seasick all at once; been so tired we actually fell asleep standing up; sat in a hole in the ground with incoming mortar fire all around us; shot at other people and been shot at ourselves; given a set of orders upon the success of which people's lives depended, or felt that particular heady rush of adrenalin when you are young and fit and confident and nineteen and you realise life is very special?[5]

I was not there and I am always very conscious of that. Equally neither we nor those who were there can go back and recreate in simplistic terms what it was like. We are the people of the 1990s, we act and react differently because of the times in which we live and our life experience and I believe we should remember that we are trying to interpret the actions and reactions of people with very different lives and life experiences, with different expectations and different priorities, all of which must be taken into consideration in the matter of interpretation of experiences.

There is another aspect too. That of the telling of the experience, the way in which it is remembered and what is remembered. I was always very conscious of this when talking with veterans of the First World War.

Early in my researches I worked with two men who had joined the regular army before 1914. One joined the Black Watch in 1912 and the other joined the Camerons in 1913. Both went to France on mobilisation and in the 1980s both were the only known survivors of their battalions. Both could remember in astonishing detail virtually every day between August and Christmas 1914; names, places, the weather, the roads, dispositions, the battles, what they felt like. After that Christmas the defence mechanism in the memory had come into play and there was left a series of isolated and often disjointed snapshots, sometimes only the ironic or amusing little incidents in a blur of this or that sector or trench, shattered village or shell-hole. It is something that I have noticed occurs particularly amongst those who were captured by the Japanese in the Second World War. On the whole, and often until recently, many have never related their experience, many do not want to, many now cannot.[6]

For those who can and want to tell us their experiences it is extremely important that we listen. These men and women will only be with us for so long. We may not like old soldiers' tales, we may not like what they have to tell us; we may not agree with the language they use, or the way they tell their experiences, but we must listen. Sanitised, academic, erudite dissertations conducted at arm's length are not enough while we have the opportunity to hear what these people have to say directly and in person.[7]

When we do listen, the telling of the experience and the way that it is told is important too. Not only are the tellers older, but they too live their lives in the 1990s. Over the last fifty years their lives have been changed by technology, by modern living standards, by television and public opinion and the many other pressures that alter our attitudes as our lives change and we get older. Too often an experience is only related because they think that that is what you want to hear. Equally experiences are strange things. You can remember that you were tired, that you were afraid, that you were cold, but there are few who can remember what the feeling was actually like. All I can say is that for those who were there the experience left them with a very special deep inner mark, and you have to ask why, fifty years on, 250,000 people marched down the Mall on VJ Day 1995 in good humour and in pride, despite their advancing years, to make one of the most courageous and definitive statements seen in this nation in a long time? The soldier's experience is important and justly deserves to be the centrepiece of our attention.

Now to move on to the experiences themselves. From the Scottish soldier's point of view I do not think that they were much different from any other soldier's experience in any given theatre of war. However, I have no doubt that the experiences of Scottish soldiers were set against a strong sense of national pride, a varied industrial, agricultural and frequently very traditional community background, where the Scottish regiments still retained an enormous popular appeal and where there was a fraternity between the men that permitted the flexibility of mutual respect to operate on the raw edges of the rank structure. Private soldier, NCO or officer, there was still a measure of, 'A man's a man for a' that.'

Casting away our armchair therefore, carefully asking the question what might it have been like to be there, we turn to the experiences themselves. It is noticeable that many of the experiences are very individual and personal and they vary widely by comparison with the type of almost standard experiences that come to us from the First World War: the Somme, Ypres, shell-holes and shattered villages.[8]

Through the experiences of the Second World War, however, run a number of common themes. There is, first, the theme of initiation and early training. These experiences are very different from those of the First World War; that war had, after all, only finished twenty one years before and the remembrance of war and soldiering was much closer than it is today, added to which there were few romantic illusions left to shatter. In addition, the TA had acted as a major source of income in the depressed inter-war Scotland and some, like my own father, were enthusiastic members who went to camp at Barry Buddon in the summer of 1939 and simply stayed in uniform until they managed to scramble aboard a boat and get away from Dunkirk in 1940.[9]

Donald Maitland, a graduate of Edinburgh University and later British Ambassador to the European Community, remembers:

Our training as potential platoon commanders built upon what we had learnt in the school and university training corps... several of our instructors had had recent battle experience. Nor were we likely to forget two pearls of wisdom from a Glaswegian sergeant instructor. The first related to mustard gas. 'Wi' this respirator ye'll be all right. But if you get it on your genial organs – aw Jings.' The second was an aphorism of universal application: 'Ye've got to cater for the BF in every section.'

In my own regiment I listened to one of my sergeants two months ago saying exactly the same words.

Then there was the departure for overseas. Again, it has to be remembered that the great majority of parents had experienced this before, but previously it had been they who were leaving. They cannot have been under any illusions. Donald Maitland once more:

> In March 1942 I went to Bridge of Allan to join the 'Dandy Ninth' – the 7/9 Battalion of the Royal Scots. I had little time ... before I was summoned by the Adjutant who said that I was to take immediate embarkation leave before joining a troopship for India ... With my parents I spent a glorious week in and around Peebles, during which our bonds grew tighter still. Paradoxically this eased the leave-taking at Waverley station. The fortitude of my parents in such circumstances never ceased to astonish me.[10]

The third common theme running through the soldier's experiences has two sides to it. Alongside the youthful exuberance and the new-found freedom from the bonds of family and an often parochial lifestyle is the overwhelming impression of amazement at the scale of events of which they were part. Again and again you will hear accounts of the awesome sights of the great convoys setting off from Liverpool or the Clyde, the accounts of the preparations for El Alamein and D-Day. There is very much the impression, 'I know that I was only a very small cog but I was there and I played my part too,' which I am sure tells us a lot about the quality of the leadership and man-management.

The 'shock of battle' has always been, and will always be, the shock of battle. There is very little that can prepare you for it completely. A young Scots Guards officer recalls one of his early battles in Italy:

> there was no sound except the subdued carping of crickets. Sergeant Stewart whispered that he thought that there was a booby-trap on the path. I advanced, revolver in hand, to examine it and found a rifle leaning against a carrier of mortar bombs. For a second it did not dawn on me these must be German. Then suddenly, and almost subconsciously, I noticed a cluster of black figures lying sheltering along the embankment at my very elbow. I spun round, yelling, 'Hands up,' in German. There was a shower of sparks as one of them fired at me. I promptly fired my revolver point blank at the nearest two who were now sitting up: and they rolled towards me screaming ... All hell now broke loose as the silence was shattered. Everyone started yelling. Other Germans sitting on the embankment, fired at us, Ex-Sergeant Hutchison finished one of them off, then, thank God, the remainder stuck up their hands.

An awful savagery now seemed to take hold of us and we rushed along the embankment shouting oaths and shooting at Germans who were lying there. I felt as if some wild animal had got me by the throat and I had to keep shouting and shooting or else my normal self would return bringing fear along with it.[11]

For the less favoured, the enormous physical endeavour of battle was followed by capture, an experience which is usually related in terms of bewilderment, anger, exhaustion and chaos. Derek Lang of the Camerons on surrender at St Valéry:

> The first moments of captivity were agonizing. The loud guttural voices of our captors, their swagger and arrogance and perhaps above all their smart turn out, which compared so noticeably with our own pitiful appearance, increased our natural despair. After a search we were hustled together in a miserable group on the promenade while they tried to round up any stragglers who were still hiding amongst the rocks. I can remember very little of the next three or four hours. I started off with a walking party but found it difficult going as I had to hold a large cotton-wool swab to my eyes and had such a dreadful headache that I hardly knew what I was doing. When the Germans discovered that I could not walk straight they put me in one of the trucks, where I passed out for the second time. When I came to it was to find that I was in some sort of temporary hospital in what had once been a casino-type hotel at Forges-les-Eaux, many miles inland. The squalor was quite appalling. We were packed together cheek by jowl. The French wounded were in row upon row on the floor of an old theatre, the British were crowded five each into balcony boxes, originally designed to accommodate two seats. I do not know whose lot was more miserable. The straw on which we lay was filthy and the smell of festering wounds almost unbearable.[12]

The next theme concerns the fact that many of the soldiers' experiences leave the impression that much of what had to be done was incredibly mundane, that it took an infuriatingly long time to get things done and that individual initiative and improvisation were all important. Lieutenant Steel Brownlie, a troop commander of the 2nd Fife and Forfar Yeomanry, records of the battle for Normandy:

> In the dark it was a matter of finding the squadron harbour, blundering about in the light of flaming wrecks and getting mixed up with other units. It took an hour or so, and then there was the business of reorganising the troop, taking on fuel, ammunition and rations from the Echelon trucks, reporting mechanical defects and seeing them dealt with by the fitters, attending an O Group to receive orders for the next day, folding and marking maps, ensuring that a member of the crew was cooking and another arranging the bedding, sometimes fitting in a wash, or a visit to a hedge outside the harbour, with a shovel and a supply of Army Form Blank, which was issued on a scale of two-and-a-half sheets per man per

day. What time would it be after all that lot? It might be one, two, or three in the morning, and the orders were to be ready to move at dawn.[13]

...and you are nineteen and it is your first battle...How many of us I wonder would be prepared to stand up and say with complete confidence that we would have been able to do it?

Improvisation was everywhere as were the 'big men', referred to by John Keegan, who were often best at it. I knew one such, who even in his eighties was still a big man. Once upon a time I was a solicitor with the great firm of Clark, Oliver, Dewar and Webster. All the senior men in the firm had been in the Highland Division together and our client lists read like the Army List of 1945. The senior partner was James Oliver, probably the most able Territorial Army soldier in the Second World War. Long-haired apprentices would stand spontaneously to attention for Mr Oliver with no idea why they were doing it. At five-foot five he was indeed a big man; and he too was part of the improvisation picture. He once showed me a scrappy sheet of foolscap paper covered with his fulsome handwriting and said, with that wonderful twinkle in his eye, 'These were my orders to the Brigade when we crossed the Rhine.' Along with these orders, which were after all for the purposes of a major military undertaking, he produced a Michelin *Guide to France and Germany* dated 1935, which was dotted with some fairly unconventional military map marking symbols which certainly were never learnt at Sandhurst.[14]

But nowhere was the spirit of improvisation more evident than in the experience of the women's services. Originally not subject to military discipline and organised on an entirely *ad hoc* basis, experiences of servicewomen deserve special mention. By 1943 the ATS boasted an active and effective force of 207,000 women who brought a resourcefulness, and a quality of humour and courage to the services. Many were ill-equipped to deal with the challenges which faced them, but face them they certainly did. I quote here from the history of the Women's Royal Army Corps, in which I had the privilege to serve for twenty-five years.

A superb example of the spirit of the ATS was the reaction of Senior Commandant Belinda Boyle to the unexpected arrival in Thurso, in Caithness, of a contingent of starving troops who were decanted by some destroyers of the Royal Navy after evacuating them from the disastrous campaign in Norway in 1940. Belinda Boyle happened to be visiting her two ATS clerks attached to the little garrison, whose commander was understandably baffled, as whatever he was geared to do it was not suddenly to produce the resources of a transit camp for the debris of a defeated army. Nor was it any use to send them on, for Thurso was a long way from anywhere and they were starving. Senior Commandant Boyle however rapidly appreciated the situation. She got hold of a sheaf of billeting forms and began amending the relevant portions for requisitioning accommodation; she requisitioned a flock of sheep, a field of cabbages and some groceries, so securing the basic supplies. For obtaining labour her personality seems to have been sufficient and no forms were needed; the Boy Scouts, called from school, rounded up the sheep, willing butchers slaughtered and jointed them, eleven patriotic housewives and the two ATS clerks made a giant stew in various coppers, buckets and a bath, and the situation was saved.[15]

Finally, for some, there was the experience of brutality, the anger, the deep sense of loss and the sheer frustration. For the prisoners of the Japanese this brutality was of a different hue and left a special mark, but it was also present in Europe especially during the long and arduous advance into Germany, when it was clear that the war was ending but the enemy simply would not give up. Many of us in this modern and very violent age dwell too long on this aspect and forget the humour and the depth of comradeship that was part of the Second World War, but for those who experienced the brutality it is difficult to erase and even more difficult to relate as just another experience. I am quoting here from another graduate of Edinburgh University, Walter Elliott of the Scots Guards. In action near Langerich he records:

I jumped off my tank and ran along the road with my platoon and then ordered Sergeant 'Joe Plush' Wilson with his section to fire at the enemy from the front, while I advanced with another section to attack the German position from the left flank. Sergeant Wilson, always the comic, was shouting, 'Splendid! Come on my men, charge!' and disappeared amid a roar of laughter in the direction of the roadblock.

Suddenly from behind there was a roar of approaching shells and we dived for the nearest ditches. The shells were right on top of us and the ground shook. There were shouts for help and the stench of cordite. Two or three lay wounded. It was clearly our own medium guns firing from behind and 'Joe Plush' lay in the middle of the road in a pool of blood. We picked him up but his eyes already had the glassy stare of the dead.

Another salvo of shells from our medium artillery descended as I found myself hugging the earth in a mixture of fury and terror. There was a shout again for stretcher bearers as the smoke cleared. I felt a blind ungovernable fury and ran down the road to where the artillery observation officer was sitting in his scout car, the turret of which was closed against his own flying shrapnel. In my anger I jumped on the top and battered on the turret with my sten gun and swore at him to stop his damned mediums wiping out my platoon.

We were glad to leave this area of burnt-out farmhouses. What we had at first taken to be a row of charred beams in a gutted cowshed, turned out to be cattle burnt alive at their halters. I could not understand why the Germans did not let them loose when they saw all the farmhouses being turned into funeral pyres in the wake of our advancing tanks. For fear of snipers every single farmhouse within range was set alight by our tanks' incendiary bullets. In a matter of minutes they were all ablaze with the tethered animals still inside, including many horses.

It nevertheless gave one a savage thrill riding through enemy territory on the back of our tanks as their guns sprayed tracer from side to side. Whenever we saw anything suspicious the infantrymen sitting on top of the tanks would also open up with their Bren guns. I remember seeing a figure in green uniform running along a hedge. He may have been a German soldier, but could have been a Russian worker, as he was by himself. Someone fired a Bren gun at him and then the whole tankload of infantry followed suit. He was completely riddled.

The tanks were machine-gunning one farmhouse we passed when an old bearded figure rushed to the door waving a large white sheet. The tanks could not stop firing in time and he gradually collapsed in the doorway – symbolic of the utter savagery of war.[16]

Walter Elliott concludes,

On Sunday 6 May I attended the Scots/Welsh Group's Service of Remembrance in the old Lutheran Church at Stade. The church was mellow red brick and had stood throughout the centuries. First we heard the National Anthem... followed by prayers. Their refrain was, 'Thine is the Victory, O Lord not mine.' The congregation replied at the end of each prayer, 'We thank thee O Lord.'

Our respected padre, the Revd David Whiteford, then stepped forward and the two commanding officers of the Scots/Welsh Group presided over the remainder of the service...

Standing by the altar each commanding officer read out the long list of his battalion's dead in a voice deep with emotion. The last post then sounded by trumpet from outside the great west door, to be followed by two minutes' silence. Then Pipe-Major Bain started playing a Scottish lament on his pipes from beside the altar – the 'Flowers of the Forest'...

His piping brought tears to many eyes as he marched slowly through the west door and down the hill, his notes growing fainter and fainter. One thought of the graves on the Italian mountains and now along the German roads. But as the piping became more distant it seemed as if the spring breezes played tricks with the lament so that it echoed back sounding more like a Scottish reel. One thought of laughter as the sun went down and of songs around the brew-can fires; of Corporal Bryson trying to catch an old hen in the Italian rain; of Sergeant-Major Lumsden roaring drunk at Hogmanay; of laughter in the back of the TCLs and of Support Company trying to pretend there was not a war on. Since the Battalion had crossed the Rhine a month before, the rifle companies had sustained the worst of the casualties with 17 platoon officers killed or wounded and 223 other ranks. Right flank had alone lost 95. During the war as a whole the Battalion had now lost 113 officers killed or wounded and 1,246 other ranks – more than in the First World War.[17]

To those of us who will only ever have the armchair experience, I leave you with these thoughts.

The Shadow of the Somme: the Influence of the First World War on British Soldiers' Perceptions and Behaviour in the Second World War

G. D. Sheffield

The early morning mist had lifted and the smoke and dust of the barrage on FONTENAY had cleared away revealing another cloudless hot summer's day, when the guns roared and thundered into action again. The barrage fell thick on the hillside beyond the village.... [when] the 1/4 K.O.Y.L.I. ...poured out of FONTENAY village and swept on up the hillside, walking behind a wall of hideous noise and smoke and dust and flame which moved forward in jumps of a hundred yards at a time ahead of them. The TESSEL WOOD feature...was taken and consolidated, and a counter-attack the same evening was successfully beaten off.[1]

This description – of infantry following a creeping artillery barrage, capturing a wood and then digging in to await the inevitable German riposte – is extremely reminiscent of the Somme in 1916. Even the unit title is correct: 1/4 King's Own Yorkshire Light Infantry did indeed fight on the Somme. Yet this passage is not referring to the First World War at all, but to fighting in Normandy in mid-June 1944. This passage illustrates that the sum of the similarities between the British armies of 1914–18 and 1939–45 far outweighed the differences. However, for many British soldiers of the Second World War folk memories of the Somme, Gallipoli and Passchendaele acted as a benchmark of the appalling nature of war, and as a result they sometimes failed to recognise just how terrible 'their' war was.

The First World War, the greatest victory the British army has ever achieved, is probably the most mythologised conflict in British history. The reality was that in the space of four years Britain raised a vast citizen army virtually from scratch. This army experienced a steep learning curve in some very bloody battles, which resulted in the creation of a highly skilled instrument of war that took the lead in defeating the Germany army in 1918, at the cost of 750,000 military deaths, with a further 250,000 dead from the Empire. The British casualty figure needs to be read in context. The losses were commensurate with the scale of Britain's role in the fighting and the issues at stake, while the losses sustained by enemies and allies were even greater.[2] Not surprisingly, a large number of Britons in the interwar years were unable to view the Western Front dispassionately. All they saw was the 'butcher's bill' of killed and wounded, and the less than perfect world of the interwar years. By 1939 the myth had emerged that the First World War had been futile, that it had been fought for nothing; it had been the 'foulest war in history'[3]; and the conduct of the war had been grossly incompetent.

The milestones on the road to disillusionment in the interwar years are well known. After a period of mental numbness lasting about ten years, there was an explosion of anti-war sentiment expressed in books, plays and films. In 1933 the British élite expressed their opposition to war in the vote of the Oxford Union against fighting for king and country, and in the same year ordinary folk followed suit in the Fulham East by-election. Two years later there was an overwhelming 'yes' vote in the Peace Ballot. Clearly, the shadow of the Somme hung over the generation growing up in the 1930s. Some of George Meddemmen's 'earliest memories' were of his uncle's stories of life as a gunner-sergeant in the First World War. Meddemmen was left with 'a horrid picture of mud and slime and death and brutality', and the outbreak of the war in 1939 'restimulated' these memories.[4]

From the perspective of the intellectual climate of the 1930s, the First World War was mere futile slaughter. For one public schoolboy, later to serve as an infantry officer and become a distinguished military historian, the idea of the serious study of war 'appeared archaic and repellent', while the pacifist ideas of Dick Shepherd, Beverley Nicholls and Donald Soper were powerful and 'plausible'.[5] George Orwell claimed that to have even the slightest acquaintance with warlike affairs, 'even to know which end of a gun the bullet comes out of, was suspect in "enlightened" circles' in the 1930s.[6]

Yet there was another view of the First World War that existed alongside 'futility' and 'never again' in interwar Britain. The First World War exercised a terrible fascination for men who had not been old enough to serve in the war. George Orwell, born in 1903, even suggested 'that part of the reason for the fascination that the Spanish Civil War had for people of my age was that it was so like the Great War'.[7] Men who were younger than Orwell, whose boyhoods had coincided with the interwar period, also became fascinated by the First World War. For some it was a matter of tradition. Patrick Hennessey, the son of a regular Indian Army officer, was raised 'on tales of the Western Front', which, interestingly enough, for him sat easily with the much older tradition of 'heroic deeds in far-flung corners of the Empire'. He was to come ashore on

D-Day in an amphibious DD tank.[8] D. A. Blake's father served in the First World War, and died shortly after the Armistice when Blake was eighteen months old. Blake was 'brought up' to believe that the service of his father and others like him was 'something very glorious, to be admired and respected'. In April 1939, with war threatening, Blake enlisted in the Territorials: 'my father did it, I couldn't not (*sic*)'. Naturally, on being commissioned in 1940, he chose to join his father's old regiment, the Royal Fusiliers.[9] Even wounded veterans could be nostalgic about the First World War. G. F. Andow's uncle lost an arm in the Great War yet would tease his nephew about joining the army, and Andow enlisted as a regular soldier in the Tank Corps in 1935.[10]

As a child Vernon Scannell, who fought as an infantryman in North Africa and Normandy, listened to his father's 'tales of the trenches'. He developed 'an almost obsessive interest' in the First World War, and this obsession was fed by things like a gift album given away with a boy's paper in which one stuck photos of trenches, guns and the like.[11] There was also a thriving nostalgia market for veterans of the war fed by part works and meetings of Old Comrades' Associations. The First World War was a terrible experience; but it was also, for many survivors, the most important and rewarding period of their lives, and this perspective was passed on to youngsters.[12] This 'counter-culture' helps to explain why so many young men proved resistant to the trend among the intellectual and social élite of revulsion against the First World War. Moreover, for an eighteen-year-old in 1939 the First World War was quite literally more than a lifetime away, and, like every other period in history, war and military life offered glamour and excitement. Many men who were to fight in the Second World War carried into that war mental images of the Great War that were quite at odds with the 'disillusioned' view which is popularly supposed to have dominated the British psyche in the 1930s.

In 1938–9 volunteers did not rush to the colours as in 1914, but none the less a steady stream of men volunteered for the Territorials when it became clear that war was likely. While some joined because of peer pressure and for summer camps, some joined to stop Hitler. One man enlisted because he believed that in 1914–18 'many men lost their lives through lack of training'.[13] Many soldiers believed that the next war would be a repetition of the last; in the words of a regular private in the Royal Tank Regiment, they 'expected more or less to carry on from the 14/18 job'.[14] Britain was in 1939 a semi-militarised society, in the sense that millions of men had served in the army in a major war and millions more adults of both sexes had experience of a total war society. Thus there was a wealth of experience upon which to draw; many veterans of the First World War had actually fought in recognisably modern manoeuvre warfare battles. In 1938–9, veterans of 1914–18 were frequently called upon to lecture on their experiences[15] and some regimental officers and senior NCOs who went to France in 1939–40 had served during the First World War. However, a vignette from a fictionalised account of Normandy neatly captures the attitude of the soldiers of the Second World War to those of the First. During a rest period, the battalion is inspected by a delegation of *anciens combatants*. With the arrogance of youth, the men regard the French veterans as 'comical little men', as figures of fun. But the British soldiers also recognised these men had something in common with them, for the French veterans 'remembered Verdun'. Other soldiers had a much more dismissive attitude towards the First World War. An officer who served in Tunisia in 1943 adopted a mocking tone when speaking of a brigadier with 'slightly First World War ideas' and criticised intelligence procedures drawn from 1914–18 experience as being best suited for 'completely static warfare'. For this officer, the First World War obviously meant 'outdated'.[16] Given that many of the key lessons of the First World War were ignored or misinterpreted by British High Command, it is not surprising that junior officers and soldiers adopted a similarly dismissive attitude towards the earlier conflict.

Many men used the experience of the First World War to identify relatively 'safe' arms of service. In 1939 the parents of one soldier strongly advised him not to accept a commission into the infantry or artillery because of the very high rate of casualties from the First World War, only for him to become a bomb disposal officer.[17] In 1939 many recruits to an artillery unit, the South Notts Hussars, had 'folk memories of the horrors' endured by their fathers in the trenches. For such men, service in an artillery or cavalry unit (and some recruits had joined in the belief that the Hussars was still a horsed cavalry regiment, as it had been in 1914–18) seemed a preferable option to ending up as 'cannon fodder' in the PBI (poor bloody infantry). One frankly admitted that 'the infantry get too close, they're real brave lads – and I thought, "Let's be in the artillery, they stand off and keep lobbing it in!"'

Ironically the South Notts Hussars were destroyed in the Western Desert in May–June 1942, firing their 25-pounder guns over open sights at advancing German tanks.[18]

Some pacifists in the interwar period seemed to believe that passing on the horrific truth about the fighting on the Western Front in 1914–18 to succeeding generations would deter young men from fighting in a future war. Clearly, this hope was not realised, but the development of the First World War myth in the 1920s and 1930s helped to change the way ordinary soldiers viewed authority and discipline. The men of 1939–45 had a much less trusting, more cynical attitude towards authority than their predecessors of 1914–18. Several reasons may be suggested for this, ranging from changes in British society[19] and improvements in education, to the presence of large numbers of articulate middle-class men in the ranks; the influence of the left-wing popular press such as the *Daily Mirror*,[20] and, not least, disillusionment with the Depression years. A comparatively neglected factor is the steady denigration of the British military leadership of the First World War in the interwar cinema.[21]

Between 1919 and 1927 British Instructional Films made a series of semi-documentaries that celebrated battles of the First World War as part of the tradition of British military victories.[22] Ten years later, many films adopted the point of view that the First World War was a tragic waste. Possibly the greatest British film ever, *The Life and Death of Colonel Blimp* (released in 1943, in time to have been seen by thousands of British soldiers in training in Britain who were not to see action until the following year) was explicit in its condemnation of the regular officer corps of 1914–18, to the fury of Winston Churchill. The 1938 film *The Dawn Patrol* showed officers of the Royal Flying Corps on the Western Front effectively being forced by High Command to condemn barely trained pilots to death by sending them out on dangerous missions: '[the] last scene shows [David] Niven, dazed by the loss of his brother and his greatest friend, welcoming a fresh wagon load of eager youngsters with his same old routine of cheerful informality...'.[23]

Although *The Dawn Patrol* portrayed the subordinate officers as disagreeing with demands of High Command, their sense of duty forced them to execute their orders – and their subordinates died as a result. While the 1930s and early 1940s were not short of traditional 'stiff upper lip' British films, the cinema, a medium of truly mass appeal, does seem to have reinforced the intellectual climate of distrust of military authority.

The habit of social deference had weakened by 1939, but it had not disappeared entirely, and the upper echelons of society still provided a disproportionate number of officers.[24] Soldiers of the First World War were perfectly capable of showing disapproval of unofficerlike behaviour, but if officers were brave and paternal their men would usually follow them.[25] The soldiers of 1939–45 were more discerning. Corporal Doug Proctor of 4th Somerset Light Infantry, 43rd (Wessex) Division fought his first campaign in Normandy in 1944. The performance of his first platoon commander and a company commander made a bad impression on Proctor, who, as a section commander, himself had responsibility for his men. Proctor later argued that in battle social class and an expensive schooling '[count] for very little. I have known NCOs who came from poor backgrounds and had accents as thick as treacle, who were far better leaders of men than some officers could ever hope to be.'

When Lieutenant Sydney Jary arrived to take over the platoon, he was very much on probation. Proctor, and his fellow corporal, Jim Kingston, would support Jary, 'but he would have to earn our respect and trust'. Jary's own account shows that he was well aware that he was on trial. Eventually, after Jary had proved his courage and leadership during several weeks of fighting, Kingston and Proctor decided that they had 'a platoon commander who was worth looking after'. After his performance during the crossing of the Seine on 25 August, Proctor wrote, Jary had 'served his "probationary" period, and...the platoon would have followed him anywhere'.[26]

This is not to claim that officers in the First World War did not also have to win the confidence of their men, and especially their NCOs – they did – but officers seem to have had a more difficult job in 1939–45 than in 1914–18. Proctor's attitude was not untypical, for in the Second World War soldiers were more questioning and less tolerant of bullshit and incompetent or gung-ho officers. A revealing little conversation appears in a memoir of the war in Italy in 1944, written by a ranker, Alex Bowlby. Ordered to wear uncomfortable greatcoats during an action, the soldiers complained:

' "This is 'ow the poor sods dressed in the last war, ain't it?" said Page. "Well, I 'ope they don't think they're still fighting the same fucking war," said Humphreys, giving his coat a vicious thump.'

Men in training camps wrote to the *Daily Mirror* or their MPs if aggrieved; in the field they sent round-robins to their officers or staged demonstrations against unpopular orders; in battle they might simply refuse to obey orders that they thought were foolish or unduly risky. Bowlby recorded an occasion when an officer leapt out of a trench during a German mortar bombardment calling for his men to follow him, but no one moved.[27] Such 'leadership by negotiation' had also occurred in the First World War, but it appears to have been much more prevalent in the Second. A particularly spectacular example occurred in 2nd Borders in Burma in late 1944. The battalion had made a bloodily abortive attempt to cross a river, and on being told that the battalion was to make another attempt NCOs simply refused and angrily returned to their men: 'It took all our powers of persuasion, while constantly toing and froing between groups of men all along the line, before we [the officers] convinced them that, with artillery support and the cover of darkness, we could make it.' It is perhaps of relevance to our theme that the battalion's catch phrase, used by both officers and men, was 'biggest balls-up since the Somme'.[28] To a greater extent than in the First World War, officers in the Second World War expected and used leadership by negotiation, and it seems likely that the portrayal of British military leadership of the First World War on the cinema screen was a factor in this.

Some officers seem to have made little attempt to adjust to the disciplinary demands of a post-Passchendaele citizen army of a democracy.[29] Others did, most famously the new breed of 'democratic' generals. Bernard Montgomery was once reported as saying that as a junior officer in 1914–18 'he had suffered under leaders and generals who were unknown, who he didn't know, who seemed as if they didn't care', and he would bring a different style of leadership to an army if he ever got the chance.[30] The care that Montgomery took to cultivate his informal style is impressive. Unlike Slim, a natural communicator whose personality is epitomised by his nickname, 'Uncle Bill', Montgomery was in reality a cold, austere man, not unlike Douglas Haig. A more junior general who thought upon similar lines was Major General 'Pete' Rees of 19th Indian Division, who reacted so violently against his First World War experience of generals that he insisted on taking numerous risks in leading from the front.[31] Not every general of the First World War was a remote figure, not every general of 1939–45 was a Monty, Rees or Brian Horrocks, and an informal approach did not impress every soldier. However, one infantry officer with much combat service could write that he did not recall 'any of the rancorous distrust' of generals or the staff so often found in the writings of First World War soldiers.[32]

The 'Shadow of the Somme' influenced the experience of the British soldier of 1939–45 in one very important way, for British strategy after Dunkirk sought to avoid a repetition of the Western Front of 1914–18. Not until

June 1944 did the British army return to Western Europe by the most direct route. One consequence of this lack of a long-running main front that absorbed the bulk of the troops was that large numbers of soldiers spent most of the Second World War training in Britain. By contrast, certain 'élite' formations moved from theatre to theatre, giving some soldiers a much more varied war than their predecessors of the First World War. It was possible for a soldier of the 51st (Highland) Division to have fought in France in 1940, North Africa in 1942–3, and France and Belgium in 1944, ending up in Germany in 1945.

The formative influence of most generals of the Second World War was the Western Front, where they had served as junior officers: while listening to the preliminary bombardment at Alamein in 1942, Major-General Wimberley of 51st (Highland) Division recalled listening to a similar bombardment at Passchendaele a quarter of a century earlier.[33] In battle, for the most part, British generals were reluctant to risk the heavy casualties of 1914–18. Montgomery's battles from Alamein in 1942 to the Rhine crossing in 1945 were mostly characterised by a set-piece approach, relying on careful preparation and massive firepower. Montgomery put into practice the methods that he had learned as a staff officer under Plumer, the commander of Second Army in 1915–18, whose methodical operations at Messines in June 1917 and at Ypres in September–October 1917 brought outstanding success. Of course, the British army of the Second World War threw up some 'thrusting' generals in the classic First World War mould, such as Ivo Thomas, of 43rd (Wessex) Division, who had served as a gunner officer on the Western Front and who earned the nickname 'the butcher' in Normandy. One can argue that by failing to take risks Montgomery sometimes caused unnecessarily heavy casualties in the long run by neglecting to seize opportunities, but the soldiers under Monty's command certainly benefited from his careful preparations for battle. This was one direct legacy of the First World War that contrasted strongly with the operations of some other British generals of 1939–45.

The British army was spared a major bloodbath like the Western Front largely because the Red Army carried out the bulk of the fighting against the Wehrmacht. Six years of fighting cost Britain 305,800 killed and missing, of whom 177,800 served in the army. In comparison with the 'million dead' of 1914–18, this seems an economical figure.[34] But a rather different story emerges when one examines the battlefield. British and Canadian battalions suffered about 100 casualties a month on average on the Western Front in the First World War. In the 1944–5 north-west European campaign, battalions suffered a minimum of 100 per month but 175 per month was not uncommon. The daily casualty rate of Allied ground forces in Normandy actually exceeded that of the BEF, including the RFC, at Passchendaele in 1917. In the North African and Italian campaigns of the Second World War the figures were about 70 casualties per battalion per month, but on some occasions casualty rates matched anything in Normandy. Moreover, these casualties were sustained by a relatively small sector of the military population. Since the First

World War, the number of riflemen had fallen steadily as a proportion of an
infantry battalion's strength, and in any war, the front-line rifleman is the
soldier most likely to be killed or wounded.[35]

The popular perception of the two world wars is that they were totally
different; that in 1914–18 the armies endured stalemated trench warfare, but in
1939–45 free-flowing mobile warfare predominated. In reality, the style of
warfare that emerged from 1916 onwards was essentially that of the Second
World War, the main difference being that in the later war technological
advances had provided effective communications and a usable instrument of
exploitation, the lack of which had prevented armies on the Western Front
from converting the 'break-in' into the 'break-through'. Even the mainly
mobile campaigns in the Western Desert between 1940 and 1942, which at
first sight do not bear much resemblance to the Western Front – although
they did to the campaigns in Palestine in 1916–18, as some of the participants
in the later campaign recognised[36] – included some essentially First World War
style battles such as the siege of Tobruk and Second Alamein. In other
theatres, where the terrain was not so suitable for manoeuvre warfare, the
similarities with the Western Front were distinctly pronounced. Static, or at
best slow-moving, attritional warfare characterised much of the fighting in
Italy (1943–5), Normandy (1944) and Holland and Germany (1944–5). Indeed,
an officer of a unit holding a static sector of the British line in Normandy in
early July 1944 used language straight from 1917 to describe the style of the
fighting: 'fighting patrols' or 'raids' gave them 'complete moral superiority over
the Germans' allowing the unit to dominate 'No Man's Land'. Ironically
enough, this unit was 45 (Royal Marine) Commando, a unit that prided itself
on its modernity and elitism.[37]

Soldiers of the Second World War were not slow to draw comparisons
between their experience and the Western Front. The First World War became
a touchstone of horror; it seems to have become an article of faith among
British soldiers of the Second World War that no matter how appalling the
circumstances they found themselves in, the men who served on the Western
Front had suffered more. Whether this was true is another matter, for the
soldiers of 1939–45 tended to believe the myth that the First World War was
the most terrible war in history rather than face the truth, that all modern wars
are terrible for those who have to fight them, especially front-line infantrymen.
When faced with evidence of heavy fighting, many soldiers automatically
reached for their folk memories of the Western Front for a comparison. On
first seeing the Normandy battlefield, Lance-Corporal Ron Garner of 43rd
Division was reminded of 'the impressions I had of the last war' not, significantly, the impressions he had presumably formed from accounts of the
fighting in France, North Africa and Italy in the previous four years. Another
private of the same division described German shelling around Hill 112 in mid-
July as 'reminiscent of the 1914–18 War'.[38] An officer serving with a Chindit
column deep in the jungles of Burma – a man who had already seen much
fighting in the Middle East and Burma – could, in his memoirs, think of no better

analogy for one battlefield than Passchendaele.[39] Many soldiers of the Second World War expressed pity, implicitly or explicitly, for their predecessors of the First World War. Mud was not exactly unknown to the soldiers of the Second World War, yet when a soldier serving with 7th Armoured Division in Normandy in July 1944 came across 'a quagmire', he wrote in his diary: 'We're in the wrong war – this must be 1917.' A few months later a sergeant of the same division, wounded in the lung, reported that the medical officer told him, 'In the First World War I would never have made it.'[40] An artillery NCO serving in Italy suffered from battleshock, and was demoted by an officer for unreliability. The NCO's comments in his memoirs mixed pity for the soldier of the First World War with anger at the leadership of the army in the Second: in the Great War, he believed, he would have been executed.[41]

The physical conditions of soldiers on campaign in the Second World War were often as gruelling as, if not worse than, those of their predecessors of 1914–18. Many soldiers of 1939–45 clung, however, to their belief that bad as things were, on the Western Front things had been even worse.[42] An infantry officer who served in Holland in late 1944 offers strong evidence of the popularity of the mythical version of the First World War. By common consent, conditions in Holland in this period were appalling, produced by the cold, deliberate flooding by the Germans, rain and the low-lying nature of the land: 'We had in two or three months the same experience as the infantry in the First World War endured for years.' In reality in the First World War a system of reliefs ensured that units regularly moved in and out of the line. Conditions in the primitive trench systems of 1914 were similar to those in Holland in late 1944, but from early 1915 onwards trenches became considerably more sophisticated. This officer mentioned a number of key factors that sustained morale, such as good logistic and medical support, letters from home, and hot food. These factors were equally important in 1914–18, and indeed the British army developed a complex and highly effective logistic infrastructure on the Western Front.

The same officer also discussed tactics used during the last months of the war. A shortage of officers and men with adequate training led to the use of 'simple tactics' which he compared to those of the First World War. Men advanced 'at a set pace behind a barrage; halted for three minutes; coloured smoke was fired indicating that you could go on...'[43] Units did indeed employ such rudimentary tactics on occasion during the First World War, as in late 1918 when the 1st Gordons was largely composed of poorly trained troops.[44] However, this is far from the whole story. A major theme in recent research into the BEF of the First World War has been the increasing tactical sophistication of the British army from mid-1916 onwards, but the tactical skill of the BEF in 1918 was virtually unknown to the fighting men of 1939–45.

The language of the British soldier of 1939–45 was coloured by the earlier war, possibly influenced by the continuing use of First World War imagery in the press.[45] While some of the slang of 1914–18 did not survive, men in both wars desired a 'Blighty one', a non-life-threatening wound that would allow evacuation to Britain or 'Blighty'. Other First World War phrases had evolved over twenty-five years. In 1914–18 the phrase 'Send it down, David' referred to rain, which the soldiers blamed on the ancient Hebrew king for some odd reason; hence soaking soldiers would shout, 'Send it down, David.' In Burma in 1945 soldiers used the phrase about shellfire, without any apparent knowledge of who 'David' was.[46]

On occasions, soldiers of the Second World War found themselves on the battlefields of the First World War. Some toured the old battlefields. In March 1940, during the Phoney War, a veteran of the First World War conducted officers of 8th Royal Warwicks over the battlefields. One found it a fascinating experience, although he was 'absolutely convinced' that similar battles would never happen again. He was killed on 21 May 1940, too soon to see the inaccuracy of his prediction.[47] A rather less formal battlefield tour occurred when 10th Indian Division advanced into Iran in 1941, when the commander of 2/4 Gurkhas, who had fought over this area as a young subaltern in 1917–18 halted the convoy to inspect his old battlefield. In 1940 some troops fought over battlefields of the First World War – on 27–8 May the 2nd Beds and Herts fought on Messines Ridge – while the advance of 21st Army Group back over the 1914–18 battlefields was a moving experience for men like Sergeant Jim Kingston of 4th Somerset Light Infantry, whose father had died in the First World War.[48] In December 1944 7th Armoured Division arrived in Ypres and found a home for a regimental cookhouse in the Cloth Hall. A particularly enterprising soldier managed to sell a gullible civilian a troop carrying vehicle. For the Belgians, it must have seemed as if the British had never been away.[49]

The 'disillusioned' war poets of 1914–18 – Sassoon, Blunden, Graves, Owen, Rosenberg – have had a disproportionate impact on modern popular perceptions of the First World War. To what extent did their poems, which emphasised 'the pity of war', influence soldiers of the Second World War? It is not difficult to find references to the poets of the First World War, but they appear almost exclusively in the writings of officers or 'gentleman rankers' rather than those of working-class soldiers. Going into battle, Alex Bowlby quoted Owen. The poet Keith Douglas, posted to divisional headquarters in 1941, recorded that the general, 'said good morning, good morning, as civilly as Siegfried Sassoon's General'.[50] Exactly what influence the poets of the First World War had on Second World War soldiers is difficult to discern. While at public school in the 1930s Peter Cochrane had picked up the 'conventional wisdom' of the Blimpish nature of the army, partly through reading Graves and Sassoon, yet on joining 2nd Cameron Highlanders in the Desert as a subaltern in 1940, the professionalism of this regular battalion forced him to revise his opinions.[51] In any case, some of the classic poetry of the First World War highlighted not just the horror of war but also the comradeship and the excitement, things that appeal to young men. Finally, the older, heroic style of war poetry was far from dead. There was a renewal of interest in Rupert Brooke during the Second World War, and in 1946 an officer concluded his battalion's war history with a wholly unironic quotation: 'If I should die, think this only of me:/That there's some corner of a foreign field/That is forever England.'[52]

Thus the Somme cast a very long shadow indeed. The soldiers who were to fight in the Second World War grew up all too aware of the horrific nature of modern industrialised warfare. Many later tried to make sense of their own experience of combat by referring to the Western Front. Yet in spite of the profound and obvious similarities between the two wars, many soldiers of 1939–45 seemed to have consoled themselves that 'their' war was not as bad as the previous one. In sum, having accepted a mythologised version of the First World War, they proceeded to mythologise their own experience of battle.

The British Field Force in France and Belgium, 1939–40

Brian Bond

The British army's experience of war in France in 1939 and 1940 was strange and its culmination unexpected; months of inactive 'Phoney War' were followed by a sudden harsh awakening to the realities of *blitzkrieg*. Alan Bennett's play *Forty Years On* captures something of the public's ambivalent reactions to a tragi-comic campaign:

> MOGGIE: We're not winning in France are we?
>
> HUGH: Well, if we are, the sites of the victories are getting nearer and nearer. Though why you expect me to know I can't imagine. I'm only in the Ministry of Information and we're always the last to know.
>
> NURSIE: It must be terribly difficult retreating. Fancy having to walk backwards all that way![1]

The experience of the greater part of this inglorious campaign has been eclipsed in the historiography by its dramatic culmination in the evacuation from Dunkirk and the myths associated with that providential escape from even worse disaster. Most of the soldiers involved knew that it had been a catastrophe, yet the survivors were welcomed back as heroes. It was all very confusing.

The campaign had begun promisingly in September 1939 with the efficient shipment of the five regular divisions to France. These were all reasonably well-equipped and trained, though defective in some crucial respects such as supporting aircraft and tanks. Major-General Bernard Montgomery's 3rd Division was probably the best, though, even so, in his *Memoirs* he was scathing about its weapons, miscellaneous transport and general preparedness for modern war.[2] A further eight Territorial Army divisions arrived in France between January and the end of April 1940. These were in general hastily improvised, with imported drafts from other units to complete establishments, short of regular officers and of varying quality. The best-trained was probably Major-General G. Le Q. Martel's 50th (Northumbrian) Division, which was unique in being composed entirely of Territorial units; Lord Gort having wisely decided that each Territorial brigade should be given a stiffening of one regular battalion. A few regular officers later expressed resentment at this association, believing that they had been let down by ill-disciplined and under-trained Territorials who retreated too easily. Three of the Territorial divisions – unofficially designated 'Labour Divisions' – were not equipped or trained to fight, but they were tragically swept into the maelstrom of the retreat and virtually annihilated. As a recent study sombrely concludes: no second-line brigade survived after its first serious contact with the enemy; and none of the second-line divisions imposed any significant delay on the German advance.[3] The sad fact was that the Field Force had not been equipped or trained for a fast-moving European war against a first-class opponent and could not be fully ready in 1940. It was heavily dependent on the French High Command and its strategy, and on fighting a static, defensive war not so very different from that of 1914–18.

Although the published source material for this campaign is vast, they are mostly deficient in dealing with the experience of war, especially from the viewpoint of the other ranks. The official history was published too soon after the event by an amateur, regimental historian and needs rewriting, but remains useful as a chronicle. David Divine, Richard Collier and others have published vivid popular accounts of Dunkirk, though Gregory Blaxland's history is much broader in scope and benefits from his personal experience as a subaltern in the Royal East Kent Regiment. Nicholas Harman injects a refreshingly critical approach, but his tone is polemical and his use of sources somewhat casual. Ronald Atkin's more recent study is based on an admirably wide range of interviews and personal recollections: it could scarcely be bettered as an impressionistic, descriptive survey but is less strong in putting the campaign in historical perspective.[4] There are numerous colourful and useful published memoirs by participants, including Ewan Butler and Selby Bradford, Bruce Shand, Anthony Rhodes and Christopher Seton-Watson, but most deal with the campaign of 1939–40 only as a first episode in their war – or indeed in their whole military career.[5]

As regards written but unpublished recollections, letters and diaries, the Imperial War Museum holds a huge collection, which has been sampled for this paper. A few of these documents are of outstanding general interest, but many are understandably too detailed, personal and limited in scope. Some soldiers simply did not have a very interesting war or, if they did, lacked ability with the pen. Again, not surprisingly, a high proportion were from the support services (Royal Signals, Royal Army Ordnance Corps, Royal Army Service Corps, and so on) where there was a good standard of literacy, but these men mostly describe life in the rear areas where little combat was witnessed.

Of more direct revelance to the present theme was the Imperial War Museum's sound recordings project on Gort's Army. This extremely useful and under-exploited source consists of expertly conducted, carefully structured interviews which follow a consistent pattern and consequently permit comparisons. Most of the interviews are also available in typescript. A few regimental histories have also been consulted, and these indicate a fascinating and under-explored field of local as well as national military history.

Here only some of the most striking facets of campaign experience can be discussed. First, let us look briefly at training during the months of phoney war. Realistic preparations for war were severely restricted by meagre resources and the urgent priority given to digging and fortifying a defensive line along the British sector of the Franco-Belgian frontier. Extremely severe weather in the early months of 1940 further restricted mobile exercises; in part because it was feared that vehicles and tanks would leave tracks which would be observed by German spotter planes.

The historians of the Coldstream Guards characteristically describe this as 'a dreary period of digging and wiring (in a dismal industrial area of France), interspersed with route marches, weapon-training and battlefield tours'. There were few incidents worth recording in the War Diary.[6] Private Victor Gilbey, of the Oxford and Bucks Light Infantry, was typical of many soldiers in experiencing very little range-shooting and no inter-arms training beyond the battalion. H. W. Dennis's battalion (the 1st East Surreys in 4th Division) only carried out humdrum and monotonous manœuvres along the border. There was plenty of weapons drill, but no firing. Major F. P. Barclay (a company commander in 2nd Royal Norfolk Regiment), summed up his battalion's experience as 'Equipment poor and late in delivery. Training all right but too orthodox, e.g. movement at night neglected. Our commanders were too rigid in their attitude. They did not learn enough about enemy tactics.' Private George Andow was a Matilda tank crew member in 4th Royal Tank Regiment, but he did very little training in tanks during the phoney war. His memoirs describe many days of idleness and whiling away the time in cafés.[7]

It was not easy to provide off-duty entertainments during the dreary winter months. There is ample evidence available on life in barns and billets (very few tents were used), relations with local civilians, the limited attractions of army cinemas and ENSA shows, and the greater ones of liquid and other forms of relaxation. Thus we find Gunner Alfred Baldwin grousing that the other ranks' living conditions and food were poor but his officers did not seem to care. The 1st East Surreys were driven to the brink of mutiny when billeted in a filthy carbon factory at Halluin and then were threatened with barrack damages for moving machinery in the interests of comfort. Yet the same witness (H. W. Dennis), who spoke some French, enjoyed good relations with local people and actually spent Christmas with a French family. Major D. F. Callander described his men (1st Cameronians) as 'very rugged indeed'. They had quite a high VD rate and were very hard to handle when drunk. He reveals that about six soldiers were staked out each night, suspended above the ground, as a punishment, even in extreme cold. As is well known, Major-General B. L. Montgomery was nearly sacked for a breezy memorandum advocating the setting up of licensed brothels, so that any soldier in need of 'horizontal refreshment' could ask directions from the military police.[8]

Apart from a few units which were given a chance to see some modest action in the French sector, there was widespread boredom over the *sitzkrieg* and even eagerness for the 'balloon to go up'. In a typical entry, Major Barclay wrote that his battalion (2nd Norfolks) were raring to go after the months of waiting.[9]

This campaign was notable for the pervasive obsession with spies and 'fifth columnists', who were said to be contacting the enemy in ingenious ways, for example by hanging out washing in semaphore fashion, shining lights at night for enemy aircraft and even ploughing give-away patterns in the fields. Doubtless there were some cases of treachery (and prevalent fears were exploited by the Germans), but many innocent people (or lesser offenders like Belgian deserters) were manhandled, imprisoned or shot. Near Tournai on 19 May, Captain J. H. Patterson (RAMC) ordered his Company Sergeant-Major to inspect a suspicious column of nuns. The CSM examined especially their feet, hands and chins and reported they were definitely female. Suspects checked by Field Security Police were treated with much less tact. With only one officer and ten NCOs per division to screen about 500 suspects a day tempers became frayed. Anthony Rhodes records a chilling conversation with the divisional provost officer responsible for providing firing parties for anyone deemed guilty. His notion of justice was 'teutonic' and there was no appeal against his decisions.[10]

The most striking single aspect of the sources consulted was that combat was experienced overwhelmingly as an air war in which the Luftwaffe enjoyed almost complete dominance. Many witnesses admit to fear and panic caused by strafing and dive-bombing, and there is almost universal grumbling and bitterness about the RAF's minimal role. Several participants saw few or no German ground troops but only glimpsed an occasional low-flying pilot. Here are just a few of the numerous comments on this aspect.

At Warlus on 21 May Major Ian English's battalion of the Durham Light Infantry was attacked by Stukas – the troops were numbed and morale plummeted. However, the survivors soon became habituated to this form of attack in which noise was the most upsetting factor. In the same action near Arras, Private Andow admits that he and his fellow crew members were frightened out of their wits and hid under their tank. Signaller S. L. Rhodda was among those who saw no German troops, and Private Victor Gilbey first encountered them as late as 27 May. Corporal E. J. Manley, in 92nd Field Regiment Royal Artillery, felt abandoned during the retreat and cursed England for causing this humiliation. At Vimy on 20–22 May he recorded that 'most of the lads are pretty demoralised and disgruntled at the lack of R.A.F. support'. Major Callander later admitted that he had felt bitter about the absence of the RAF. On 26 May Captain Patterson counted hundreds of enemy aircraft so that 'the sky was black with them', but he was charitable towards the poor showing by the RAF. On reaching England the troops in his train screamed at RAF personnel in another train 'Where have you been?', but the latter only cheered.[11]

When the German onslaught began early on 10 May 1940, the Field Force's leading formations advanced quite efficiently to the line of the River Dyle in Belgium as agreed with their allies, and for a few days the line was held with deceptively little enemy action. Once the retreat began, however, due to the German breakthrough further south, confusion and even worse soon occurred. Only a static or slow-moving campaign had been planned and communications (in all senses) soon collapsed under the unexpected strain. There had been sufficient motor transport for a planned advance but not for an improvised retreat on roads congested with refugees. There was soon an acute shortage of fuel. The British had relied almost entirely on the Belgian public telephone system, which now collapsed. Radio communications had been restricted for security reasons and now proved useless, as did field telephones. Thus orders could not be transmitted with any confidence and, in effect, confusion reigned.

Typical of many Territorial units' experience was that of 1st Buckingham-shire Battalion, which formed part of 145 Brigade, 48 (South Midland) Division. The battalion was based in Aylesbury and contained a high proportion of local men led by their civilian employers. They mobilised and trained near Newbury before sailing to Le Havre in mid-January 1940. Then they moved up to the dreary mining area of Wahagnies, where the men were mostly billeted in farm buildings and spent most of their time digging what proved to be useless anti-tank ditches and marching on *pavé* roads. On 12 May the battalion moved up by lorry to the battlefield at Waterloo, where there was desultory skirmishing with German infantry. After the retreat began, four days were spent defending the canal line and units began to get mixed up; in a night march the transport and weapons were separated. The colonel collapsed through overstrain and had to be replaced; his successor was killed a few days later, as was also the adjutant. By 25 May the battalion had been pulled back to provide the sole defence of Hazebrouck whence GHQ was about to depart. On 27 and 28 May the battalion, whose scattered outlying companies lost touch with headquarters in the town centre, put up a valiant defence against an armoured division attacking from the west. With ammunition running out in a hopeless battle between infantry and tanks at close range, and with cellars crammed with wounded men, the new commander Major Elliot Viney had little option but to surrender. From a fighting strength of about 600, 100 were killed or missing and between 200 and 250 taken prisoner. Approximately half the battalion, mostly from companies on the perimeter, escaped back to England.[12]

The metaphorical 'fog of war', which enveloped 1st Buckinghamshire Battalion as it moved back from Waterloo to its nemesis at Hazebrouck, affected all units of the Field Force in varying degrees. To the inevitable friction resulting from inter-Allied misunderstandings and contradictory orders, there was added the appalling and pitiful chaos caused by huge crowds of refugees. But when every allowance is made for these hazards of war, many other ranks and junior officers were surely right to complain that they were never adequately 'put in the picture'. This was a lesson that Montgomery took to heart: trust the troops and try to keep them informed about your plans and what is happening.

Private A. F. Johnson, a driver with 101 Company RASC, was among those relieved to find the Royal Navy in charge when he reached the coast at La Panne. He had not seen his section officer for eleven days. 'Ignorant troops are a liability,' he wrote. 'Muddle and incompetence reigned.' Gunner Alfred Baldwin was more scathing: not only were his battery officers absent; even the sergeants forgot what they were supposed to do. He could not understand why the sergeants did not gather their own gun crews and bring them to the beach as a party. There were only sixteen men in Gilbey's section after it had destroyed its transport, and they marched the final nine miles to La Panne. But in his experience the military police kept firm control on the beaches.[13] Thus, though there were honourable exceptions, we have to imagine not an orderly retreat in disciplined units, but rather a myriad of small *ad hoc* groups, all too often without their officers, marching in irregular, spasmodic stages towards the sea, sustained by a vision of evacuation to 'Blighty'.

For the most part the weather during the retreat was sunny and very hot. Exhaustion from constant, bewildering movement and lack of sleep was bad enough, but some groups also suffered from lack of food and even worse, water. For example, S. L. Wright's battalion of the Green Howards had had no food for forty-eight hours and so reached Dunkirk as 'a starving, disorganised mob'. Major Cocke's signals company in II Corps found lack of water to be its most acute deprivation. One desperately thirsty RASC driver, Stan Smith, thought he had found relief in a store of lemonade, only to find he was drinking shampoo and hair oil – it was a barber's shop! Other groups liberated cellars full of wine, to the detriment of their marching stamina.[14]

Nearly all accounts stress the troops' utter exhaustion, caused by lack of sleep, and uncertainty, punctuated by moments of terror. Major English calculated that he had had ten hours' sleep in the first week (from 10 May), and only seven in the second. He summed up his experience as 'Almost total disorganization and chaos really, mainly due to total lack of information about the enemy. Constant movement and lack of sleep reduced one's efficiency to about half. Everyone was just so tired.'

On the retreat from the Dyle, 2nd Battalion Coldstream Guards had fought a hard battle at Pecq, and by the night of 23 May had only two officers per company on duty. When ordered to keep marching overnight, Captain Pilkington recorded:

> Weariness came down like a board on our heads. As I marched my eyes kept closing, and I knew that I was walking about like a drunkard. After a little, nausea overcame me, and I fell out for a few minutes. When we halted for a ten minutes' rest I lay flat on the cold wet road and found a few minutes' relief.[15]

German conduct of the campaign was marred by two appalling massacres of British prisoners, as well as several smaller-scale atrocities and others narrowly averted by the intervention of senior officers. On 27 May in a hamlet inappropriately named Le Paradis some ninety men of the Norfolks surrendered to SS troops of 2nd Infantry Regiment Totenkopf Division. They were kicked, beaten and led into a paddock to be mown down by two heavy machine-guns. The wounded were finished off with bayonets, but two miraculously survived to publicise the incident, one getting his account published as *The Vengeance of Private Pooley*. As a consequence the regimental commander was tried and executed.[16]

On the following day an even more notorious massacre of a similar number of prisoners, from the 2nd Royal Warwicks, was perpetrated near Wormhoudt. SS troops from Sepp Dietrich's division shot several prisoners out of hand and the rest were herded into a barn. When a British officer complained that there was no room for the wounded to lie down, hand grenades were thrown in and orders given for any survivors to be shot. Once again, a few badly wounded men survived and escaped to publicise the incident, but in this case the officer allegedly responsible, Wilhelm Mohnke, has never been brought to trial.[17]

The excuse of the SS units involved was that they were retaliating for British massacres of German prisoners during the fighting round Arras on 21 May. No specific charges were made, but Nicholas Harman resurrected the allegations in 1980 in his book *Dunkirk: the Necessary Myth*. Harman was deliberately imprecise, so his passing reference to the possibility of British war crimes was too vague to be established or refuted. Indeed elsewhere in the book he provided an alternative explanation: namely that some 400 'missing' German prisoners might have been drowned in a French ship sunk while carrying them to England.[18]

Two Durham Light Infantry soldiers, interviewed in the Imperial War Museum's project, recalled that German prisoners were safely handed over on 21 May, but their accounts are unfortunately inconsistent. George Self (an NCO with 8th Durham Light Infantry) admits that at Warlus his comrades lost their heads and shot about six enemy soldiers after an apparently dead German had shot a British soldier in the back. But, under Captain Walton's orders, Self says he escorted the remaining prisoners back from Warlus and handed them over to the French. The other witness, somewhat confusingly named Howard Sell (at the time transport officer with 6th DLI), affirms that the prisoners were disarmed and taken to a collecting point for the military police. They were given food and water, and he personally saw them put on lorries.[19] While it is clear that some German prisoners, or soldiers intending to surrender, were killed in the heat of battle, it seems most unlikely that any were murdered in cold blood after they had surrendered and been handed over to higher authorities. Nothing has come to light to lend any credence to SS allegations in mitigation of their own terrible actions, which were to be repeated on a far more horrific scale in other places throughout the war.

How serious was the breakdown of the British army's discipline in the march to the coast, the defence of the Dunkirk perimeter and, most especially, during the evacuation from the ports and the beaches? Critical accounts, such as those of Nicholas Harman and, to a lesser extent, Ronald Atkin, play up some of the much-quoted incidents concerning Anglo-French friction and even open hostility, notably over crossing the canal line, where the British took control and insisted that all vehicles be destroyed. There were also certainly some cases where individuals – of various nationalities – lost their nerve or attempted selfishly to jump the queues for the boats. As regards Anglo-French differences, it must be remembered that the Field Force had government authority to embark from 26 May and that the Royal Navy had been preparing for the likely emergency over the previous week. The French forces had no such orders until early June and even then their naval resources were meagre and unprepared. Understandably there was anger on the French side when it emerged at the Supreme War Council meeting on 31 May that 150,000 British troops had been evacuated to date but only 15,000 French. On a wider view, however, the sad reality was that the alliance was breaking up so that the British and French viewed the evacuation from completely different standpoints.

Discipline was mainly dependent on the degree to which units remained coherent bodies receiving clear orders from their own officers and NCOs – though it must be added that on the beaches some brave officers were successful in calming panics and exerting authority over miscellaneous groups of soldiers. Here are some contrasting experiences in the final phase of the campaign.

We have already seen that the 2nd Battalion Coldstream Guards was nearing exhaustion by 23 May. On 28 May the battalion was ordered to march more than 50 miles that night from Roubaix to the coast. The men had covered 32 miles over congested roads and in torrential rain when they were picked up by lorries and taken to the perimeter defences. The next four days were spent in intense fighting to hold a 2,000 yard sector of the Bergues–Furnes canal in which the battalion's strength was reduced to 200 men. An acting captain in the 2nd Battalion East Lancashire Regiment, Marcus Ervine-Andrews, won the VC for conspicuous gallantry defending the Canal de Bergues on 31 May. Beating back a dawn attack, he killed 17 Germans with his rifle and many more with a Bren gun. Only 8 of his 85 men survived. During these days a procession of stragglers streamed through the Coldstreams' lines, but they also witnessed the inspiring sight of two platoons of the Welsh Guards, remarkably clean and well turned out, marching across in formation. Colonel Carter recalled that 2nd Battalion Sherwood Foresters did not lose a single man in a continuous forced march of 50 miles. The battalion remained united until it boarded the *Fair Maid of Perth* at Dunkirk, when it was split up and the men's rifles taken away. This was infuriating. Major Callander's Cameronians also retained their weapons, and their cohesion; indeed so impressive were they that on the beaches they were joined by a huge Welsh Guards sergeant who had been separated from his own battalion. Callander was also outraged when, once aboard the ship, the men's weapons were confiscated and they were separated from the officers. On reaching England Callander recorded sardonically that he realised the situation must be desperate when he heard British civilians cheering their soldiers.[20]

Many witnesses admitted to the fear they experienced on the beaches, but most expressly rebut the allegations of widespread indiscipline. In general their evidence reinforces the impressions derived from contemporary photographs and newsreels of the long serpentine queues patiently waiting their turn for a passage home. Private Rhodda is typical in writing that he saw no panicking but 'only a bit of a scramble for the boats'. When his own signals officer tried to carry excess kit on to the beaches he was overruled by a naval officer. Alfred Baldwin, who had bravely swum out to fetch a dinghy but then failed to get aboard, was one of the many who paid tribute to a young naval officer who stood in the water for hours holding off with a revolver troops who would otherwise have rushed and swamped the boats.[21] Apart from sheer exhaustion, it was surely the prospect of getting home safely which enabled the majority of the soldiers to endure the dangers and disappointments on the beaches, at Dunkirk and the smaller coastal resorts as far as La Panne. Curiously, however, several evacuees did not realise they were homeward bound even when taken aboard. Major Callander, for example, was surprised to discover that his Cameronians had been shipped directly to England rather than to a more westerly French port whence they could rejoin the fight. Private Johnson was another escaper who, waking aboard HMS *Greyhound*, was amazed to see the white cliffs of Dover.[22] These instances serve to underline the point that many of the other ranks had little idea of where they were and what was happening: perhaps this accounts for some of their later attempts to bring order and meaning to their experiences during these hectic days by writing about them.

The Field Force's precipitate retreat and evacuation was clearly not a victory, but the magnitude of the Allied catastrophe and its implications were muffled for the British public by censorship and the promulgation of 'necessary myths'. The ambivalence fostered by the conflict between personal experience and propaganda is reflected in soldiers' recollections. On the criteria of casualties and prisoners, the abandonment of vehicles and all heavy weapons and the disorganisation of the higher formations, their commanders and staffs, the campaign had clearly ended in disaster. As Major English reflected: there had been a lot of gallantry and steady fighting, 'but we hadn't been very effective as an Army'. On the other hand, the survivors were surprised and heartened by the heroes' welcome most of them received; there was delight at being free in an English summer after the recent nightmare; and a dogged confidence that with better equipment and training they would be more than a match for the Germans. Colonel Carter summed up a widespread reaction when he remarked, 'It *was* a defeat... there you are. [The] men were angry but not demoralized.'[23] What probably remains puzzling to later generations is the almost universal, but hardly rational, confidence in ultimate victory after this catastrophe following soon after the débâcle in Norway.

As this brief survey has suggested, the British army's experience of war on the Continent in 1939 and 1940 was more varied and interesting than the obsessive attention paid to Dunkirk might suggest. Though kept too much in the dark by politicians and the High Command, this was a reasonably well-educated, literate and inquisitive army composed mostly of civilians in uniform. Individuals in all ranks, from private to general, recorded their impressions in great detail, or displayed remarkable powers of recall when interviewed many years later.

From the military historian's viewpoint the main conclusion must be that the other ranks and junior officers paid the price for the government's belated acceptance of a continental role with inadequate equipment and tanks, an almost total lack of close air support, unimaginative training still based essentially on First World War attitudes, and mediocre leadership. The speed and decisiveness of the Allied rout came as a terrible shock to those who understood the situation, resulting in some bitter criticisms of the politicians and generals held responsible. Understandably too, though it now seems rather shameful, much of the blame was transferred to Britain's recent allies, France and more especially Belgium, for the latter's sudden announcement of a ceasefire on 28 May. But there was also a sentiment of profound respect for the Wehrmacht, and particularly for the Panzer divisions and the Luftwaffe. The experiences of 1940 exercised a lasting influence on generals such as Alan Brooke and Henry Pownall, making them very reluctant to open a 'second front' in Western Europe even in 1944.

For the general public a combination of rigorous censorship, abetted by voluntary self-censoring, skilful propaganda and Churchill's defiant rhetoric, succeeded to a remarkable degree in concealing Britain's precarious position and the unlikelihood of her ever being able to defeat Germany, given the Nazi-Soviet Pact and American isolationism. Nevertheless, King George VI doubtless expressed the irrational optimism of many of his subjects in his often-quoted remark in a letter to his mother: 'Personally, I feel happier now that we have no allies to be polite to and to pamper.'[24] Thus the underlying interpretation of this peculiar 'victory' was that it signified an escape from continental commitments, and hence of the terrible prospect of repeating the experience of 1914–18.

This all-too-brief foray into the British army's experience of war in 1939 and 1940 leaves a strong impression of the great range and interest of personal accounts for this campaign, and later campaigns, in the war, which have scarcely been exploited by historians, compared with research using comparable sources on the First World War. It is to be hoped that this book will stimulate much deeper and more comprehensive research than has been attempted in this chapter.

The Real *Dad's Army*: the British Home Guard, 1940–44

S. P. MacKenzie

The power of television is still sometimes underrated by 'old-fashioned print people' (to paraphrase Marshall McLuhan) such as academics. How people view the past, as well as the present, is increasingly conditioned by what they see on the box. This is certainly true of the Home Guard, mention of which (in Britain at least) immediately conjures up images of Captain Mainwaring, Lance-Corporal Jones, and the other characters of the television comedy series *Dad's Army*.

As Jimmy Perry, one of the programme's creators, recalled: 'The strange thing was that the Home Guard had never been discussed for twenty years before I got the idea for *Dad's Army*.'[1] The great success of the programme, however, which has been run and rerun on BBC1 and BBC2 more or less continuously since 1967, has meant that mention of the Home Guard generates instant public recognition even among those born long after the war.[2]

Dad's Army is also the point of reference for many veterans of the Home Guard. Tellingly, one of the first questions asked in interviews conducted by the staff of the Imperial War Museum sound archive is the extent to which experiences matched what interviewees later saw on television – an approach which always draws a ready response.[3] Without *Dad's Army*, who is to say that the Home Guard would not appear as obscure to non-specialists as does (say) the Volunteer Training Corps of the First World War?[4]

When people think of the Home Guard, they immediately think of *Dad's Army*. Television, in short, has shaped popular memory. In this case, though, no great harm is done; to a considerable degree, the real Home Guard actually matches – with due allowance for comic exaggeration – at least some of the perceptions we have acquired from television.

What sorts of perceptions? In terms of structure: the obvious presence of First World War veterans; civilians from a variety of classes and backgrounds coming forward; and volunteers being organised and led by a local notable (in the case of *Dad's Army* a bank manager). In terms of outlook and behaviour: the great seriousness, even zeal, with which volunteers took themselves in

their anti-invasion role, combined with a tendency to think, speak, and sometimes act on the basis of a civilian, rather than military, attitude toward orders
and authority.

For despite War Office involvement, the Home Guard was very much a self-
made force, organised and run by volunteers at the local level from the ground
up, so to speak, rather than being the result of a thought-through, government-
imposed, top-down plan (as in, say, the ARP organisation). The bottom-up
nature of the Home Guard had enormous implications for its social make-up,
outlook, expectations, and evolution. The rather chaotic origins of the Home
Guard – or Local Defence Volunteers (LDV), as it was known until Churchill
personally substituted the term 'Home Guard' in July 1940[5] – go far towards
explaining the somewhat anarchic nature of the force.

Though far from unprecedented, the idea of a volunteer anti-invasion force
was unattractive to both military and civilian authorities well into the first year
of the war. In an age where the profession of arms had become increasingly
technical and sophisticated, what place was there for the enthusiastic amateur?
A case could be made for the pre-war development of civil defence services,
but almost nobody, even after the outbreak of war in September 1939, could
see the utility of civilians in arms (as opposed to volunteers or conscripts
trained in the army).[6]

All this changed drastically in April–May 1940 in the public mind, with the
opening of the German campaign in the west. The seizure of Denmark and
Norway, followed by the success of German thrusts into France and the Low
Countries, created an atmosphere of considerable alarm. How could the
enemy have achieved such stunning victories within a matter of weeks?
What worried people most were the reports – usually exaggerated – of German parachutists, sometimes said to be disguised as nuns and often assisted by
fifth columnists, seizing important points ahead of the main advance. Who
was to say that the Nazis, assisted by their friends, were not about to descend
on Britain itself?

As editorials, letters in the press, and the reports of the Ministry of
Information's home intelligence division in this period indicate, the public
was increasingly certain that a fifth column existed and that an airborne
invasion of Britain was imminent. 'The situation in a few places', as a home
intelligence report of late May 1940 noted, 'has become slightly hysterical.'[7]
Even *The Times* by 11 May was warning its readers that German parachutists
'might speak English quite well. Some might be sent over in civilian dress to
act as spies. The general public must be alert.'[8]

One of the main consequences of this public alarm was strong lobbying
efforts by a variety of MPs, peers, newspaper columnists, members of the
British Legion and other patriotic individuals, all intent on getting the government to 'arm the people' and set up a local defence force to meet the
parachutist/fifth column threat.[9]

The government, almost as badly informed as the public as to what was
really happening on the Continent, was equally alarmed. On 27 May, for

example, the chiefs of staff presented a paper to the War Cabinet arguing that 'ruthless action should be taken to eliminate any chances of Fifth Column activities'.[10] Reports of German parachutists dressed as policemen and nuns were repeated as gospel in Home Office memoranda, while Lord Croft (made Under-Secretary of State for War when Churchill assumed the premiership on 10 May) estimated that 100,000 German paratroopers could descend at any moment anywhere in the British Isles.[11]

Both the Home Office and the War Office, however, had their doubts about responding to the invasion threat by 'arming the people'. Who knew what chaos might ensue if, as one far from atypical journalist was advocating, 'every householder, every head of family', took up arms 'to protect his own homestead'?[12] If every Tom, Dick and Harry started carrying around a loaded weapon and thinking of himself as a local patriotic fighter, the potential for mayhem and manslaughter would be incalculable. It was for this reason that the War Office politely declined offers by various individuals to form local defence bands in April and early May 1940, and that the Home Office felt compelled to issue a press release explaining that: 'It would not be correct for country gentlemen to carry their guns with them on their walks and take flying or running shots as opportunity offered.'[13]

But in the crisis atmosphere of spring 1940, people were evidently ready to ignore the authorities and take matters into their own hands. By the first week of May local worthies up and down the country – municipal officials, company directors, retired officers, the local squires, and so forth – were forming anti-fifth column and parachutist defence columns which were arming themselves with whatever weapons were at hand. By 12 May the Ministry of Home Security was alarmed enough to send off a teleprinter message to all regions requesting confirmation that 'bands of civilians were forming all over the country and arming themselves with shotguns, etc. for the purpose of detecting and dealing with German parachutists.'[14]

Under such circumstances both the War Office and Cabinet rapidly reversed their position on the need for an officially sponsored local defence force, and plans were hastily drawn up between 11 and 14 May to deal with what one staff officer involved described as the 'burning problem' of what to do 'before civilian residents on the East Coast took matters into their own hands'.[15] With time at a premium and military staff resources already stretched to the limit in coping with the army's troubles, it was decided that the LDV would have to be organised, county by county, by the lords-lieutenant and regional commissioners, and by the volunteers themselves at the local (i.e. platoon and company) level. The sense of urgency was such that on the evening of 14 May, scarcely forty-eight hours after Cabinet approval for the Local Defence Volunteers was given, Anthony Eden (then Secretary of State for War) was making a radio appeal for men aged between seventeen and sixty-five to register at their local police station and join the 'Local Defence Volunteers'.[16]

All well and good, in so far as people were now being given an official outlet for their patriotic urges. Unfortunately, Eden, like everybody else in the War

Office, vastly underestimated the scale of public enthusiasm for such a force and badly misinterpreted what people expected it to be like. The original plan seems to have been that the LDV would serve as a sort of part-time special constabulary, on the lookout for parachute landings and suspicious activities.[17] A more aggressive role seemed out of the question, given the arms and training needs of the army (especially after Dunkirk). From the perspective of the volunteers themselves, however, the LDV was supposed to be an armed force, ready to pounce on the invader wherever he landed.[18]

All else being equal, the War Office ought to have had its way. Its staff, after all, were the professionals with greatest knowledge of what was and was not feasible. Several related factors, however, soon made it apparent that the LDV would not simply 'Look, Duck, and Vanish.'[19]

First, the number of men whose expectations of an immediate supply of arms and equipment (beyond the issue of LDV armbands) were not being met was very large indeed. The planners had anticipated a maximum of about 150,000 volunteers; even Lord Croft thought that at most 500,000 men would come forward; but by the end of June 1940 the number of registered enthusiasts had reached 1,456,000 and was still climbing.[20]

Second, a significant number of the volunteers were veterans of the First World War. The actual number could vary radically from one unit to another, but the overall proportion – estimates during the war varying from 40 per cent to 75 per cent[21] – was impressive. And as self-styled 'old sweats', these men thought themselves fully prepared to meet the Hun again as long as they were given arms and equipment.

Third, whether ex-soldiers or not, the volunteers were classified as civilians until the invasion came, and hence were not subject to the restrictions on freedom of expression contained in King's Regulations. Hence any grievances could be aired in public without fear of a court martial.

Fourth, the organisational instructions for the LDV in a sense merely gave official encouragement to a process that was already underway: defence bands generated from below rather than above. In Leeds, for example, in a manner replicated in countless locations and organisations, the local golf club turned itself into an LDV unit simply by electing the club captain as commander and going on night patrols of the course with clubs and sticks.[22]

And finally, especially among the more senior LDV administrative appointments, there were a lot of men of real influence. MPs, members of the peerage and retired senior officers abounded. Of the ten central London LDV zones, for instance, four were run by retired lieutenant-colonels, one by a full colonel, two by retired brigadiers, and one each by a retired major-general, lieutenant-general, and full general.[23] The Home Guard unit for the Palace of Westminster included 95 MPs and 17 peers, many of whom, like old generals such as Sir Hubert Gough who were determined not to fade away, were prepared to use their influence and position to lobby the relevant authorities.[24] As Lieutenant-General Sir Henry Pownall, appointed inspector-general for the Home

Guard in June 1940, recorded in his diary, there were bound to be difficulties caused by the 'masses of retired officers who have joined up, who are all registering hard and say they know much better than anyone else how everything should be done'.[25]

And it worked. The War Office quickly reversed its position over the summer of 1940 and made strenuous efforts at least to *appear* to want the Home Guard to become a front-line fighting force. Tactical instructions were modified to make the force more aggressive, for example, while strenuous efforts were made to find old equipment and purchase as many surplus .300 rifles and machine-guns as possible from the United States.[26] The Home Guard had made it abundantly clear that it took its anti-invasion role very seriously indeed, and for the next four years Churchill in particular (along with a wide range of ministers and generals) took care to try and make sure that the Home Guard did not feel slighted in any way.

This concern over appearances manifested itself in a variety of forms. The manufacture and issue of a series of special sub-artillery weapons (Northover Projector, Blacker Bombard, Smith Gun);[27] the inclusion of the Home Guard in the December 1941 National Service Bill (whereby civilian workers began to be conscripted into the Home Guard, along with the civil defence services, less because of a need for manpower than because it was feared that leaving the Home Guard out would create the impression that it was not being taken as seriously);[28] the huge public-relations effort devoted to 14 May anniversary parades;[29] and also the extreme cautiousness with which the authorities approached the question of standing down the Home Guard (an event which took place in late 1944, many months after even the remotest possibility of an enemy raid had disappeared – but at about the time even the true believers in the force had given up hope).[30]

It may not be going too far, indeed, to suggest that in many ways the Home Guard manufactured itself according to its own image. As General Pownall wrote: 'The H.G. are voters first and soldiers afterwards...What they think they need, if they say so loud enough, they will get.'[31]

Another, related, aspect of the seriousness with which the volunteers of the Home Guard took themselves was their sheer zealousness. 'Our enthusiasm verged on the fanatical,' one volunteer recalled.[32] 'Morale was very high,' as another put it concerning the common desire to get to grips with the enemy. 'We genuinely looked forward to the Germans coming.'[33] In the summer of 1940, there was no alternative to letting the Home Guard organise and run itself on a local basis with minimal instruction and oversight. But left to their own devices, local units – or at least local commanders of the Captain Mainwaring type – could sometimes go overboard in terms of preparing to meet the Nazi invasion.

It was one thing, in lieu of issued rifles, for sections and platoons to arm themselves with whatever they could lay their hands on (anything from shotguns to cutlasses and bags of pepper);[34] it was quite another to assume that Home Guards had unlimited powers in dealing with the fifth column threat in their particular area. Acting on this assumption, which was quite common, caused a good deal of inconvenience and occasional tragedies.

Zealous volunteers went on night patrols and set up roadblocks, conducting impromptu identity checks on their own authority. Courting couples, air-raid wardens, firemen, even soldiers and policemen, were stopped and challenged. Nobody, it seemed, was above suspicion. 'One man,' the *Daily Express* reported on 15 August 1940, 'said that he had been stopped twenty times on a journey of eight miles.' Not even the mighty were immune from high-handed treatment: Lord Gort, for instance, former commander of the BEF, found his car commandeered one night in August by a Home Guard.[35]

This sort of situation formed the basis for more than one *Dad's Army* script. 'Cases have been reported', the CO of the 7th Herefordshire Home Guard Battalion noted in June, 'where Army Officers have been stopped and asked for identity cards by LDV who themselves could not establish their own identity.'[36]

The darker side of all this zeal arises from the fact that trigger-happy Home Guards managed to wound or kill at least a dozen innocent motorists.[37] 'I have lost two officers in the course of ten days,' the CO of the 2nd Anti-Aircraft Division wrote to GHQ Home Forces on 12 June, 'one shot dead in Eastern Command and one shot and dangerously wounded between Nottingham and these headquarters'.[38]

Orders were issued to put a stop to this sort of thing, and the War Office Home Guard Directorate and GHQ Home Forces made successful efforts to impose greater regulation on the force over the winter of 1940–41 (involving the issuing of standard battledress uniforms, new administrative regulations, and a formal commissioning process for Home Guard commanders).[39] Yet even after the Home Guard was quietly purged of some of its more rambunctious elements, and a barrage of instructions issued on how battalions should operate and be run (along with regular adjutants and quartermasters to oversee them), many units continued to behave as if they, rather than the War Office, had the final say in Home Guard affairs. This applied especially to matters of weapons and role.

Though orders were issued that Home Guards were not to manufacture home-made weapons (due to a number of accidents), everything from *ersatz* armoured cars to soup-tin grenades continued to be constructed in some areas. These were units in which it was felt not enough (or the right kind of) weaponry was being provided by the War Office.[40]

Similarly, though the operational plans produced by GHQ Home Forces in 1941–2 called for the Home Guard to engage in essentially static, 'nodal point' defence – that is, defending villages and towns – various units, either completely on their own initiative or through creative interpretation of ambiguous instructions from higher authority, developed operational plans of a different kind.

Some Home Guards wanted to become mobile, even armoured. They trained for the role and equipped themselves with assorted civilian vehicles converted for military use. Machine-guns and Home Guard sub-artillery, designed for local manhandling, were sometimes modified by such units (again without authorisation) so that they could be towed or mounted on cars.[41]

Other Home Guard units, inspired by the Soviet example, wanted to become guerrilla fighters. One Edinburgh company commander, for example, later related how he had worked hard to turn his unit 'into a body of first-class guerrillas or "banditti" rather than second-class regular soldiers'.[42]

When GHQ Home Forces learned of such activities stern counter-orders were issued. Motorised Home Guards charging hither and yon as the mood took them would impede vital army traffic, while a guerrilla mentality could lead to the abandonment of static defences. But the Home Guard was so large (well over 1.5 million men at any one time), and the administrative structure so limited (especially above battalion level), that some units were able to go their own way pretty much unmolested down to the end of the war. Disguises and unorthodox tactics were pursued enthusiastically in inter-unit exercises (again often despite orders to the contrary); while among the mobility enthusiasts there were one or two units which went from mere armoured cars to acquiring (through various means) a few obsolete tanks.[43]

It is important to stress that by no means all units favoured such activity. Variety, indeed, was a characteristic – and inevitable – feature of a force so home-grown in character. The outlook and training of Home Guard units varied a lot according to region, whether in town or country, and on the whims of commanding officers (and the men themselves). A Home Guard moving from one part of England to another could find himself in a different world; though the practice of congregating in pubs after parades and exercises seems to have been well-nigh universal.[44] But this – plus the fact that neighbouring units often regarded each other with almost as much suspicion as the enemy[45] – only underscores the essentially civilian and rather parochial nature of the force (certainly in comparison to the regular army).

Even the issue of class supports this. As already mentioned, the Home Guard was essentially organised through social rank. This was true not only in the country – where what can loosely be termed the squirearchy (or what the *Daily Worker* called the 'Hunting-Shooting-Fishing Oligarchy')[46] tended to be put in charge by the lord-lieutenant or simply assume command over the lesser folk – but also in towns and cities, where officers of Home Guard factory units tended to be management while the men in the ranks were workers. The 5th City of London (Press) Battalion, for instance, was composed of men from *The Times* and commanded by the newspaper proprietor John Astor.[47] When the LDV was organised in John Lewis's in Liverpool, the NCOs and officers were the floor and department managers;[48] Charles Lightfoot, a 'well-known businessman' in the glass industry, 'automatically became the commandant [sic] of the Home Guard in Manchester'.[49]

The social gulf between officers and men in the Home Guard, however, should not be exaggerated. In areas where there was an over-abundance of retired senior officers (e.g. Kensington-Belgravia or Chelsea) they could and did serve quite happily in the ranks.[50] Both officers and men in particular units often came from a variety of social and professional backgrounds; and at root, as one instructor put it, a platoon could be 'just a bunch of chaps, that's all'.[51]

More significantly, differences in social rank appear not to have translated into full-scale military deference. Especially in 1940, but also in later years (in more indirect forms), men in Home Guard uniform not only questioned or challenged authority outside the force – the police, the army, the War Office, and so on – but also within it. Instructions could be queried or ignored, after all, with relative impunity, and really objectionable orders could be countered by simply resigning (sometimes *en masse*). To judge from the extant correspondence files of Home Guard units, disputes could go on for months and sometimes years. 'I am in the odd position', as one volunteer explained in a letter, 'of saying exactly what I like to my officers. They swallow it most generously.'[52]

Everything dealt with thus far figures, to a greater or lesser degree, in *Dad's Army*. Though some former volunteers were offended at the way the force was portrayed, for others the scripts – with due allowance for comic exaggeration – had the ring of truth to them (hardly surprising in view of the fact that Jimmy Perry and David Croft, the creators of *Dad's Army*, had both been in the Home Guard).[53] But there are also aspects of the Home Guard experience where art, or at least entertainment, does not mirror life.

The fictional Home Guard platoon of Walmington-on-Sea matches the image fostered during and after the war of a force that was primarily rural, based in small towns and villages.[54] Yet in point of fact the Home Guard was largely an urban phenomenon. Urban centres were where the majority of the population lived, and city dwellers were just as keen to do their bit as country folk. Moreover, urban industrial workers, already organised in semi-regimented fashion on the job, could simply transform themselves into factory Home Guard platoons and companies. In consequence, the size of urban Home Guard units was much greater than those in less heavily populated areas.[55]

It also tends to be forgotten that from 1942 to 1944 a significant proportion of the Home Guard – around 300,000 – were in fact conscripts: men in reserved occupations, not liable for call-up into the army, who were directed to join the Home Guard as a result of the 1942 National Service Bill. Some directed men appear to have taken to life in uniform (up to forty-eight hours per month), but by no means all were enthusiastic, especially as the threat of German invasion continued to recede.

Trade unionists were particularly annoyed at what they saw as over-zealous attempts to prosecute in the civil courts conscripts who did not turn up for parades. From the perspective of the old volunteers and regular officers working with the Home Guard, such as Major-General L. A. Hawes, such men 'judged their military duties by civilian standards...that is to say by factory regulations, trade unions, home life, daily civil contacts, and by the knowledge that they were unpaid and their punishments limited'.[56] By November 1943 friction had developed to the point where Walter Citrine, General

Secretary of the TUC, publicly attacked the 'savage penalties' levelled against workers who sensibly wanted to avoid 'useless drills'.[57]

There was, in short, a reluctant Home Guard – not least in connection with one of the duties the Home Guard increasingly undertook: manning anti-aircraft batteries. As the manpower needs of the field army grew, Anti-Aircraft Command came increasingly to rely on the Home Guard (as well as the ATS) to replace men needed in the infantry through the manning of searchlights and gun and rocket batteries, and by May 1943 nearly 112,000 Home Guards were assigned to ack-ack sites.[58]

This was important duty. Unfortunately, it was also fraught with problems. Creature comforts on the more remote sites were few, relations between the Home Guard regular crews were often bad, and conscripted Home Guards in particular were so lacking in motivation that General Sir Frederick Pile, GOC AA Command, considered them next to useless. 'They never turned up when they were wanted,' he later noted, 'and, in fact, many units learned not to expect them.'[59]

A third aspect of the Home Guard experience that is mostly ignored in the world of *Dad's Army* is the role of women. From 1940 down to early 1943, despite vigorous lobbying efforts by Dr Edith Summerskill, MP, and others, women were officially excluded from the force. 'We were never really recognized as the Home Guard,' as one Leeds woman recalled.[60]

Yet approximately 50,000 women in this period, according to a contemporary estimate, served in an unofficial Home Guard auxiliary function as clerks, typists and cooks. Rather ironically, by the time the War Office consented to the formation of an official women's Home Guard auxiliary of 80,000 in 1943, there was very little interest. By March 1944 only 28,000 women had enrolled.[61]

The Home Guard is unique among the forces discussed in this collection in that it never saw action. Conjecture, however, is hard to resist. What if Hitler *had* invaded in 1940 or 1941? How would the Home Guard have fared?

Any answer to such a question must be, of course, highly speculative; but given what is known about the relative strengths and weaknesses of the forces involved the likely 'Home Guard experience' in combat would have been bloody indeed. Units in the invasion area, largely untrained and very poorly equipped (especially in 1940), would have had to confront highly trained, well-armed, and battle-hardened professionals. They might have slowed down a German advance, but the cost would have been very high – especially given that the German government refused to recognise members of the force as legitimate combatants (which would have meant few prisoners would have been spared). As a volunteer of the Isle of Wight Home Guard concluded: 'We'd have put up a good show for about half a day. And that would have been the end of us.'[62]

As it was, the Home Guard – which encompassed roughly 20 per cent of the adult male population not in the army or civil defence services[63] – was spared this ordeal. Though contributing a very real service to the British war effort through maintaining morale and freeing up regulars in Anti-Aircraft Command, the Home Guard retained a curious 'grown-ups-playing-at-soldiers' innocence. 'We all lived out our fantasies in this strange way,' Jimmy Perry reflected – for which viewers of *Dad's Army* and members of the real Home Guard can both be thankful.[64]

The British Soldier on the Home Front: Army Morale Reports, 1940–45

J. A. Crang

The popular image of the British army during the Second World War is largely concerned with events overseas and the pictures that most readily spring to mind are of battle-hardened troops fighting through the jungle of Burma, the desert of North Africa or the bocage of Normandy. Whilst these episodes undoubtedly reflect an important element of the soldier's experience of the war, they do not, however, tell the whole story. What the popular image tends to overlook is the fact that the majority of the British army spent the greater part of the war in Britain. A vast number of anti-aircraft, home defence, field force, logistical, training and administrative units were based in the United Kingdom, either defending the country against attack or supporting the army abroad, and it was not until after D-Day that the majority of troops were serving overseas. Indeed, for much of the war more than 1.5 million troops, which represented well over half of the army, were stationed in Britain and even in the spring of 1945 about a million, which at that time was around one-third of the service, were still serving at home.[1] For most soldiers, therefore, the experience of war was not one of daring deeds at the sharp end, but rather of a sedentary existence in camps and depots across the country polishing their brasses and wondering why they were there. This essay focuses on the experience of British soldiers on the home front during the Second World War and, through an analysis of army morale reports, attempts to build up a picture of the factors that influenced their morale.

Army morale reports were, it seems, compiled occasionally in the early period of the war. At least one document, dated September 1940, still exists. Written by the Directorate of Military Intelligence in the War Office, it provides a summary of the prevailing mood of the troops and draws on the views of serving soldiers and officers as well as those of civilians who were deemed qualified to comment.[2] It was not, however, until the beginning of 1942 that army morale reports were produced on a regular basis. Several factors were involved in this new development. To begin with, the War Office became increasingly concerned about the morale of the troops, which threatened to undermine the military efficiency of the army. From the outset there was little enthusiasm for the war amongst the conscript soldiers who made up the bulk of the army and their priority was to get the conflict over with as quickly as possible. Yet here they were, the majority of them, sitting around in Britain taking little active part in the war effort and depressed by the string of humiliating defeats that the army had suffered in the early years of the war; a problem compounded by what was seen as a scurrilous campaign in the popular press against the incompetence and blimpishness of the 'brass hats'.[3] What was more, the military authorities had little means of assessing the morale of the troops in any comprehensive fashion. Certainly, there were theories about what was wrong: Sir James Grigg (Secretary of State for War) thought that at least part of the explanation for the mood of despondency amongst the soldiers lay in what he described as the 'pansydom' of the inter-war years with its emphasis on privileges rather than duties to the state.[4] There was, nevertheless, no machinery in place for surveying and analysing the morale of the army or co-ordinating any action on the basis of judgements that were made.[5] The influence of the Adjutant-General, Lieutenant-General Sir Ronald Adam, was also important in the efforts to monitor morale more effectively. A man with deep concern for the welfare of the citizen soldier, Adam was prepared to utilise a variety of new, and sometimes radical, techniques in order to maintain the morale of the army, which was a matter of particular importance to him as the military authority responsible for personnel matters. 'This war is going to be won or lost on morale,' he argued in a frank statement to his senior colleagues in the War Office. 'We are too apt to leave the problem alone. Morale is a psychological problem like sex, and therefore the Britisher is almost ashamed to talk about it.'[6]

For all these reasons, in January 1942 the first in a regular series of army morale reports on the army at home was produced by the adjutant-general's department and in March a special Morale Committee was set up in the War Office to monitor them and co-ordinate action on the basis of the assessments.[7] The reports were compiled by Major John Sparrow (Deputy Assistant Adjutant-General and subsequently Warden of All Souls College, Oxford) from a variety of sources: divisional, district and area commanders' reports; censorship reports of soldiers' letters; courts-martial statistics; home intelligence surveys by the Ministry of Information; and enquiries addressed to Professor John Hilton, who gave a weekly broadcast talk from the BBC to soldiers and their families between May 1942 and May 1943.[8] The reports were produced on a quarterly basis and circulated to the Army Council and contributing commanders. Similar surveys were drawn up for the army overseas.[9] In his record of the War Office's activities in this field, Sparrow defined morale as comprising 'all those things which make the soldier more, or less, keen to carry out his job of soldiering, and readier, or less ready, to endure the hardships, discomforts and dangers that it entails'.[10]

One obvious factor which was identified in the reports as influencing the morale of the troops was their general living conditions. In the early stages of the war, and particularly after the withdrawal from Dunkirk, makeshift camps sprang up around the country and there were many complaints from soldiers who found the conditions hard to endure after the comforts of civvy street. The morale report for September 1940 drew particular attention to the inadequacy of the messing arrangements, the poor standard of accommodation and the lack of basic amenities in outlying areas. 'The food is a monumental scandal,' complained one soldier in Scottish command. 'One slice of bread for breakfast and one slice for tea – we are continually hungry.'[11] Another lamented: 'The bell tent in which we sleep leaks ... Haven't had clean clothes for 3 weeks or bath ... If we want a wash or shave we have to go half a mile to wash in a *ditch*.'[12] 'In the mornings', wrote a further sufferer, 'my blankets and ground sheet are clammy with wet, but an hour ago the Engineers started to erect the huts, so perhaps by this time next week we shall be in the dry.' He was to be disappointed: '*Shattering* news,' he continued, 'the huts being erected by the sappers are for themselves.'[13] For some, all this proved too much:

> What I have to put up with, such as bad grub, wet clothes and blankets, no beds to sleep in, a wet tent that leaks like a sieve ... I tell you that they have broke my spirit so that when I go to bed at nights I could sob my heart out ... The first chance I get I shall desert, never fear.[14]

As the war went on and camps became better established and more Nissen huts were provided, so living conditions improved. The messing arrangements, in particular, seemed to have advanced considerably and there was a good deal of praise for the new Army Catering Corps.[15] According to morale reports in 1942–3, however, there continued to be pockets of dissatisfaction. It was found that in Eastern Command, where the decision was taken relatively late to house large numbers of troops, the standard of accommodation left much to be desired.[16] There were also complaints from Anti-Aircraft Command, manning guns or searchlights at isolated sites, about the lack of drying areas, the shortage of quiet rooms and the quality of lighting available in their huts.[17] To make matters worse, the arrival of large numbers of American troops placed extra pressure on the accommodation available and eventually resulted, it was said, in a quarter of a million British soldiers being forced to sleep on the floor without beds.[18] 'Shoddiness', wrote one soldier in December 1942, 'is the word to describe conditions in the men's billets and messes in many units.'[19]

The military authorities did tackle some of the problems which came to their attention. In Eastern Command, enquiries were made into the provision of accommodation and steps were taken to improve the ventilation in Nissen huts by the installation of extra dormer windows.[20] As for Anti-Aircraft Command, the scale of drying areas was increased, investigations were conducted into the establishment of quiet rooms and efforts were made to enhance the quality of lighting by the more effective use of methylated spirits in the Tilley lamps.[21] Arrangements were also put in hand to explain to the troops the reasons for particular hardships they were forced to endure, especially those caused by the arrival of the Americans.[22] Whilst complaints about living conditions did not cease, the improvements that were made to the accommodation available and the gradual reduction in the number of troops in Britain that needed to be housed, meant that by the final stages of the war the morale reports indicated that these matters were no longer at the forefront of the soldier's mind.[23]

With so many troops based in Britain, often engaged in fairly routine tasks, another aspect of army life which the reports highlighted as influencing morale was the availability of recreational facilities. In the early stages of the war, and particularly in the wake of Dunkirk, many camps were without such facilities as canteens and entertainments, and there was a good deal of dissatisfaction on the part of soldiers who found themselves bored and restless in the absence of the leisure pursuits of civilian life. The morale report for September 1940 made special reference to the lack of spare-time activities in the more remote parts of the country. 'I think this place is getting on everyones' nerves,' wrote one soldier from Scottish Command. 'It would be alright if there was something to do.'[24] Another pleaded: 'Time lags dreadfully monotonously on our hands...it is an immense relief when we go to bed and forget for a few hours.'[25] 'If there is no sign of leave starting in the near future,' warned a third, 'I expect to see a real riot.'[26]

Efforts were made to improve the recreational facilities for the troops, with voluntary organisations such as the YMCA helping out where they could and public-spirited individuals also encouraged to participate: one lady (in Rotting-dean) opening her garden to soldiers for a short time each day and various hosts and hostesses organising small supper parties, subject to some discretion in the choice of their soldier guests.[27] Morale reports for 1942–4, however, identified continuing sources of complaint. One particular problem concerned the canteen provision of the NAAFI. The managers, it was said, lacked initiative; the premises were considered drab; there was a shortage of sweets, chocolate, razor blades and good-quality cigarettes at pre-budget prices; and the tea was rotten.[28] The variety shows provided by ENSA were a further subject of criticism. The entertainers, it was reported, often did not visit remote areas and when they did perform the soldiers complained – with surprising prudishness – that the acts were too low-brow and suggestive in tone.[29] There were also representations from Anti-Aircraft Command about a shortage of wireless sets and film shows which, when they did take place, featured out-of-date films.[30] 'The Army', concluded the morale report for May–July 1942, 'still suffers from being bored and "browned off".'[31]

The War Office did try to deal with these complaints. Enquiries were made into the running of the NAAFI canteens and there were some attempts to improve the service available – NAAFI tea, in particular, being found to taste infinitely better when dispensed from tea-pots rather than urns.[32] There were also efforts to weed out ENSA shows that were of poor quality and greater encouragement was given to service performers to entertain their own units.[33] In Anti-Aircraft Command, extra wireless sets and film equipment were provided and the Ministry of Supply was approached over the availability of the latest Technicolor films.[34] Although grumbles about recreational facilities did not cease, the steps that were taken to upgrade the amenities and the smaller number of troops to cater for, meant that by the end of the war morale reports recorded a considerable improvement in these matters.[35]

Whilst consideration was given to improving the recreational facilities available, there were various aspects of the soldier's working life that the reports identified as being important to the morale of the army. On entering the service it was clear that the wartime soldiers preferred to take part in hard and realistic training rather than waste their time on what, to the civilian mind, seemed like pointless rituals. This was particularly emphasised by the morale report for September 1940:

> Close contact with troops reveals continuously the soldiers' preference for service in a unit which makes continuous, hard, and in his judgement, useful demands on his mental and physical powers, to service in what he describes as a 'heap', even if life is incomparably softer.[36]

Yet to his frustration what the soldier's day-to-day routine often seemed to entail was not a rigorous preparation for active service, but merely a series of unnecessary parades and excessive 'spit and polish', most of which appeared to have little relevance to the war effort. 'The present-day soldier', the report observed, 'is very quick to differentiate between War efficiency, and "spit and polish" efficiency.'[37]

Some attempts were made to improve the working life of the soldier and rid it of some of its more outmoded practices.[38] Dissatisfaction, however, remained. The morale report for February–May 1942 argued that whilst the strategic necessity of keeping large numbers of troops at home was generally understood, there was still a longing for more active training and a continuing resentment of many of the rituals that had to be performed. 'The average soldier', it was noted, '...particularly objects to "spit and polish" where its application serves no apparent purpose.'[39] What all this helped to bring about was a dangerous tendency towards apathy amongst the troops. The morale report for May–July 1942 warned that

> though there are no positive signs in the Army at home of war weariness, apathy, or the tendency to ask 'What difference does it all make?'...the troops, in their present state of mind and feeling, might well fall victims to propaganda which aimed at producing such an attitude. It is reported that it is not uncommon for soldiers to ask lecturers whether Hitler has not done a good deal for the trade unions in Germany, and similar questions which reveal that they have no idea of the kind of enemy they have to face.[40]

The military authorities took action to remedy some of these problems. To begin with, efforts were made to make training more lively and realistic, and to this end field-firing exercises were encouraged and a new battle drill was introduced to simulate the conditions of combat; the most bloodthirsty troops, it was said, being those from the mobile laundry units and the Royal Army Medical Corps.[41] Some of the rituals of the service were also modified. Instructions were issued that pay parades could be expedited by paying companies on a platoon basis; that church parades should be less formal occasions; and that sick parades be replaced by an 'MO's hour'.[42] It was further laid down that standards of 'spit and polish' ought to be 'suited to the occasion' and observed in a manner which the men could 'understand and respect'.[43] Underpinning this, attempts were made to inspire some crusading fervour amongst the troops by the organisation of weekly current affairs discussions designed to convince them of the objectives for which Britain was fighting.[44] Whilst there continued to be some complaints about unnecessary 'eyewash',[45] the measures that were taken to improve the working life of the troops, combined with the increasing prospect of active service overseas, ensured that by the later part of the war any tendency towards apathy referred to in earlier morale reports seems to have faded from the minds of the soldiers.

A further area of army life that the reports identified as being crucial to morale, and one linked to the working conditions of the troops, was the relationship between officers and their men. In the early stages of the war the relationship left a good deal to be desired. The wartime soldiers were more class-conscious and better educated than their predecessors and were to look much more sceptically on the privileges of rank and the autocratic style of leadership that had characterised the pre-war army. This was particularly highlighted by the morale report for September 1940, which noted that the new recruit was 'quicker than before to resent what he calls "the old school tie" stand-offishness when conditions lend themselves to closer comradeship'.[46] Yet rather than making the necessary adjustments to their relationships with the soldiers, not only was it apparent that many young officers openly flaunted the privileges of rank, but they also displayed a complete lack of interest in the welfare of their men. 'There is almost everywhere', it was observed, 'a failure to recognise the full significance of the cliché, "The men must come first." '[47]

The army did recognise these problems and tried to inspire more mutual respect between the ranks.[48] However, the relationship between officers and men remained a cause for concern. Two aspects of the problem were identified by the morale reports. First, it was clear that whilst the privileges accorded to officers were generally accepted if they were the result of military necessity, those for which there seemed to be little justification, apart from the assumption that gentility was concomitant with an officer's status, were keenly resented by many of the soldiers. The morale report for May–July 1942 noted:

The morale and fighting spirit of the Army as a whole would be enhanced if the ordinary soldier could be reassured that differentiations due to social tradition and the subordination involved in military discipline do not imply a fundamental conflict of interests. Anything, on the other hand, that strengthens his belief in the existence of a fundamental gulf or barrier between himself and his leaders has an immediate and marked adverse effect on morale.[49]

The second, and more important, element in the strained relationship between officers and their men was the poor standard of man-management on the part of junior officers. Whilst most soldiers were prepared to concede the necessity of officers leading a somewhat detached life from their own and following even the worst leader, it was clear that they were going to require tactful and sympathetic handling if their willing co-operation was to be maintained. But whilst it might have been thought that the new wartime officers, all of whom had spent a period in the ranks, would bring a new degree of sympathy and understanding to their relations with their men, it was ironic that many of them seemed to go to great lengths to distance themselves from their subordinates. 'Several commanders report', observed the morale report for February–May 1942, 'that young and recently joined officers are "woefully ignorant" of man-management and lack interest in their men.'[50]

In trying to account for the poor standard of man-management several observations were made. One contributory factor, it was suggested, lay in the continuous cross-posting of officers and men that prevented the build-up of any sense of *esprit de corps* in a unit.[51] This was a particularly sensitive matter for the Scots, Welsh and Irish. 'Our only trouble', wrote one officer of the Northern Irish Horse in the report for February–May 1942, 'is the lack of Irish officers, there are only about 5 of us here now; I am sure it is wrong – when there are so many Irishmen in the Regiment. The English officers we get these days are terrible idiots.'[52] Another explanation concerned the quality of officer material that was available to the army. Initially it was believed that the problem lay in the fact that many of the new wartime officers came from the grammar schools, rather than public schools, and were thus not immersed in the traditions of leadership. 'Most candidates for commissions', observed the morale report for November 1942–January 1943, 'now have not a "background" which helps them to adopt the right attitude towards their men.'[53] But as the war went on, it was found that those who had the right 'background' were also at fault. 'It is, perhaps, significant', continued the report, 'that one of the Commanders who register complaints in this matter is Commander of a Guards Brigade Group: presumably the regiments under his command draw upon promising sources, yet he finds that "many young officers on joining have no idea of man-management".'[54] Thus, the report concluded, the shortcomings of young officers were not only caused by a shortage of candidates from the traditional sources, but were also the result of 'selfishness and negligence on the part of young officers from whatever background they are drawn'.[55] A further explanation, it came to be recognised, lay in the attitude of commanding officers of units, the majority of whom were regulars. Not only was it said that they had a poor understanding of the mentality of the wartime soldiers, but also that a good many of them were simply not interested in their men and therefore did not ensure that their junior officers took an interest in them either.[56] The morale report for November 1943–January 1944 noted that 'it is upon C.O.s and company commanders that the chief responsibility for the state of man-management must rest'.[57]

Whatever the reason, morale reports portrayed a catalogue of mismanagement on the part of young officers. 'It is evident', lamented the report for May–July 1942, 'that there is still a deplorably large proportion of officers who fail to care properly for their men's welfare and to inspire their men's respect.'[58] This situation, it was argued, had several important consequences. First, it created a worrying gulf between officers and men that threatened to undermine the very solidarity of the army. 'The problem', warned the morale report for August–October 1942, 'is largely one of officering: the troops are ready enough to feel friendliness, respect, and admiration for the right type of officer... The troops' letters show, however, that such a relationship is far from universal.'[59] What was more, the poor relationship between officers and their men, combined with what was seen as the sham of senior officers and government ministers conducting grand, formal inspections of the troops without a word of encouragement to the long-suffering other ranks, led to a sense of alienation from the higher military authorities. 'To judge from the tone of his letters,' noted the report for May–July 1942, 'the ordinary soldier does not fully identify himself with the Army; he looks with detachment upon it and those who control it, and thinks of those in authority, whether political or military, as his governors rather than as his leaders.'[60] Indeed, it was further suggested that the poor relationship between officers and men threatened to undermine the existing political system since references to high-handed officers in the soldiers' letters were nearly always followed by allusions to 'capitalism', 'democracy' and 'Russia'.[61] The morale report for February–April 1943 concluded that 'one selfish or "stand-offish" officer produces more potentially disaffected soldiers than six communist agitators'.[62]

The War Office tried to improve the relationship between officers and men. For a start, some of the privileges of rank were restricted. Officers, it was instructed, were no longer permitted to reserve hotel bars and lounges for their own exclusive use;[63] the use of military transport by officers for recreational purposes was limited;[64] and attempts were made to equalise the living conditions on troopships, which were a particular source of grievance to the rank and file.[65] In addition, efforts were made to improve man-management. New officer-selection methods were introduced which included psychological tests to assess a candidate's capacity to manage his men;[66] the syllabuses of officer cadet training units were modified to incorporate more training in man-management skills;[67] commanding officers were encouraged to improve man-management practices in their units;[68] and a system of weekly 'request hours' was instituted when men could approach their officers informally over any matter that was troubling them.[69] Steps were also taken to reduce the cross-posting of officers and men by the decision to use designated 'lower establishment' divisions as pools of draftees.[70]

Alongside this, attempts were made to bridge the gap between soldiers and the higher military authorities. Formal inspections of troops by senior officers, it was recommended, should be kept to a minimum;[71] shortened and simplified versions of army council instructions were to be displayed in all units;[72] a new wireless programme, entitled *The War Office Calling the Army*, was launched to explain the reasons for current regulations;[73] and a suggestion-box scheme was introduced which, at the very least, allowed troops to let off a bit of steam: one commanding officer, it was reported, receiving just two suggestions, both of which he described as 'facetious'.[74] Although there continued to be some complaints about the relationship between officers and their men, and the gap between the troops and the higher military authorities was never really bridged,[75] the measures that were taken to restore a sense of mutual respect between the ranks, and a growing confidence in the ability of army commanders through their performances overseas, ensured that by the later stages of the war a greater degree of solidarity seems to have returned to the army.[76]

Whilst the relationship between officers and men came under scrutiny, another issue which the reports identified as being important to the morale of the army was the social status of the soldier. The wartime soldier, as a recent civilian, was acutely aware of his social status and wished to be looked favourably upon by the general population. This was affirmed in the morale report for September 1940 which noted that the soldier was immensely grateful for any 'interest' or 'sympathy' shown towards him by the public, as long as it was done without a show.[77] Yet despite the efforts of the military authorities to build up the image of the army in the public mind,[78] it was apparent that as the war progressed the soldier's social status left much to be desired.

Several reasons for this were identified. To begin with, the army was said to receive less favourable publicity when compared to the other services, and the wireless and the newspapers were full of stories about the exploits of the RAF, or the Canadian and American armies, rather than the British soldier.[79] 'The opinion exists', observed the morale report for January 1942, 'that the Army does not get a fair deal from the Press.'[80] Another matter concerned the army's serge battledress which compared very unfavourably with the more glamorous collars and ties worn by the RAF and the Americans. 'If I had the merest chance of joining the American Army,' commented one envious soldier in December 1942, '...I should have a good uniform and not a sack.'[81] Perhaps the most important concern, however, was the soldier's pay. For a start, the average soldier's basic pay was low in absolute terms – amounting to only 17/6 a week in 1942 – and thus did not allow him much of a social life; one commander calculating that on a typical weekly budget a man could not save for his leave, walk out with a girl or have a night out in the nearest town with his friends, unless he smoked fewer than twenty cigarettes a day or cut out his tea and buns.[82] But what made matters worse was that in comparative terms the ordinary British soldier was paid less than his counterpart in the RAF, or in the Canadian and American armies, and was also much worse off than many wartime civilians. The morale report for May–July 1942 recounted that:

A soldier is continually reminded of these contrasts by letters from home drawing attention to the financial position of neighbours who in peace-time were no better off than himself, by the proximity of civilian workmen on jobs where both soldiers and civilians are employed, and by daily contact with overseas troops and the R.A.F. in shops and public houses. In the words of a R.S.M.: 'The soldier really begins to grouse when he goes into a pub and finds that he can only afford a single glass of beer, whilst the munition worker slaps down a £1 note and asks for whisky or gin.'[83]

This situation led to some hostility towards American troops, in particular, whose spending power tended to push up prices in areas in which they were billeted, and it also prompted some antagonism towards civilian workers who resorted to strike action.[84] Above all, however, it damaged the soldier's social status; and this usually came home to him every time he went out for an evening, perhaps in search of female company, and found himself regularly shunned in favour of his more affluent rivals.[85] 'Women', observed the report for May–July 1942, 'almost invariably prefer the society of the R.A.F., civilian workers, U.S. and Dominion troops, to that of the British soldier, because he has less money to spend on them.'[86] What all this brought about was a worrying lack of self-respect on the part of many troops. 'Over recent months', the report concluded, 'it has become increasingly evident that the ordinary soldier suffers from a lack of respect for himself as a member of the Army.'[87]

The War Office did try to improve the army's public image. The public relations machinery was overhauled and much effort was spent in trying to promote the service through newspaper articles, broadcast talks and films such as *Desert Victory* and *The Way Ahead*.[88] Collars and ties were also granted to soldiers once the supply situation had eased, and although some thought that they might be a ploy to tempt men to stay on in the regular army, they were generally welcomed.[89] Moreover, several increases in the soldier's pay were announced by the government during the remaining war years.[90] Whilst there continued to be some discontent over disparities in pay, the steps that were taken to cultivate the social status of the soldier, along with the increasing success of the army overseas, ensured that during the second half of the war the troops appear to have recovered some of their self-respect.[91]

A further factor that was identified by the reports as being vital to the morale of the army, and one that became increasingly urgent as the war went on and soldiers faced periods of extended separation from their wives and families, was home worries. Several worries were singled out by the morale reports. One such anxiety concerned the financial plight of soldiers' families. During the war a soldier's wife was entitled to a family allowance for the upkeep of herself and her children, and for the wife of a private soldier with two children this amounted to 43/- a week in 1942.[92] This, though, was considered to be a grossly inadequate sum, particularly in view of spiralling wartime inflation. 'Letters passing between soldiers and their families', noted the morale report for May–July 1942, 'show clearly that there are many soldiers' homes in which the allowance does not provide for the necessaries of life at the present cost of living.'[93] Another problem was the soldier's concern for his children when his wife was ill or pregnant. Before the war the task of looking after children in these circumstances would usually have been undertaken by a husband, a close relative or a neighbour. But with husbands away, relatives scattered across the country and neighbours engaged in long hours of war work, this proved to be very difficult and children could often be left unattended. The morale report for May–July 1943 drew particular attention to 'the difficulties of soldiers' wives who are temporarily incapacitated and have families to look after'.[94] Perhaps the most pressing concern for the soldier, however, was the fidelity of his wife. During the long periods apart it was not unusual for wives to seek some male companionship and the presence of large numbers of glamorous, well-paid Canadian and American troops created a good deal of anxiety on the part of British soldiers; the Americans, in particular, gaining a reputation for not always being 'chivalrous and restrained' in their relations with women.[95] 'A worry which is constantly sapping the morale of a great part of the Army', observed the report for May–July 1942, 'is due to the suspicion, very frequently justified, of fickleness on the part of wives and "girls".'[96] The morale report for August–October 1942 continued in similar vein:

> Wives who should, presumably, be a source of comfort seem to be in many cases the reverse [and] letters from wives who are not equal to the test of prolonged separation do perhaps more than any other single factor to undermine the soldier's morale.[97]

All this of course put a great strain on many marriages. 'A really alarming number of soldiers' homes', warned the report for May–July 1943, 'are being upset, if not actually broken up, by matrimonial difficulties.'[98]

The military authorities did try to remedy some of these concerns. As far as the financial circumstances of soldiers' families were concerned, the government's announcement of increases in family allowances alongside rises in soldiers' pay helped to relieve the problem, but the War Office itself sought to ease the situation by consulting with the Ministry of Pensions in an attempt to expedite the complicated 'war service grants' scheme which existed to supplement family allowances in cases of particular hardship.[99] To help alleviate the difficulties faced by a soldier's family when his wife was ill or pregnant, representations were made to the Ministry of Health over the provision of extra 'home helps' for wives and the creation of hostel places for children when their mothers were unable to look after them. The Ministry of Labour was also approached with a view to releasing relatives from war work to help look after soldiers' families and making available additional staff for the Soldiers', Sailors' and Airmens' Forces Association which sought to provide assistance wherever possible.[100] As for the infidelity of soldiers' wives, the military authorities were somewhat at a loss as to what to do. The initial reaction was to approach the archbishop of Canterbury for spiritual guidance on the matter, and consider an appeal to the women of Britain to exercise some moral restraint by enclosing a special leaflet on infidelity in a new issue of family-allowance booklets; a suggestion which, on reflection, was thought to be inappropriate for a matter of such delicacy.[101] Some practical steps, however, were taken to mitigate the problem. First, a reconciliation scheme was introduced, involving voluntary welfare workers, which sought to patch up differences between husband and wife before a family was broken up without good cause.[102] Second, an army legal aid scheme was launched, with the assistance of serving lawyers, to tender advice to soldiers on a range of civil matters and begin divorce proceedings if that was the only option available.[103] In addition to this, steps were taken to smooth the granting of compassionate leave to soldiers with particularly pressing problems.[104] All these measures did help to ease soldiers' home worries and there was some evidence of an improvement in these matters; a process helped by the departure of many Canadian and American troops, which reduced fears over infidelity in particular.[105] But many domestic anxieties could only really be resolved by the return of soldiers to their families and thus, according to the morale reports, they were to remain a pressing concern for soldiers right up until the end of the war.[106]

Together with home worries, another consideration identified by the reports as being important to the morale of the army, and again one that became more urgent as the war went on and victory came closer, was the soldier's post-war prospects. The wartime soldier saw his military service as very much a temporary interlude before resuming his civilian life and he was naturally anxious about the sort of future he might expect once he was released from the army. Several aspects of this were identified in the morale reports. The majority of soldiers, it was reported, were not interested in a 'new Britain', but were simply concerned that they should return to a decent house and secure employment in the post-war world.[107] 'The ordinary soldier', noted the morale report for November 1943–January 1944, 'is not pre-occupied with political theories, but he is intensely interested in "post-war planning" in so far as it affects his home, his job, and the prospects of his family.'[108] A number of soldiers, however, were said to favour more radical change and the creation of a 'Utopia'. The morale report for May–July 1943 warned:

> In the background of these visions looms Russia ... that country is evidently looked upon as a paradise of freedom and justice, with a benevolent Government which provides unlimited scope for the individual, and 'Joe' Stalin is a figure among political leaders second only in affectionate regard to the Prime Minister himself.[109]

Whatever their hopes for the future, a generation of soldiers who had been brought up to believe that their fathers had been 'sold out' after the First World War wanted some statement about their post-war prospects. The report continued:

All that the soldier seeks, it seems, is reassurance; he does not want Utopia, and if he is satisfied that his needs are really being considered sympathetically by those in authority he will care little about 'plans'. On the other hand, unless he does receive this reassurance there is a danger that a large proportion of soldiers ... will feel bitterly resentful towards those in authority and be very ready to listen to irresponsible political extremists.[110]

Alongside their fears about the post-war world, soldiers were also naturally anxious about their eventual demobilisation from the army. There were particular concerns, it was recounted, about the order in which troops would be released from the service; about the provision of training to prepare them for their peacetime occupations; and about whether wartime women workers would retain jobs that should be made available to men.[111] Again, they were eager to have some indication that these matters were being considered. 'The one strong universal feeling', concluded the report for May–July 1943, 'is that it is desirable that a clear statement should be made about demobilization priorities at the earliest moment.'[112]

The War Office did seek to reassure the troops about their post-war prospects. The government's publication of various white papers on reconstruction in the later stages of the war helped to relieve the problem, but the military authorities themselves eased concerns by facilitating debate on these issues in the troops' weekly current affairs discussions as a means of keeping them in touch with developments.[113] As for demobilisation, again the government's publication of its release arrangements in the final period of the conflict soothed worries. The War Office, however, also played a role by announcing a new education scheme designed to help prepare soldiers for civilian life.[114] These measures did something to reassure the troops, but, according to the morale reports, there was a good deal of scepticism about whether the government would be able to provide enough homes and jobs for servicemen after the war and soldiers continued to worry about their future prospects right up until the end of the conflict.[115]

A final factor that emerges from the reports as playing an influential role in the morale of the troops, and one that formed a backcloth to many of the issues discussed, was the progress of the war overseas. From the beginning of the conflict it was apparent that the soldier wanted to feel part of a winning team and the morale report for September 1940 confirmed that 'the highest state of morale in a man is achieved through the consciousness that he is a member of an efficient corporate body'.[116] In the early years of the war, however, the string of defeats suffered by the army overseas meant that this was not occurring. These defeats, it should be noted, did not have a disastrous impact on the morale of the army. Events in the Far East, it was said, did not really register with the troops because they were too far away, and although the Middle East was closer to home they were reported as being simply 'puzzled' by developments there.[117] Nevertheless, whilst there was always confidence in an ultimate victory, the military reverses contributed to a lack of pride in the army and also led inevitably to contrasts being drawn with the performance of the Russians. 'The effect of Russian successes and our own reverses', warned the morale report for February 1942, 'has been to make men take an increased interest in the Russian system of discipline and Government.'[118]

The military victories that the army enjoyed in the second half of the war did, however, lift the spirits of the troops considerably. Admittedly, the victories did not solve all problems. There was, it was reported, a belief amongst some soldiers at home that they were not really associated with the successes abroad and therefore could derive little comfort from them.[119] A number, it was further noted, became anxious that the war would end without them having been on active service overseas.[120] There was also a tendency, it was said, to underestimate the British contribution to the Allied war effort and to regard the Russians as 'the sole saviours of Europe'.[121] Despite these reservations, military victories none the less restored a measure of esteem in the service. 'The Army', announced the morale report for May–July 1943, 'is "on the map" and knows it.'[122] Indeed, by the time of D-Day the overall morale of the army was much improved compared with the situation earlier in the conflict. 'There is little doubt', concluded the report for February–April 1944, 'that during the period covered by this Report the morale in the Army at home has risen to a higher level than at any time during the present war.'[123]

Having discussed some of the factors that were regarded as important in shaping the morale of the army, it is necessary to look briefly at how valuable the army morale reports are as a means of assessing these matters. Certainly, the morale reports have their shortcomings. As the War Office itself was aware, the sources upon which they were based did not always provide a perfect insight into the army's morale. The reports from commanders, for instance, could not, by their very nature, give more than a fleeting impression of the views of ordinary soldiers and might well have been influenced by the need to impress higher authority. Censorship reports of soldiers' letters depended on the circumstances and surroundings of the individual writer and tended to feature whatever issue happened to be uppermost in his mind at the time of writing, rather than what he thought of most constantly or felt most deeply. Courts-martial statistics were problematic in that they generally applied to only a small percentage of the army and were distorted by habitual offenders, whilst the home intelligence surveys by the Ministry of Information were considered to be of limited value because they dealt mainly with civilian rather than military morale. As for troops' enquiries to Professor John Hilton, they were inclined to indicate areas of special complaint rather than more routine day-to-day matters.[124]

The interpretation of the sources in the War Office is also open to scrutiny. It is possible, for instance, that in some respects John Sparrow tended to be optimistic in his assessments of morale since he could not have been entirely sure on whose desk the reports might land. In other respects, however, it is arguable that he erred on the side of pessimism. In general terms, the reports tend to emphasise the dysfuntional elements in army life and underplay the sources of satisfaction.[125] There is little or nothing, for instance, to be heard about the comradeship or the humour which must have kept the troops going through the most adverse of circumstances. More specifically, his frequent warnings of the troops' obsession with Russia have been criticised for being too alarmist.[126] Certainly, some troops were politically radicalised as a result of the class system they encountered in the army and the author Anthony Burgess vividly recalls a Welsh sergeant confessing at the end of the war, 'When I joined up I was red. Now I'm bloody purple.'[127] But, as Sparrow himself admitted, few soldiers had genuine communist sympathies,[128] and his reports have a tendency at times to paint a picture of dangerous political discontent when all that was apparent was a natural 'bolshiness' on the part of the troops to the circumstances of army life.

Although the morale reports have their shortcomings, they are, however, a valuable adjunct to other historical sources, such as the individual testimonies of soldiers lodged in the Imperial War Museum or the opinion surveys conducted in the army by the Mass-Observation organisation. For a start, they are based on a fairly wide range of sources and thus provide a useful overall view of morale. They are also reasonably consistent in that they were produced at regular intervals, using broadly the same format, and thus allow for an analysis of morale over time. Moreover, even if they do not always provide an entirely accurate assessment of morale, they are, nevertheless, important because they reveal what the military authorities *thought* was the state of morale of the army at any particular moment, regardless of how valid that judgement ultimately was.

What, then, in conclusion, do the reports tell us about the morale of the army on the home front? First, they reveal that the morale of the army was a complex matter determined by a number of factors – ranging from the quality of the NAAFI tea to the competence of officers, to the loyalty of wives, to news from the battlefronts. Second, they illustrate that morale did improve over time and, whilst the reports never admitted any great cause for alarm, it was clear that by the end of the war substantial progress had been made from the low point of 1942.[129] Third, they suggest that the morale of the army, at least at home, was largely conditioned by the fact that the soldier remained very much a civilian at heart who, in the memorable words of one morale report, continued to look upon the army as 'a body of a different caste, a large part of whose function consists in badgering the soldier about, for reasons which he does not understand'.[130] And one final point: it is perhaps worth noting that whilst a good number of Colonel Blimp figures undoubtedly remained in the wartime army, the establishment of the morale machinery in the War Office, and the steps that were taken to improve morale, demonstrate just how far the army had developed as a social institution during the war and how it had become, in many ways, a more enlightened organisation than that which had existed at the outset.

Part Three
Commonwealth Contingents

"Twas England Bade Our Wild Geese Go': Soldiers of Ireland in the Second World War

Ian S. Wood

'Twas England bade our wild geese go, that small nations might be free.
But their lonely graves are by Suvla's waves and the edge of the grey North Sea.
Oh, had they died at Pearse's side or fallen with great Cathal Bru,
Their names we would keep, where the Fenians sleep, in the shade of the foggy dew.

The 'wild geese' celebrated in this famous republican ballad of the 1916 rising were of course Ireland's soldiers of fortune, whose loyalty to the defeated Stuart cause took them across the sea to offer their swords to Europe's Catholic monarchs. Their exploits and their social structure as a sizeable military community in exile have been explored in some fine work.[1] Kipling used the phrase in his 1918 tribute to the Irish Guards and, at the time he wrote it, many thousands of Irishmen had given Britain faithful service on countless battlefields. They did again, from both sides of a partitioned island's border in the Second World War, and not simply because England, in any coercive sense, 'bade them go'.

North of that border, military virtue is a central ingredient of the self-image of Loyalist Ulster and paramilitary bodies like the Ulster Defence Association and the Ulster Volunteer Force constantly celebrate it in their literature. The fiftieth anniversary of the D-Day landings was marked by the UDA in its magazine, not just with tributes to the bravery of Ulstermen, but with an article addressed to the then Taoiseach or prime minister of the Irish Republic entitled, 'What did you do in the war, Albert?', which was of course a bitter diatribe against his country's neutrality in the Second World War.[2]

When war was declared in 1939, recruitment in Northern Ireland to the armed forces was not a great success. Since there was no conscription there, it must be accepted that this might also have been the case in mainland Britain if the Chamberlain and Churchill governments had had to rely on voluntary enlistment, but when registration of the province's population for food rationing was announced, the prime minister Lord Craigavon had to give assurances that it would not be used as the basis for any military call-up.[3]

Unemployment in Northern Ireland grew in the opening months of the war as important overseas textile markets were lost, and Loyalist workers with bitter memories of how Protestant Ulster's sacrifices on the Somme had been followed by the hardships of the interwar period feared for their jobs if they enlisted.[4] Even attestation for Home Guard service needed to be backed by official reassurance that it would not involve any commitment to overseas service.[5]

Concern at the situation was widely felt but could not easily be voiced publicly. Sir David Lindsay Keir, vice-chancellor of Queen's University, Belfast, wrote in late 1940 to Harold Wilson, joint secretary to the London Cabinet's Wartime Requirements Committee, to express his unease at the Ministry of Information's 'Go To It!' poster campaign: 'Those who are loyal are, as you say, disheartened by there being nothing to "go to". Among the other (i.e. the Catholic) community extremisms of all kinds, e.g. Communism and the IRA, are flourishing on industrial depression. I have even been told of young Protestants being attracted into illegal organizations.'[6]

Stormont government propaganda, backed up by cinema newsreel output, sought to represent the province as being totally behind the British war effort,[7] while avoiding the contentious issue of conscription there. The debate on its extension to Northern Ireland is alluded to in the official volume dealing with its role in the war,[8] but with important gaps later to be filled by Robert Fisk.[9]

Early in 1940 the Chamberlain government rejected Lord Craigavon's case for conscription, backing away from the likely reaction both of the province's nationalist minority and of the de Valera government in Dublin.[10] Craigavon's successor, John Andrews, raised the issue again in May 1941, only to meet with a similar response from Churchill. Charles Wickham, the Royal Ulster Constabulary's inspector general, who accompanied his prime minister to Downing Street, argued that a call-up, to be viable, would have to apply to both communities and predicted that nationalist resentment would be heightened by the large number of Protestants in reserved occupations who would escape any liability to service.[11] With an IRA campaign under way, republicans would have exploited any imposition of conscription upon their community anyhow, so the matter was shelved until Churchill resurrected it in correspondence with Roosevelt once large numbers of American troops began to arrive in Northern Ireland in 1942.[12]

Nearly 40,000 volunteers for war service none the less came forward in Northern Ireland, and for those who opted for army service it was a natural step to enlist in famous regiments which had traditionally recruited in both the nine historic Ulster counties or within the sub-state created by partition in 1920. None of these were ever sectarian in recruitment, so Catholics could join them within the province, or by crossing the border from Eire, which the de Valera government never prevented them from doing.[13]

Enlistment by Catholics in nationalist areas of Northern Ireland was never a decision to be made lightly, least of all in a tense year like 1942 when the crown forces operated martial law in some areas after the hanging in September in Belfast of the IRA volunteer Tom Williams.[14] Well before that, as Brian Moore's fictional account of the city at war reminds us, even to wear Civil Defence or ARP uniform could, in the minority community, provoke derision or overt hostility.[15] Yet Northern Catholics did volunteer, though in what proportion of the total who did so remains hard to quantify.

The regiments they joined varied in their traditions of enlistment. The Irish Guards had always recruited from the other side of the border as well as from the expatriate Irish in Britain, and the official history of Northern Ireland at war refers to a detachment of the regiment going into action in Belgium in 1940 'with some Ulstermen in the ranks'.[16] Others remained predominantly Protestant, like the two battalions of the Royal Ulster Rifles which landed in Normandy on D-Day. The 1st Battalion had converted to an airborne unit and the 2nd Battalion went ashore not far behind it. Among its ranks were three Magill brothers from Sandy Row, a famously Loyalist area of Belfast, and a Major Hinds, who won the Military Cross and later became a deputy grand master of the Orange Order.[17] No unit from Northern Ireland, however, was exclusively Catholic or Protestant and what cements the morale and identity of any fighting unit goes far beyond mere religious allegiance.

War service, however, did not resolve the problems of a deeply divided society, something that came home with chilling clarity to Sam McAughtry, a Belfast Protestant who has written vividly of his experiences in uniform. After demobilisation he was approached in a pub in his own Loyalist area of north Belfast by an acquaintance who had seen out the war in the police. After a drink he handed McAughtry application forms for the force's B-Special Reserve, with the words, 'That's in case there's a real war.'[18]

In the Free State, or Eire as it became in 1937, however, there were many who continued to serve in the British army. Partly this was because of proud family traditions but it was also because the new state's Defence Forces were, from the conclusion of the Civil War in 1923, kept small for both financial and political reasons. An event which confirmed the original Free State's policy towards the army was the March 1924 mutiny over the rate at which soldiers, especially officers, were being discharged from service. Many of these were 'old IRA' men, as opposed to anti-Treaty IRA members, who had accepted the 1921 agreements with Britain out of loyalty to their leader Michael Collins.[19]

The crisis had clear political overtones but it was defused by the Cosgrave government. None the less, it acted as a deterrent to any expansion of the new state's army, even when de Valera and the anti-Treaty party in the form of Fianna Fáil took power in 1932. They in turn suspected some elements within the army of sympathy for General Eoin O'Duffy's paramilitary and authoritarian 'Blueshirt' movement which, though it failed politically, brought serious street violence into the Free State's politics.[20]

Training levels and combat efficiency remained poor and it has been claimed that 'the country was almost defenceless when the Second World War broke out'.[21] Even a rapid build-up of strength failed to resolve the problem of low pay and, in 1942 with the Defence Forces' strength on the 40,000 mark, a private soldier's weekly pay of less than thirteen shillings was eaten into by compulsory deductions for haircuts, laundry and social welfare facilities, the last of a very basic nature.[22]

With British army service offering substantially better pay as well as opportunities for travel and overseas service, the outbreak of war in 1939 accentuated a pattern which was already set. One sergeant in the Irish Guards put it well at the height of the Anzio battle in 1944 when an English-speaking German prisoner asked him why, as a citizen of a neutral state, he was fighting for Britain. His reply was, 'They've fed me for seven years. Now I'm earning my keep.'[23] Well before then British recruiting depots were trying to establish whether Irishmen offering themselves for service had in fact deserted from their own country's Defence Forces.[24]

G2, the Irish Military Intelligence Branch, was also aware of this problem, and in 1945 drew up its own estimates of those who had left the state's Defence Forces:

> There are at present almost 5,000 non-commissioned officers and men of the Defence Forces in a state of desertion or absence without leave. Of these approximately 4,000 are absent for more than 12 months, many of them for as long as 3 or 4 years. There is little doubt that the majority of them are or have been serving in the British forces or are in civilian employment in Great Britain and Northern Ireland.[25]

When some of these absentees who had survived their war service began to return home after 1945 the question was raised in the Dáil and elsewhere as to whether they deserved to be penalised for what they had done, for example by the loss of such welfare benefits as were available in post-war Eire. De Valera, however, was prompt in rejecting the idea.[26]

Such men inevitably had a variety of reasons for joining the British forces, but for a significant number of them doing so was a reaction to the fact of their state's neutrality and the blanket censorship that went with it. The Irish press was absent from all the major war theatres and the many deaths in action of Irishmen were edited off the relevant pages of the newspapers. Frank Aitken, the defence minister in de Valera's government, even felt that such notices could be interpreted as propagandist if families of the fallen were allowed to insert the familiar words 'Greater love hath no man, that he lay down his life for his friends.'[27]

This atmosphere created a sense of isolation which the late F. S. L. Lyons captured memorably in his account of the years of 'the Emergency', as the war years have always been called in Irish writing:

> It was as if an entire people had been condemned to live in Plato's cave, backs to the fire of life and deriving their only knowledge of what went on outside from the flickering shadows thrown on the wall before their eyes by the men and women who passed to and fro behind them.[28]

Escape from the cave was a necessity for many thousands of Irishmen, especially the young and the brave, as well as democrats and anti-Fascists influenced perhaps by the same feelings articulated by Louis MacNeice in his angry poem 'Neutrality'.[29]

The first Irish regiment to see action in the Second World War was the Irish Guards. Raised early in the century, it had from its creation a nationalist tradition within its ranks, at least in the form of sometimes vocal support for constitutional Home Rule. On 22 July 1914 John Redmond, leader of the Irish Home Rule party, emerged with his deputy John Dillon from crucial all-party talks in Buckingham Palace on the Ulster crisis. As they passed Wellington Barracks where the 1st Battalion was stationed, they were recognised and cheered to the echo by hundreds of men crowding up to the barrack railings. This overt breach of discipline went unpunished by a commanding officer who confined his response to the following coded admonition: 'You are all, or nearly all, racing men and like a good bet. Back what horse you like but keep your tips to yourself.'[30] All bets were off only too soon as the battalion met its bloody baptism in the killing fields of Belgium and France.

Twenty-five years later, after some demanding service in troubled areas like Egypt and Palestine, another generation of Irish Guardsmen went into battle again as the regiment's 1st Battalion was committed to action in Norway after the German invasion of 9 April 1940. Over 700 officers and men left King's Cross with a crate of beer in every compartment for a convivial journey to Greenock, where Arctic clothing and troopships awaited them as the fight intensified for control of some major Norwegian ports. They arrived as the British and French front in central Norway began to collapse and in time only for a planned attack on Narvik to be abandoned.

On 10 May, as news came through of the German attack on France and the Low Countries, they were ordered south, on board an elderly Polish passenger liner converted to a troop carrier, which was quickly intercepted by German aircraft on 14 May. As the ship caught fire and ammunition started to explode, the battalion sustained its first serious casualties, including the lives of its commanding officer, second in command, adjutant and three company commanders. Losses would have been far worse if good discipline had not been maintained by all ranks on deck, even when it became clear that loss of power made it impossible for any lifeboats to be used.

Everything depended on the Royal Navy's ability to align a ship alongside with gangways and ropes, but this took time, during which Father Cavanagh, the battalion chaplain, played a key role. 'Realizing that somehow he had to keep the men occupied, the priest started to say the Rosary, and with bared heads on a burning ship in the Arctic Circle men said the prayers that they had learned long ago in the quiet churches and farmhouses of Ireland.'[31] This helped the battalion forget its extreme peril until rescue was at hand, and 694 men were ultimately taken off safely in the space of sixteen minutes.

Amidst these events, the regiment's 2nd Battalion had embarked at Dover on 10 May for the Hook of Holland, already under threat from the German invasion. The battalion chaplain kept a diary and recalled a hasty meal eaten as their ship rolled in heavy seas: 'Immediately afterwards I went below to hear confessions as, if we were not torpedoed on the way, people seemed to envisage a sort of Gallipoli landing.'[32] These fears turned out not to be justified, but as German air attacks on the Hook intensified, it became clear that part of the battalion's brief was to guard the safe embarkation of the Dutch royal family.

Casualties mounted before they could re-embark and the chaplain, in a diary entry for 14 May, recorded some of them:

> My first memory of that morning is of going, in a lull between air-raids, with Pipe-Major Cosgrove, the chief stretcher-bearer, to identify the corpses of eight Irish Guardsmen who had been taken the night before to the mortuary in the Dutch camp near a coastal battery. We identified the poor white contorted bodies with their staring eyes making them look still alive.[33]

A further 200 casualties were to be sustained by the 2nd Battalion once it was redeployed in France prior to its evacuation from Boulogne.

Other Irish regiments took their part in the retreat to Dunkirk, the 2nd Royal Inniskilling Fusiliers leaving its beaches with barely 200 men who had survived the retreat, while the 2nd Battalion of the Royal Ulster Rifles and the 1st Battalion of Royal Irish Fusiliers were also there, along with anti-aircraft artillery units recruited from Northern Ireland.

After Dunkirk controversy developed over whether Irish and Ulster battalions should be grouped into 'Irish brigades'. Such a term gave a message that not all Ulster Unionists wanted to hear, for its historical lineage went back to the early days of the 'wild geese' in exile who had served England's enemies. Some of the acrimony centred upon the London Irish Rifles, a Territorial unit which in the interwar years had recruited substantially among young Irishmen from the new Free State who had found work in London. John Andrews, Northern Ireland's prime minister in 1942, saw a sinister agenda here and wrote to Attlee early that year objecting to what he called a policy 'calculated to obliterate or blur the distinction between the belligerency of Northern Ireland and the neutrality of Eire'.[34]

Seen against this kind of Unionist suspicion, the formation of at least one all-Irish infantry brigade was no small achievement and in fact owed much to Churchill's support for the idea. An existing brigade was simply renumbered in January 1942, and its English battalions replaced with the 1st Royal Irish Fusiliers, the 6th Royal Inniskillings and the 2nd London Irish Rifles.[35]

Intensive battle training for the new brigade followed and by late October 1942 it was ready for its part in the British and American landings in North Africa, a huge operation which would not have taken place at all had not Roosevelt been swayed by Churchill's advocacy of it as a way to tighten the noose around the Axis powers in the Mediterranean and to support the British Eighth Army's advance from El Alamein. Eisenhower took overall command with grave doubts about what he saw as a diversion of essential resources from the opening of a Second Front in Europe itself.[36]

The Allies, it has been said, 'landed in French North Africa ill-adapted to tackle anything but the most limited resistance'.[37] Three separate forces went ashore almost 700 miles apart at Casablanca, Oran and Algiers, and so movement inland by them was slow compared to the rapidity with which formidable German units were brought in to secure a line between Bizerta and Tunis. This has been viewed by some historians as an act of folly on Hitler's part, but he had to make a quick decision in response to the first landings and his instincts were almost always to fight rather than give up territory.

This the Irish Brigade quickly found out for themselves when they made their first contacts with the enemy on the windswept and barren area around Goubellat. 'There they first heard the zip of the bullet, the quick stutter of the Schmeisser – the whine of the shell, followed by its bark, and that most bloodsome crump of the mortar, the most formidable of the lot. And all fired with malicious intent.'[38] Here the Irish battalions took their first battle casualties since 1940, casualties that began to mount once, in January 1943, they were ordered up to the Bou Arada area to attack 'Two Tree Hill'. It was a key artillery observation point overlooking a plain ten miles wide and twenty-five miles long, and was described as 'a blighted spot, steep-sided and covered in small rocks and scree, virtually unapproachable without the goodwill of the residents'.[39]

The 6th Royal Inniskillings were savagely bled when they attacked there on 13 January, impeded by mud and driving rain as well as relentlessly accurate German fire which killed and wounded over a hundred of them. Then as pressure for more attacks grew from London and Washington, it was the turn of the 2nd London Irish Rifles in the same intractable area. Late at night on 19 January they fought their way up the main Bou Arada ridge and managed to dig in there, only to realise that, without anti-tank guns, they were expected to try to hold it against heavy German Panzers. A survivor recalled what followed in the battalion diary:

> There was no delay at all before the storm burst and the German counter-attack came in like a visitation from the angels of Hell complete with chariots of fire. The force and vigour of the onslaught was matched only by its audacity. True disciples of Rommel, the tank commanders rode in sitting in their turret tops armed with Very pistols and star shells to guide them. They simply charged along the ridge from one end to the other, leaving their Jagers to pick up the bits later. Driving straight over the top of the feature in the starlight they went on and down, and over the road itself in an ever widening torrent and our defence dissolved into fragments before them.[40]

Fighting of even greater ferocity lay ahead for the brigade among the rugged hills, almost impassable tracks and razor-back ridges around Medjez-el-Bab, the Siegfried Line of Tunisia in German eyes, which had to be breached before the

Allies could advance on Tunis itself. After it was finally taken on 7 May, the British corps commander under whom the Irish Brigade had served revisited the area with Nelson Russell, their brigadier and after viewing the terrain and the German defences, simply asked him: 'How on earth did they do it?'[41]

An equal share of the Tunisian ordeal fell to the 1st Battalion of the Irish Guards, who also sustained heavy casualties in a campaign in which German units like the Hermann Goering division were allowed to exploit all their mastery of the arts of fighting, retreat and regrouping under attack. Only quite remarkable acts of individual bravery could, sometimes, neutralise these enemy qualities, like that of one of the battalion's bren-gunners, Lance-Corporal Kenneally, who was awarded one of the war's most celebrated Victoria Crosses. Already wounded in the battle for Hill 212 on 29 April 1943, he fought on, held up by a comrade and refusing to hand over his weapon. 'His extraordinary gallantry in attacking single-handed a massed body of the enemy and breaking up an attack on two occasions was an achievement that can seldom have been equalled.'[42]

The terror and frenzy of this sort of close-quarter action under concentrated fire could put a high premium upon the qualities of non-combatants like the Irish Guards' Catholic chaplain, Father Brookes. He had served on the Western Front a quarter of a century earlier and remained with the regiment until he took Holy Orders in 1925. Amidst the carnage of 29 and 30 April he seemed to be everywhere, blackthorn stick in hand, tending the wounded and giving the last rites to the dying.[43] Pipers and drummers doubling up as stretcher-bearers could also perform conspicuous feats of stamina and bravery. An adjutant in the battalion recalled one of them, Hickey, helping him to treat the wounds of a fellow officer: 'Hickey and I had to stuff field dressing after field dressing into a hole the size of a teapot in his back, after he had been shot in the shoulder in the early stages of the attack.'[44] The officer died of his wounds and Hickey had to have a leg amputated shortly afterwards when he was hit by a mortar shell on the forward slope of Hill 212.

The Irish Guards left behind them in Tunisia all too many crosses in the ground inscribed with the words 'Quis Separabit', the regimental motto, and the 38th Brigade's losses had been so severe that, even as new drafts arrived from Britain, there was talk within the British command of disbanding the London Irish Rifles because they had suffered so heavily. Russell, as brigadier, used all his influence to oppose this, arguing that the brigade's Irish identity must be preserved even if it meant drawing upon replacements from non-Irish units with men who had some Irish ancestry. An Irish stepfather would be good enough, he argued, and the spirit of his three battalions would do the rest.[45]

Russell never forgot the bravery of his battalions during this brutal initiation. When he was required to hand over his command nearly a year later, his war diary for 18 February 1944 reverted to his memories of Tunisia:

I turned over one poor chap on a rocky, bloody crag on Tannagoucha. He was facing the right way, the last round of a clip in the breech and three dead Germans in front of him. His name was Duff. After all is over – and the remainder of the Empire is understandably irritated with Ireland – I hope these countless Duffs, from both the North and the South, and in all three services, will be remembered.[46]

After Tunisia, the 38th Brigade was reassigned to the British 78th Division and committed with other Irish battalions to the invasion of Sicily in July, an undertaking that was to suffer from the distances involved, multiple layers of command vying for resources, and from the competing egos of commanders on the ground. The 2nd Royal Inniskilling Fusiliers and the 1st London Irish Rifles went in ahead of the 38th Brigade and played a key role in support of British parachute troops who seized and held the vital Primo Sole bridge on the road to Catania.[47]

Indecision and poor co-ordination of effort by the Allies' air and naval forces allowed the Germans to regroup and indeed reinforce their troops from across the Messina Straits. Severely outnumbered but skilfully led by Kesselring, parachute and Panzer units mounted a tenacious defence of their positions in the north-east corner of the island, making chaos of Montgomery's plan for an advance up the coastal plain around Catania in which, as he had put it with some condescension to his allies, the Eighth Army would be his sword and Patton's Seventh United States Army, operating inland, would be his shield.

The plain of Catania, dominated by Mount Etna, scorched dry by the sun of high summer and broken up by jagged outcrops of rock and dry river beds, proved to be a cauldron for Montgomery's troops, including the five Irish battalions who fought there once the 38th Brigade was committed to the battle. The Scottish poet, Hamish Henderson who was also there, wrote his own description of this unyielding terrain and the fear it generated:

> O, with prophetic grief
> mourn that village that clings
> to the crags of the west
> a high gap-toothed eyrie.
>
> For the tension in the rocks
> before sunset will have formed
> a landscape of unrest
> that anticipates terror.
>
> A charge has been concealed
> in the sockets of these hills.
> It explodes in the heat
> of July and howitzers.[48]

Reminiscent of some of its hill actions in Tunisia was the Irish brigade's storming of the town of Centuripe, perched on a precipice but controlling the only road to the west of Etna. It was a position the taking of which cost many more Irish lives and Montgomery, later shown the place where two Inniskilling companies had scaled a cliff under German fire, described the whole operation as being 'impossible'.[49] Sacrifices like these did not, however, prevent Kesselring being allowed to ship most of his troops across the Straits of Messina to hold southern Italy against an anticipated landing there by the Allies.

The Italian campaign which followed, because of the terrain involved and the limited role for tanks, consumed infantry at a rate comparable to the worst fighting on the Western Front in 1916 and 1917, yet it became in many ways for the Allies, it has been said, a matter of 'strategic diversion on the maritime flank of a continental enemy, the "Peninsular War" of 1939–45'.[50] It was a campaign too, like Sicily, dogged by some bitter rivalry over crucial command decisions among the Allied armies. Yet, from the start, generals' dispositions and battle plans were limited by how far metalled roads would carry infantry before they faced goat tracks to take them to a 'front' barely measurable on maps where, in all weathers, concepts like advance or retreat quickly dissolved into the most localised and brutal attritional combat.

December 1943 found the battalions of the 38th Brigade along the line of the Moro river and seeking to move north of it to the Sangro against tenacious German opposition and atrocious weather. Particularly severe fighting centred upon the contested village of St Vito, where Canadian troops relieved a 1st Royal Irish Fusiliers Battalion close to breaking point. A Canadian officer recorded his impression of what sustained combat had done to them as well as his own premonitions of what lay ahead.

It was pelting rain when I went forward with the Fusiliers' Intelligence Officer to see what he could show me. Long lines of soaked and muddy Fusiliers wound their way past us, moving to the rear. Their faces were as colourless as paper pulp and they were so exhausted they hardly seemed to notice the intense shelling the coastal road was getting as they straggled down it.

But I noticed, as I had never noticed before. The rancid taint of cordite seemed to work on me like some powerful and alien drug. My heart was thumping to no regular rhythm. It was hard to draw breath, and I was shivering spasmodically though I was not cold. Worst of all, I had to wrestle with an almost irresistible compulsion to stop, to turn about, to join these deathly visaged men who were escaping from the battle that awaited me.[51]

The courage of young soldiers in war has never been a bottomless well to be endlessly drawn upon, and battle-induced trauma can come to any unit exposed for too long to the terror and frenzy of battle as well as to exhaustion and lack of sleep. It clearly overtook the 1st Battalion of the Irish Guards at Anzio, an operation which may have been an opportunity wasted to shorten the Italian campaign. The Allies' landing there in January 1944 might, with more resources and bolder leadership, have breached the German's Gustav Line, isolated their position at Monte Cassino and opened the road to Rome itself. Instead the British and American landing force found itself besieged within a confined bridgehead from which it only broke out after four months and at appalling cost.

Anzio was a killing ground which destroyed the 1st Battalion, though it had retrained and been reinforced after its losses in Tunisia. In Naples they were piped on to their landing craft for Anzio to tunes like 'The Minstrel Boy' and the 'Wearing of the Green', but what lay ahead was a grinding, claustrophobic struggle which their chaplain, Father Brookes, described at the time as worse than anything he had experienced on the Western Front a generation earlier.[52]

The bridgehead secured after the initial landing by the Allies in late January 1944 had, until Mussolini came to power, been an almost impenetrable malarial swamp. In a prestige reclamation project the regime had drained it for agricultural use, but heavy rain like that of early 1944 could quickly bring water back to the surface and reduce much of the area to a quagmire. Crucial to control of it and to the prospects of any breakout that would put the Allies on the road to Rome were a road and railway which intersected the bridgehead. They were flanked some eight miles inland by a feature which became known on unit maps as the Gullies, a confusing sequence of deep, muddy, overgrown channels and ravines, some forty feet deep on average. A perfect defensive position, the Gullies were laced by Kesselring's men with every sort of lethal anti-personnel device. These included the dreaded miniature Schuh mine, easily concealed and capable of ripping a man's foot off at a touch, and any movement was tracked by ceaseless fire from German tanks and artillery ringing the bridgehead's perimeter.

It was in the Gullies that the Irish Guards over a seven-week period fought a series of battles of a horror that can still drain the imagination half a century later. At one point they alone held the enemy at bay. For three days and four nights the depleted battalion held the Gullies against continual attack. It was a savage, brutish, troglodyte existence, in which there could be no sleep for anyone and no rest for any commander.

> The weather was almost the worst enemy and the same torrential rain, which sent an icy flood swirling around our knees as we lurked in the Gullies, would at times sweep away the earth that covered the poor torn bodies of casualties hastily buried ... isolated by day and erratically supplied by night, soaked to the skin and stupified by exhaustion and bombardment, surrounded by new and old corpses yet persistently cheerful, the Guardsmen dug trenches and manned them till they were blown in and then dug new ones, beat off attacks, changed their positions, launched local attacks, stalked snipers, broke up patrols, evacuated the wounded, buried the dead and carried supplies. The bringing up of supplies each night was a recurrent nightmare. Carrying parties got lost, jeeps got bogged and, as the swearing troops heaved at them, down came the shells.[53]

As in Tunisia, prodigies of courage were performed by battalion medical orderlies and stretcher parties, whose work never ceased as constant enemy sniping and close-range shelling took their toll. 'What I remember most', said one officer later, 'is the long strain of hanging on all day to hear the list of casualties every evening, to see the stretcher-bearers livid with fatigue staggering past with their load, a dirty Red Cross flag held aloft as a precarious appeal.'[4]

Those still surviving in the Gullies knew they were living off borrowed time when the order came through that the battalion, or what remained of it, was to be relieved by the Duke of Wellington's Regiment. Their commanding officer watched in awe as a ragged, gaunt, mud-caked column passed him on its way to the rear. 'As the tall Guardsmen filed out,' he wrote, 'leaving us the heritage of death and desolation they had borne so long, a peculiar sense of isolation struck us. In all the long crucifixion of the beach-head, no positions saw such sublime self-sacrifice and such hideous slaughter as were perpetrated in the overgrown foliage that sprouted in the deep gullies.'[5] The memory of these men stayed with the commanding officer of their replacement unit and many years later he made sure that his son at Sandhurst made the Irish Guards his first choice for a commission.[6]

The full muster of the battalion when it landed at Anzio had been 1,080 officers and men, but when they were shipped out on 7 March the survivors numbered just 267. Many were still too drained physically and mentally even to celebrate St Patrick's Day and Anzio proved to be the end of the battalion's war. Enough men, however, were fit in June, after the liberation of Rome, to form a composite company to be received by the pope at the Vatican for a special blessing upon them.[7]

Parallel to the grinding struggle to achieve a breakout from Anzio was the protracted onslaught on the German positions on Monastery Hill above the town of Monte Cassino. The Irish battalions of the 38th Brigade were at least rested out of the line before being committed to battle there and were able to celebrate St Patrick's Day ahead of their movement orders. The 6th Inniskillings seem to have led the way the night before with a party and ceilidh that took on a life of its own. Their battalion war diary records a flow of St Patrick's Day greetings sent out by telegram, as the revelry increased, to King George VI, the duke of Gloucester, colonel-in-chief of the regiment, the Roman Catholic and Church of Ireland primates of Ireland, numerous generals and, not least, the Irish Taoiseach or prime minister, de Valera. Many of these greetings were acknowledged and reciprocated, but his reply is not on record.[8]

In the light of what followed, it was just as well that the party had been a good one, for the brigade's deployment around the villages of St Angelo and Monte Cairo, under constant fire from German parachute troops dug in around the monastery, proved to be a season in Hell itself. 'Daylight movement invited death. Cairo village was so accurately registered by the German artillery that survival was a miracle; even answering a call of nature could attract the attention of the Grim Reaper.'[9] Digging in on rock-hard ground was near to impossible, as was burial of the dead, and staying alive could depend on improvised sangars, or stone shelters, which might, if their occupants were lucky, blend in enough with the terrain to deceive enemy gunners.

One survivor of the fighting at Monte Cairo later wrote of his time there with the Inniskillings:

We were many times better off than the battalion directly facing the monastery, where throwing the contents of his lavatory tin out of the back of his sangar earned for many a man a hail of mortar bombs. Just try sharing a rock hole about the size of two coffins with three others, then, in a prone position, lower your costume and fill a small tin held in the hand. The stink of excrement competed with the smell of death in every position. Bowels will not wait for nightfall.[60]

While it fell to other troops, notably the Free Polish Corps, finally to storm the monastery position itself, the 38th Brigade's role was an important one. The pressure its attacks put upon the enemy in the Rapido and Liri river valleys, and its eventual breach of the Gustav line west of Cassino, helped to make the German position there untenable and materially assisted the Polish attack once it was launched. Once again the cost was heavy though accompanied by a whole clutch of citations, not least for the brigade's Kerry-born Catholic chaplain, Father Don Kelleher, who was awarded a Military Cross. His charismatic bravery was vividly recalled years afterwards:

On the night when Cairo village received Evensong from the German guns and while every living soul in that heaving rubble lay flat and prayed, a figure walked calmly among the shell bursts carrying a dying man in his arms to shelter. No one begrudged Father Don Kelleher, the Kerryman, that MC and he probably got a ghostly pat on the shoulder from old St Benedict himself, peering from Heaven to see what we were making of all his architecture.[61]

Victory at Cassino opened the way to Rome's liberation by American troops on 4 June, and six days later a representative 38th Brigade group paraded in St Peter's Square for a papal blessing which may well have been received by more than one Orangeman, for sectarianism meant little to men recruited from both sides of a disputed border as well as among the expatriate Irish in Britain. At the Pope's request the brigade's pipers and drummers in their distinctive caubeens and saffron kilts then formed up in the square to entertain him and some Irish priests with traditional airs and marches.[62]

From Rome a long march lay ahead for the Irish battalions as Kesselring's men regrouped once more behind new defensive positions. One of the most formidable of these lay along the River Po, an obstacle German propaganda claimed would destroy the Allies. Long after the Po was crossed at heavy cost in April 1945 one Church of Ireland rector's son, who had fought there as a young lieutenant, described the Eighth Army massing for the attack:

The Irish turned up in force. Not only were there battalions of the Royal Inniskillings Fusiliers, Royal Irish Fusiliers and the London Irish Rifles carrying the fighting tradition of the historic Irish Brigade, but there were the North Irish Horse and thousands of Irishmen from the 'neutral' South who had infiltrated every unit in the country – all keen to get into the fight while applauding de Valera's astuteness in keeping them out of the war.[63]

Events in Italy leading to the Po crossing had of course been overshadowed by the Normandy landings in June 1944 and the protracted battles to achieve a breakout from the area initially secured by the Allies. Nothing in the extended training of British troops prepared them for the ferocity of the resistance they would meet in the confined bocage country of Normandy's farmlands, with their narrow lanes and high, dense hedgerows which gave perfect defensive cover to hardened German units as well as to some fanatical fighters from the SS and Hitler Youth. Some British battalions broke under the horror of it and had to be disbanded or redistributed to other units, and even seasoned formations like the 51st Highland Division were criticised by Montgomery for what he claimed was loss of momentum.[64]

The strain of course was potentially severe on units without combat experience, which was the case with most of the Irish units committed either to the landings or the subsequent fighting. None the less, they showed qualities that were to single them out for special praise and, of course, for special assignments, like Montgomery's attempt to seize Caen within twenty-four hours of D-Day. The 2nd Battalion of the Royal Ulster Rifles, who had gone ashore on 6 June with the British 3rd Division, lost heavily in an attack across rolling cornfields and under concentrated German fire from prepared positions around the village of Cambes. Six miles inland, this was a key to opening the road to Caen itself, but the battalion's first major attack, which cost it 200 casualties, failed in its objectives and served as a brutal warning of what lay ahead.[65] Yet again, an all-Ireland sacrifice of life was involved, for George VI, when inspecting them before D-Day, had commented on the number of Eire men serving in an Ulster battalion.[66] Amongst the two Irish Guards battalions in Normandy, the 2nd converted to tanks and the 3rd serving as infantry, few had seen combat before.

The 3rd Battalion was soon in action, but the worst fighting it experienced in Normandy was without a doubt in early August when it got orders to clear an area between Montchamp and Estry in support of units from the 15th Scottish Division. Heavy German shelling had already devastated the area and when they moved up, casualties from other battalions choked their regimental aid post. Under heavy fire they moved to higher ground in order to hold the ridge beside the village of Sourdeval as a launching point for an attack upon nearby enemy positions. The village was already a ruin, stinking of corpses and Camembert cheese one Guardsman recalled.[67] It was from this imagined vantage point that the 3rd Battalion, with minimal artillery support, was flung into one of the most ill-judged actions of the Normandy campaign, in which German parachute troops had a clear field of fire from prepared positions against attackers who had little but uncut corn for protection against a torrent of Spandau and 88-mm gun fire.

Much bitterness followed upon this disastrous attack and survivors' accounts make it sound grimly similar to countless ill-fated infantry attacks a generation earlier on the Western Front. The moment the leading companies crossed the crest of the ridge the ground around them was scythed by German fire, and three-quarters of 2 and 4 Companies were killed or wounded by the time they reached their first objective, a stream that intersected fields below the ridge before the ground rose to the enemy's positions.

> The wounded lay where they fell or, if they could, crawled to the shelter of the standing corn, but the burning phosphorus of the mortar bombs fired the dry stalks. The companies choked and coughed their way through a swirl of smoke and flame and the wounded watched the flames creep on to them. Some dragged themselves painfully out of the corn back to the open root fields. The stretcher-bearers shed their equipment and worked frantically to shift the unconscious till they too dropped wounded in the corn.[68]

Late in the afternoon of 12 August, Brigade HQ sanctioned a smokescreen to be laid so that survivors could be pulled back and the following day went in burying in makeshift graves those of the dead who could be retrieved.

Heavy casualties were also the lot of the Irish Guards' 2nd Battalion in Normandy as its tank crews quickly found out what the lethal defects of their vehicles meant in close combat with Tiger and Panther tanks manned by SS units with experience of the Russian front. Montgomery himself, in a rousing address to the battalion on 12 August, although burnt-out Sherman tanks were in full view as he spoke, extolled the superior virtues of British and American equipment. German Panzers, he urged them, could be disabled from the flanks and were unreliable mechanically.[69] For ordinary crews and tank commanders like those of the Irish Guards, the problem of course was that every German tank could destroy a Sherman, while only one Sherman out of four could penetrate German armour with its turret gun. 'In the Normandy tank battles the Allies defeated the Germans because they could afford to lose six tanks to every one German.'[70] The 2nd Battalion of the Irish Guards lost 175 tanks in Normandy, a figure better converted to crews burnt to death as their 'Tommy cookers' were targeted by superior German fire power.

Weary and savage months of fighting lay ahead for the Guards and for other Irish and Ulster units across France and Germany, and in Italy too, before hostilities ended in Europe. The casualty lists grew, but so too did the citations for valour; yet the meanness of the politics of Irish partition festered still in ways that could mock the courage of those whose war has been the concern of this chapter. The possibility of the Free State, or Eire, actually providing as many or even more recruits for Britain's war than Northern Ireland became a matter of obsessive anxiety to Unionist politicians. With the war still on, the Stormont government asked all the service departments in London to comb through their files to quantify the national origins of those serving. The exercise, it has been shown, was futile because Irish citizens who went to Northern Ireland to enlist could easily be incorporated within the province's own enlistment statistics, while many Irishmen in Britain could join up from British addresses or, if resident there in 1939, be liable for call-up anyhow.[71]

Ultimately the British Dominions calculated that over 32,000 men 'born in Eire' were serving in the British army at the end of 1944, while an estimated 38,000 from Northern Ireland volunteered for all three services,[72] but the controversy over comparative numbers was unworthy of men whose readiness to serve and suffer alongside each other left little room for the politics either of partition or of sectarianism. Yet the issue of Northern Ireland's contribution to the war and attempts to use it against the Irish Republic resurfaced inevitably after 1969. Politicians in the Republic could be as mean of spirit as Ulster Unionists and in 1983 Fianna Fáil, then in opposition, bitterly attacked the Irish army being represented at a Remembrance Day service in Dublin.[73] At that time Dublin's own memorial to Ireland's war dead was little visited and badly neglected, a situation few public figures seemed concerned to criticise.

Since then the memorial has been restored and the ground it occupies at Islandbridge has been beautifully landscaped. It was rededicated by the Irish Taoiseach in a simple ceremony in April 1995 and all parties in the Irish Republic and Northern Ireland, including Sinn Fein, were there to hear his tribute to the fallen.[74] Originally the memorial was raised to honour those killed in the First World War but de Valera declined to open it in 1932. That was a time when it was 'politically correct' for Irish politicians to remember only the Republican dead. After 1939, however, their number was swiftly overtaken by those from both sides of the border or of Irish descent who once again proved ready to offer up their lives in a British war. This chapter has sought to tell at least something of their story.

'If I Fight for Them, Maybe Then I Can Go Back to the Village': African Soldiers in the Mediterranean and European Campaigns, 1939–45

David Killingray

In the official plans to commemorate the fiftieth anniversary of VE Day in London one group of ex-servicemen were not included. African troops, according to the British government, did not fight in Europe and therefore were not entitled to be represented in the celebrations. This ill-informed policy was denounced and corrected by a number of people. African troops had fought in Europe during the Second World War, most notably in Italy; also they made a major contribution to the important victories over the Italians in Abyssinia in 1940–41 and in the North African theatres which prepared the way for the invasion of Italy. Criticism of this Whitehall insensitivity resulted in some representation being granted, but not in keeping with the extent of Africa's contribution to the defence of Britain's imperial interests during the war of 1939–45. By contrast, the French government invited all the franco-phone African heads of state to Toulon in August 1994 to celebrate the landings in Provence fifty years before, which had helped to liberate France from German control.[1]

*

The subject peoples of European empires invariably have been drawn in to metropolitan wars. Black loyalists during the American revolution, slaves in redcoats, *tirailleurs* from West Africa, recruits from the north Indian plains, all fought in European wars overseas and also in Europe. By the beginning of the twentieth century new sources of manpower in the African colonies offered a useful military substitute for hard-pressed imperial powers, not merely for distant wars overseas but in European theatres. The French led the way, but other countries followed.

All European colonial powers raised local indigenous troops in their African colonies. These small, lightly armed forces were recruited from supposed martial races, often 'loyal aliens' and people from the periphery of the colony, and were commanded mainly by white officers. The role of colonial troops was to defend the borders of the colony but principally to pacify the country and then maintain internal security. Black men were employed by white men to put down black men! They might also be employed in the defence of a neighbouring colony. The policy was cheap and practical. From the mid-nineteenth century onwards the French had used West African soldiers in imperial wars; the British had a much larger reserve in India, what Salisbury called that 'dusky barracks in an oriental sea', which provided an imperial fire-service for Asia and also Africa. It was but a brief step for the French to involve West African *tirailleurs* and for the British to use Indian troops in the campaigns of the Western Front by 1914.

The root of French policy of employing West African troops in European theatres was the growing difference in population and thus manpower resources between France and Germany. In 1910 General Mangin suggested that West African troops could form *la force noire*, a substitute garrison to relieve French troops in North Africa for service in defence of France. Conscription was introduced in West Africa in 1912. However, by late 1914 French West African troops were being employed against the Germans.[2] Altogether 215,000 colonial troops served in France, of whom 175,000 were from Africa. As soldiers in a modern war they acquitted themselves well, although their suffering in the winter led the authorities to withdraw them to the south of France at the beginning of that season of the year, a policy known as *hivernage*. During the First World War African troops were also employed in the Mediterranean and Black Sea campaigns, and as occupation forces on the Rhine in 1919–20. Conscription was extended by the law of 1919, and in the interwar years African colonial forces saw service as garrison troops in North Africa and the Levant.

The military forces in Britain's African colonies – essentially the King's African Rifles (KAR) recruited in East Africa and Nyasaland, the quasi-federal Royal West African Frontier Force (RWAFF), and the Sudan Defence Force (SDF) – were intended solely for service within Africa. The Indian army was used as an imperial force. Climatic, political and practical arguments were advanced against using African troops in a European campaign and against a European enemy. Nevertheless, in 1916–18 a 20,000 strong South African Native Labour Corps was recruited and deployed behind the lines in France.[3] Towards the end of the war senior officers were talking seriously about emulating the French and recruiting a large black army for imperial service. Peace arrested such an idea, but it was revived when Britain faced acute military manpower shortages in the immediate post-war years, again in 1940 and once more after 1946.[4] Fascist Italy recruited a large African army in the 1930s, reputedly 200,000 men many of whom were mercenaries, which was regarded as a serious threat by the other imperial powers. A large part of this colonial army was a labour force. In the bloody war of 1935–41 many Italian *askaris* either proved unreliable and simply faded into the countryside or deserted to join Ethiopian guerrillas.[5] One oral story from the East African campaign of 1940–41 is of two Italian *askari*, arms raised in surrender, greeting soldiers of the Gold Coast Regiment in a northern Gold Coast language. They had gone on the *hajj*, become lost in the Sudan and wandered into Italian territory, where they had been swept up into the colonial army. In the well-established pattern of mercenary armies the prisoners simply changed their uniform for that of the GCR. Belgium had an army of over 20,000 men in the Congo, the Force Publique, which was primarily used for internal policing. When Belgium collapsed in mid-1940 the Force formed the largest part of the army then available to the government in exile.[6]

Twentieth-century warfare required a large army of non-combatant labourers to support the military in the field, to dig trenches, build roads, move supplies, as dock workers, and to guard prisoners and war material. In the First World War pioneer labour for the Western Front was drawn from all over the world, including from North and South Africa and from China. The 'little wars' in tropical Africa in the late nineteenth century had relied upon thousands of human carriers, men, women and children, to move supplies up to the front line in regions where horses and cattle could not operate because of disease. During the East African campaign, 1914–18, hundreds of thousands of porters were forcibly enlisted by the colonial powers to serve as carriers in harsh and deadly conditions; disease, starvation and official neglect resulted in the death of around 100,000 people.[7] In the Sinai theatre against the Turks, the British used 300,000 Egyptian *fellahin* in their labour corps. The Second World War also made a heavy demand for labour to support the military. In that conflict, however, Africans were enlisted into pioneer battalions destined essentially for non-combatant duties, although the shovel was often joined by a rifle in front-line conditions.[8]

*

When war broke out in September 1939 the borders of France were guarded by 80 divisions, including 10 colonial divisions of which 7 were from black Africa. French West Africa alone provided 100,000 soldiers of whom around 80,000 were serving in France by mid-1940. When France asked for an armistice in June 1940, 9 per cent of the French army consisted of African soldiers, a considerably higher percentage than in the First World War. African troops were among the French forces evacuated from Dunkirk. The rank and file of de Gaulle's Free French army, from 1940 to 1944, was largely African with North Africans forming a substantial element; de Lattre's invasion force in southern France in August 1944 included 20,000 black Africans, one-fifth of the total force. Altogether throughout the war the Vichy and Free French recruited 200,000 soldiers from sub-Saharan Africa, many of whom saw service in the Mediterranean or European theatres.[9] Some men volunteered for military service but many were caught by selective conscription. 'We had been forced to go into the army – forced to go to France,' said one old soldier, while another recalled that 'if I fight for them, maybe then I can go back the village'.[10] The Belgian Force Publique, which numbered 23,000 men in 1940, nearly doubled in size by 1943. The Force saw action in Abyssinia in 1941, and from March 1943 to July 1944 detachments were under the British Middle East Command in Egypt and the Levant, largely employed on garrison and other non-combatant duties. Rumours spread, either purposely or gratuitously, among Italian and German prisoners-of-war in Egypt that the Belgian colonial troops guarding them were cannibals; the filed teeth of some soldiers seemed to compound this idea which, so it was reported, helped discourage prisoners from attempting to escape.[11]

Britain's African colonial forces were intended only for service within Africa. The size and equipment of the various military forces reflected that perceived role. In 1939, even with some pre-war preparation, the KAR, RWAFF and the SDF were essentially lightly armed infantry; they lacked adequate transport or the necessary support services for involvement in a modern war, even one fought in Africa. Most of those services had to be created from scratch in the early years of the war at the same time as rapid recruitment expanded the African armies. Wartime recruitment was not the voluntary process it was often projected to be. Soldiers were enlisted via indigenous rulers who were compelled to act as recruiting agents. Large numbers of men were forced in to the army, often 'just sent' by their chiefs who were given quotas. Bildad Kaggia, a Kenyan recruiting clerk, said of the chiefs: 'They used every method from persuasion to force to reach their quotas...the chief used the power of conscription to get rid of anybody they did not want in their locations.'[12] 'The orders I had received', said a Masutho ruler, 'were to send out chiefs and ask them to call their men together and then pick out a number of young men.'[13] Several men were interviewed by Michael Crowder in Botswana: '*All* explained...it was not their choice to go to war but they were forced to answer the call. In Serowe they were put inside the big kraal at the *kgotla* and told they were to go to war. In Lobatsi they were kept [behind a] fence.'[14]

Official recruitment drives sent men scurrying for cover 'in the bush' or across nearby frontiers. When insufficient men were forthcoming, particularly semi-skilled and literate men for the newly created specialist corps, emergency legislation empowered the colonial authorities to introduce conscription. This became more urgent as the decision was taken in 1941 to use African troops as military labour in North Africa and the Middle East. In the course of the war, Britain's African armies expanded greatly so that by 1945 more than half-a-million men, drawn from all parts of the continent, had been through the ranks. These forces served as combat troops in all theatres, but most notably in Abyssinia, Madagascar and then, in 1944–5, in Burma against the Japanese. In addition African troops also provided a vital labour role in those campaigns, but this was particularly so in the North African, Levant and Italian operations. The majority of British African troops in the Mediterranean area were employed in non-combatant roles, which, on the line of supply, meant often being exposed to enemy fire with only limited cover.

In the Second World War the South Africa government recruited over 120,000 non-white service personnel, over one-third of the total Union Defence Force. Of these 76,000 were black; about 40,000 black and coloured troops served outside the Union. Most were employed in the non-combatant Cape Corps and the Native Military Corps (NMC), collectively known as the Non-European Army Services (NEAS). White opinion in South Africa was sharply divided over recruiting blacks into the armed forces. Smuts's government adamantly opposed armed black units and only reluctantly agreed to this when they were moved into combat zones. Men of the Cape Corps and NMC served in North Africa and Syria, where 'they kept the lines of communication open, fed thousands of mouths, dug many miles of trenches, loaded and unloaded tons of essential war materials and carried many wounded to safety'.[15] In 1941 recruiting of troops began in the three High Commission Territories (HCTs) for service in the Middle East. Proposals by South Africa that these soldiers should be integrated with their own forces were rejected, but Pretoria's demand that High Commission troops should be non-combatant was initially accepted, although by 1942 High Commission troops in North Africa were integrated with British units and trained in combatant roles such as anti-aircraft gunners.[16] Altogether the three territories sent more than 28,000 men for service in the African Auxiliary Pioneer Corps (AAPC), who saw service in North Africa, the Levant and also Italy.[17]

*

First-hand published accounts of African soldiers' experiences of war service are limited and relatively rare. Most recruits were non-literate and only a small handful of Europeans who served with African troops recorded their service.[18] Some information on daily life can be found in the official newspapers that were produced for African soldiers, for example *RWAFF News*, published in Cairo and Bombay, and *Indlhovu*, which was distributed to South African and High Commission troops in North Africa. Similar material was used in official publications directed at an African audience and intended to encourage recruiting and maintain morale, also broadcast on the radio, and published in the English and vernacular language newspapers of many colonies. Inevitably this was information carefully selected and doctored to present a positive picture of African military service, and for the most part it needs to be treated with great caution. Metropolitan and national archives contain a great deal of material on Africa's war effort, but much of this was generated by officials and thus represents a view from above.

Where there are comments on the African rank and file they tend to be statements by European officers; only occasionally is the African voice to be heard.[19] One place where it was expressed was in the letters written home by African soldiers that remained in the hands of the censors, or went to African rulers. Good collections of such letters are in the National Archives of Ghana, Accra, and in the National Archives of Botswana.[20] Undoubtedly there are other collections of similar letters in archives elsewhere in Africa and, of course, in private hands. Wartime propaganda generated a large number of photographs of African soldiers, but at the same time individual soldiers either took their own photographs or had them made in studios in cities such as Jerusalem, Cairo and Rome. A small number of African autobiographies have included wartime experiences, most notably that by Bildad Kaggia.[21] A rapidly wasting source is the oral evidence of ex-servicemen. A small amount of this was collected in East Africa by Shiroya in the 1960s and since then by other researchers elsewhere in the continent. A collection of oral and written material was gathered for a BBC Africa Service series on 'Africa and the Second World War', broadcast in 1989, and use is made of that material in this essay.[22] A good portion of this splendid written, oral and visual material, covering all aspects of the war service and experience of Africans, came from personal and family collections, a source that no doubt offers a rich cache that has been hardly touched by researchers.

*

The French used their colonial infantry as front-line troops in all their campaigns during the Second World War, on the Western Front, in Syria, North Africa, in the invasions of Europe and in the liberation of France in 1944. African casualties were high, particularly in the Battle for France in 1939–40; figures vary widely but an official figure records a total of nearly 16,000 *tirailleurs* killed and missing in that campaign.[23] The non-combatant role of many British African troops meant that they suffered fewer casualties, although they were often in the unenviable position of working in the front line but without any means of defending themselves. Both the French and the British African troops suffered from the winter climate of Europe and the Mediterranean, even the troops from the High Commission Territories who were recruited in the belief that they would be better able than East and West Africans to withstand the cold weather. African memories of the casualties of the First World War affected recruitment in the years 1939–45. The horrendous death rate of carriers in the East African campaign in the First World War served as a powerful deterrent to enlistment, as did the loss of over 600 men of the South African Native Labour Corps when the troopship *Mendi* was sunk in the English Channel in November 1917. Officials were cautious in releasing casualty figures, especially those from the loss of ships carrying African troops and prisoners-of-war.

For many French colonial troops on the Western Front the war was short and bitter. Tuo Lieroulou's *tirailleur* regiment was moved from Toulon to the Somme to reinforce the front line against the German attack in mid-1940:

> We marched to the Somme, section by section, carrying our backpacks. We were ready for them, but we didn't know until the war started who was stronger... The war lasted nearly two months at the Somme. It was there that they took us. They were stronger than we were. They dropped bombs on us.

Another soldier from Côte d'Ivoire recalled that the fighting 'was very confused – people were running. We were trying to find a place to shoot from that would be safe. Bombs were dropping. Everyone was for himself.'[24] Sekongo Yessongui's memories were of digging foxholes and killing Germans in the process of advancing and retreating in 'seven days of fighting'. 'It was hot at the front', one veteran remembered, and 'when it was over, practically everyone was dead – all but a few of us. Then Pétain lowered the flag.'[25] In the process of retreat and defeat many African soldiers were caught by the advancing Germans. Soro Nougbo's unit retreated because 'the Germans were too many for us. It was very serious – so many dead. The Germans found us when they crossed the river. Our job was to collect the corpses – but then we ran too.'[26]

In the Syrian campaign and in North Africa, French colonial troops fought against each other on behalf of the rival Vichy régime and the Free French. This confusing situation was recalled by Namble Silué, who was with the Vichy forces in Syria:

> There was war – against de Gaulle. They said to us, 'Tomorrow we go to Damascus for manoeuvres.' De Gaulle's men attacked us. The Camerounians attacked us. We were amazed because they weren't Germans. They were black. Why were they attacking us? They had never told us we were fighting against the French.

And after the fighting was over the captured Vichy African troops were re-enlisted in de Gaulle's Free French forces.[27] As part of de Gaulle's army they were equipped by the Americans and deployed in Tunisia to fight the Germans. There, Nanga Soro told Nancy Lawler, 'What I will never forget is the infinity of corpses – *Sénégalais* and French. I found some whose souls had not yet left them, but I knew they were going to die. You could not eat because of the smell of the bodies. The blood.'[28]

Few of the Africans enlisted by the British had any idea of what battlefield conditions in North Africa were to be like. A Seychellois soldier later said that 'we were like cattle being led to the slaughter house'.[29] In the confusion and retreat before the German advance into Egypt in 1941, many African soldiers were wounded and captured. At Bardia, in December 1941, a black South African, Corporal Berry Gazi, volunteered to crawl towards the German lines with a Red Cross flag to request a ceasefire while the wounded were removed from the battlefield. He said:

> I proceeded to crawl towards the enemy under very heavy fire, shells were bursting around me and bullets were whining over my head. I cannot say what my feelings were at that moment...I managed to reach the enemy without being wounded and delivered the message. I returned safely to our lines.

He was awarded the Military Medal for his bravery.[30] Towards the end of the war Sergeant Ontwetse M. Molefi, wounded and in hospital, wrote to his chief: 'I never get sleep out of the pains from my knees. I now fell (*sic*) that I am a cripple who will not do anything in life.'[31]

South African and High Commission troops took part in the operations in Italy from late summer 1943 onwards. For example, Swazi troops landed at Salerno in September 1943 and were reported, admittedly by their own comrades, as playing football at Anzio 'even in the heat of battle'.[32] In the battlefront conditions of Italy, some HCT pioneers were converted to combatants. As one man told Tshekedi Khama: 'Even the war is progressing well. We are in Italy consisting of 8 units with each equipped with heavy guns. These guns can be used to shoot down aeroplanes hence there is nothing to fear.'[33] Most High Commission troops continued in labour roles. Conditions were particularly harsh in winter for the companies employed on the front line as they had

> to contend not only with the weather but with the enemy. Many of them whose work is to carry ammuniton, food and clothes to the men in the line can only work under cover of darkness because of enemy shelling and mortar fire. This is a very arduous occupation for night after night with heavy loads on their shoulders they clamber up steep and dangerous slopes in rain and snow, deliver their loads and return (very often carrying wounded men) to their little bivouac tents in the early hours of the morning.[34]

Swazi soldiers at Anzio worked under constant bombardment from German heavy guns, sleeping at night in dugouts away from the billets. One anonymous soldier's letter commented on the work as being

> very much routine...mixed with tense moments when we were subjected to bombing shelling and one time we had to stand by for break-through. As weeks passed and still our advance was stayed we were called upon to carry arms and rations up to the now tired fighting troops who had the highest of mountains to assail, some of the mountains took us $13\frac{1}{2}$ hours to climb...[35]

French African troops spearheaded the invasions of Corsica and Elba in June and September 1944. The fighting to take Elba was brief but fierce and the Germans and Italians strongly resisted the landings. 'It was four in the morning,' recalled Ditemba Silué. 'We all disembarked now. The war was heating up. Many were dead – on the ground. To get off the boat, there was a long rope. The water was waist-high, we had to hold on the rope or drown. We were being bombarded.' Tuo Nahon said, 'The Germans started firing at us while we were still in the water. All the same, we captured them. The landings were difficult. When you left the boat – to get on the sand – they were shooting at you. Six were killed on the shore of the sea. We were lucky.'[36]

In August 1944, two months after the D-Day landings, the French and Americans launched Operation Anvil, the invasion of Provence. A substantial part of the French force was composed of North and West African troops. Again German resistance was strong but short-lived. Kiwalte Kambou remembered that as the 9th Infantry Division advanced towards Toulon the fighting was sometimes 'combat with bayonets – man to man'.[37] West African infantry were also involved in the fighting to drive the Germans out of France. The winter of 1944 was bitterly cold and West African soldiers suffered greatly. 'It was the first cold for me,' recalled Yeo Kouhona. 'We were well dressed, but it was so cold – ice everywhere...We were sent south to the French (*sic*). We were happy...Some were furious when we left though, because we had started the war and it was almost over now and we were being replaced by French troops who had been afraid. We wanted the Africans to win the war.'[38] De Gaulle's policy was to advance France's political position and it was decided that the army to invade Germany should be exclusively white, a policy known as *blanchissement*. Thus French African troops were removed to southern France for the winter of 1944–5.

*

As the Allies advanced across France and Germany colonial prisoners-of-war were freed. The largest number of captives were West Africans taken with the fall of France in mid-1940.[39] German behaviour towards African prisoners was often harsh. Some captured and wounded soldiers were shot out of hand, for example at Chasselay Montluzon and Erquinvillers, both near Lyons. A similar fate nearly befell Léopold Senghor, the future president of Senegal, who was captured at Charité-sur-Loire. Only the intervention of a French officer saved his life. Crowded in makeshift camps in northern France and Germany, the African prisoners suffered from poor and inadequate food and often brutal treatment. Escape for black or brown prisoners was difficult. After the armistice some African soldiers were released; a handful joined the resistance. For many captivity continued for another four years, either in camps or as forced labour on farms and defence works. At the end of the war, repatriation was slow and the cause of considerable discontent among African soldiers. West African *tirailleurs*, recently freed from captivity and *en route* for Dakar, mutinied in a transit camp near Liverpool and then more dramatically at Thiaroye in Senegal in December 1944. There were other but less serious incidents of unrest in camps in southern France.

The Italians and Germans captured several thousand African prisoners from the British and South Africans in North Africa. The welfare of these captives was disregarded by both Italians and Germans. In contravention of the Geneva Convention, African prisoners were forced to handle war supplies and to unload ships even during air raids. A few escaped, some to perish in the desert or from mines. One or two attempted sabotage, one spectacular case being of a South African who made a small bomb and detonator and blew up a German freighter in Tobruk harbour. As Axis control of North Africa crumbled, some prisoners-of-war were shipped to southern Europe. A few died in ships that were torpedoed. Conditions in camps in Europe were if anything worse; Babenhausen and Chartres were particularly notorious for members of the NEAS. Nzamo Nogaga, held at a camp near Munich, was kept alive by Red Cross parcels. 'Personally the only dead meat I did not eat was dog's flesh,' he wrote. 'I ate horse, donkey, cat, anything to keep life in me. The reason I did not eat dog's meat was because no dog came my way!'[40] An attempt by the Italians to make a propaganda film, showing the superiority of whites over blacks with the former world heavy-weight boxing champion Primo Carnera in a ring against Kay Masaki, a large Zulu prisoner, ended in spectacular disaster when Masaki knocked out Carnera. Both the Germans and Italians used African prisoners for wartime films, in one instance resulting in loss of life.

*

Army life pitched African soldiers into a range of new experiences. Many recruits came from rural areas and had little contact with the products or methods of the industrialised world. Tarmac roads, machinery, canned food, motor vehicles, not to mention the sea, large ocean-going ships and the weapons of modern warfare, were either unknown or concepts which they barely grasped. Military service introduced soldiers to new technologies, ideas, foodstuffs, peoples and relationships, regular wages, and the welfare services of the modern army. This started in training camps with the issue of unfamiliar uniforms, boots, and a system of discipline geared more to industrial time than to the more relaxed notions of work schedules determined by agrarian life. In training camps sanitary and hygienic ideas were also taught and enforced by punishment. Here men were also given instruction in number, perhaps received orders in an alien language such as French or English, and learned how to assemble, look after, and fire a rifle of precision design. Soldiers ate new foods, especially once they went overseas, many learned how to handle shovels, drive and maintain lorries, and also how to read in their own language, in French or English. And the army had a welfare and medical system, not altogether new to those who had been migrant labourers in the mines, but certainly to those fixed in their rural backgrounds. In addition there were new places seen and ideas absorbed. The full extent of the impact of these experiences on nearly a million men drawn from all parts of Africa is neither well documented nor fully explored. At a time when the research might have been undertaken, for example by anthropologists in the immediate post-war years, their interests were directed elsewhere. A few historians and political scientists have drawn some conclusions about the way in which the war raised the expectations and political awareness of some African soldiers and how this influenced the nationalism of the post-1945 period, but in most instances this has not been supported by adequate evidence.

One thing that needs to be stressed strongly is that African armies of the Second World War were not necessarily very different from the armies raised in the modern industrial states. In the latter recruits were predominantly young men, many of them drawn from rural areas where there was a traditional hierarchy of status and authority, and the army introduced them to a variety of new practices and skills. African and European armies had their own hierarchies and command structures where non-commissioned officers were of primary importance to the ranks. The military authorities exercised a strong measure of social control over soldiers, conscious that large numbers of men congregated closely together were potentially volatile and needed firm but wise discipline. A major difference between European and colonial armies was that in the latter most officers were white. Race thus was a likely point of conflict, and it was recognised early on in the war that it was vital, as far as was possible, to employ officers who would work within the acknowledged racial notions that underpinned the colonial order but without descending to the kind of overt racist behaviour widely seen in South Africa and the white settler colonies. Black and coloured South African troops resented racial abuse from officers and responded by deserting, strikes and more violent forms of protests. A few unpopular officers were murdered by their own men, something that was not peculiar to African colonial armies. Among African troops desertion rates were relatively high while they were in the colony of origin; men tended to desert at times of harvest and when orders came to move overseas. On active service it was much more difficult to desert successfully; colour and race were markers. Strikes and mutinies were more frequent than is indicated by the current meagre literature, as indeed was also the case with European troops. Even the official record remains mute about many minor incidents among African troops, particularly in North Africa, some of which had to be suppressed with force and resulted in the loss of life.[41]

The several hundred thousand African troops in Mediterranean theatres were housed in a variety of often makeshift encampments. In most cases the first camp was tents. Bildad Kaggia said that his 'depot was in the desert and the headquarters was housed in a tent. For the first time I slept under canvas and tasted life in the desert...where sandstorms were a daily occurrence. Cooking and working in the office were sometimes very difficult.' When his unit moved to a new and more permanent camp in Ismailia it was, he wrote, a 'paradise' fully provided 'with hot water taps, modern baths and kitchens. We regarded Ismailia as more than a home...'[42] Most camps were sited away from urban areas; this was partly the necessity of wartime operations, but it was also official policy to limit contact with local people and to minimise the usual kind of conflicts that develop between soldiers and civilians.

*

The main use for Africans was as labourers in uniform under military discipline. They should be 'pioneers first, soldiers second', argued Henry Gurney in assessing the East Africa Military Labour Service in late 1942. 'Our object', he said, 'should be to provide reliable and disciplined labour under all conditions, and military training should be limited to the bare minimum, otherwise the men will regard themselves as soldiers rather than Pioneers. On this basis, the training should last four months, one of which should be devoted entirely to Pioneer duties.'[43] Although many African soldiers were specifically recruited for service as pioneers, there was among them widespread resentment at labouring work which was not seen as proper soldiering. In the East African campaign it led to a serious mutiny in the KAR. A South African soldier who had served in North Africa, Sergeant R. Moloi, caught something of that resentment when he said:

> My people kept complaining that there is no war. They were told that they are joining the war – that they are going to see the front line. We do not see the front line – we are only digging these holes for the Arabs. But I told them they have to do what they are told and then we did that. I understood it was a good help during the war – that is why they stopped the Germans.[44]

The work schedule for African labourers was often hard and gruelling. It involved road, railway and aerodrome construction, dock labour, and serving the supply line of the Eighth Army. The climate was often severe, hot in summer and cold, especially in the Levant and Italy, during the depth of the winter. South Africans, with mine labour experience and recruited on a contract basis for the NMC, worked through the Syrian winter of 1942 to build a railway and two tunnels as well as a sea wall. East African and Indian Ocean island military labour 'got trapped in the treadmill of dock work' so that one of the most common entries in the war diaries for Seychellois pioneers is 'Men working at the docks'.[45] Military demands on labour invariably meant long periods of intense hard work with intervals of idleness. To Charles Arden-Clarke, the Resident Commissioner for Basutoland who was touring the Middle East, this spelled potential trouble. He reported: 'I came across instances of companies which appeared to me to have been kept too long at heavy labour under hard conditions and without change or respite to the detriment of their morale and discipline.'[46] The nature of the work did not change and for those African troops serving in Italy, and even a few in Yugoslavia, the winter inflicted considerable suffering. In an attempt to maintain morale and to help with further recruiting, the colonial authorities arranged for visits by African chiefs to the men at the front. Invariably these were stage-managed events and only compliant chiefs were used, often those who had or hoped to receive the King's Medal for Chiefs. One chief from the central Nyanza District of Kenya, in a speech to the troops in North Africa, told them:

> If you were asked to go ahead into Europe, I want the name sent to me of any man who is afraid...All people at home are wanting to join up...I have seen in the desert tours the destruction of war, we do not want that destruction to come to our villages.[47]

A large part of military labour work was dangerous. Units were often under shellfire or bombed from the air. Handling ammunition and petrol, and working with heavy machinery, often in precarious conditions, exposed men to the danger of accident and injury. For African pioneers who arrived in North Africa during the fraught days of 1941 and were involved in supplying a buckling front line, there was a high risk of being killed, injured or captured. Mauritian troops told a raw Seychellois company recently arrived at Mersa Matruh, that 'flocks of birds come by day and night to lay their eggs here'. For Samuel Accouche, also from the Seychelles, the usual shelter in an air raid was a thick-walled cement cistern. However, on one day someone else took his place. 'It was on that day', he said, 'that the German planes turned up. They dropped their bombs and one fell thirteen and half feet from the cistern. The shock lifted the cistern from the ground and threw it onto the place which we used for cooking and the debris pierced the cistern in such a way that it looked like a tea-strainer.'[48]

*

Most military service involves long periods of enforced idleness and boredom for soldiers. A major problem for the military authorities was what to do with African labourers sitting about in camps with little or nothing to do other than routine maintenance tasks. Official reports on African pioneers regularly warned that indolence would lead to unrest. A typical example is the following rather clumsy report on Seychelles troops 'hanging about' at Mersa Matruh:

> The native personnel are now beginning to feel the strain of being kept, as it were, hanging about so long. It appears to me some very fine material both fighting and working is being wasted by the men being kept for so long as it were on their own doorsteps and subject to influences from which it would be desirable to get away. I would therefore respectfully urge your excellency to do everything in your power to have the company moved either to a theatre of war or another sphere of activity.[49]

As the war moved out of North Africa and into Europe, with most African troops behind in support roles, the situation became more acute. By 1944–5 the sickness, heat and boredom of the North African camps had generated deep resentment among African pioneers. Unrest was endemic, and small incidents had a potential to generate greater discontent. Junior officers such as Captain Atkinson, often best able to feel the pulse of the men, reported that 'mass disobedience has occurred among all African troops. It is not being hushed up now – it cannot be – and indeed it is a most important fact to be faced. Successful African mass disobedience is a product of this war.' He went on to describe the anger over the slowness of repatriation and said that men did not believe stories of shipping shortages.[50]

The strains of long absence from home compounded by wartime conditions fuelled soldiers' grievances. Soldiers' letters, often written on scraps of paper, even lavatory paper, due to shortages of supplies in the front line, give a good idea of the concerns and anxieties that chafed sores on to the minds of men away at the war. The particular focus of interest of African soldiers was slightly different from that of European troops, but in essentials they were very similar concerns – over wife and marriage, children, relatives, whether a job would be kept, and care for property and possessions. Many of the letters were written by literate men for their non-literate comrades, often in exchange for a payment. Many of the letters dwell on a similar range of topics. For men from a rural and largely self-sufficient economy, land, cattle and livestock were vital. A Tswana private serving in the Middle East bitterly wrote: 'We are fighting here and they're taking women's field and their tobacco. I want that field. A Kalanga does not own any land.'[51] The fidelity of wives was a constant concern of soldiers, and many letters asked relatives or chiefs to enquire into the behaviour of a wife or neighbours. Private Rahakwena worried that he had not heard from his wife, particularly as he had been paying money to her: 'Chief I am asking myself the following question every day, is that person whom I regard as my wife present or has she decided to venture for green pastures in other areas... I am saying this because I am not receiving any letters from her...'[52] For another soldier news from home brought what he hated to hear: 'since my departure my wife has found another man and they are in fact staying together... This causes me great pain when taking into consideration the fact when I departed she was a decent wife of mine.'[53]

Another cause of concern was whether the compulsory and deferred payments of money from a soldier's wage were being received by the right person and used in accordance with the soldier's wishes. The pay received by pioneers varied from ninepence to over one shilling a day with West and South Africans receiving higher rates than men from East Africa. The authorities arranged for money to be paid to wives or other relatives and also for a sum to be deducted as savings so that a soldier had money to return home with on demobilisation. Private Manaheng Lesotho had put down his uncle's name to receive money and to ensure that his wife was cared for, but the uncle had not replied and he heard reports that 'my wife goes about in poor clothing'.[54] Family disputes figure in some of the letters, over ploughs, the use of oxen, access to wells and water, who has the responsibility for the education of children, and the behaviour of relatives. One soldier told his chief that his younger brother was troublesome: 'he beat up mother and robbed her of £12,' and other offences, and therefore, 'Chief I now feel that the only alternative of dealing with this boy is sending him this side as a recruit.'[55]

In all respects the military authorities gave less consideration to the interests of African troops than that given to British soldiers. In matters of leave, food, uniform, accommodation, privileges, there seems to have been a widespread and unquestioned acceptance by senior military men that Africans would not receive or be considered on the same terms as European soldiers. Of course, this was already evident from different levels of pay, something of which many Africans became all too aware when they reached the Middle East and served alongside other Africans and Europeans. Many of the officers who served with African troops argued for greater concern for their men. The disparity of treatment was highlighted by the failure to give home leave to African soldiers. 'During the first three years an African was lucky if he got as much as one week's leave a year,' wrote Barber after the war.[56] For many High Commission troops in North Africa and Europe, home leave only came after 3–4 years service. When the Germans surrendered, two drafts of HCT troops were in South Africa awaiting shipment back to the Middle East. They simply refused to embark. The authorities were reluctant to use force and in the end had to accept the demands of the soldiers.

*

Local leave was granted to African soldiers but this was often organised. Men did go unaccompanied into Cairo, Alexandria and other Egyptian cities, but the military authorities were keen to exercise a measure of control over their activities. There were official leave camps in Cairo, Beirut, and Jerusalem. A European officer reported:

> as much liberty as possible is given on the off duty days. Parties are taken on lorries to visit other Bechuana Companies in the vicinity and others are marched off to a neighbouring town where, at first, the focus of interest was the ancient Roman ruins...Many of the men are keen on learning English and classes are often held most evenings after work.[57]

For many Africans the cities of Egypt were the first large urban centres that they had seen. One South African wrote: 'Arrived at a great city called Kei Road [Cairo]. It is very much bigger than the place in this country that has that name. It stands beside a river much wider than the Kei.'[58] Men who came from the tropics were struck by the aridity of North Africa: 'Egypt is a very bad country...it has never rained not a single drop. No tree, no grass is seen here, just bare sand. All the daily man's necessities are found in this great Capital – Cairo.'[59] They also commented on the extent and degree of poverty among the Egyptians.

Many African soldiers, particularly those with mission-school backgrounds, were eager to visit Palestine during local leave. This was one reason why a leave camp was established near Jerusalem. The military certainly used the

Christianity of soldiers as one way of exercising social control; padres, church services, and the distribution of bibles and religious books all contributed to filling the idle hours and meeting the interests of men. Similar policies were pursued for Muslim soldiers. However, it is important to stress that African soldiers brought their religious faith with them and that they often set their own agendas for worship and religious practice. For example, one European missioner reported that the Bechuana in the Middle East 'have brought their Christianity with them. Each company has its own church, with a leader, a band of communicant members and a larger number of catechumens.' They 'engage in bible teaching, collect money for church work at home'.[60] European and African clergy, appointed as padres, conducted visits to the biblical sites of Palestine. Bildad Kaggia from Kenya recalled his excitement at the prospect of going to Jerusalem: 'I shall never recapture the feeling that I had that morning when I boarded the train heading for Jerusalem, the "Holy City", the city of David, the city paved with gold.'[61] Kaggia and other soldiers were met at the station and taken to the leave camp where they were issued with a book specially written for soldiers entitled *Walks Around Jerusalem*.

The cities of the Middle East and Italy offered unimagined opportunities to African soldiers. Possession of a uniform and money, and in the company of other similar men, they went to places and did things that they would certainly not have done in the towns of their home colonies. Most men record the respectable pursuits and places, riding on trolley buses, going in lifts, the variety of shops and buildings, and visits to the zoo. Peter Ansah, a pupil teacher from the Gold Coast who joined the army in 1944, wrote of the 'privilege' of visiting Cairo and Alexandria: 'I had the privilege of seeing the Cairo zoo, the piramids [sic]. It was my first time of seeing elephant. At the piramids I also saw the Spinphix [sic].'[62] Soldiers in Italy visited the local towns but for some the highlight was to go to Rome. Jachoniah Dlamini of the Swazi Regiment was in Italy in 1943–4, and he recalled:

> Generally, I would go to towns for sightseeing. I had the chance to take ten days' leave in Rome, where I visited the Vatican, and was blessed by the Pope. He told us, 'I thank you for not bombing Rome because it took time to build and it is a place of religion.' I was very impressed by the Pope. I thought Rome had too many people and was too big, the people unlike the Arabs, treated us very kindly.[63]

*

The cities of the Mediterranean area also offered bars and women and the chance to meet and mix with soldiers drawn from all over the globe. It is difficult to quantify the impact that this had on the perceptions and expectations of African soldiers, many of whom came from rural societies subject to well-established patterns of authority and custom. It certainly worried Europeans, who thought such contact would pose a possible threat to the future colonial order. A South African colonel told recently freed black prisoners-of-war in Europe, 'You have been in strange lands. You have seen strange things, and even found yourselves in positions different from those found in the Union, therefore do not try to be a European Gentleman, but instead be a gentleman according to your customs and culture.'[64] His anxieties were probably well placed. African soldiers did mix with soldiers of other races and had contact with local peoples where they were stationed, all of which helped to challenge ideas of racial hierarchy that were an intrinsic part of the colonial order, especially in South Africa and the white settler colonies, and which shaped the colonial armed forces. As one Kenyan soldier wrote:

> I am very annoyed with the Europeans we have now, they are no good at all – they are very bad indeed. Here in Egypt there is no racial discrimination, we feed on the same food, eat at the same tables and so with dress-ing...our uniform is the same as that worn by officers. Here all races have the same privileges, and we got rid of Kenya Europeans.[65]

Parsons quotes a Kamba corporal who wrote of his experiences in the Middle East:

> There is no difference between black and white here. We go to the bar together, among white girls, who much prefer us to them. They long to be with us all the time and although we used fear them at home, we are playing with them here. They cost sh.5.[66]

Bildad Kaggia, in his autobiography, wrote about racial animosity and the discrimination practised by white soldiers, but he also had a good deal to say about good relations between the races.

The military authorities were particularly worried at African soldiers having contact with local women who were light or white skinned. This was partly concern to prevent venereal disease and also to limit tension with local communities, but mainly fear at an assault upon the highest peak of colonial racial exclusivity, the sanctity of white womanhood. This intimate contact between Africans and Europeans generated an urgent correspondence between Pretoria and London; South African political pressure led the War Office to agree to remove black troops from Italy as soon as militarily convenient.[67] This was done in November 1945 when HCT troops were returned to labour and garrison work in Palestine. Fights occurred regularly between rival groups of soldiers (and here regimental loyalties were important) over access to women. When race was put into the equation the fighting was more bitter. 'We used to fight with the South Africans at least three or four times a week,' recalled Joe Culverwell, a 'coloured' volunteer from southern Rhodesia who was in the Argyll and Sutherland Highlanders. 'Because this was the first time – particularly in Egypt – that we mixed up with white girls. You know, they sent out millions of WACS from England, and the South Africans didn't like it.'[68] All kinds of stories were put about to discourage Africans from going with prostitutes and local women. As Jachoniah Dlamini said, off duty we generally played football, engaged in traditional Swazi dancing or played *injuba*, a traditional African game that some westerners liken to chess. On some occasions we went into nearby towns, but we were warned by our officers to stay away from the local women as we were told that the Germans had poisoned the women, which would result in your penis dropping off if you engaged in sexual intercourse with them.[69]

It is unlikely that these scare stories deterred soldiers. African soldiers had recourse to white prostitutes, when it was possible, throughout the Mediterranean area; some Africans in Britain for the victory parades in 1946 certainly went with white prostitutes.[70] Official entertainment for African soldiers was also regulated. Blacks were banned from films and live shows which had white women in any state of undress or in any 'compromising' relationship with a non-white male. There were also attempts, particularly by the South Africans, to keep out of African hands magazines that might depict white women as sexual objects, and also even innocuous journals such as the London *Picture Post* which contained occasional illustrations of African-American soldiers enjoying hospitality in white homes in Britain.[71]

It is now amusing to read the verbal contortions of the semi-official historians of the South African war effort in describing black male/white female relations in Italy. They wrote that this caused 'problems...where unsophisticated NMC men were for the first time exposed to all the temptations of Continental European life... Their background was primitive and left them unable to cope with the emotional and moral stresses to which they were subjected in a Europe at war. Months were to pass before the problem was satisfactorily solved.'[72] The South African military attempted to regulate closely the movement of African soldiers, but with great difficulty in places where there was no official racial discrimination. Contact between former black prisoners-of-war temporarily in Britain in 1945 and local white people should be reduced to the absolute minimum by placing certain areas out of bounds, warned South African officials: 'In the absence of an official "Colour Bar" in England, entertainments of the NE (Non-Europeans) at clubs, such as the Donoughmore Club, will disturb the perspective of such personnel making it difficult for them to readjust themselves to South African conditions when they return to the Union.'[73]

*

As has been said, race shaped the African colonial armies. The French tended to be slightly more colour blind, or practical in certain aspects of their colonial policies, and from the end of the nineteenth century their colonial forces included a small number of black officers from either Africa or the Caribbean. Indeed, it was possible as a black French citizen to climb by merit to the top of the colonial administrative system, as did Félix Éboué, the governor of Chad and ally of de Gaulle in 1940. Black officers served in both world wars in

European campaigns. By contrast the British had a clear racial divide between officers and the rank and file. Officers with the African colonial forces, unlike those in the Indian army, were seconded for duty to Africa. They carried a King's Commission and by army regulations this was restricted to those of 'pure European parentage'. This was temporarily suspended for the duration of the war in 1939 thus allowing black Britons and also Africans to become commissioned officers.

In wartime only a handful of blacks were granted a commission, only two being given to Africans, both from the Gold Coast. The War Office wanted to revert to the pre-1939 situation once the war was over; most senior military men had deeply rooted ideas of white racial superiority and argued that Africans, and non-Europeans generally, lacked the innate qualities required to be an officer and that in any case black men would not command respect from the white rank and file.[74] So, effectively, the highest rank that an African could hope to gain in the British African colonial forces was as a senior NCO, and even they were required to address white NCOs as 'Sir'. In the pre-1939 years this discrimination passed with little comment from soldiers, although it was occasionally aired by literate nationalists. The expansion of the armed forces in wartime brought many literate men into the ranks, sent them overseas where some occasionally encountered black men with pips on their shoulders (Kaggia mentions the impact of meeting black US officers in the Middle East), and fuelled feelings of resentment at the discrimination which pervaded the military system, most especially focused on questions of rank and pay. This resentment was constantly present among literate soldiers in the Mediterranean theatres of war. One or two Africans (in all cases, I think, having one white parent) served in the British army during the Second World War in professional capacities as doctor or engineer. The Royal Air Force, the younger service and less hidebound, enlisted and commissioned a steady stream of Africans, almost all from West Africa, from 1940 onwards. Most who served in European operations were with Bomber Command and a few became prisoners-of-war.[75]

Some of the earlier historians of nationalism were keen to identify wartime service by Africans as a catalyst for post-war political change. It was a reasonable assumption that service overseas and exposure to a range of new experiences would raise the levels of expectations, including the political horizons, of African soldiers. Clearly this did happen, and the men who left Africa for the Middle East in 1941 and returned home four years later had drunk from a remarkable cup of experience. Political awareness and expectations were increased for many men, but there does not seem to be a great deal of evidence (more for East Africa than West) that those experiences were then channelled into nationalist politics. The vast majority of Africans who served in the Mediterranean and Europe, and also in Asia, returned to their rural homes and did not participate in national politics. There are instances where they did, but these are exceptional given the large number of men who served in the African colonial forces.

In the Middle East, North Africa and Europe Africans rarely encountered political ideas or activists. In all probability contacts with US African American troops were at a social rather than a political level. Ironically, political ideas were more likely to come from the Army Education Services or from meeting conscripted British soldiers who had a more robust and critical view of class and racial distinctions. The political impact of the war on African servicemen is clearly a subject that needs to be looked at more carefully. The many letters from African soldiers that have been examined do not contain anti-colonial ideas but rather the concerns that exercise most men who are forcibly absent from home for a long time. The expectations aired in the letters are not about the overthrow of an alien régime, although a handful refer to the need for political change, but to personal ambitions related to career, income, and family. Admittedly these are expectations that could be translated into a political agenda, but that does not emerge from soldiers' letters home.

*

The end of the war and demobilisation brought into sharp focus further racial differences over the more favourable treatment given to white soldiers compared to Africans. Whites were needed for reconstruction and their demobilisation was given precedence over non-whites. Large numbers of Africans languished in camps in the Middle East and Africa. Jaconiah Dlamini ended his war in Italy:

> Our officers came and told us the war was over, it was sometime in 1945, the exact date I cannot remember. We returned home in January 1946, having embarked from Ancona and sailing to Egypt... We spent more than nine months in and around Ancona, when we were leaving to return to Swaziland, the town people all came to say farewell. That is they were shouting... siSwati for 'goodbye'... Up to this day, I never received any payment for fighting in the war.[76]

The farewell greeting by the Italians gives some indication of the close relations enjoyed by the Swazi troops with local people. The comment about non-payment is a frequent complaint still to be heard from ageing African ex-servicemen. Even with deferred pay, most soldiers returned home with very little to show for their long years of absence, hard work and danger. The average sum saved *per capita* by Gold Coast troops was little more than £30, and the GCR contained a good number of literate men who received higher pay in the various service corps. For Sergeant Muliango there was the added grievances of supposed promises broken and yet a further turn to the white-settler screw. He wrote in 1989:

> I remained in North Africa until the end of the war when our regiment came back to Northern Rhodesia. To our surprise the government never gave us [black soldiers] any money or farms as promised. However, our

white soldier friends were awarded farms and money. For example, Colonel Bruce Miller and Colonel Gidings were given farms in Kabwe, and another Colonel whom I have forgotten was given in Choma, Southern Province.[77]

Towards the end of the war a small number of African soldiers were selected to go to Britain to help supervise the repatriation of African prisoners-of-war from liberated Europe. Among them was Bildad Kaggia who, in his autobiography, provides two chapters on his experiences in Britain. He describes flying-bomb raids, his contact with British people including clergy and missionaries, and attending Labour Party meetings. In Britain, Kaggia and his companions received British rates of pay and their 'pay book had "U.K. Pay" on them, something many European officers did not like to see'. However, when they were demobilised and repatriated via France to Alexandria in November 1945, their pay was immediately reduced: 'To me', wrote Kaggia, 'the reversion to the old, low pay was an unbearable humiliation. We resented the depot and warrant officers who spared no time in reminding us that we were no longer in England. I longed to be on my way home, to escape the ridicule and humiliation.'[78]

Kaggia's time in Britain was the realisation of a hope 'of acquiring experience in a country which was almost legendary to me'. More soldiers came after the war as contingents in the victory parade in London. Among these was Major Seth Anthony, one of the two Gold Coast men commissioned in the war, who had seen service as a combatant officer in Burma. The African units in the parade are captured on film and there are numerous photographs of individual groups of soldiers smartly turned out in their uniforms. Joseph Mulenga of the Northern Rhodesia Regiment came to London, as he told an interviewer from the BBC: 'War had ended... Fourteen of us were chosen to go to London', via Kenya, where

we stayed for three weeks, rehearsing for the [victory] parade. Many ex-servicemen came from southern Rhodesia... Tanganyika and Kenya ... We arrived in Liverpool and went straight by electric train to Kingstone Gardens [?] near Buckingham Palace, where our camp was ready for our arrival... Our stay in London was a special treat... We did our march past before King George Six and Her Majesty the Queen of England – Queen Elizabeth and other dignitaries.[79]

*

The war of 1939–45 took more than three-quarters of a million men from Africa for service with European armies. Many of them remained in the colonies in which they were recruited but several hundred thousand served overseas in Western Europe, the Mediterranean area, and also in Asia. Although large numbers were initially enlisted as non-combatant labourers, during the course of the war they learned military skills and took an active role in combatant duties, manning artillery and serving as front-line troops in a variety of capacities. The labour role of Africans was vital at certain times as it freed better-trained Europeans for other duties. British official war historians largely ignored the men from the villages and towns of Africa. However, their contribution to the war against the Axis powers was not forgotten by those Europeans who served with or alongside them. On every 11 November, small groups of Africans, steadily depleting each year, often ragged but proudly displaying their medals, gather at poignant war memorials in the centre of Nairobi, Accra, Maseru and other African cities, to remember their companions who died during a war that was not theirs or of their making. Fifty years after the end of the war these African ex-servicemen deserve to have their contribution to the war more fully recognised and their history better recorded.

'Matters of Honour': Indian Troops in the North African and Italian Theatres

Gerard Douds

> A man's destiny is his own village,
> His own fire and his wife's cooking,
> To sit in front of his own door at sunset,
> And see his grandson and his neighbours' grandson
> Play in the dust together.[1]

T. S. Eliot's lines but they mirrored the thoughts of many a sepoy, naik and halvidar. Precisely who were those Indian troops who fought so far from home in the course of the Second World War; specks in the Libyan wastes, dots on the frozen bastions of the Gustav Line? A clear vision issued from Wavell in his eulogy to an Indian army 'in which all creeds and races of Indians served together with British in mutual trust and concord'.[2] This ideal was robustly maintained on the public platform. In the course of a lively debate within the Indian Legislative Assembly on 19 November 1943, the motion was carried that key services of the Indian army should be thrown open and should not remain 'the monopoly of a few privileged classes'. For the government of India, C. M. Trividi, Secretary to the War Department, accepted the resolution on the basis that such a situation already existed.[3]

In reality recruitment policies favoured traditional martial 'types', as is revealed in an intriguing 'Note on Recruitment Policy', dated 21 July 1943, by the vastly experienced Secretary to the Military Department at the India Office, General G. N. Molesworth. His historical perspective emphasised the targeting of 'virile races...Sikhs, Muslims, Rajputs, Dogras, Pathans, Jats'. The rejected regions included the populations of Bengal, Bihar, Orissa, part of the United Provinces and Central Provinces, Madras and Bombay: areas which, as Molesworth put it, had entered into 'the suffocating penumbra of Victorian and Edwardian peace'. The special value of the Gurkhas as 'foreign mercenaries' came to be recognised at the time of the Mutiny. Exceptions to his large list of disqualifications were Madrasi Sappers and Miners, recruited from what were referred to as 'special stud farms' near Bangalore, and Mahrattas, who had enjoyed a considerable reputation throughout the First World War. The so-called 'voluntary system' employed during 1914–18, ultimately masked 'press gang methods' which drained the 'martial classes...to the last drop'. 'In other words', explained Molesworth, 'we exhausted Fortnum and Mason without tapping Marks and Spencers and Woolworths to any great degree.'[4]

Molesworth's exposition of recruiting priorities has to be seen within the larger context of pressures, notably from the United States, dramatically to expand recruitment in a sub-continent of some 400 million souls. To the end, he steadfastly maintained that while some 'new' classes might have to be recruited post-1939, the true Indian figure for 'really good troops' remained static at 850,000, excluding Gurkhas. In acknowledging the incapacity of critics in Britain and the United States to grasp the strength of these limitations, Molesworth pressed the case that 1943 recruitment had gone beyond the safe limits of recruitment of fighting men, pushing for quantity at the expense of quality. But as British representatives in Washington pointed out on 12 June 1942, a very real need existed to correct 'false impressions' that HMG 'are and have been reluctant to see, for political reasons, the Indian Army expanded'.[5] In a war fought for individual and collective freedoms, it did not do to make too much of fighting élites. Major Yeats-Brown was upbraided by an American officer beneath the slopes of Monte Cassino: the Indian recruits were a 'fine bunch', why were there not more of them? Why not apply conscription 'and get four million or even eight million soldiers'? Yeats-Brown countered discreetly that 'India would starve if she were mobilised for total war; it would be impossible to take more men off the land, because she is not a mechanised nation, especially in agriculture'.[6]

In the event, India came to assemble an army of close to 2 million engaged in the Allied cause. On 1 September 1939, troops in India numbered 43,500 British and 131,000 Indian. Following a sleepy start, expansion brought, by the end of 1940, a rate of recruitment approaching 20,000 a month; by the end of 1941 it was 50,000 a month. Additionally, the quixotic Indian princes ruling populations of around 90 million contributed their Indian States Forces, some of which had had an aspect of Gilbert and Sullivan, with recruitment rising to 76,000 by August 1941. At the close of 1941 the Indian army stood at 900,000; 300,000 having gone overseas. By the end of 1942, recruitment stood at beyond 1,800,000, with some consequent departure from Molesworth's 'martial types' particularly with an extension to southern India, where literacy rates were generally higher.[7] Major questions revolved round expectations of these volunteers, whose capacities seemed to be so largely determined by perceived historical reputation and record. They would be called upon to fight very far from home, in generally alien and hostile terrains, and at times when their homeland faced imminent invasion, 'stripped of trained soldiers and munitions of war to sustain the fight in the West'.[8]

Doubts about the reliability of Indian troops began at the top. Wavell confided to his diary, on 24 June 1943, his frustration with Churchill's obsessions that 'the Indian Army was likely to rise at any moment; and he accused me of creating a Frankenstein by putting modern weapons in the hands of sepoys, spoke of 1857, and was really most childish about it'. Wavell's written and verbal reassurances went for nothing. The PM retained 'a curious complex about India and is always loath to hear good of it and apt to believe the worst. He has still at heart his cavalry subaltern's idea of India; just as his military tactics are inclined to date from the Boer War.'[9]

In the matter of equipment and training, the vaunted 'volunteer' army of the First World War had been left to run down in the interwar period, its equipment mildewing, fast on the way to becoming, as graphically expressed by the late Professor Gallagher, 'a force of screwguns and mules, incapable of taking on any serious opponent'.[10] At a time when the military budget accounted for more than 40 per cent of the total expenditure of the government of India, local taxpayers could not be expected to spend more while Britain retrenched. With an eye on the deteriorating international situation, the Congress Party had put forward proposals, in 1937–8, for the expansion of the Indian army, its mechanisation and the development of 'absurdly small' naval and air arms.[11] The government's sticking point over growth had distinct political overtones, a long-standing British reluctance to commission Indians. Not till 1932 did India inaugurate its own cadet-training establishment at Dehra Dun. Only 290 officers had emerged by 1939, after which the emergency produced a large influx of Indian officers. But the slow input in the early stages had its effect. By 1946 only three Indian officers had reached the rank of brigadier.[12] The limitations of officer training available at Dehru Dun made a particular impression on Jawaharlal Nehru, who had no doubts as to cadet smartness on parade:

> but I wonder sometimes what purpose this training serves, unless it is accompanied by technical training. Infantry and cavalry are about as much use to-day as the Roman phalanx, and the rifle is little better than a bow and arrow in an age of air warfare, gas bombs, tanks, and powerful artillery. No doubt their trainers and mentors realise this.[13]

What sort of political risk did the new generation of Indian officers represent? In General Chaudhuri's recollection, discussion of politics was always taboo in the military cantonments and officers' messes, 'yet no Indian, either in the officer cadre or among the men, was unaware or uninterested in these developments'.[14] Opportunities for social intercourse were more or less non-existent, even for the new Indian officers hand-picked for demonstrable qualities of loyalty. D. K. Palit, joining an Indianised unit at Peshawar in the late 1930s reminisced: 'never, during my ten months in the station, was I asked to so much as a cup of tea by my own British CO or Company Commander or any other British officer'.[15] With the coming of war and the availability of only a few hundred Indians in an officer-corps of several thousand, selective Indianisation had to be ditched and massive expansion launched. It was transparently clear to political India that the war emergency had produced the volte-face. 'We have been playing a losing hand from the start in this matter of "Indianisation",' confessed Auchinleck in October 1940:

> The Indian has always thought, rightly or wrongly, that we never intended the scheme to succeed and expected it to fail. Colour has been lent to this view by the way in which each new step forward has had to be wrested from us, instead of being freely given. Now that we have given a lot we get no credit, because there was little grace in our giving.[16]

The subversive, S. C. Bose, set out to convince Ribbentrop in May 1941 that not only Indian nationalists but also Indian officers 'listened very attentively to the news broadcasts from Berlin'.[17] There were grounds for dissatisfaction to be played upon. The unimaginative political clauses of the 1938 Chatfield Committee precluded any shared responsibility between British and Indian leaders for the disposition of Indian troops or even common problems of defence policy generally.[18] By April 1941 Bose had fired up the Italian minister in Kabul with a projection that intensified propaganda, to include Bose's own broadcasts, would result in the mass desertion of entire Indian divisions.[19] Arguably the ultimate test for the loyalty of the Indian officer corps arrived with the aftermath of the Congress Rising of August 1942. The confidential report of commander-in-chief India on 22 November 1942 settled for a cautious but comfortable assessment:

> the feelings of the Indian Commissioned Officers are difficult to evaluate; many, as is understandable, take a keen interest in the political problems of the day, the outcome of which must profoundly affect their own futures. It is also probably true to say that the bulk of I.C.O.s who have come from civil life would welcome a self-governing India; at the same time they have, as a class, behaved very well...no cause for dissatisfaction.[20]

By and large, the commander-in-chief's assessment was borne out.

Arguably, recruiting other ranks carried more risk, particularly when deviation from Molesworth's 'martial types' came to be considered. With a population of around 400 million, a monthly average of 66,000 recruits could be just about sustained, though by later 1942, with the virtual exhaustion of Molesworth's favoured reservoirs, a discernible decline set in both in physique and intelligence.[21] In March 1940 the India Office felt sufficiently wary, as to the mood of Indian troops, to curtail their reading. Delhi was instructed accordingly: 'as it is possible to obtain vernacular Indian papers here please send by airmail list of papers unsuitable for Indian troops'.[22] There were grounds for concern in the gist of an article written for the *Statesman*, in the early stages of the war, by the erratic Bengali literateur, N. C. Chaudhuri: 'a majority of Indians showed a deep-seated inhibition to all news favourable to the Allies. They received reports of Allied successes, exploits and power with mental reservation, but showed themselves over ready to give the Germans more than their due.'[23]

The likelihood is that Chaudhuri was influenced by sentiment in Bengal, not a favoured area for recruitment and the cradle of the initial Indian political renaissance. Following the fall of France, prominent Bengalis would portray Hitler as a reincarnation of Vishnu; allegedly, German tanks flew the Kapidvaja, the 'Monkey Banner' which bedecked the chariot of Arjuna, legendary hero of the Mahabharata.[24] Indian provinces could show a marked disparity in their responses to the war situation. Governor Haig of the United Provinces made much, as early as October 1939, of 'large scale and violent anti-recruiting meetings'. Governor Craik, responsible for the Punjab, pressed the Viceroy to mount an immediate recruitment drive in order to exploit enormous general enthusiasm so apparent in the Land of the Five Rivers.[25]

Once recruited, the transportation of Indian troops to the far-off theatres of war in the West presented German propagandists with opportunities. Major Holder of Skinners Horse was nonplussed by the general exit from his regiment, prior to embarkation, of its many followers, grooms, cooks, water-carriers and their fellows. On investigation,

> it transpired that the Germans had a flourishing fifth column in India which passes the word round the bazaars that any ship leaving India would be promptly torpedoed, and everyone knew of the submarine successes that they were having in the Channel and the Atlantic. The Indian officers were aware of this but they refrained from telling us about it lest they could be accused of spreading alarm and despondency.[26]

On the day that Holder sailed, the establishment made up with 'unemployables', Lord Haw Haw broadcast to India precise details of the Indian 5th Division's shipping arrangements with the baleful rider: 'none of them will arrive at their destination which is Port Sudan'.[27]

Clear instances of mutiny occurred relatively soon after the commencement of hostilities. In the forefront of resistance were Sikh companies. An alarming case blew up in Egypt in November 1939, brewing among the RIA Service Corps of the 4th Indian Division. GOC Egypt's report pinpointed 'Sikh soldiers at the root of the trouble'. In line with the accompanying analysis of J. P. Morton, Indian Police, the emphasis lay upon the influence of 'subversive propaganda...there exists within the Cairo bazaar an Indian civil population...a medium of subversive discussion'.[28] A monthly intelligence summary, issued by Army HQ India in September 1940, struck an alarmist note in its assessment of the problem: 'there is absolutely no doubt that at the back of the Sikh trouble there is a widespread subversive organisation for getting at the Army. Some of the leaders and workers in the organisation have been arrested, but there is clear evidence that many other secret communist workers remain.' The summary went on to rake up the First World War Ghadr conspiracy: 'Since the Communist-Ghadrite party is composed almost entirely of Sikhs, it is natural that the Sikhs of the Army should be the first to be affected by subversive influences.'[29] The more mundane, and not implausible, explanation, only briefly referred to in Morton's investigation, referred to the objections of the disaffected Sikh motor transport unit to unloading vehicles, which they understood to be 'coolie' work. Contributory grievances, also deep-felt, concerned hair-trimming to facilitate the wearing of steel helmets. This last concern resurfaced in Hong Kong in December 1940, when Sikh soldiers of the heavy artillery unit mutinied over the helmet issue. In the immediate aftermath, harsh sentences were imposed before wiser counsels prevailed and a decision was published on 18 February 1941 not to impose steel helmets on Sikhs.[30]

The major mutiny broke with the refusal of the majority of the Sikh squadron, Central Indian Horse, to go overseas from Bombay in June 1940. Due to an unexpected delay, the squadron had been shunted into a siding for a day and a night. Thus an opportunity presented itself to four Sikh political activists, imbued with 'incendiary peasant communism', to exploit unrest due in part to the legacy of a highly inadequate commanding officer. An official report linked the mutiny to communal tension in the Punjab. Understandably anxieties followed from the publication of Muslim League proposals for Pakistan, and its implications for the historic integrity of the Punjab. Viceroy Linlithgow's telegram to the India Office referred to 'a concerted effort...to play upon fears of troops by spreading rumours of terror of modern warfare and certainty of either being killed by bombs or drowned at sea on the way to scenes of action'.[31] That what formal mutiny there was came to be directly associated with Sikh subversives had been anticipated. As a British officer from a Sikh regiment forecast in uncompromising terms: 'in war, under stress, there is no one like the Sikh; work him almost to death and he is magnificent. But relax, give him leisure and a chance to recuperate, and he will start to intrigue.'[32] The Central Indian Horse Mutiny was severely dealt with; sixteen ringleaders were sentenced to death, a further ninety-five soldiers were given sentences of ten years and upwards.[33]

Given their disproportionately high recruitment into the Indian army, Muslim sensibilities were taken seriously. In July 1941 General Dill laid particular stress on a Mediterranean strategy which would deny the Axis powers the whole resources of Africa. The clinching argument highlighted the enormous repercussions which a British withdrawal might have on the Muslim world, stretching right through to India and thus prejudicing 'the good will on which we were so greatly dependent'. Muslim loyalties within the Indian army were not to be put under too much pressure. Although Hitler's Directive No. 30 identified the 'Arab Freedom movement' as 'our natural ally against England', the rhetoric foundered both on a contemptuous attitude towards non-Europeans, the hallmark of Nazi ideology, and a pressing political need to sit well with Turkey, the only land access for Germany to the Middle East.[34] Although the Afghan minister in Berlin was ready to talk about the subjected Islamic peoples and '15 million Afghans suffering in India' awaiting German liberation, Islamic leaders within India held aloof. Only after the fall of France did a hitherto stalwart of the British-Indian connection, the Aga Khan, meet secretly with Prince Max Hohenlohe in Switzerland in July 1940. Prince Max was made privy to a number of curious arrangements and proposals. The Aga Khan and the former Khedive of Egypt would be taking champagne together on the first night spent by Hitler at Windsor Castle. The Aga Khan's followers and several younger maharajas were standing by to assist in the administration of India after the Axis victory. However, with the conversation turning to the question of a financial subvention for the Aga Khan, interest faded and instructions followed from Ribbentrop to break off contacts with him.[35] If suspect maharajas existed, they kept their heads down. Only a single picket of the Gwalior Lancers went over to Japan in the course of the Arakan counter-offensive in February 1944.[36] In the European theatres, Indian States Forces held up, notably with the heroics of the Jaipur men in forcing Casolino Ridge on the jagged summit of the Apennines.[37]

A confident assessment of the level of Indian troop morale in the North African campaigns is difficult to arrive at, given the destruction of virtually all censorship reports drawn up by the Middle East Intelligence Centre. The surviving fragments from June and July 1940 emphasise, as perhaps they would, the positive content in servicemen's letters. Since the correspondence was two-way, letters of exhortation from India are cited: 'You are lucky to have the opportunity to show your courage. Do your best to gain repute.' From Egypt to India: 'We do not hope to come back soon; not before the death of Hitler. We are determined to finish him off this year.'[38] The author-ities calculated that morale had moved up following the entry of Italy, 'and the war being brought much nearer to Indian troops, they are forgetting what grievances they may have had. They appreciate the possible chance of fight-ing and increase in work relieves monotony.' Prominence is accorded to the gung-ho letter, the pleasure to be derived from giving the Italians 'big kicks ... Italian bombers have been over every morning between 4 and 6 o'clock but they are useless. They could not hit a door if they had hold of the handle.' Some uncertainty, likely to have been more general, is revealed in the corres-pondence of an Indian officer in the aftermath of the fall of France: 'I can't understand the French packing it in as they have done and I thought they hated the Germans sufficiently to go on fighting anywhere – they must have been in a pretty rotten state to come to that.'[39]

The censors fought shy of reaching general conclusions: 'in view of the reticence to express their opinions shown by the majority of writers, it is difficult to appreciate their exact outlook'. However, satisfaction and apprecia-tion were expressed concerning rations, broadcasts, and the absence as yet of real hardship on field service; 'there is no doubt in ultimate victory'.[40] Report-ing later from the Italian front, Major Yeats-Brown testified to impressive provisions for the support of Indian troops:

> special broadcasts have been arranged for them in eight languages from Cairo, Baghdad, Teheran, and Beirut ... their weekly paper, the *Fauj-i-Akh-bar* (Army Newspaper) is published in five languages and has a circulation of 30,000 copies. The bi-weekly *Jang-i-Khabren* (War News) is published in eight languages, and prints more than 120,000 copies. There are six other publications for Indian troops.[41]

The commander-in-chief India's assessments on Indian troop morale revealed at least something of the special strains imposed by service so far from home. In the course of the report for November 1942, acknowledgement is made that 'Axis broadcasts materially helped to sustain an atmosphere of suppressed excitement...more or less indiscriminate sabotage and arson, labour agitation and intense pamphletering'.[42] Concern for families at home weighed on troops abroad. The commander-in-chief's assessments for November 1943 and February 1944 raised the old distinction between units maintaining 'excellent' morale and many subsidiary units, 'particularly in those enlisting non-martial classes', where morale remained 'of variable quality', this last a recurring phrase. It was conceded that 'the soldier is greatly disturbed by the economic position of his family'; that living costs were high and that rates of pay were condemned by all ranks as inadequate for the support of a family. Though the Congress rising of August 1942 blew over, the next year returned the spectre of famine and with it added anxieties for the troops abroad, particularly with regard to rationing and the victimisation of wives. The August 1944 assessment squarely faced the paucity of leave programmes, but with more than a hint of complacency: 'the infrequency with which the Indian soldier can obtain leave to visit his home tends to upset his contentment and encourage desertion. So far as the conditions of his service are concerned the Indian soldier has very few complaints: the standard of life in the Army is as high and generally higher than that to which he was accustomed.'[43]

By and large, the confidence of the government of India in its troops was borne out. The steadfast character of Indian soldiers, both in North Africa and Italy, is a recurring theme within the recollection of well-placed witnesses, British and Indian. By tradition, Indian troops trained far from home and were hardened to long absences.[44] Thus the work of S. C. Bose, in the European theatres, to recruit from Indian prisoners-of-war enjoyed very limited success. In a colourful televised sketch of Bose, a force of 50,000 is projected to invade India; only some 16,000 prisoners-of-war became available, of whom 4,000 allegedly came over to Bose's Indian Legion in Germany. None saw front-line action.[45] Such statistics appear on the high side: as late as August 1942 the Indian Legion in Germany had no more than 320 men undergoing training. A move to send Bose to Egypt, in September 1942, to convert captured Indians, never materialised.[46]

A retrospective government of India review of May 1945 confirmed the long-standing attempts by Axis agencies to undermine the loyalty of Indian captives, habitually singled out for special attention. The conclusion drawn is one of comprehensive failure. An Axis-formed unit under Italian auspices mutinied as soon as it became apparent that it had to fight against the Allies. The unit was disbanded. In France a number of prisoners-of-war discarded German uniforms and joined the Maquis. The review records an approximate number of 13,000 Indian prisoners in Europe for 1944; some 3,000 had accepted service with the Germans, the great majority under *force majeure*. A positive construction was placed on this not insignificant number of converts:

'dressed up in Axis uniforms, Indian soldiers have in the vast majority of cases seized at the first chance of escaping from Axis masters'.[47] The assured loyalty of the Indian army persisted as a central theme in newsreels for the reassurance of the British public at war. Prior to the 1942 Congress Rising, Pathé Gazette projected a feature on 'Indians at Buckingham Palace'. A contingent from the 4th Indian Division was presented to Their Majesties in the inner courtyard; 'splendid warriors' fresh from triumphs in Libya and Tunisia ran the effusive commentary. In the aftermath of the revolt, newsreels played down the true extent of the disturbances, far and away the most serious since 1857, and represented Indian troops 'as loyal as ever, symbolic of the determination of India as a whole. In France, Libya and Burma, Indian troops have given their lives fighting for a cause which they know to be just. They represent the spirit of India, not the Congress mob.'[48]

The battle experience of Indian troops would be severe. Absence of modern equipment produced major disabilities. The 1938 Auchinleck Report concluded that, judged by modern standards, the Indian army was unfit to take the field against land or air forces equipped with up-to-date weapons.[49] With war declared, the government of India confirmed its readiness to send its troops overseas

> if they could be equipped with the modern weapons and equipment appropriate to the conditions, their own scale being quite unsuitable against a first class power. The proportion of artillery to infantry was comparatively low; it was mostly British, but armed with obsolescent guns. There were no anti-aircraft guns, few anti-tank weapons, and there was a lack of modern light machine-guns, mortars and carriers. A start had been made with modernisation, but much of the cavalry was still horsed, and animals provided the greater part of the Army's transport.[50]

Steps were taken but a major setback lay in wait. 'After the disaster of Dunkirk, India had shipped to Great Britain so much of her reserves of rifles and other weapons that when new Indian divisions were raised later there were grave shortages in their equipment,' recalled Field-Marshal Slim. 'Thus when my division went to war in 1941, some 500 of its soldiers in the transport companies were virtually unarmed. They were combatants and should have carried rifles, but alas, there were no rifles for them.' The unarmed drivers fell easy prey to looting Arabs, who left nothing but the main members of the chassis lying on the sand.[51] To cope with the fast Ford cars of the Iraqi machine-gunners, Slim had clamoured for armour. It arrived:

> a high ugly machine with, perched on top of it, a turret like an old-fashioned bee-hive from which poked a Vickers machine-gun, its sole armament . . . its large wheels had tyres of a size used by no modern vehicle. The body of the car was twenty years old; its engine five. Both had led hard lives. Now, in the hot early summer in 1941 in Iraq, an armoured car regiment, equipped with these museum pieces, had joined my Indian division.

In later operations against Persia, Slim came to curse the Persian batteries of modern 155-mm guns, 'which so easily and so far outranged our old 18-pounders'.[52] As survivors of campaigns in both North Africa and Italy recall, the lack of anti-tank weaponry posed acute problems for Indian infantry defending against armour. Frequently only mortars were available and there were always major transport problems.[53]

Italian and German equipment would present still more formidable challenges. At Gallabat on the Sudan-Abyssinia frontier, Slim's nineteen old Gladiator biplanes were up against forty superior Italian fighters and a number of bombers. Slim's faith in the quality of his Indian troops was amply confirmed by events at Gallabat, which saw British units routed by Italian bombing while Baluchis, Garhwalis, and a section of Indian sappers and miners stood firm.[54] Early on in the desert campaign, the 4th Indian Division encountered the fearsome German 88-mm dual-purpose gun; 'its flat trajectory and terrific hitting power made it a deadly weapon'.[55] The superiority of German medium and heavy tanks was manifest. Bereft of their own armour, Indian units relied on speed, initiative, and a fair measure of daring. Thus the Central Indian Horse, with trucks but no tanks, regularly operated behind enemy lines in North Africa with a view to drawing out enemy tanks into positions vulnerable to Allied 25-pounders.[56] In no encounter was the equipment gap as sharply illustrated as the stand of the 3rd Indian Motor Brigade at Meikili in April 1941. The constituent regiments were neither equipped nor trained for desert warfare: 'two of the regiments had given up their horses only six months before; none of them had any armour; all had had to supply drafts of trained men to other units; they were 40 per cent below strength in light machine-guns and each regiment had one anti-tank gun instead of forty-two'. Meikili lay across Rommel's major attack route; armoured formations engaged the Indians ferociously. At three points over the next forty-eight hours German emissaries requested a surrender, the third message coming from Rommel personally. With its headquarters and other elements overrun, part of the brigade broke out; 300 prisoners were taken on the way home by Major Rajendrasinhji who became the first Indian Sandhurst-trained officer to win the DSO. The resulting 48-hour delay in the German advance enabled the 9th Australian Division to reach Tobruk. In the calculation of Indian official historians, had Tobruk fallen, 'the Axis had a fair chance of reaching Alexandria before Alamein was even thought of'.[57] On such single episodes, entire campaigns may turn.

Arguably, the Italian campaigns produced even more severe tests of Indian morale, tenacity and raw courage. Climate took a firmer hand. Philip Mason mused that 'something in the dry, stimulating air of the desert seemed to bring out the best in the combination of British and Indian'. In the recollection of an American veteran of the Italian campaigns, fierce winter conditions made Indian troops, who had enjoyed the highest of reputations in the desert, less effective.[58] This is strongly contested by representatives of the Rajputana Rifles who point, with justification, to their regiment's valour at Cassino and elsewhere.[59] In forcing the Gustav Line, the 8th Indian Division in late 1943 encountered Italy's General November. As Indian sappers toiled to bridge the bitter, swollen River Sangro, a freezing spate wiped out their progress. Conditions over that winter stretched the Indian Medical Services to the limit: 'troops who had never known extreme cold now fought in frozen foxholes, patrolled in slush and sleet, waded icy rivers, slept in snowdrifts, bivouacked in blizzards'. For three horrendous weeks in February 1944 Gurkha and support units were pinned down on the deadly slopes of Hangman's Hill below Cassino, blasted by blizzards, forty yards from the German positions. Ever resourceful, Indian troops on the lower Adriatic front, when cut off by a detachment of German ski troops, used their greatcoats as the Romans once used their shields, and sledged to safety.[60]

In this savage theatre, Indian troops, traditionally trained, could reflect that mechanisation was not always the answer. The Indian units forcing the Trigno found their supply problems solved by the arrival of the 13th and 14th Indian Mule Companies. The 4th Indian Division before Cassino organised on a mule-pack basis. However, the Indian mule complement had been internationally enlarged with mixed results, mischievously described in an official history as 'a heterogeneous assemblage of French, American and Italian mules of diverse training, habits and temper'. Whether intractable or not, mules earned the undying gratitude of Indian troops, as daily carriers of food, water and ammunition along paths exposed to incessant shell and mortar fire. They were looked after. Though one officer did report that the Sikh tendency to treat mules like potential Derby winners produced higher casualties among the drivers than the mules.[61]

The contours of the Italian peninsula produced more formidable natural barriers than were generally encountered in North Africa: high crests, verticalsided gorges and heavily wooded trackless valleys. The veteran war correspondent, Martha Gellhorn found the mountains of Italy 'horrible; to attack always against heights held by well-entrenched and well-trained enemy troops is surely the worst sort of war. Nothing can help the infantry much in mountains: Germans dug into the stone sides of these cliffs can survive the heaviest shelling. Tanks cannot operate.'[62] German excellence in exploiting such terrain for defensive purposes was sharply illustrated by the construction of 'hedgehogs', anchor strongholds, on the eastern flank of the Gustav Line:

the villages had been transformed into fortresses, with shelters 20 feet deep, machine-gun nests and connecting tunnels proofed against the heaviest shelling. Houses in key positions had been reinforced with concrete. Pillboxes had been built. Often different floors or different rooms in the same building were converted into separate strongpoints, so that if assailants broke into one side of the house, the other side could still be defended. Escape tunnels connected the houses, and the storming of one strongpoint usually left the victors under fire from a nearby redoubt.[63]

Indian infantry were faced with an endless sequence of miniature Stalingrads. Their opponents were the toughest yet. In North Africa and the Middle East, French, Iraqis, Persians and Italians had given place to the more formidable Afrika Korps. Now, Indians were pitted consistently against German military élites. The 15th Panzer Grenadier Division held Cassino town, seasoned veterans of a dozen battlefields; their instructions came direct from Hitler. At every turn during the harrowing drive north, Indian infantrymen had to overcome the pride of the German army, the 1st German Parachute Division, an élite of fanatical Nazi conviction. Sophisticated and deadly landmines, notably the little Schuh mine, undetectable in non-metallic casings, were broadcast in millions before defensive positions. A sepoy prodding with a bayonet was dead. A further devilish device was the Nebelwerfer, a multiple mortar operated by remote control.[64]

On several counts, Indian troops excelled in this grim setting. Major-General Tuker of the 4th Indian Division was, above all, a specialist in mountain warfare. The 'cerebral general', his training methods were judged to border on genius.[65] As an artist of standing, his etcher's eye captured fine gradations of perspective and the lie of a battlefield, lost on a conventional commander. The troops of the 10th Indian Division, so successful in the Tiber valley, had endured years of waiting in Syria, Cyprus and Palestine. Throughout, mountain warfare was the training diet; 'warfare which was flexible and individual', which understood 'the values of high ground and dead ground, of observation points and hidden approaches, of unobtrusive infiltration and deep penetration'.[66] The extent to which Indian infantry responded is sharply documented in an official history, which speaks repeatedly of 'cat-eyed, soft-footed' Indian raiders. In the Arno valley, confronted by the strong-points of the Gothic Line, Indian reconnoitring craft facilitated the penetration of these formidable defences. The legacy of generations of warfare on the Frontier was exploited, 'keen sight, silent movement, quick decision...military qualities inherent in the blood of men whose ancestors have been soldiers for a thousand years'. As a North-American 'Indian', serving with a Canadian scout patrol testified of the Gurkhas: 'Boy, are they good? I thought I knew a bit about tracking, but I can't teach these boys anything.'[67] Italy tested Indian fighting qualities to a new pitch. The 8th Indian Division was the only formation to smash through the charnel house of Cassino. No doubt a Hogmanay at Pisa, spent with the Argyll and Sutherland Highlanders, was a tonic for this famous division but it would also require further reserves of endurance. When German emissaries presented themselves at Allied headquarters to accept the victor's terms, the game was not quite over. The 6th DCO Lancers sped far up the road to Austria and intercepted old enemies, the 1st German Parachute Division. One British officer, two Sikhs and six Jats arranged the surrender of 11,000 men.[68]

How is the very considerable success of Indian troops against high odds to be accounted for? Field-Marshal Slim came to feel that an Asiatic army, in which he included both Russian and Indian armies, enjoyed several natural advantages over Western counterparts. With his Russian peer, the sepoy enjoyed a nature that 'inclined him to courage and patriotism... and his normal standard of living left him content to be without many things Western soldiers regarded as indispensable necessities – a great military advantage'.[69] Mason has been more precise in locating the framework of discipline and organisation, essential to victory, within Hinduism itself and traditional social structures: generations of warriors proud of their courage, for whom cowardice is the ultimate disgrace.[70] In essence, this last is a replay of Molesworth's 'martial races' theme and a good deal had happened in the deserts and mountains to undermine it. The old caricatures of educated Indians as effete, garrulous and non-officer material, coupled with 'martial types', bellicose but unable to lead, took a knock.[71] Mason himself testifies, as do official histories on so many occasions, to the impressive accomplishments of those in the basement of caste, say the conduct of the untouchables of the Madras Sappers.[72] With his eye on contemporary history, General Sinha recalled the fighting tradition of south India and the sanguinary exploits of the 'Tamil Tigers'.[73]

Realisation of the wind of change came with Auchinleck's command. The segregation of Indian officers in 'Indianising' units was abandoned: 'now all Indian Army units (except Gurkha battalions) took both British and Indian officers. Their pay was now equal and there was no longer any question of their replacing the old Viceroy's Commissioned Officers, the subadars and jemadars who had been raised from the ranks.' Of course Auchinleck's move away from recruitment of 'martial types' may be explained by the exhaustion of their stock, but his direction of recruitment towards landless labourers and even the defiled Chamars, the leather workers, points to an understanding of and sensitivity towards criticisms from political India, unheard of in any previous commander-in-chief. Unprecedented tours of military installations and training centres took place for the benefit of elected members of the Indian Legislature. The Indian army of 1945 became an incalculable asset after Independence.[74]

To what extent were the Indian troops who saw it through to VE Day new men? Initial doubts as to the suitability of Indian troops for Europe had focused upon the multiplication of specialist cadres required for a technician's war. Bravery and discipline might be conceded, but how would the limited horizons developed in the Indian countryside serve? The critics were answered on all counts. The survivors 'were not the raw recruits, the simple peasants, who had fared forth three, four, or five years before. They had travelled thousands of miles; they had seen diverse peoples: they spoke foreign languages; they had matched their manhood against the greatest aggressor and had not been found wanting.'[75] Above all, British reputations as representative of the guardian caste, would not be untarnished. As the remarkably frank Rifleman Bowlby recalled of his Italian experiences, a county battalion had been brigaded with Indian troops:

the whole Battalion had cut and run from a German tank attack. The Indians recaptured the ground. Once the Brigade was out of the Line the English Brigadier formed both Battalions into a square, the Indians on the outside, with their weapons, the county regiment inside without weapons. He then told them what he thought of them.[76]

How much did the Indian troops know of Britain itself? Surviving files for the visits of Indian soldiers to the UK only refer to immediate post-war arrangements. A mere handful of troops, drawn by ballot from Middle East garrisons, was able to participate. The scheme was dogged by War Office pressure, eventually successful, to close it on grounds of cost. A curious itinerary started with Whipsnade and included the Ford Motor Works, ice hockey and football. Once consulted, Indian participants successfully pressed for the inclusion of three days in Scotland.[77] Indian troops did hold together in adherence to conviction and the resolution of 'matters of honour', in Philip Mason's inspirational phrase. They were moved by a remarkable sense of camaraderie. The essence of this bonding is powerfully expressed by General Chand Das: 'comradeship built up under fire is greater than blood relationship'.[78] But were the bargains kept? In November 1993, an immaculate group of much-ribboned Indian ex-servicemen gathered at the sparsely furnished Milan Day Centre in Southall, west London. To a mood of despondency had been added fury and fear. 'I am a proud and loyal man,' contributed Sergeant-Major Rajinder Singh, 'we had so much faith in this country. In the war, I thought it is time to help Britain to save democracy and fight fascism. They don't remember what we did...Today in Britain, a fascist has won an election. Can you imagine how we feel?' The gathering took place in the aftermath of the BNP victory at Tower Hamlets.[79] Time to recall the favourite refrain of Indian soldiers nailed to the grim slopes of Monte Cassino:

> Oh, bury me at Cassino
> My duty to England is done.
> And when you get back to Blighty,
> And you are drinking your whisky and rum,
> Remember that old Indian soldier,
> When the war that he fought has been won![80]

Mr Wu and the Colonials: the British Empire's Evacuation from Crete, 1941

Angus Calder

In the small hours of the morning of 1 June 1941, Brigadier Howard Kippenberger, with the 20th New Zealand Battalion, was determined that whoever got left behind on Crete, it would not be the 230 men he had selected to go. Around the tiny beach at Sphakia, off which Royal Navy vessels waited to take up evacuated Commonwealth troops, a mob of stragglers, chancers, deserters and men whose luck had run out surged with such energy as was left them after their trek across high mountains with scant access to food or water. The cordon keeping them away from the ships was provided by survivors of the 22nd New Zealand Battalion, who had fired the first shots in the Battle for Crete on 20 May, just twelve nightmarish days before. They were determined to get off themselves, and they had orders to shoot if necessary. Grimly, Kippenberger counted his men through, grieving for the scores of soldiers he had had to leave behind to surrender to the Germans when daylight came.[1]

A few hours before this, Colonel Bob Laycock, of the Commando formation called, after him, Layforce, had ordered his brigade HQ to embark after squaring the officer in charge of the beach. With him went his intelligence officer, Captain Evelyn Waugh. Layforce had airily been told to push their way through the rabble to join their leader. They couldn't do it, and men who had fought bravely as rearguard to the Commonwealth retreat across Crete fell into German hands next day.[2]

That the British evacuation from Crete marked a major military setback was never seriously denied. News of it came at a bad time for British morale. Would Britain be the next target for airborne invasion? The latest German air raids had been especially heavy, notably on London, where Churchill had wept over the ruins of the House of Commons. People were not to know that the Luftwaffe was going east for Barbarossa, Hitler's attack on the Soviet Union, nor that, far from coincidentally, the fact that as many as 18,000 Commonwealth troops were taken off Crete while 12,000 remained alive as prisoners could be attributed to an easing off of the daily attentions of the Luftwaffe there too. Churchill himself was downcast and highly critical in private of Wavell, the British commander in the Middle East, and Freyberg, who had led the force on Crete. News on 22 June that the Nazis had struck towards Russia lifted a pall of gloom, and in the new phase of war, Crete was comfortably forgotten, except by relatives of men killed or taken prisoner there. But there was never any attempt to romanticise it as a second, smaller version of Dunkirk exactly one year later. The evacuation, as John Keegan has put it, was 'a shaming culmination to a benighted battle'.[3]

After the war was over, numerous able writers on the British side picked at the scab. The British official historians were cautiously defensive; a Panglossian line emerged that holding Crete would have involved a deleterious diversion of forces from the desert battle against Rommel, so that the forced evacuation had been a blessing in disguise. The honest and acute author of the official New Zealand history, Dan Davin, was freer to be critical – after all, ultimate responsibility for the débâcle rested with Wavell, who had failed over several months before the battle to supply and reinforce the British garrison properly, and Churchill, who had insisted on Crete's defence when this had simply not been practical. The Poms had let the Kiwis down. However, several later authors of a string of detailed books about the battle have inclined to think that Crete might have been held quite easily if this, that or the other strategic or tactical decision had been taken.[4] After all, as was finally admitted, the Bletchley code-crackers had supplied Wavell and, through him, Freyberg, a fortnight before the battle, with ULTRA information, derived from slips by Luftwaffe ENIGMA operators, which revealed exactly, in detail, the plan of campaign which the Germans actually followed.[5]

My purpose here is not to attempt fresh judgement on the battle, but to consider front-line morale on the Commonwealth side in so far as this can be reconstructed from first-hand accounts and from secondary sources using these. 'Mr Wu' was the pet-name accepted by Evelyn Waugh in his correspondence with Lady Diana Cooper, wife of one of Churchill's ministers, whom he in turn jokingly identified with 'Mrs Stitch', an all-powerful society hostess in his fiction. Their lifelong platonic friendship epitomises Waugh's fascination with the British ruling class into which, as the son of a literary man, he, most painfully and regrettably, had not been born. A sergeant-major had called him 'Wuff' and play with his name – pronounced 'Wauch' in Scotland, whence, via his grandfather, it had come – was one of the smallest amongst many tribulations suffered by this talented but unhappy man in the course of war service almost from the start till very near the finish, volunteered for despite quite advanced age – he was thirty-six in 1939.

In his trilogy *The Sword of Honour*, published between 1952 and 1964, the protagonist Guy Crouchback's war matches his creator's closely. The fall of Crete is the subject of some fifty virtuoso pages in *Officers and Gentlemen*, the second part of the trilogy, which was published in 1955 after Waugh had had the chance both to influence – since Davin consulted him – and to read the official accounts of the battle. The novel's details relate very closely, incident by incident, to a long 'Memorandum on LAYFORCE' written by Waugh after his taste of action on Crete. Hence we have in Waugh's novel first-hand testimony adjusted to secondary sources. He could not have taken his fiction far beyond well-known facts even if he had wanted to. His description of the state of affairs in Crete is remorselessly pessimistic. Now that the trilogy is regarded as a modern classic, it is bound to determine what many readers know, or think they know, about the battle.[6]

The other significant writer on Crete in May 1941 was Dan Davin, a subaltern in the New Zealand Division. His two quite striking short stories about the battle are unimportant compared to the massive volume, *Crete*, which he contributed to the New Zealand series of official war histories. Published in 1953, the year when his fellow-countryman Hillary scaled Everest, this must be regarded as the 'classic' historical account.

Waugh attached himself to the class of Englishmen who saw 'colonials' from the Dominions as absurd provincials, earnest, ill-mannered and ignorant of the necessity of social hierarchy. On Crete, he was not attuned to be aware that the New Zealanders, by their actions, were writing national epic. As viewed by Davin, and others among or close to the New Zealand Division, Crete was a grim story but not a negative one. The New Zealanders, volunteers to a man, had been gallantly blooded in the rout which had driven the Commonwealth forces out of Greece. Now they bore the brunt of exceptionally ferocious fighting in Crete. Superbly trained in advance, they emerged awesomely battle-hardened by their reverses – 'reverses', as Davin says, 'the more readily assimilated because the fighting man and his officers rightly believed that they were not due to their own failings in the field'.[7] A quarter-century before, New Zealanders had suffered through British blunders in the agony of Gallipoli. Now, not so far away, they rubbed in the point that, especially with the bayonet, they were the best soldiers in the world – a claim which Rommel himself endorsed as they helped thrust his Germans back in the Western Desert, on to Sicily, up to Monte Cassino, where the doughty Kippenberger lost both his feet, finally to Trieste, where they pre-empted takeover by Tito's Yugoslav partisans.

One man, Waugh's, account of morale on Crete as shameful must be balanced against the testimony of others to the exceptional courage and steadiness of many, perhaps most, of the New Zealanders. Perhaps in these different perceptions we seem to have in little the contrast between a rising nation and a decaying imperial ruling class.

On 6 April 1941, Hitler sent German forces into Greece to cut short the Greeks' successful resistance to his ally Mussolini's attempted invasion and check the British, who had depleted their forces in North Africa to send support to their valiant ally. He had to secure his southern flank in advance of his invasion of the Soviet Union. Operation Marita was swiftly successful. Within three weeks the Swastika flew over the Acropolis. Of the Commonwealth contingent, 12,000 were left behind, dead or in captivity. Those who were evacuated arrived in Crete without most of their equipment. Even blankets and cooking utensils were in short supply.

The island was idyllic in spring. Men relaxed after their campaign, bathed, drank wine and enjoyed the friendship freely offered by the local population. But Crete would be hard to defend. 152 miles long, 35 miles wide at greatest, its population of something over 400,000 was mostly settled on the northern coastal plain with concentrations at Canea, the capital, in the west, and Heraklion, the main port, further east. The only serious road on the island ran along this coast, where there were now airfields at Malame, Retimo and Heraklion. The harbour at Suda Bay, near Canea, could handle only two ships at a time. From the autumn of 1940, the British had garrisoned Crete, but it was over 400 miles from Egypt, no fighter cover was normally possible, and supplying troops on the island was awkward and, if the Luftwaffe intervened, very risky for the Royal Navy. Lack of sustained attention from headquarters in Cairo was compounded by the fact that six different commanders in as many months, were in charge on Crete before, on 30 April, Wavell handed it over to the reluctant General Freyberg of the NZ Division. Inadequately armed before the influx from Greece – which included thousands of Greek soldiers without rifles – the British force was now encumbered with drivers without vehicles, gunners without guns, and scraps of units.

Defeat could subsequently be attributed to any one of several key shortages. The RAF flew a few Hurricanes and Gladiator biplanes against the Luftwaffe, but removed the last half dozen, after constant attrition, on 17 May, leaving the sky open to dive bombers and low-flying Messerschmitt fighters. Anti-aircraft guns were in short supply. So, crucially, was wireless equipment – a lack exacerbated as telephone lines were severed in the fighting. Twenty-three inferior tanks were insufficient. Even entrenching tools were in short supply.

To the *in situ* Creforce of 5,200 men were added about 20,000 from Greece, about half of these of little or no military significance, and nearly 3,500 reinforcements from Egypt. The grand total was imposing, but they faced attack by crack German troops with total control of the air.

Hitler had been listening to those who wanted to use the airborne army created for Goering's Luftwaffe by General Kurt Student. It had been deployed in support of ground attack in Norway and the Netherlands, fanning wild fears in Britain. Now what could it achieve against islands? Malta, the most valuable target, was strongly defended, and Hitler approved Operation Merkur against Crete. 22,000 soldiers were committed, most to descend by parachute or glider. An airforce of 280 bombers, 150 Stuka dive-bombers and 200 fighters was in support. Though the techniques of German parachuting were crude and contributed to appalling losses, such men as survived would be well armed and supplied. The aim was to capture the island's three airfields, starting from Malame in the west, and roll back and squeeze the Commonwealth defenders along the north coast road.

Despite having softened up the defence with days of continuous air attack, the Germans still failed to capture a single airfield on 20 May. Retimo and

Heraklion remained in Commonwealth hands till the end of the battle, so the British and Australian troops defending them were never defeated and never went into retreat. Those at Retimo, however, became prisoners, and many of the soldiers from Heraklion were lost at sea after the navy had taken them off. The decisive fighting was in the west of the island. Crucially, on the night of 20 May, the New Zealander Colonel Andrew VC, commanding the defenders of Malame airfield, withdrew his men after expected reinforcements had failed to come up. The Germans had suffered so heavily that they expected defeat, but next day were able to seize and use the airfield. A belated Commonwealth counter-attack on 22 May failed. Growing steadily in strength, the Germans drove east as planned. Canea was subjected to an air raid reminiscent of Guernica. The Commonwealth defence got so tangled and confused that any coherent account of what happened is intrinsically suspect. By 26 May, Freyberg was convinced that the island was untenable. A stream of men south over the White Mountains, which rose to 8,000 feet, had begun, heading for the tiny harbour at Sphakia. The road stopped short several kilometres from Sphakia village and the last ordeal was to scramble down a goat track to the little beach. Rearguard fighting continued to the end, but the traffic from 26 May was all one way.

No one ever denied that the spectacle was pitiful, though the Latin phrase *via dolorosa*, which features in several accounts including Freyberg's, attempts to accord the trek such sacred dignity as might be attributed to exhausted men who had done their best against odds. The road was steep, footwear disintegrated, wells were few, food hard to come by, dysentery rife. That the navy, operating only at night, was able to take off 18,000 men makes the Dunkirk evacuation seem almost a dawdle. Why didn't the Germans find some way, easy surely, of cutting the road off and the British to pieces?

Davin's description of the *via dolorosa* can stand for many grim accounts:

> The natural savage grandeur of the mountain road was overprinted with the chaos of war. Every yard of the road carried its tale of disaster, personal and military. The verges were strewn with abandoned equipment, packs cast aside when the galling weight had proved too much for chafed skin and exhausted shoulders; empty water bottles; suitcases and officers' valises gaping their glimpses of... pullovers knitted by laborious love in homes that the owners might not live to see again; steel helmets buried in the dust...[8]

Most men – though not brave Captain Waugh – were wary in the daylight, hiding in gullies and caves and groves from the Luftwaffe. Even at night, fear was intense. Kippenberger, leading the 4th NZ Brigade in relatively good order, on a gruelling night-march to a new rearguard position, stopped, unsure of which fork to take, and turned his torch on a map. From the bank above voices chorused, 'Put out that fucking light!' As Kippenberger recalled it, deadpan, 'a man rushed up and kicked the torch out my hand. I stood up and seized him by the throat, throttled till he started to choke, and threw him

down. I then stated that if there were any more such talk I would open fire.'
Silence followed.[9]

The Cretan débâcle remains controversial not because of the indisputably
humiliating retreat to Sphakia, but for other reasons. It was, as Len Deighton
has put it, 'history's only defeat by an unsupported parachute army'.[10] The
word 'unsupported' is literally true because the Royal Navy smashed an
ancillary German attempt to land men and equipment from the sea. Student's
men suffered such heavy losses – 4,000 dead, 220 out of 600 transport aircraft
knocked out – that Hitler would never permit another such operation.[11]

The Cretans, furthermore, were the first people conquered by the Nazis to
start resistance against the Germans from the moment they arrived. Formed
by centuries of militant resentment of Turkish rule, traditions latterly revived
in recalcitrance against the reimposition of monarchy in Athens prompted
extraordinary heroics, not just by ill-armed men but by women and children,
priests and monks. At the village of Platanias, a New Zealander saw dead
parachutists. 'That they had fallen to the villagers was gruesomely obvious.
These Cretan women were already widowed from the fighting in Albania
[against Mussolini], and they made short and ghastly work of any German
who fell into their hands.'[12] The Germans responded in a style which came to
typify their conduct in eastern theatres of war.

Resisted by locals at Kastelli, they revenged themselves on 27 May by
shooting about 200 hostages in the town square. While the British, in this
savage little campaign, disposed unceremoniously of Germans who attempted
to surrender,[13] the systematic German atrocities against civilians, which continued during years of occupation, were of a different order of barbarity. This
point may be taken either to supplement or to modify Omer Bartov's
implication, in his study of *Hitler's Army*, that the barbarisation of German
soldiery originated on the Eastern Front.[14] What he describes happened
spontaneously on Crete – a self-righteously brutal response to a hostile
population deemed to be savages themselves – though Hitler had so admired
the spirit shown by the Greeks against Mussolini, and so idealised these
descendants of classical heroes, that he had ordered, before his campaign
against them began, that all Greeks taken prisoner should be released as
soon as an armistice was signed.[15] It seems that *übermenschen* became *unter-
menschen* in the minds of German soldiers as soon as they first saw a German
corpse with eyes pecked out by crows and assumed that vengeful villagers had
torn them out.

It does seem that if the British had done more to arm the Greek soldiers
evacuated to the island, and had furnished local people with weapons, the
German attack would probably have failed. In this case, Wavell's lack of
attention to the island before the attack might make him seem the author of
defeat. Other candidates for that title include Andrew, who made the fateful
decision to withdraw at Malame on 20 May when the Germans thought
themselves beaten; Brigadier Hargest, who failed to support Andrew timer-
ously; and Freyberg, held to have misread an ULTRA message for his eyes only

so as to reinforce his own obsession that he must position men to defend the beaches against a main thrust by the Germans from the sea. No one blames Admiral Cunningham, whose RN ships were grievously knocked about by the Luftwaffe as they defended the island, reinforced it, and finally took troops off it.

But it is hard to hold anyone on the spot truly culpable granted the conditions under which the Commonwealth troops fought the battle. Leaving aside the smoke rising from Suda Bay and bombed Canea, the stench of corpses in the sun and the stink of wounds assailed by flies, the lack of efficient communication by telephone and wireless, and inability to use in daylight what little transport was available, made co-ordinated command impossible. On 26 May, Brigadier Puttick, commanding the NZ Division, had to walk four miles to Freyberg's HQ, then four miles back, delivering an essential message. The Luftwaffe dominated proceedings completely. As the first British official historian says, 'It is difficult to think of any instance in the history of warfare where a force has been so pinned down and paralysed at *every level*, from the rifleman in his slit trench... to the staff officer at Force Headquarters waiting for information that does not arrive or planning movements that will not be carried out...'[16]

Tolstoy would presumably have welcomed news of this reduction nearly to absurdity of evidence to support his view, in *War and Peace*, that generals do not really control battles. Had the morale of the half-starved Commonwealth troops disintegrated completely, the enormous condescension of history might have been prevailed upon to forgive them. Lieutenant Roy Farran, a twenty-year-old British tank commander, was horrifed at Malame when the commander of his lead tank was killed just ahead of him. His own machine soon packed up in a bamboo field, under a 'swarm' of Messerschmitt 109s. 'The terror of the aeroplanes', he would write, 'had turned me into a frightened, quivering, woman...I lost my head. I was so afraid that I could have burrowed into the ground.'[17]

But as Farran should have known, Cretan women were amongst those on the island who remained uncowed by the endless snarl of planes. So was Captain Waugh. Different men, different units, reacted very differently to the same conditions. One case in point, already alluded to, involves Colonel Bob Laycock, a dashing, intelligent, upper-class soldier whom Waugh admired almost unreservedly.

Layforce had been sent from Egypt to reinforce Crete in a gesture by Wavell too little and too late. The first contingent, without their commander, turned up in Suda Bay on the night of 23–4 May. Laycock and Waugh, detained by shipping difficulties, arrived with the rest on 26–7 May. Freyberg already knew that the battle was lost. Layforce consisted of commandos, recruited for daring strikes by sea against enemy-held coastlines. On Crete its members were consigned to rearguard infantry duties remote from their training and aspirations. However, they seem on the whole to have performed these effectively. They were, bar Germans, the freshest troops on the island.

As 'last on', they were to be 'last out', even though Freyberg insisted on priority for fighting units. On 31 May, the last night of evacuation, they were listed to embark after the New Zealanders and Australians who were with them in the rearguard, before only the Royal Marines. But Laycock and his staff flitted before all the New Zealanders were on the ships. Though only a handful of his Layforce men managed later to push through the increasing chaos on the beach and get out of Crete, and Layforce itself was soon disbanded, Laycock's career was unimpeded. He rose to be a general, to head Combined Operations Command, to be knighted. Before he died in 1968, his affable and distinguished personality had predisposed inquisitive historians not to pry too zealously into a scenario where he had not only disobeyed orders – 'Layforce to embark after other fighting units but before stragglers', as Waugh had noted – but abandoned brave subordinates to their fate.[18]

Others, who had fought longer and harder, reacted differently to the painful moment when it became clear that the navy could accept no more battering and thousands would have to be left on Crete. Kippenberger protested vehemently when told that he must leave his second in command on the island and only agreed to get off when told that Freyberg had expressly ordered it. He agonised in detail over which 76 of his 306 men should stay, and succumbed, changing his mind when a deputation of subalterns came to him asking to remain themselves in place of the officer he proposed to leave in charge, who, as they pointed out, was married.[19] On 29 May, Freyberg himself had been most reluctant to leave before his rearguard, but if he had been captured, ULTRA could have been compromised, and he bowed to orders and necessity.[20] At much lower level, men sacrificed themselves. Michael Woodbine Parish, a very young gunner acting as General Weston's liaison officer on a motorcycle had no romantic illusions – he later wrote that Waugh had 'clearly portrayed' in his novel the 'shameful' retreat to the south coast. He was ordered by Weston himself to embark, but volunteered at Sphakia to carry rations up to the Maoris in the rearguard. The Maoris got away, but Woodbine Parish was taken prisoner.[21]

Evelyn Waugh's original view that the war was or should be a crusade against the Nazis and their wicked Soviet allies had been challenged by the boring and at times farcical time he had spent in khaki. In 1955 he would still feel able to dedicate *Officers and Gentlemen*

TO

MAJOR-GENERAL

SIR ROBERT LAYCOCK

that every man in arms should wish to be

but the night of 31 May in Sphakia seems to have completed a long process of disillusionment.

Like his protagonist in *The Sword of Honour*, Guy Crouchback, Waugh had been captivated by the old-fashioned traditions of the Royal Marines, who became the 'Halberdiers' in his fiction. But he never rose above, and sometimes fell below, the rank of captain which he soon achieved. In August 1940 he embarked with the Marines on an abortive attempt to capture Dakar from the Free French. Influence in high places – he knew Churchill's crony Brendan Bracken – helped secure his transfer to the Commandos, where he mingled with aristocratic clubmen whose arrogant, casual, expensive ways he partly idolised, partly laughed at. His second chance of participation in glorious warfare came when 8 Commando, stationed in Egypt, raided Bardia along the coast. It was supposed to be full of Italians. There were none there.

If Dakar and Bardia had been farcical, Crete was something worse than that. When Laycock asked him on the ship which took them back to Egypt how he had found his first battle, his reply was, 'Like German opera – too long and too loud'. He wrote to his wife afterwards: I have been in a serious battle and have decided I abominate military life. It was tedious and futile and fatiguing. I found I was not at all frightened only very bored and very weary... The thing about battle is that it is no different at all from manoeuvres with Col. Lushington [of the Marines] on Bagshot heath – just as confused and purposeless.[22]

From disillusionment, Waugh went on to suffer petty vexation, quarrelling with and sneering at superiors in a succession of desk jobs. There was a very bitter moment in 1943 when Laycock, ordered to take his Commando to Italy, left Waugh behind. Eventually he was seconded for six grim months to the British Military Mission in Croatia headed by his old drinking companion Randolph Churchill, the great man's son. When VE Day came Waugh, some weeks out of uniform, mused: 'I remember at the start of it all writing to Frank Pakenham that [the war's] value for us would be to show us finally that we are not "men of action". I took longer than him to learn it.'[23]

If Waugh was not a 'man of action', his physical courage, bordering on death-wish, determined his reactions to the Cretan disaster. Arriving at the very moment when defeat had been acknowledged as inevitable, he was the harshest possible judge of the behaviour of exhausted men who had been fighting Germans. He became tired himself, though not as tired as they were, and suffered hunger like them, but he still could not allow for the effect of many days under bombardment and strafing on the nerves of commanders and men. His 'Memorandum on Layforce' records instance after instance of what he took to be cowardice. The 'first indication' is the arrival on the ship which brought Waugh with Laycock into Suda Bay of a 'stocky, bald, terrified' naval commander gibbering with weariness and panic. 'My God it's hell... Look at me, no gear. Oh my God, it's hell. Bombs all the time. Left all my gear behind.' Shortly afterwards, on land, they met Lieutenant-Colonel Colvin, whose crack-up would provide Waugh with the model for Major Hound's in his novel, but who 'did not seem particularly nervous that night', and then Freyberg – 'composed but obtuse'.

After they reached Layforce HQ, Laycock sent Waugh forward to give Colvin his orders. 'At one point in our journey, General Weston popped out of the hedge. He seemed to have lost his staff and his head.' He reproved Waugh for producing a map to show him where Laycock was. 'Don't you know better than to show a map? It's the best way of telling the enemy where headquarters are' – a remark baffling in its inconsequentiality. Weston tried to hitch a lift from Waugh back to Laycock, but Waugh insisted on going forward to find Colvin. 'I used to command here once', Weston said wistfully. Waugh's habitual insolence towards superior officers – probably the main reason for his frustrations in the army – was never more cruelly and successfully applied.

Back in Egypt Colvin's obsession with discipline had plagued the unorthodox Waugh. Now Waugh found him a total wreck – hunched under a table in his HQ in a farm building 'like a disconsolate ape'. When the Luftwaffe took their usual break for lunch, Colvin emerged, and 'still looked a soldierly figure when he was on his feet'. But as the aeroplanes returned he lay rigid with his face in the gorse for about four hours. At night, in a panic, while Laycock was absent, Colvin ordered Layforce HQ to withdraw and took men on an all-night march through mobs of retreating stragglers. 'All the officers', Waugh noted cynically, 'seemed to have made off in the motor transport.' Colvin marched insensately on. 'Nothing but daylight would stop him. The moment that came he popped into a drain under the road and sat there.' Waugh found Laycock and brought him to Colvin, 'still in his drain'. Laycock relieved him of his command – 'You're done up. Ken will take over from you.'[24]

Laycock, polite to this unfortunate officer, showed a consideration which Waugh refused to concede. Colvin was obviously shell-shocked. When Laycock was negotiating his flit from the island, General Weston, who had at first charged him with the task of surrender next day, changed his mind – 'it was foolish to sacrifice a first-class man' – and allotted the role to Colvin. What became of Colvin is not clear. He may have embarked with Laycock, who sent the surrender instruction back to George Young, still with the Commando rearguard. The editor of Waugh's diaries thought that, like Hound in the novel, he 'disappeared'.[25]

Waugh was consumed with guilt over Laycock's flit and his connivance in it. This may explain his obstreperous mood when he returned to Egypt. He went to see his friend and future biographer Christopher Sykes, who had a job at GHQ, Cairo, and when Sykes asked about Crete, overboiled with anger:

> He said that he had never seen anything so degrading as the cowardice that infected the spirit of the army. He declared that Crete had been surrendered without need; that the officers and men were hypnotised into defeatism by the continuous dive bombing which with a little courage one could stand up to; that the fighting spirit of the British armed services was so meagre that we had not the slightest hope of defeating the Germans; that he had taken part in a military disgrace, a fact that he would remember with shame for the rest of his life.

As Sykes notes, such extreme views, repeated to others, 'including people he met for the first time', can have done his prospects in the army no good at all.[26]

In *Officers and Gentlemen*, begun eleven years later, Waugh of course refracted what had become his settled view of the battle through his fictional characters. Colvin became Brigade-Major Hound – less senior, rather more sympathetically observed. Laycock was eliminated, since his fictional counterpart Colonel Backhouse breaks his leg in transit to Crete and takes no part in the battle. Something like Laycock's flit, though, is attributed to Ivor Claire, a Household Cavalry officer serving with the Commando who deserts his soldiers at Sphakia and is spirited from Egypt to India by the resourceful Mrs Stitch lest word of his disgrace might ruin him. Guy Crouchback has grown especially fond of Claire – he is outrageously effete, but a wonderful horseman, and seems possessed of spiritual depth. He is shocked and shamed by Claire's desertion, but connives in the shielding of him by the aristocratic Mrs Stitch, defending one of her own class against rule-bound vulgarians. Guy himself, left behind on Crete, escapes with *ad hoc* company in an open boat and barely survives a terrible crossing to Egypt. (In historical fact, hundreds managed to get off in similar ways.) His sacrificial courage clearly relates to Waugh's wish that he himself had stayed honourably behind rather than going with Laycock, just as his disillusionment with Claire echoes the shattering of Waugh's own hopes of a chivalric, crusading war in the company of choice, aristocratic, spirits.

Walking through the chaos of Crete, Crouchback at one point encounters his old comrades, the 'Halberdiers', now in the rearguard. Royal Marines were actually on the island, but commanded by Weston whom Waugh despised – this is a dreamed-up encounter. The Halberdiers are shining lights of stiff morale, completely cool under bombardment. Guy wistfully asks to be taken back – after all, he points out, they have incorporated a company of New Zealanders, who, Colonel Tickeridge has told him, 'rolled up and said please may they join in our battle – first class fellows'. After Tickeridge replies, 'No can do' – Guy is under a different command – Guy watches the Halberdiers execute a withdrawal in which 'everything' is done correctly, before he plods on to rejoin his Commando.[27]

So Waugh, in his artistic resolution of his Cretan experiences, did acknowledge that not all morale was bad, and, in a brief but telling reference, recognised the special qualities of the New Zealanders.

Perhaps the New Zealanders did so well because Crete, as a terrain, was less alien to them than it was to their Commonwealth comrades-in-arms.[28] In the Antipodes, New Zealand lay in Mediterranean latitudes. Its people depended on agriculture and livestock, like the Cretans, and unlike the urbanised British and Australians. New Zealand's mountainous terrain, especially rugged in South Island, perhaps made the White Mountains which were crossed to get to Sphakia seem not quite so daunting to its soldiers.

The first attacks on Crete on 20 May, in the western sector, were borne almost entirely by the NZ Division – 'the Div' as its members called it. Four battalions of Hargest's 5th Brigade were stretched along about five miles of the north-west coast. Some three miles to the east, there was the newly improvised 10th NZ Brigade, of which more later. The 4th NZ Brigade was towards the outskirts of Canea. So New Zealanders were first in the fighting, and virtually the last off in the navy's evacuation ships. In between, two of them won VCs.[29]

Correspondingly, questions of blame for the defeat revolve mostly around New Zealand commanders. Bernard Freyberg was a huge man physically, VC with DSO and two bars in the First World War, when he had been copiously wounded. Churchill described this durable hero as the 'Salamander of the British Empire'. He was perhaps, as Waugh thought, 'obtuse', and suffered from an incapacity to deal sternly with faulty subordinates, but he communicated, as one historian puts it, his own 'Homeric gusto' to the infantry he led. Waugh saw him at dusk outside the cave in the gorge above Sphakia where had set up headquarters, 'saying goodbye to New Zealanders who were leaving that evening. Some had photographs of him which he signed.' Cultivated, somewhat romantic, brilliant at training men, Freyberg had a 'relaxed attitude towards formal discipline'.[30]

However, he was not so popular with his senior subordinates. He was not quite a real New Zealander. Born near London in 1889, he had been taken to New Zealand as a small child, had trained as a dentist and territorial soldier there, but then departed, as a ship's stoker, to see the world. After his First World War heroics, he had settled in Britain as a professional soldier, deeply aggrieved when he was retired on health grounds in 1937. When the New Zealand government in 1939 accepted his own offer of his services and put him in charge of their volunteer expeditionary force, 'the Div', he was an Anglicised outsider coming in above Brigadier Puttick, who had played a leading part in forming and training the force. On that fateful 20 May, Puttick commanded in the Malame area. Under him there was Brigadier Hargest of 5th Brigade, like Freyberg a Gallipoli veteran, but latterly a Conservative MP in the New Zealand parliament. Freyberg had tried to have him rejected in 1939 as too old and without recent military experience, but political pressure had prevailed. Brigadier Inglis was, like Kippenberger, a lawyer in peacetime – 'quick thinking', but also 'pugnacious and opinionated' – the Judas who eventually spoilt Freyberg's reputation with Churchill when he got to London soon after the Cretan disaster.[31]

All these men, though, had one thing in common with Freyberg. They had distinguished themselves in the trench warfare of 1914–18 and were not quite able to adjust to a new style of war in which air power might be decisive and mobility was imperative. As Dan Davin tells us, such veterans receded from senior positions as 'the Div' fought on, so that 'in its prime' in Africa and Italy it was 'officered by men in their twenties and thirties, at least at battalion level'.[32]

However, the way 'the Div' had been raised guaranteed that even in Crete there was *rapport* across the ranks. New Zealand had a population of only

1,630,000 people, including 90,000 Maoris. Commitment to the British cause was total. The smallest of the Dominions rationed its copious supplies of home-grown food so that Britain might be fed, agonised over the Mother Country's ordeal in 1940, and committed the same proportion of its expenditure to the war effort, becoming, on a par with Britain, the most heavily mobilised of all combatant countries. Nearly a tenth of the population entered the armed forces, 50,000 served overseas. Proportionately, more New Zealanders were killed in the war than from any other part of the Empire.[33] Under this overarching, national dedication, 'the Div' fostered local loyalties.

'The Div' had three brigades, each with three battalions. The Maori battalion was attached to any one of these as occasion prompted. It was subdivided into units approximately corresponding to 'tribal' areas. In each brigade, likewise, the first two battalions were drawn from the Auckland and Wellington areas of North Island and the third from South Island. As Davin writes:

> The territorial divisions thus described became more difficult to maintain as the war went on, but to the end – at battalion level especially – they were an important factor in pride and morale: officers and men in a given unit tended to know each other from civilian life – from school, the office or factory, university, the rural district, work or the playing field – or a combination of any of these. Since men and officers alike were drawn from an egalitarian population, social class as a distinction or sanction barely existed...

Ties to New Zealand, so many thousands of miles away, remained strong: 'the troops were as zealous to sustain their standing in the eyes of family and friends back in, say, Southland, as a provincial Rugby football side would be or the All Blacks [NZ national rugby team] on tour'.[34]

Whether or not primary group loyalties are seen as always, in general, everywhere, the main reason why soldiers keep on fighting in such a hell as Crete became, there is no doubt that in the particular case of 'the Div' they were crucial. The closeness between Freyberg, Kippenberger and their men made strict, formal discipline inappropriate. Thinking of John Keegan's 'triad' of factors in morale, 'coercion' was relatively unimportant compared to the 'inducement' of sustaining local and family traditions of service and courage. If 'narcosis' featured, the adrenalin-rich rush of pride, comparable to the emotions experienced on and around the rugby field, could be as potent as drugs or alcohol. The very fact that New Zealand was a small nation paradoxically made for big courage – and there were, in Keegan's parlance, some very big men providing inspiration to others in 'the Div'. They didn't come bigger than Freyberg or cooler under fire than Kippenberger.[35]

Sandy Thomas, a junior officer on Crete, thought his platoon there 'a grand crowd – wonderful company for the past year since I, a very raw second lieutenant of twenty years had welcomed them all into the mobilisation camp at Burnham, New Zealand. Most of them were from my district and were either miners, bushmen or farmhands. There were also those who, like me, had been country bank clerks, or shop assistants.' Such men, Thomas noted, did not take kindly to being told to withdraw from the Malame area. 'They had seen so many of the enemy dead that their morale was quite unshaken by the terrific air attacks by day. Man for man they considered that they could lick the German despite his superior weapons and equipment. Their fathers had made a name in the first war for ruthless and invariably successful night attacks.'[36]

New Zealand troops made fun of the Luftwaffe, capturing the coloured strips used by the Germans to signal supplies from the air, and collecting for their own use what was duly dropped. High spirited when cheerfulness was in order, they stuck by each other when things got bad. In a characteristic cameo, three New Zealand sappers are on the road to Sphakia. The officer, Phil, has bad feet, torn to shreds by walking, and tries to order the other two to leave him behind to his fate. They refuse. As Sapper Trethowen recalled, 'We had been through a lot together and we were staying together.' The three eventually embarked at 12.45 a.m. on 1 June, amongst the very last to go. Whatever his failings as politician or brigadier, Hargest, like Freyberg and Kippenberger, was loath to accept evacuation ahead of some of his men. As his 5th Brigade marched as best they could down the rocky path to Sphakia to embark, it was seen that 'All had shaved. They wore helmets and haversacks. Every man carried a weapon.' Lesser mortals 'got on their feet to watch them pass. There were murmurs in the darkness, "It's the New Zealanders, the New Zealanders."'[37]

The young English tank officer Roy Farran had experienced early in the battle that awe which New Zealand morale eventually evoked in many outsiders. As he took his tank up towards Galatas on 20 May, after the airborne attack had started, he 'passed several New Zealand positions, where the troops stood up and gave us the "thumbs up" sign with a grin. Just to look at their confident, smiling faces was good for the spirit.' Later that day, told to co-operate with New Zealanders in an attack on the Galatas cemetery, he hid his tank among trees and walked down to 19th Battalion HQ to report to Colonel Blackburn for orders. 'I had some difficulty in finding him, firstly because we were interrupted by an air raid, and secondly because he was out in the forward defence lines potting at a sniper. I ran along to him with my head ducked, bullets whistling all round, until I noticed that ducking did not quite seem to be the fashion in this part of the world. With a tremendous effort, I tried to appear brave and walked up to him. He muttered, "Just a moment," out of the corner of his mouth, and only turned when he had brought the sniper tumbling from a neighbouring tree.' Later in the campaign, he watched Maoris falling back with the Germans close behind, 'but not a shadow of fear showed on their smiling, copper faces. As they passed my tank, they winked and put up their thumbs. Some fifty yards behind the rest came two Maoris carrying a pot of stew across a rifle.'[38]

At Galatas again, on 25 May, Farran was able to help New Zealanders stage a counter-attack which would figure in history books as a little epic of spontaneous group heroism. About four that afternoon, Farran was ordered to take his two tanks up to co-operate with Kippenberger, currently in charge of the improvised New Zealand 10th Brigade, whose headquarters had been in Galatas.

Around the events which followed, the first British official historian, Buckley, felt entitled to deploy purple prose unsuitable for other parts of his sorry story. The New Zealanders, 'men of English blood from the land of the Southern Cross', matched, he said, the spirit of Alfred the Great's Saxons against the Danes at Ethandune, with that heroism 'that has so often snatched victory from defeat'.[39] Unfortunately, Kippenberger snatched no more than the briefest of respites, if that, from ongoing defeat, and his own quite dry account should take precedence as we try to imagine the counter-attack.

His 10th Brigade represented in microcosm the higgledy-piggledy, patchwork character of the whole Commonwealth force in the west of the island at this time. When battle had been joined on 20 May, he had had under his command 2,300 Greek troops in two battalions whom he thought were mostly useless. The 20th NZ Brigade numbered 650, but could not be used without approval at divisional level. The NZ Divisional Cavalry, 190 men, were untrained in infantry work, though well disciplined. There were 1½ platoons of NZ machine-guns, and one battery armed with three Italian 75s without sights. There was, finally, the First Composite Battalion of 750 men, gunners and members of service corps mobilised as infantry, but no good, Kippenberger thought, except in defence. By 25 May, many of these men were coming round to the opinion that they had done enough fighting. Generally ill-equipped, the brigade had only about six digging tools per company.

A picturesque young Englishman, Michael Forrester, an officer in the Queen's Regiment attached to the Greek Mission, had turned up on a visit and decided to stay for what he called 'the party'. He had managed to rally many of the Greeks and on 22 May had led a charge of about 100 Greek soldiers and Cretan villagers, including women and children, 'yelling like Red Indians', which had startled German attackers into precipitate retreat. But the Composite Battalion was demoralised, and General Puttick, in overall charge of the sector, decided on 23 May that the still battleworthy 18th Battalion should reinforce Kippenberger.

On 25 May, the Germans attacked the 18th in front of Galatas. The village was lost. A first counter-attack by a force including the padre, clerks, batmen and 'everyone who could carry a rifle', failed. The 20th Battalion, now ordered up, found the Composite Battalion 'nearly all gone', and as the Germans continued to press, 'suddenly the trickle of stragglers turned into a stream, many of them on the verge of panic'. Kippenberger rallied them in person, 'shouting "Stand for New Zealand!" and everything else I could think of,' and managed to place many under the nearest officers and NCOs – 'in most cases the men responding with alacrity'.

In the new line which he constructed on a ridge to the west of Galatas, the band of the 4th Brigade stood next to the Pioneer Platoon of the 20th

Battalion, the Kiwi Concert Party of entertainers intended to relieve the *longueurs* of army life, and A Company of the 23rd Battalion. The 20th Battalion was to extend this line. Kippenberger's memoirs state that he ordered Farran to go into the village to reconnoitre and blaze away. In Farran's own account, the young man, bitter about signs of impending defeat and anxious to help the New Zealanders whose wounded he had seen, bandages on heads, arms in slings, as he drove up, 'asked permission to go in first, alone'. Two more companies of the 23rd arrived while Farran's tanks were shooting up the village. Kippenberger told the commanders that they would have to retake the village, with the two tanks. 'Stragglers and walking wounded were still streaming past. Some stopped to join in.' So did a noted New Zealand sportsman, Carson, a brave officer now leading only four men. When Farran reappeared his lead tank had two wounded men in it. Volunteers were asked for to replace them, and the chosen two were taken off for ten minutes' training.

Then came the charge, in gathering darkness. The tanks set off downhill, followed by the *ad hoc* infantry, who began running and shouting in a 'terrifying crescendo'. Kippenberger listened to the din, augmented by gunfire. By the time it was clear what had happened in the village, 'we had lost', he notes drily, 'both tanks, Farran was wounded and in each company some thirty men were hit. Two of the subalterns who had led the charge, Sandy Thomas and Rex King, were badly wounded...' So far from feeling triumphant over the recapture of Galatas, Kippenberger felt 'more tired than ever before in my life, or since'. He walked to report and consult at Brigade HQ, where the commanders agreed that Galatas must now be abandoned, and everyone brought back before morning to the line established by Kippenberger on the Daratsos ridge.

Next day it turned out that the Composite Battalion had retreated, 'for its own reasons'. The game was clearly lost, and Kippenberger was 'unashamedly pleased' to get Freyberg's order to move towards Sphakia.[40]

So the recapture of Galatas achieved, at the time, nothing of consequence. As years wore on, though, it came to encapsulate the improvisational flair, raw courage and volunteer morale of 'the Div' on Crete. The outsider Farran's account, published in 1948, turned it into pure Hollywood.

Farran claims that 'about three hundred' New Zealanders volunteered to replace the wounded men in his lead tank. Later, lying with wounds on his legs and right arm in the inadequate shelter provided by a low stone wall, he prays for the New Zealanders to arrive. 'They came up the main street in a rush, but were met by a hail of machine-gun bullets on the corner. Several went down in a heap, including the Platoon Commander. I shouted, "Come on New Zealand! Clean 'em out New Zealand!"'

The wounded platoon commander was Sandy Thomas. Someone, according to Farran, called out that Germans were on the roof. Farran saw Thomas 'lift himself up on his elbows and take careful aim with his pistol. The German machine-gunner came tumbling down the slates on to the street below. It was an astonishing shot in such a light and at such a range.'[41]

Astonishing, indeed, if it happened. One suspects that memories of screen westerns had got mixed up in Farran's mind with events in Galatas. He and Thomas should have published the same story, since both were captured by Germans in the improvised field station and both were flown to Athens for treatment, successful in both cases, of gangrenous wounds; they confabulated in transit. But Thomas's account, published three years later, while equally filmable, is quite different.

He and his comrades, suddenly ordered up to Galatas, sheltered from a lone German aeroplane and first see Kippenberger from the ditch where they have flung themselves, 'a slight figure, pipe in mouth, standing unconcernedly on the road in contempt of danger'. Rapidly drawn into the brigadier's plan of attack, they find themselves lined up on a terrace just 200 yards from the village. 'It occurred to me suddenly', Thomas writes, 'that this was going to be the biggest moment of my life.' As the infantry followed Farran's tanks, 'the whole line broke spontaneously into the most blood-curdling shouts and battle cries. The effect was terrific. One felt one's blood rising swiftly above fear and uncertainty until only an inexplicable exhilaration, quite beyond description, remained.' Elsewhere, Thomas, quoted by Davin in his official history, analysed that epic yell as a composite of Maori-style *hakas* chanted as rugby matches commenced by school and college teams throughout New Zealand. An NCO, also quoted by Davin, described it as 'the most ungodly row I ever heard...cat-calls and battle cries, machine-guns, rifles, hand grenades all going on at once'.

As the New Zealanders surged on yelling and firing at random, a tank, Farran's lead, flew back towards them in retreat. 'The shouting stopped. From the turret a frenzied man screamed, "Let me through...For cripes sake run for it − the place is stiff with Germans."' This may have been one of the eager volunteers, though Thomas does not say so. Thomas, threatening the driver with a revolver, got him to turn the tank round. 'The maniac in the turret' leapt out and began to flee through the ranks. Thomas knew the effect on morale could be disastrous and that his 'duty as an officer' was to shoot him, but he could not bring himself to do it, and was relieved when a private soldier did the job for him and 'looked grimly around his friends. No one said anything [Thomas goes on]. We just all moved on again, quieter now, but I think the better for that.' The tank sent back soon stood wrecked by the village square, its driver dead. Thomas decided, seeing this, that 'Action, quick action was essential. I decided to charge...' As one man the New Zealanders marched across the square and were soon in hand-to-hand combat with the Germans. When someone shouted, 'Look out, that bastard on the roof,' according to Thomas he fired along with several others at the helmeted form above which was dropping a grenade. This wounded Thomas in the back at the same time as his thigh was hit by a bullet...But he did remember hearing Farran in his 'very English voice' calling out, 'Good show New Zealand...come on New Zealand.'[42]

Davin's official account inevitably smooths discrepancies into a single heroic flow. From a letter by Captain Bassett written soon afterwards, he conjures up the 'big man' figure of 'that great lump of footballing muscle William Carson, with a broad grin, licking his lips before the counter-attack and saying, "Thank Christ I've got a bloody bayonet."' Davin salutes the brief affray as 'one of the fiercest engagements fought by any New Zealand troops during the whole war'.[43]

When peace came, after Mr Wu had returned to his writing-desk, the surviving members of 'the Div' went back to their rugby clubs. Their war stories barely interconnected with his. New Zealanders had not been immune to onsets of strange behaviour under Luftwaffe attack such as had appalled Waugh in Colvin and Weston. By 27 May, Kippenberger saw the strain telling on his officers. As German fighter planes harried them, one 'continued digging until he could not climb out of his pit'. Another 'appeared walking very fast, with odd automaton-like steps and quivering incessantly. He made a great effort to control himself, offered cigarettes to Jim [Burrows, second in command] and me, and continued his move at high velocity.'[44] To reduce the story of Crete to a straight contrast of colonial courage and British cowardice would be absurd. But Dan Davin's countrymen could pull pride from defeat. Evelyn Waugh's, convinced by his account, will think of it as silly, sorry show. It ain't what you see, it's the way that you see it.

'If this war isn't over, And pretty damn soon, There'll be nobody left, In this old platoon…': First Canadian Army, February–March 1945

Terry Copp

Oh, what with the wounded
And what with the dead
And what with the boys
Who are swinging the lead
If this war isn't over
And pretty damn soon
There'll be nobody left
In this old platoon.[1]

This verse, from an anonymous Canadian, captures the feelings of the men who toiled in the rifle companies of the American, British and Canadian armies in the last months of the war. General George Patton expressed the same view when he analysed the imperfect victory in the Ardennes in January 1945. While he believed that Montgomery's inability to understand that the proper object of military operations was to destroy the enemy's army had contributed to the failure of the counter-offensive, the real problem was the nature of the Allied war machine. 'We have to push people beyond endurance…', Patton wrote, 'because we are forced to fight with inadequate means.'[2] The Allied armies were short of replacements and ammunition and desperately needed more divisions.

The manpower crisis, which had forced the Canadian government to send 16,000 conscripts overseas and led Churchill to order the call-up of a quarter of a million additional men in January 1945, was the product of a long series of miscalculations. The Allied plan for the invasion of north-west Europe had always been dependent upon the ability of the Red Army to continue operations which occupied the energies of most of the enemy; the Allies had simply not created a force large enough to confront the German army. Indeed British (and thus Canadian) preparations for Operation Overlord were strongly affected by the overall war policy of Great Britain, which was based on the desire to defeat the enemy by means other than direct conflict with the German army in north-west Europe. In the spring of 1944 the British-Canadian component of the Allied forces available for the invasion included formidable strategic and tactical air forces, a naval commitment of unparalleled power and a small army which was especially deficient in infantry.

The 21st Army Group was to have an establishment of some 750,000 men by 1945, but there were only nine infantry divisions (including two Canadian) available. Fifteen per cent of the total troops wore infantry badges,[3] but the designation 'infantry' should not be confused with actual commitment to battle in a rifle company. A standard 1944 infantry division contained 915 officers and 17,247 men, but less than half were infantry and of those only 4,500 served in the 36 divisional rifle companies.[4]

The British had long since determined, as the Germans in north-west Europe were soon to discover, that artillery was to be the army's principal weapon and fully 18 per cent of the troops in the bridgehead were gunners.[5] The planners also believed that the 14 armoured brigades allotted to 21st Army Group would play a prominent role in cracking the German defences. It was hoped that these two 'arms', together with the 2nd Tactical Air Force, would ensure that the campaign did not become a bloody replay of the Western Front in the First World War.

Viewed in hindsight the Normandy campaign went much better than the planners had hoped. Not only were the Germans unable to bring substantial reinforcements from the Eastern Front (the Russian summer offensive which began two weeks after D-Day was to cost the Germans close to one million casualties), but poor intelligence and the Allied deception scheme 'Fortitude' kept large elements of the forces available in the west away from Normandy until it was too late. But, and it is a very large but, the discipline and determination of the German defenders meant that the Allied infantry were required to attack, occupy and hold small parcels of ground under circumstances which fully paralleled the horrors of the fighting on the Western Front in the First World War.

Modern memory has a firm image of 'suicide battalions' and long casualty lists in the First World War, but we are not accustomed to thinking of Normandy in these terms, perhaps because of the relatively short duration of the campaign (88 days) and the overwhelming victory which climaxed the battle. A single crude comparison will help to make the point. During a 105-day period in the summer and autumn of 1917 British and Canadian soldiers fought the battle known as 3rd Ypres which included the struggle for Passchendaele. When it was over our forces had suffered 244,000 casualties or 2,121 a day.[6] Normandy was to cost the Allies more than 200,000 casualties or 2,354 a day, and 70 per cent of these casualties were suffered by the tiny minority of men in infantry rifle companies.[7]

The fighting in north-west Europe placed a burden of almost unbearable proportions on one small arm of the Allied armies – the infantry. British (and thus Canadian) planners had not, despite the lessons of Italy, prepared for this eventuality. Well before D-Day Montgomery had been warned that reserves of infantry replacements were dangerously limited. The War Office had already calculated that 'at least two infantry divisions and several separate brigades might have to be disbanded by the end of 1944 for lack of reinforcements'.[8] By early July the character of the Normandy battle had brought this crisis to hand and infantry battalions were frequently operating well below strength. By early August the Canadians, for whom infantry casualties were running at 76 per cent of the total,[9] were reporting a deficiency of 1,900 general duty infantry,[10] or the equivalent of four battalions of riflemen. By the end of August shortages had reached the staggering figure of 4,318.[11]

This situation, which was only slightly less serious among British divisions, forced Montgomery to cannibalise the 59th (Staffordshire) Division. On 14 August he told Alanbrooke, 'My infantry divisions are now so low in effective rifle strength they can no longer – repeat, no longer – fight effectively in major operations. The need for action has been present for some time, but the urgency of battle operations forced me to delay a decision.'[12]

Perhaps enough has been said to indicate something of the context within which the infantry fought in Normandy. Let us turn to the impact which the battle had upon the men who fought it as measured by the phenomenon of 'Battle Exhaustion'. Needless to say, battle exhaustion, the preferred term for neuropsychiatric casualties, was largely an infantryman's problem. More than 90 per cent of the known cases were among the infantry. The large majority of individuals diagnosed as suffering from battle exhaustion exhibited what the psychiatrists described as acute fear reactions and acute and chronic anxiety manifested through uncontrollable tremors, a pronounced startle reaction to war-related sounds and a profound loss of self-confidence. The second largest symptomatic category was depression with accompanying withdrawal. Conversion states such as amnesia, stupor or loss of control over some physical function, which had made up a large component of those described as 'shell-shocked' in the First World War, were rarely seen in the Second World War.[13]

The most commonly used method of measuring exhaustion rates was the NP ratio, which measured neuropsychiatric casualties in relation to total non-fatal battle casualties. Experience in the Mediterranean among the British, American and Canadian units suggested that a NP ratio of 23, more than 1 in 5 of non-fatal casualties, was normal for infantry divisions involved in intense combat.[14]

During the worst weeks of the Normandy campaign Allied psychiatrists had talked about a battle exhaustion 'crisis' which was threatening the military effectiveness of their armies. All of the infantry divisions in action since D-Day, the 3rd British, 3rd Canadian, 50th Northumbrian and 51st Highland, were in dreadful condition and were evacuating battle exhaustion casualties by the hundreds. Those divisions that joined the campaign during the prolonged stalemate were in a similar condition by late July. Major D. J. Watterson, 2nd Army's Adviser in Psychiatry, described the crisis in his monthly report:

The high optimism of the troops who landed in the assault and early build up phases inevitably dwindled when the campaign for a few weeks appeared to have slowed down. Almost certainly the initial hopes and optimism were too high and the gradual realization that the 'walk-over' to Berlin had developed into an infantry slogging match caused an unspoken but clearly recognizable fall of morale. One sign of this was the increase in the incidence of psychiatric casualties arriving in a steady stream at Exhaustion Centres and reinforced by waves of beaten, exhausted men from each of the major battles. For every man breaking down there were certainly three or four effective men remaining with their units.

Swings of morale often tended to overshoot the mark and this happened during the first half of July. Thereafter men settled to their new appreciation of this War of Liberation, discarding their notions of marching through welcoming and gay French villages, replacing them by more realistic appraisals of a brave and skilful enemy, of battered towns and of necessary days, perhaps weeks, of grimly sitting down and holding under mortar fire, cloudy skies, rain and mud.

Finally in the last week of the month a noticeable steadying and bracing of morale occurred so that the subsequent breakthrough South of Caumont by our own army, the long strides of the Americans into Brittany and the pursuit of the enemy by the Russians through Poland and the Baltic states caused no sudden inflation of false optimism but rather a sober satisfaction that the hard fighting ahead would bring its own similar rewards. With this background the incidence of exhaustion and neurotic breakdown in the army may be assumed to have reached and passed its peak.[15]

Watterson, writing in early August, was far too optimistic. Battle exhaustion rates declined in the Second British Army after Operation 'Bluecoat' because most divisions were out of combat. The First Canadian Army, including the 51st Highland Division, remained caught up in the attritional battle north of Falaise throughout August and endured even higher levels of exhaustion casualties than it had in July.[16] The pressure on everyone eased in September, but October brought a new battle exhaustion crisis especially for the two Canadian infantry divisions involved in the battle to clear the approaches to Antwerp. In the struggle for the Breskens pocket the NP ratio averaged 17 and rose to as high as 25 in two of the brigades. The divisional psychiatrist Major Dick Gregory described some of the 421 exhaustion cases evacuated to his casualty clearing station as men who now lack 'the morale or volition to carry on'.[17]

The foremost cause of this seems to be futility. The men claim there is nothing to which to look forward to – no rest – no leave, no enjoyment, no normal life and no escape. The only way one could get out of battle was death, wounds, self-inflicted wounds and going 'nuts'.

The division [they believe] had no interest in them except to get blood from a stone in order to bring glory to others.[18]

Neither the military planners, frightened by the shrinking pool of infantry reinforcements, nor the psychiatrists, professionally committed to reducing the NP ratio, were very happy with either the quantity or predictability of neuro-psychiatric casualties. The military authorities could and did, on occasion, issue directives attempting to forbid soldiers from breaking down. The major impact of such orders was to make life difficult for the regimental medical officer and to make it nearly impossible for historians to compile accurate battle exhaustion statistics. Shortly after General E. L. M. Burns assumed command of I Canadian Corps, regimental medical officers were ordered to be very strict and hold all NP cases until it was certain that they could not be returned to their units, and regimental COs were told that battle exhaustion was their responsibility and if it occurred in the coming action 'it would be taken as a reflection upon the ability of these officers'.[19] General Guy Simonds, commanding II Canadian Corps, never went quite that far, but it is evident that he had little patience with the policies of the First Canadian Army relating to battle exhaustion.[20]

The psychiatrists in the British and Canadian army, fully supported by their American colleagues initially took a different approach. If battle exhaustion was a psychoneurosis, i.e. 'an emotional disorder in which feelings of anxiety, obsessional thoughts, compulsive acts and physical complaints, without object-ive evidence of disease, in various patterns, dominate the personality,' then their training indicated that the neurotic individual must have been predis-posed to neurosis by childhood experiences. If enough attention was paid to screening combat units for predisposed individuals, then the NP ratio should be significantly reduced.

By the autumn of 1944 most psychiatrists with front-line experience and certainly most regimental medical officers were beginning to understand that the emotional breakdowns they were labelling 'battle exhaustion' were the product of battle. Many individuals who gave no indication of predisposition were cracking under the pressure, while soldiers who could readily be labelled neurotic were surviving. It was clear that every man had his breaking point and that it was group cohesion as much as individual reserves of courage that kept men going.[21] The only realistic way of reducing battle exhaustion was to lower the intensity, cost and duration of combat. As the 21st Army Group prepared for the Rhineland campaign of February–March 1945, British and Canadian front-line psychiatrists got ready for another battle exhaustion crisis.

*

From November 1944 to February 1945 Montgomery's Anglo-Canadian Army Group was not involved in any major offensive operations and there was time to consider ways of improving the nature of the Allied war machine which might limit physical and mental casualties. The major source of new ideas was No. 2 Operational Research Section, which operated under the direction of the Scientific Adviser to the 21st Army Group, Brigadier Basil Schonland. Together with his Canadian deputy Omand Solandt, Schonland, a South African physicist, had built a team of brilliant young scientists who looked at operational problems without the blinkers imposed by service loyalties, traditions or untested assumptions.[22]

The OR section had examined most of the major weapons systems in use in Normandy. The chief problem was undoubtedly 'the location of enemy mortars, which were causing appalling casualties and proving almost impossible to deal with'. This was a classic OR challenge, a problem 'midway between the technical and the operational'. Major Michael Swann interviewed battalion medical officers from four different infantry divisions and found that all agreed in placing the proportion of mortar casualties at above 70 per cent of total casualties. He found that divisional counter-mortar staffs tended to under-estimate the number of mortars and Nebelwerfers opposite them, noting that a German infantry division possessed as many as 57 81-mm mortars and between 12 and 20 of the 120-mm type. Panzer divisions were equipped with about half these numbers. In Normandy, the German army had also provided a regiment composed of 54 six-barrel Nebelwerfers on the scale of one per division. Swann estimated that to bring the problem under control divisions might need to obtain between 60 and 80 hostile mortar locations a day.[23]

Swann and Schonland were able to persuade the 21st Army Group to develop a more systematic counter-mortar programme and to establish 1st Canadian and 100th British Radar Batteries. These units, equipped with anti-aircraft, gun-laying (GL MKIII) radars, were tested in the field during 1944 and used with remarkable effectiveness to suppress and destroy enemy mortar fire in the Rhineland.[24] The OR section also played a large role in the introduction of mobile radar control posts, which resulted in a new degree of accuracy for medium bombers employed in support of ground troops.[25] These important achievements provided some assistance to the men who launched Operation Veritable on 8 February 1945, but much remained to be done before the problems of the Allied 'war machine' could be overcome.

The OR section had concentrated most of its energies on efforts to gauge the effectiveness of Allied heavy bombers in support of the land battle and on the role of the Tactical Air Force. Their careful study of the use of heavy bombers in Normandy and the Channel ports had demonstrated that attacks on enemy forward defence lines caused relatively few casualties and not much material damage to dug-in positions. Bombing was accurate, in the sense that nine-tenths of a bomb load 'usually fell within a circle of 1,000 yards in diameter', but since 'quite often the centre of the pattern is wrongly placed' the risk to Allied troops remained high.[26] Heavy bomber support in the Rhineland was confined to administrative and gun areas, sparing the troops the dangers of short bombings but removing the heavies from a direct support role.

OR analysis of fighter-bomber activity had provoked a bitter controversy between the army and air force after the OR section had established beyond all reasonable doubt that Tacair claims of tank and other ground target destruction were wildly exaggerated. The RAF, which knew how inaccurate its bombs and rockets were from its own OR studies, refused to accept the army's evidence and more importantly refused to re-examine methods of employing fighter-bombers in close support.[27] This attitude and the predictably poor weather of the late winter months meant that tactical air power played a marginal role in the battles of February and March.

The OR section had also continued pre-invasion studies of the performance of Allied and German armour, confirming the most pessimistic views about the inferiority of Allied tanks.[28] Their work documented what every crew member knew: the Sherman was dangerously vulnerable to all calibres of German anti-tank guns. The statistics were stunning. Sixty per cent of Allied tank losses were the result of a single shot from a 75-mm or 88-mm gun and two-thirds of all tanks hit 'brewed up' when hit. German armour-piercing shells almost always penetrated and disabled a tank; the armour offered so little protection that the only way to survive was to avoid being targeted.[29]

The contrast with German tank casualties was especially striking. Only 38 per cent of hits from the Sherman's 75-mm gun penetrated German armour and both the Panther and Tiger frequently survived one or two penetrations. The sloping frontal armour of the Panther and the 75-mm self-propelled guns prevented penetration of three-quarters of all direct hits.[30]

The OR section could offer little hope of any improvement in tank armour. Attempts to install extra armour plating had been fruitless. 'In no recorded case in our survey,' the report noted, 'has the extra outside appliqué armour resisted any hit.' It was possible to limit the proportion of tanks brewing up by carrying ammunition only in armoured bins, but while this practice could save lives it would require more frequent withdrawal of tanks for resupply.[31]

Major Tony Sargeaunt, the OR section's tank warfare specialist, accepted anecdotal evidence that tank-tracks loosely welded on to the front of a Sherman might deflect some shots but the main impact was on morale. Tank crews believed it worked and this helped to sustain them in battle.[32] Sargeaunt recommended that the army concentrate on providing a better gun 'to make German tanks more vulnerable rather than attempt to decrease our own vulnerability'. In practice this could only mean increasing the number of tanks equipped with the 17-pounder gun. But if the proportion of Sherman 'Fireflys' was increased the effectiveness of armour squadrons in an infantry support role would be limited because 'Fireflys' only carried armour-piercing ammunition. Sargeaunt concluded that Allied armour would remain highly vulnerable 'until better methods of spotting tanks and anti-tank guns are found'.[33] In close country like the Rhineland battlefield the enemy could maintain a homogeneous defence, holding fire until ranges were short and flanking fire possible. Tanks would have to stay back and function as mobile artillery firing from dead ground or risk destruction. The lightly armoured M10 self-propelled anti-tank guns were at even greater risk, but M10 crews had long since abandoned a close support role.

The limitations of Allied air power and armour were even evident to all who cared to study the battlefield and the consequence was an even greater reliance on the principal Allied weapon, artillery. Unfortunately Allied artillery had failed to live up to expectations. The 25-pounder and its 105-mm counterpart, the basic gun of the divisional field artillery regiments, fired a shell which had to strike within 3' 6" of a slit trench to transmit a shock wave and only direct hits caused casualties. Given the dispersed nature of forward defences, operational researchers calculated that even very heavy concentrations could only offer temporary neutralisation. The difficulty was that existing methods of predicted fire resulted in errors of range and line, which frequently meant that the mean point of impact was off by 100 to 300 yards. The psychological effects of artillery concentrations were lost if casualties were not even a possibility.[34]

Medium artillery fired a heavier shell with an increased zone of lethality for exposed troops, but it was subject to the same limitations as the 25-pounder with regard to accuracy and the need for direct hits. Until proximity fuses were available and new methods of directing fire systematised, the only answer was to increase the number of guns employed in supporting infantry attacks.[35] If there were enough guns neutralisation could be achieved because the zones of lethality overlapped, minimising the significance of inaccuracy.

For Operation Veritable an extraordinary fire plan employing guns of all calibres was employed. Described as a 'Pepperpot', the fire plan was based on OR advice about the probability of increasing casualties by multiplying the number of lethal fragments in the barrage. Wooded areas like the Reichswald Forest would further increase effectiveness by creating air bursts with added tree fragments.[36]

The 'Pepperpot' certainly helped XXX British Corps break into the Reichswald on 8 February 1945, but once through the crust of German defences the battlefield was quickly transformed into a killing ground for anyone who was not dug in. Since it was up to 51 Highland, 15 Scottish and 53 Welsh to move forward, clearing the forest row by row, the field dressing stations were quickly swamped with British wounded.

XXX Corps had set aside no. 35 field dressing station to deal with battle exhaustion casualties and Major John Wishart, the corps psychiatrist, was quickly overwhelmed. In a 1990 interview Dr Wishart read his 1945 reports and relived his memories of dealing with '1,000 odd psychiatric casualties' in the space of two weeks.

Conditions at the field dressing station quickly deteriorated and a second psychiatrist was sent for. Previous policies of sedation, rest and return to unit seemed impossible to implement. Wishart reported that the largest group of patients

consisted of men who had been wounded previously – these made up quite 50 per cent of the total, and in one afternoon ten out of twelve examined fell into this category. Their story was largely the same – wounded in Normandy, evacuated to UK, three to six weeks in hospital, a month in convalescent depot, leave, a month's training, and out here again; generally it was the first hard battle since their return. The wounds had varied in character and severity – one man for example had been four months in hospital with a fractured elbow, but many seemed to have had small, multiple flesh wounds that hardly justified so long a period out of action, and during this period they had enjoyed all the 'glamour' attached to being wounded in the invasion. Some were cases which, had they occurred in civil life, would have had attention from their panel doctor and perhaps a week off from work. I took the opportunity of asking numbers of them what their philosophy had been towards being killed or wounded prior to their injury, and the answer was nearly always the same: 'Never thought about it.' And when asked what their philosophy was now, the general reply was: 'I know what it's like now – I know what can happen.' The majority said they had had no nervous complaints prior to being wounded, but some had complained of headaches and insomnia or battle dreams while in hospital, or reported sick with these symptoms after leaving hospital, and a few had apparently been wounded while behaving in an unstable manner in the field.

The second group was composed of young, immature boys experiencing their first severe action. On the whole, this group looked younger than their age, and quite a number had only recently begun to shave or had never yet had to shave. They experienced grief reactions at the sight of dead and mangled friends, and it was noticeable that many of them were of very poor combatant temperament, and often rather below average in intelligence. Some of them gave a history of 'nerves' in the blitz period. One said he joined the army because he was 'fed up with living in the shelter with his mother and her nerves'.[37]

The first week of combat brought XXX Corps through the Reichswald to the ruins of Cleve. The weather, terrain and determined resistance of the troops of the First Parachute Army had turned Veritable into an experience veterans compared to the horrors of Hill 112 and the Scottish Corridor in Normandy.

The simple truth is that the Allies' 'numerical and material superiority', 'powerful naval forces' and 'tremendous air forces' meant little at the tactical level. What really mattered on the battlefield was that the Allies were seldom able to concentrate enough troops to obtain the kind of force-ratios necessary to overcome a well-equipped, well dug-in, defending army. Air power allowed the Allies freedom to manœuvre behind their own lines but could not change the fundamental realities of the battlefield. The enemy had to be drawn into combat, not simply struck by high explosives. The Germans could not trade space for time, so Allied advances were inevitably met with fierce counter-attacks. It was in these clashes that the army of the Third Reich was destroyed.

Veritable had originally been planned as the northern arm of a giant pincer movement intended to trap and destroy the German forces west of the Rhine. General Bill Simpson's Ninth US Army, built up to a strength of twelve divisions, was to cross the Roer on D-Day+2, advance to the Rhine and then north to meet the British and Canadian forces, but the enemy opened the Roer dams, creating an impassable barrier and postponing Operation Grenade for two weeks.

German reserves had arrived in the Cleve sector in time to prevent XXX Corps from achieving a breakthrough but at a tremendous cost to both sides. XLVII Panzer Corps, with the 116th Panzer Division and 15th Panzer Grenadier Division under command, followed standard German practice by launching a counter-attack over a wide front. By the evening of 12 February the counter-attack had collapsed and the 'fighting power' of the 15th Panzer Grenadier Division was 'broken'.[38]

The actions of XLVII Corps in the counter-attacks of 10–12 February and a similar operation mounted against the Canadians with Panzer Lehr Division a week later require comment. Military historians have frequently criticised Allied battle doctrine, but have been curiously reluctant to examine the German army's operational art in specific battlefield situations. When such studies are undertaken they will show that the German army was destroyed in counter-attacks similar to the ones directed at the British and Canadians in the Rhineland. Immediate counter-attacks are not a very good idea if your opponents are expecting them and if their most effective weapon is *observed* artillery fire.

*

First Canadian Army commander Harry Crerar decided to introduce 2nd Canadian Corps into the battle on XXX Corps' left flank, permitting the exhausted British divisions time to regroup. The 2nd and 3rd Canadian infantry divisions took over a front less than 2,000 yards wide. The 3rd Division, nicknamed the 'water rats', had spent the first week of Veritable in Buffalo LVTs clearing the flooded villages along the Rhine. Now on dry or at least drier ground, they were ordered to seize the wooded ridge near Anne of Cleves's Moyland Castle.

As all historians of military operations know, terrain is one of the most important primary documents available for study and much can be learned from walking the Rhineland battlefields.[39] Unlike the state forests Moyland is a natural mixed wood growing on the broken slopes of a low ridge. On 15 February it was defended by two battalions of Panzer Grenadiers, who were supported by artillery and mortar fire including guns firing from the east bank of the Rhine. Guy Simonds, the Canadian corps commander, wanted Moyland cleared quickly so that the corps' flank would be secure for the second phase of Veritable, now developing as a new set-piece attack code named Blockbuster.[40] The 7th 'Western' Brigade, with one armoured regiment, was assigned to the task. Simonds appears to have believed that the enemy would not risk being cut off and would withdraw to Kalcar, but Hitler had once again ordered that no fortified area could be evacuated without his permission and the First Parachute Army followed his orders.

There are many accounts of the battle of Moyland Wood and all agree that this was one of the most difficult engagements fought by the Canadians in north-west Europe. After two days of combat the Germans withdrew the Panzer Grenadiers and replaced them with a fresh battalion of paratroopers. The killing went on for another twenty-four hours until 7th Brigade's reserve battalion, a new armoured regiment and teams of Wasp flame-throwers, six with each company, burned the enemy out of the woods one square at a time. German losses will never be known but Canadian casualties, 485 killed and wounded and 120 battle exhaustion evacuations,[41] forced Simonds to withdraw 7th Brigade from the battle.

Canadian psychiatrists saw scores of patients who were as burnt out emotionally as the charred bodies in the forest.[42] When the corps psychiatrist Burdett McNeel reviewed the cases he noted that 'at least one-third...give a history of being previously wounded or "exhausted". 3rd Cdn Inf. Division estimate their rate of repeaters as much higher than this figure.'[43] Most of those evacuated as battle exhaustion casualties were too emotionally spent to be returned to unit. Wishart sent less than 20 per cent of the British back and the Canadians, who had always been more cautious, asked fewer than one in ten to face another day's combat.[44]

The struggle for Moyland Wood and the Goch–Calcar road was difficult enough, but Operation Blockbuster was even worse. The 4th Canadian and 11th British Armoured Divisions tried to force their way through the position known as the Hochwald 'lay back'. Armoured regiments bogged down in the mud and tanks brewed up all across the battlefield. The 11th Armoured suffered 446 non-fatal casualties in 72 hours, 99 of them due to battle exhaustion. The 4th Canadian must have held its cases within the formations and then evacuated them as 'sick', for it reported the same number of exhaustion cases out of 1,100 'wounded'.[45]

The Battle of the Rhineland was the last great attritional battle fought by the 21st Army Group in the Second World War. It was fought at the sharp end by ordinary young men from Britain and Canada who had no ambition to conquer the world or even to save it for democracy. By 1945 all that was left was some personal pride and the feeling that you shouldn't be the one to quit and let your buddies down. The enemy, now fighting on their own soil seemed to the Allied soldiers to be implacable bloody-minded bastards, all too ready to die for an evil cause in a war which was already lost; or to surrender minutes after they had killed your best friend and faced their own extinction. The politeness of post-war discourse among military historians should not be allowed to obscure the attitudes of 1945 or we will fail to understand what war does to soldiers.

British soldiers were particularly bitter about their lot in life. The 21st Army Group's adviser in psychiatry insisted that 'the prospect of further service in south-east Asia is now accepted by troops as in the main inevitable', but he urged the army to take steps to 'debunk the horror picture many still have of the Far East'.[46] This was official optimism. The initials BLA, for British Liberation Army, now stood for 'Burma Looms Ahead' and the prospect only pleased those who wished to continue as professional soldiers. The Canadians knew that they need not serve in the Pacific unless they volunteered to go, so their future seemed a bit brighter, if only the enemy would face reality.

In February 1995 newspapers in Canada and in the UK marked the anniversary of the bombing of Dresden. In almost every article reference was made to the 'fact' that the raid was particularly cruel as the war was almost over. Dresden was bombed on the eve of Moyland Wood, four long bloody weeks before the last surviving German soldier retreated to the east bank of the Rhine. There were no newspaper stories about the agony of the Rhineland battle. Our collective memory of the war has yet to find room for stories of the endurance and survival of ordinary soldiers.

Part Four
Seven Armies in Europe

'No taste for the fight'?: French Combat Performance in 1940 and the Politics of the Fall of France

Martin S. Alexander

The Battle of France had not run half its course before accusations of cowardice in the face of the enemy and defeatism among the rank and file of the French army, along with 'fifth column' treachery, began to fly.[1] More than a month before the Franco-German Armistice of 22 June, French leaders were busily preparing alibis and identifying scapegoats.

Shockingly, it all emanated from the very top: with the notorious, self-serving report of 18 May 1940 written by General Maurice Gamelin, supreme commander of French land forces, which alleged that:

> The French soldier, yesterday's citizen, did not believe in the war. His curiosity did not extend beyond the horizon of his factory, office or field. Disposed to criticise ceaselessly all those having the least authority, encouraged in the name of civilisation to enjoy a soft daily life, today's serviceman did not receive the moral and patriotic education during the years between the wars which would have prepared him for the drama in which the nation's destiny will be played out...The regrettable instances of looting of which our troops have been guilty at numerous points on the front offer manifest proof of...this indiscipline...Too many failures to do their duty in battle have occurred, permitting the enemy to exploit local successes, to turn the flank of the most gallant defenders, to wreck the execution of the leaders' concept and know-how. The rupture of our dispositions has too often been the result of an every-man-for-himself attitude at key points, local at first, then quasi-general.[2]

In hindsight, there is something grotesque in Gamelin spending his final days on active service drawing up an indictment of the common French soldier. The report was intended for Edouard Daladier, prime minister from

April 1938 to 21 March 1940, also minister for national defence and war, and thus politically responsible for the army and for French defence and rearmament, since June 1936. But it was overtaken by the pace of politics even as it was conveyed from Gamelin's headquarters at the Château de Vincennes to the Ministry for National Defence on the rue Saint-Dominique. When the document reached the ministry, Paul Reynaud, the head of the government, had used the political flap triggered by the German breakthrough on the Meuse to move Daladier to the foreign ministry, and assume the direction of the defence department himself.[3]

The episode well illustrated how seriously civil–military quarrels and political feuds prevented the creation of a well-oiled, efficient French war administration.[4] Indeed, conflict between French soldiers and statesmen smouldered before 1939, impeding good Franco-British relations and hindering strategic planning.[5] Reynaud's assumption of the powers of the minister for national defence and war signified Daladier's eclipse.[6] As a result, the remaining time in command of Gamelin, who had been Daladier's military protégé since 1936, was to be measured in hours. On 19 May the general was tersely informed by Reynaud that his services were no longer required.[7] Twenty-four hours later, in a fateful gesture, Reynaud installed Gamelin's erstwhile predecessor, the 73-year-old General Maxime Weygand, in his place.

The recriminations of ambitious and jealous generals can form, however, no more than a starting point if we are to re-evaluate the combat experiences of French troops in 1940. Gamelin's charge-sheet against his own soldiers accused them of 'too many failures to do their duty' and 'a quasi-general attitude of every-man-for-himself'. The army had assumed 'the outward form, in some instances the veneer' of strength and efficiency. But experience revealed that the essence was lacking 'particularly in certain major reservist formations, despite eight months in a state of war'.[8] Gamelin's comment underlines the need for any serious study of the French army's quality in 1940 to ask whether it used the Phoney War respite to good effect.[9]

But it also highlights the diversity, the heterogeneity, within Gamelin's forces. French units possessed highly uneven levels of modern armaments, training, leadership and morale. On the eve of the advance into Belgium and Holland as part of General Henri Giraud's Seventh Army, Captain Pierre Dunoyer de Segonzac felt great confidence in his 4th Armoured Cuirassiers. This was a pre-war 'active' unit, equipped with the fast, well-armed SOMUA S35 tank. But, he noted pensively, 'I knew that very few of our regiments had such good war material or level of training.'[10] The Canadian historian John Cairns, writing in the 1950s, acknowledged that some French units could not be proud of their conduct under fire. But there was, he added, another side to the coin. The

French Army [of 1940] was nothing if not diverse. Not every division belonged to the reservist, undertrained and underequipped Series B formations, or to the 'apartment janitors' of the Maginot Line... 'trailing about the corridors in slippers, playing innumerable games of cards under the silent guns, yawning from morning to night'. The whole French Army had not simply waited to be captured or bolted before the enemy.[11]

As Pétain's Vichy régime replaced the Third Republic, most senior political and military figures welcomed Gamelin's shift of the explanation for collapse away from the army high command. Indeed, French leaders breathed a sigh of relief as attention was diverted on to the performance of rank-and-file troops who had actually faced the enemy.

The war of words surrounding 1940 has been more protracted than any imaginable 1914–18-style trench stalemate. Yet among the many battles fought in the courts and parliamentary commissions of enquiry, among the volleys of memoirs and apologias, no campaign was more vigorously waged than the one by senior officers to deflect all criticism away from their ilk – with the sole exception of Gamelin. Responsibility was to be displaced on to civilian society and the pacifists, for producing an inadequate, irresolute and unpatriotic soldiery – and on to the parliamentary régime, the parties, for producing political leadership that they condemned as a contradiction in terms. The commanders, many soon to be Vichy ministers, found no shortage of *responsables*. And of these, strange to say, none save Gamelin was from the top ranks of the armed forces.[12] Vichy's cabinet of 1940–41 contained numerous generals and admirals. In a stream of pamphlets, radio broadcasts and speeches, these ascribed the defeat to a clutch of erstwhile Republican politicians, functionaries and officers, along with such predictable targets as factory workers, trade union militants, socialists and the Parti Communiste Français.[13]

But former parliamentarians were not Vichy's only scapegoats. Though most senior commanders (except for Gamelin and Corap) escaped reproach, rank-and-file civilians-turned-soldier, the reservists especially, did not. The army blamed some of its own for what had gone wrong in 1940. Vichyite generals rounded on Republican ones. And both sorts rounded on the other ranks, the common 'citizen-soldier'. What surfaced was an unedifying 'them and us' syndrome: 1940 left not just a defeated army, but a deeply divided one.[14]

Defeated generals and untried admirals were, ironically, among the biggest political winners in 1940; they became members of France's new ruling class.[15] They helped make the rules for the first round of enquiries into what had happened. The chief suspects were 'Republican' commanders, and the 'defective' citizen-soldiers of the army's rank and file. Mireau and Broulard, the murderously cynical generals of Humphrey Cobb's novel and Stanley Kubrick's film of 1957, *Paths of Glory*, allowed their ambitions to prevail over natural justice and professional integrity by court-martialling their surviving troops, after the defeat of an assault on a German strongpoint known as 'The Ant Hill'. In reality, too, French generals between 1940 and 1942 were ready to blame failure on their men. As in the fictionalised account of events in 1916, politics and personal careerism predominated in the analysis of military failure.

Amongst Vichy's military villains, the role of principal scapegoat was reserved for Gamelin. At the Riom trial in 1942, Pétain's court sought to prove that he treasonably neglected French national defences before the war. Another conveniently available man on the spot was General André Corap. The luckless commander of the exposed and ill-supported Ninth Army behind the Belgian Ardennes, Corap faced a desperate situation the moment battle was joined in May 1940. His forces, mostly infantry and lightly mechanised cavalry, were hopelessly mis-matched against the eruption of the panzer divisions of Rommel, Kempf and Kuntzen across the Meuse at Dinant, Houx and Monthermé.[16]

On Corap's right wing in 1940 (literally, as it would turn out) was General Charles Huntziger, commanding Second Army. He, too, was a 'man on the spot'. It was divisions of the Series-B reserve under Huntziger's authority, General Lafontaine's 55th and General Baudet's 71st Infantry, badly armed and ill trained, with only two regular officers per regiment, which collapsed under the more southerly armoured onslaught launched at Sedan by Guderian's XIXth Panzer Corps. The posting of the two poorest divisions in his entire army at the crucial Sedan hinge was Huntziger's decision, made without pressure from the chain of command above him: General Gaston Billotte (1st Army Group), General Alphonse Georges (the theatre commander), or Gamelin. Visited in October 1939 by the former French military attaché to Warsaw, who had come to report on the alarming bravura of the panzers' performance in the east, Huntziger smugly retorted: 'Poland is Poland...here, we're in France, General.'[17] Later in 1940, at Vichy, Pétain would show no trace of professional embarrassment in appointing Huntziger minister for war.

When re-examining the French soldier's combat effectiveness and 'spirit' in 1940, it seems worthwhile to ask what his German adversary made of him. Hitler was soon informed that the 'whole French Army had not simply waited to be captured or bolted before the enemy'.[18] The Führer wrote on 25 May to Mussolini (who would not enter the war until 10 June) that:

> Very marked differences become apparent in the French when their military ability is evaluated. There are very bad units side by side with excellent ones. On the whole, difference in quality between the active and nonactive divisions is extraordinarily noticeable. Many of the active units have fought desperately, the reserve units are for the most part obviously not equal to the impact of battle on morale.[19]

Though Pétain, Weygand and even Gamelin lacked the integrity to resist claiming that their soldiers in 1940 fought less well than those of 1914, plentiful German evidence suggests otherwise. Wehrmacht witnesses testify to the ferocity of the fighting.

*

Methodologically, in seeking to assess French 'taste for the fight', it is crucial to note the varied composition and quality of major formations. The principal active units (20 infantry divisions based in metropolitan France in peacetime, 13 of which were fully motorised, 3 mechanised cavalry divisions, 4 mainly-horsed cavalry divisions, reinforced since September 1939 by 5 North African infantry divisions and one colonial infantry division) were well trained, equipped and led. Less capable, with far fewer career officers and NCOs, were the 20 Series-A reserve divisions. Weaker still were the Series-B reserves (where the rank-and-file were typically 35 years old or more, and which came last in the pecking order for modern weaponry). We should surely acknowledge spatial and temporal divergences: the theatre of north-eastern France and Belgium was a large one, and the Battle of France lasted for six weeks. There were marked qualitative differences between French corps assigned to one part of the front and those assigned to another.[20] Even so, glib generalisations do not pass muster: German accounts reveal instances of ferociously brave French defence in the campaign's final days.

The remainder of this essay seeks to probe French battlefield performance, by focusing on action at 'the sharp end' in three different phases of the 1940 campaign and on three different sectors. A first case, illustrated by German sources, comes from the fight on the Meuse for the river crossings, from 13 to 16 May. The second is the encounter battles on the Aisne and Sambre during the westward advance of the Germans between 18 and 25 May. The third comes from the Wehrmacht's resumption of a general offensive southwards, to force the French lines on the Somme and Seine and finish the campaign, from 5 June to the Armistice seventeen days later.

Most general histories which mention the breaching of the Meuse suggest that the Germans sliced through the French defences like hot knives through butter. At Sedan the river crossing gave Guderian's XIXth Panzer Corps little difficulty.[21] But the Germans met serious problems at their two more northerly bridgeheads. At Monthermé, Kempf's 6th Panzer Division and Kuntzen's 8th Panzer Division (forming Reinhardt's XLI Panzer Corps), initially made no headway against a resilient French defence. This action assumed a fateful significance: for it led Georges, the French commander-in-chief of the north-eastern theatre of operations, to disregard the danger posed by Reinhardt. No orders were issued to any French reserves to move up in support of the units of Corap's Ninth Army resisting on the river line at Monthermé. As a result, a catastrophe began to unfold when the first-line defences did crumble under the weight of vastly superior attacking forces. For there was nothing behind the French front to mount a counter-attack or even impede the break-out and exploitation by Reinhardt's tanks and panzer grenadiers.[22]

At Dinant, the Meuse crossing by General Erwin Rommel's 7th Panzer Division (part of XXXIX Panzer Corps) was also fiercely opposed by the French – and, once again, by units of the often-maligned Ninth Army. Accounts of the action on this sector are readily accessible from Rommel himself, and from one of his panzer company commanders, Captain (later Colonel) Hans von Luck.[23] At the first attempts to push the division's two infantry regiments (6th and 7th Panzer Grenadier) across the river in inflatable boats 'all hell broke loose'. French heavy artillery was skilfully sited and accurately ranged on the river. Its shells, supported by small-arms fire, stopped the Germans at once. Well screened on the wooded slopes west of the Meuse, these French defenders were impervious to counter-fire from the German tanks and field guns.

A crossing was won only when Rommel assumed personal command of the 2nd Battalion of 7th Panzer Grenadier Regiment. He ordered houses up-wind of the crossing point to be set alight to create a smoke-screen. With this second wave Rommel got across and, recounts Luck, 'it became possible to form a small bridgehead in the teeth of the French, who defended themselves bravely'. During the night of 13–14 May the divisional engineers ferried tanks over to the toehold the infantry had established. The attack resumed on 14 May. Yet even with their defensive river line now breached, the French did not flee. Rommel was in the thick of the action once more, as Luck recalled: 'His command tank was hit and the driver put it in a ditch. Rommel was slightly wounded, but hurried forward on foot – in the midst of enemy fire ... It made a strong impression on all the officers and men.'[24]

Rommel's division now broke out from its bridgehead and moved westwards into open country. Its tank regiment, supported by a special engineer unit, forced a breach in the French line a mile and a half deep. On the night of 16–17 May units of 7th Panzer advanced through Avesnes and on 17 May were able to seize undamaged bridges across the River Sambre. At this juncture the Germans claimed the rewards due to the side with both the initiative and the element of surprise. One of Rommel's 'unorthodox orders' emphasised the enemy's confusion and instructed his units to 'take advantage of it'. The French on the Sambre 'were caught completely unawares by our impetuous advance and retreated, to some extent with signs of disbandment. *La guerre est finie, je m'en fous*, we heard, shouted by some French soldiers.'[25] From this point on, Luck's account of the advance across northern France assumes a more familiar cast – not so much a battle as the mopping up of a beaten foe, one degenerating at times into an unmilitary rabble. On 18 and 19 May, pushing west between Cambrai and the St Quentin canal, the division's reconnaissance battalion and Luck's tanks were 'involved again and again with the flood of retreating French soldiers, who in their panic mingled to a large extent with the civilian population'.[26] These scenes are confirmed by the diary of René Balbaud, a French infantry reservist. His unit had briefly moved into Belgium near Maubeuge, but was now north of Rommel's advance and retreating from Valenciennes, past Lille to Poperinghe, eventually to become one of the French soldiers evacuated at Dunkirk. Digging in for a temporary stand on 16 May, he noted the 'Never-ending columns of refugees winding their way along – I would never have believed so many people lived in Belgium!'[27]

Meanwhile French armoured cavalry had been defending resiliently in the battle at the Gembloux Gap on the Belgian plain. There the tanks, armoured cars and mechanised dragoons of General René Prioux's mechanised cavalry corps, the 2nd and 3rd DLMs (Divisions Légères Mécanisées), clashed with General Erich Hoepner's XVI Panzer Corps (3rd and 4th Panzer Divisions). These formations (3rd DLM having only formed in February 1940) fought well against experienced enemies (3rd and 4th Panzer being formed in 1936 and 1938 respectively). And the Germans were stopped in their tracks for two days.[28]

However, the second stage of the campaign offered a qualitatively different experience of war. It saw hastily assembled French formations striving to improvise defences and stem the retreat. Now French troops and their officers faced a truly terrifying test of military fortitude, leadership and morale. For everything was unexpected: from the need to fight in such chaotic conditions at all (the very antithesis of French 'methodical battle' doctrine, the *bataille conduite*) to the locations themselves, the rivers, canals, hill-slopes and villages of Picardy, Artois and Champagne that pre-war planning had intended to be far in the battle's rear.

From the fighting in the open country of northern France, we will examine two actions. The first is the destruction of Captain Dunoyer de Segonzac's squadron of the 4th Armoured Cuirassier Regiment on the edge of the Mormal Forest on 18 May. The 4th Cuirassiers belonged to the 1st DLM, which had been despatched towards Breda on 10 May with Giraud's Seventh Army, in an effort to link with the Dutch.[29] The 1st DLM was perhaps the best, certainly the longest-established, mobile warfare unit in the French army. It had been assembled as early as 1933. General Jean Flavigny, its pre-war commander, was one of France's acknowledged armoured-war theorists. The division enjoyed unofficial 'élite' status and had frequently staged exercises to show off the French army's best modern material and tactics to foreign missions.[30] What the 1st DLM might have accomplished in any reasonable combat situation in 1940 may be judged by the fact that further north the 2nd and 3rd DLMs, despite poor deployment by Prioux, their corps commander, gave good accounts of themselves and handed Hoepner's panzers a 'drubbing'.[31]

But reasonable combat situations were few and far between in the experience of Allied soldiers in 1940. Indeed the conditions in which the 1st DLM, now led by General Picard, engaged the enemy were disastrous. Hastily called to turn about, the regiments and squadrons were separated from one another and from their integral divisional artillery, motorised infantry (*dragons portés*), signals units and higher command. They were ordered south one after another, as the French command sought to improvise a response to the German break-out across the Meuse. On 15 May Prioux learned that the 'sorely tried' 1st DLM, 'still without being placed under my orders', was 'coming down from Holland into the First Army's zone'. Still the higher command failed to concentrate its armour into a really powerful counter-attack force.[32] Dunoyer de Segonzac's squadron, detached from its regiment, made the move on flat-bed railway wagons from Duffel, south-east of Antwerp on 13 May, to Charleroi. It spent 14 and 15 May on the Charleroi canal, awaiting a panzer attack. None came. On the 17th fresh orders moved Segonzac's squadron again at maximum speed, this time on its own tracks and tyres, westwards up the Sambre valley, to Le Quesnoy, a small town west of the Mormal Forest. That day it made what amounted to a forced march, covering 100 kilometres along roads clogged with refugees. At Le Quesnoy, Segonzac discovered the rest of the 4th Cuirassiers, last seen in Holland. The unit was ordered to seal the 'enormous breach' which had opened up between the southern flank of French First Army (General Georges Blanchard) and the retreating remnants of Corap's Ninth. The mood was sombre: the cuirassiers had been pushed from pillar to post for a week. Officers and men were sleepless and exhausted, tanks worryingly low on fuel – and all 'without our even having fired a shot'.[33] Segonzac received more instructions: push eastwards through the Mormal Forest and take up defensive positions at Berlaimont. Given a support company of mechanised infantry, Segonzac had his command tank and 4 platoons, each of 5 tanks, a total of 21 SOMUAs. As they passed through the forest a demoralising sight greeted them: the hulks of abandoned Chars B1 bis, the most powerful tank on either side in the 1940 campaign. These had simply run out of fuel. It was symptomatic of the logistical chaos which plagued all Allied efforts to regroup and reform a solid defence.

Once in action, however, Segonzac's force performed well: its superior fire-power quickly drove off the German advance guard. By nightfall on 17th, Berlaimont had been secured. But three French tanks had been destroyed. The fuel reserves of the remainder were low once again. Segonzac regrouped in a forest clearing, and personally set off by motorcycle to seek fuel. After several hours he found and commandeered a bowser, and also re-established contact with regimental headquarters. The latter ordered him to retire to the west of the forest, occupy Jolimetz, and 'hold it to the last'.[34]

The village's inhabitants had already fled. Now reinforced by a company of Moroccan *tirailleurs*, Segonzac had 'time enough to organise a well-thought-out defence'. The Germans appeared at 9 a.m. on 18 May, launching a combined-arms assault with tanks, artillery and infantry. The fight was ferocious, with the French mounting several counter-attacks in the early afternoon. But the German pressure was unrelenting, pushing in the French positions so that by evening the fighting was in the streets. 'It was at dusk on this 18th of May' that the indomitable Segonzac finally 'felt myself beaten'. Only about ten men survived; all but one of his tanks was destroyed.

> This fine squadron, of which I was justly so proud, whose men were brave and good soldiers, whose tanks were modern and well deployed, had been wiped out. I felt stupefied, like a long-distance runner who, believing he's well prepared and knowing he's good, sees his rival fly past him at the approach of the winning-tape.[35]

From this account, published in 1971, two poignant aspects stand out. One was Segonzac's admission that, thirty years after the shattering experience of seeing a squadron that he had led, trained and honed for two years annihilated around him in a single day's fighting, he still had no idea of the enemy's identity or strength. It was a mystery what manner or size of opponent had destroyed him in this savage little 'battle without a name' on 18 May 1940. 'I never had any more of a picture of the battle at Le Quesnoy as a whole', he confessed, 'than did Fabrice del Dongo at Waterloo.' The other poignancy was Segonzac's enduring uncertainty as to whether his unit had deserved to be destroyed that day. His memoir records his struggle over three decades, endlessly re-living the events, to come to terms with the disaster that had so swiftly, so brutally, overwhelmed his force. Thirty years on, he still asked himself whether they had been beaten fairly and squarely, honourably, by a superior opponent. For otherwise, how could such a well-led, well-equipped, well-trained and courageous unit have been simply swept away? 'Plainly we encountered panzers intoxicated by their stunning coup in the Ardennes, but did they enjoy the sort of qualitative and quantitative superiority to account for our defeat?'[36] Three of his officers and many of his men had been killed outright; others were wounded.[37] 'But from the very fact that they measured up so splendidly to what was required of them, our defeat seemed all the more humiliating. And so the doubts crept over me; they were doubts about me and doubts about my country.'[38] Dunoyer de Segonzac would, of course, found the leadership training school, the Ecole des Cadres at Uriage, be dismissed by Pierre Laval's Vichy ultra-collaborators in December 1942, go underground and escape through Spain to join the Gaullists at Algiers. He, at least, was one of the fortunate French junior officers. His 'soldier's experience of war' in 1939–45 would end a great deal more happily than it began, for in 1944–5 he fought in the liberation of Alsace, commanding the 12th Dragoon Regiment in Marshal Jean de Lattre de Tassigny's Free French First Army.[39]

More typical, however, was the long imprisonment in German prisoner-of-war camps which defeat brought for French veterans of 1940. Such would be the fate of Lieutenant Lucien Carron, commander of a section in the 3rd Battalion of the 6th Infantry Regiment, 44th Infantry Division. Stationed at Neuwiller in lower Alsace when the German offensive began, the 44th was redeployed by rail and road to Rheims between 16 and 18 May. It then moved forward by motor transport and forced marches to the Aisne–Oise canal, deploying from Pontavert to Moussy-Verneuil, between Soissons and Neufchâtel. Carron's war in 1940 was a short series of shapeless, unplanned, frightening actions against the panzer drive to the coast, as his units became part of the defensive flank Gamelin sought to improvise along the southern edge of the panzer corridor.[40]

The 28-year-old Carron's section consisted of 25 non-commissioned officers and men. All were reservists, aged between 26 and 40. Only one had ever been under fire. Yet, as his men dug in to defend the bridge over the Aisne at Oeuilly, Carron was confident, 'really sure that whatever happened they would not give way. Their looks, when they met mine, told me: they would hold.' Indeed, in spite of the 3rd Battalion having only 400 men, and these forced to spread themselves along a frontage of over eight kilometres, units such as Carron's 'did hold, and a lot longer than had ever seemed possible'.[41]

Soldiers such as these, with their courage and tenacity, did not let France down. What undid the stoutest resistance was the way Allied generals lost their grip on the campaign. Some officers had seen this happening, and understood all too well its significance. 'It was more and more apparent', noted Prioux on 17 May, 'that... the enemy was imposing his will on us and that we had lost the operational initiative.'[42]

As this essay has tried to suggest, Gamelin's indictment of the rank-and-file combatants on 18 May 1940, endorsed afterwards by the Pétainists, was disingenuous, cowardly and well-nigh obscene. France's soldiers did not let the generals down: rather the reverse. What was fatal was the speed and daring of the German advance. The Germans got 'inside the Boyd Loop' or decision cycle of French corps, army and army-group commanders. It forced them into hasty improvisations, uncoordinated counter-attacks, and obliged them to deploy units where their frontage would unavoidably be too long for success-ful defence. Trained to expect positional warfare, entire French divisions were confronted by the disintegration of their logistical support arrangements. Consequently, stalwart resistance repeatedly fizzled out through the exhaustion of ammunition supplies, or fuel stocks, or both.[43]

Compounding these problems was the battle's unpredictability and thus its uncontrollability. Many French commanders were bewildered. Some were paralysed. The British chief of the imperial general staff, Field-Marshal Sir Edmund 'Tiny' Ironside (a massively built man, 6 foot 4 inches tall), visiting the French commanders in the north on 20 May,

> found Billotte and Blanchard at Lens (First Army), all in a state of complete depression. No plan, no thought of a plan...Defeated at the head without casualties. *Très fatigués* and nothing doing. I lost my temper and shook Billotte by the button of his tunic. The man is completely beaten.[44]

Others too, including Georges, suffered emotional and physical collapse.[45]

The fog of war descended, blinding commanders. Brigades, divisions and even entire corps blundered about without orders, ignorant of the situation on their flanks. Headquarters were overrun, or were forced on to refugee-clogged roads to avoid capture. The battle acquired a fluidity from 17 May to 4 June that had never been expected, still less planned for. In its wake came a disintegration of command, control and communications. This disintegration was all the more complete because of the dearth of powerful two-way wireless sets, and because headquarters relied on telephone land-lines. These were lost to service as the German advance forced staffs to flee from their châteaux. Exacerbating the confusion and breakdown of command, some generals were captured – such as Giraud, just after he was reassigned from his own army (the Seventh) to take over Corap's retreating Ninth. Fifteen were dismissed. Others suffered accidents – including Billotte, who died in a messy, unmilitary fashion, somehow emblematic of the Allied débâcle, after his car crashed on a road thronged with refugees.

This second phase was fought largely as a myriad of local battles, and purely tactical defensive actions. Among the chaos there were some notable counter-attacks that demonstrated the bravery and competence of parts of the French army. Two involved tanks of Colonel Charles de Gaulle's newly formed 4th DCR (Division Cuirassée de Réserve) at Moncornet on 17 May and north of Laon on 19–20 May, which for three days running pushed back the Germans' southern flank-guard. 'I've just got out of a long and hard scrap which went *very well* for me,' wrote de Gaulle to his wife on 21 May; 'it was a full-scale fight and – a rare thing since this war started – it went very happily for us.'[46] At one stage de Gaulle had even bottled up Guderian's command troop in a wood, coming within a whisker of capturing the celebrated panzer general.[47] Another aggressive counter-thrust which seriously discomfited the Germans was mounted near Rethel by de Lattre de Tassigny's 14th Infantry Division (admittedly a pre-war 'active' formation).[48]

Impressive as operations, these French manœuvres were too disconnected and too late to have any overall strategic impact. The campaign on the Allied side was out of control, units going into action in penny-packets. Battalions fought cut off from regimental headquarters; regiments were isolated from divisional command; divisions were lost by their army staffs, then sometimes rediscovered, broken up, reassigned, and finally commandeered by the next senior general to appear at their headquarters.[49]

Upon such a disastrous canvas, it is perhaps not strange that few scholars have troubled to sketch the operational and tactical setbacks that the French did inflict on the Wehrmacht. General histories of 1940 have given still less credit to the fierce French resistance that met Wehrmacht formations *after* the Germans resumed their offensive towards the south on 5 June. This was the final, vain, stage of the Allied defence, mounted along the so-called 'Weygand Line'. This position did, it is true, give the Allies a continuous front once again – for the first time since 15 May. But it lacked solidity. It had been hastily improvised along the lower Seine, the Somme, the Aisne and eastwards towards the extremity of the Maginot Line. It had become not a matter of whether France would be defeated – but only a question of when, and on what terms. All the same, the Germans did not find the going easy, despite their now-decisive superiority in men and material.

To defend the entire front from Normandy to Switzerland, Weygand had the equivalent of about 60 battleworthy French divisions remaining (compared with over 90 that started the campaign). These were supported by Major-General Victor Fortune's 51st Highland Division, Major-General Roger Evans's 1st Armoured Division and other units of the so-called 'second BEF' – a corps under Lieutenant-General Sir Alan Brooke hurriedly impro-vised on the basis of the lines-of-communication troops left in Brittany and Normandy after Dunkirk. The rest of the Allied forces available on 10 May had now been lost. Once again, however, German accounts modify the popular perception that this phase of the campaign was little more than a mopping-up operation admixed with some early-summer tourism.

Recounting the resumption of general offensive operations after 5 June, Major-General F. W. Von Mellenthin writes that the German attack progressed well in Picardy and the lower Somme, especially for Guderian's panzer corps. This rapidly advanced into the lower Seine basin, threatening Rouen.[50] The elation felt by the German mechanised troops is confirmed by Luck, whom Rommel had now promoted to lead the reconnaissance battalion of 7th Panzer Division:

> On 5 and 6 June we advanced in 'open battle order' across the flat terrain, avoiding the main roads, along which the civilian population and retreating elements of the French Tenth Army were moving south. We reached the Somme and took possession of its bridges, surprisingly intact...On the far side of the Somme we came upon resistance, the Weygand Line... and...we came under heavy artillery fire.[51]

On 7 June Rommel's units reached the northern bank of the Seine near Rouen. Though 'huge clouds of black smoke' hung over the city, showing that 'the Luftwaffe had done quite a job', the French had remained disciplined, ensuring that every bridge across the river was demolished. This dashed German hopes of a rapid crossing and a drive into the Cotentin peninsula. Instead 7th Panzer was ordered to swing west and advance to the Channel coast north of Le Havre. Yet between dawn and dusk on 9 June, Luck's panzer reconnaissance battalion could advance only five kilometres. They encountered 'strong anti-tank defenses, against which we had nothing to throw in'. Leading the bulk of his division on the small port of St Valéry-en-Caux, where German intelligence had detected British units awaiting evacuation, Rommel also encountered 'stiff resistance'.[52]

Luck's battalion was detached to the south, where he bluffed the mayor and the French garrison commander at Fécamp into surrendering the port to avoid the destruction of the town and its famous monastery. The grateful abbot rashly said he wished to offer Luck's men a bottle of Benedictine each, for sparing the cellars. 'The abbot paled when I told him the strength of my unit, 1,100 men. But he kept his word.'[53] The dénouement of this part of the battle is well known, Rommel capturing the entire Highland Division and Fortune, its singularly misnamed commander, at St Valéry.[54] But what is notable here is that the German advance, with only some forty kilometres to cover to the coast, was no sightseeing jaunt – rather a hard-pounding slog in the teeth of Franco-British resistance which was quite at odds with the strategically hopeless situation of the Allies.

In the battle to smash through the Weygand Line further east, the German offensive progressed even less smoothly – and this not because of the lack of any revivifying *consommation* to compare with that enjoyed by Luck and his men. Panzergruppe Kleist, recounts Mellenthin, 'tried in vain to break out of the bridgeheads at Amiens and Péronne; the French troops in this sector fought with extreme stubbornness and inflicted considerable losses'.[55] On the night of 9–10 June, Panzercorps Schmidt crossed the Aisne, west of Rethel.

This was, however, where General de Lattre's 14th Infantry Division, supported by several reconstituted French mechanised units, had moved up to launch their counter-attack:

> There was fierce fighting on 10 June; the country was difficult with numerous villages and woods which were strongly held by the French...On the afternoon of the 10th French reserves, including a newly-formed armoured division, counter-attacked from Juniville against the flank of our panzers, and were driven off after a tank battle lasting two hours...On 11 June Reinhardt beat off several counter-attacks by French armoured and mechanized brigades.[56]

Mellenthin concluded from this phase of the campaign that Guderian's success and Kleist's failure resulted from a difference in methods. 'The attacks of the latter from the Amiens and Péronne bridgeheads demonstrate that it is quite useless to throw armour against well-prepared defensive positions, manned by an enemy who expects an attack and is determined to repulse it.'[57] Mellenthin's own formation (General Meyer-Rabingen's 197th Infantry Division) formed part of the German First Army. It saw limited action in Lorraine, attacking part of the Maginot Line at Puttlinger, south of Saarbrücken on 14 June. After piercing the French static positions, the division force-marched some thirty-five miles a day. Assigned Donon, the highest summit in the northern Vosges, as its objective, it wheeled left into the rear of the Maginot Line. By 21 June the division had passed through the front of another one 'which had been pinned down with heavy losses'. French defenders 'had blocked the roads by felling trees' in the densely wooded hills, 'and his artillery, snipers, and machine-guns took full advantage of the excellent cover'. The 197th was, by nightfall on 22 June, a mile from the peak of Donon. A day of 'slow, bitter fighting' ended when Mellenthin received a telephone call from the corps chief of staff, Colonel Hans Speidel, conveying the news of the Armistice.[58]

Certainly some French troops cracked under pressure – but so did other Allied units, Dutch, Belgian and even British. And so, on occasion, did the Germans. Moreover, 'French reservists [as Jeffery Gunsburg has commented] broke under their baptism of fire in 1914 too. The reservists at Sedan faced overwhelming superiority in air and ground firepower; it was not surprising that many panicked, although some fought to the end in their casemates.' General Jean Etcheberrigaray's Series-B 53rd infantry division, part of Corap's Ninth Army, gave a respectable account of itself in unpropitious circumstances when fighting in open terrain against Guderian's panzers west of the Meuse.[59]

Moreover the German offensive's second phase, on the Somme on 5 June and on the Aisne four days later, saw fierce French resistance. 'There's a rumour', noted Georges Sadoul as his unit briefly rested at Sully-sur-Loire on 16 June, 'that the division is to defend the Loire, like it defended the Somme and the Oise. But with what? Our division is practically annihilated. Our two infantry regiments are down to the strength of a company...We've even lost our general...and many of our lorries are missing.'[60] This had been the French soldiers' last stand. As a recent history of Franco-British relations in this era of *mésententes* has acknowledged:

> for a few days the spirit of the French Army suddenly recovered...morale was high. The German combination of tanks and dive-bombers which had previously played havoc was now faced with determination and some success. It was a remarkable recovery, too often omitted from accounts of the campaign; and it was achieved almost without help from the British. This last stand came from within the French Army.[61]

A measure of the fighting's ferocity comes from the losses. In a communiqué of 4 June 1940, the Germans listed casualties for 10 May to 3 June as 10,000 killed, 42,500 wounded, and 8,500 missing. The communiqué of 2 July put casualties through to 25 June as 27,000 killed, 111,000 wounded and 18,500 missing. Hitler's men did not find France a walk-over in 1940. Most of these losses fell among the German mechanised and motorised units, especially in the early stages of the campaign. Each division in Hoth's XV Panzer Corps (Hartlieb's 5th Panzer and Rommel's 7th Panzer) lost some 50 officers and 1,500 men.[62] (Rommel himself admitted to his division suffering 1,600 casualties, and claimed 97,648 prisoners-of-war captured.)[63] Furthermore, each division had seen about 30 per cent of its armoured vehicles knocked out. Heavy losses in weapons resulted, in this part of the theatre, from frequent combat with Allied tanks.[64]

Allied casualties were heavier still. This has been attributed, in part, to the fact that German mechanised units, light on manpower, were frequently pitted against Allied infantry. The French formed by far the largest national component of the Allied coalition. They bore the brunt of the fighting across the theatre as a whole. Their casualties were greatest in absolute terms: some 120,000 killed, 250,000 wounded, and 1,500,000 prisoners during the six weeks. Many more French troops became German prisoners-of-war after the Armistice.[65] French losses during the battle were higher, proportionately, than those for a comparable period in the fighting at Verdun in 1916.[66]

*

In conclusion, one must mention the vexatious matter of comparisons between the performance of the French in 1914 and 1940. It is tempting to believe that because France survived the Schlieffen Plan, but not *Sichelschnitt*, the French troops must have 'gone rotten' by 1940. Yet a comparison should emphasise how narrow is the divide separating triumph from disaster in military matters. In 1914 France was pushed to the very edge of defeat even by Moltke's badly adapted and poorly directed perversion of Schlieffen's Plan. French strategy was at least as ill-judged and amateurishly executed in 1914 as in 1940.[67] The headlong rush into Lorraine embodied in Plan XVII was arguably rasher than even the Breda Variant of April 1940. Joffre made almost every mistake in the book – and a good few that were not. His field commanders were found sorely wanting – so much so that his dismissal of 140 officers in the rank of brigadier and above in the first month of war became known as the 'mass graveyard of the generals'.[68]

Yet Joffre was fortunate – certainly more fortunate than Gamelin (who, in 1914, had been Joffre's chief of operations). For Joffre's war moved at the speed of horses and foot soldiers. Tactical setbacks could be, and were, put right by means of more rapid redeployments by rail from quieter parts of the front. The creation of General Maunoury's Sixth Army east of Paris on the eve of the crucial Battle of the Marne was a case in point.[69]

A generation later, however, the French faced a German mechanised and motorised spearhead able to cover 100 kilometres or more in a day. By 1940 railway redeployment was too ponderous, too time-consuming, in the face of the bold and deep penetrations of the panzers. Joffre had time to recover from a bag-full of initial blunders; Gamelin, who actually made fewer mistakes, did not. One of Gamelin's most junior officers, Lieutenant Carron of the 44th Infantry Division, wrote his memoirs after repatriation from Germany, expressly to place on record that 'our troops in this war...have not been unworthy of their great forebears of 1914–18. Like them, they would have done well, if they'd been given time.'[70] John Cairns, characteristically, offered wise words of caution on this approach in one of his pioneering explorations of the French defeat written some forty years ago. He pointed out that comparisons between 1914 and 1940 were 'difficult, even impossible, and that whatever purpose they served, it was not that of demonstrating that the generation of 1940 was unworthy of its fathers. An army in defeat is unlikely to look very well at any time.'[71]

Aphorisms can be seductive; they may mislead more than they enlighten. Their seductiveness lies in their simplicity. One that sprang to mind in exploring this subject was the saying that there is no such thing as bad soldiers – only bad officers. We should be properly cautious, as critical historians, after the necessarily selective, limited reappraisal in this essay of the evidence of French fighting performance and leadership in 1940, before asserting that the case 'proves' this attractively straightforward adage. All the same, there *were* bad officers in the French armies of 1940 – albeit these were usually inert, unfit, reservist colonels and ageing, demoralised generals. Most scandalously of all, the generals seem to have held the other ranks in such low regard that they blamed them for what were really the senior officers' own failures to organise and command properly, rather than praising the men's courage and fortitude under enemy fire.[72] And beyond doubt the rank and file of most French units had an unwarranted confidence in their commanders that seems bitterly ironic in light of these commanders' derogatory commentaries. Most unedifying of all was the aftermath at Vichy: Pétain joining in the excoriation of the 1940 *poilu* whilst making ministers out of his inadequate, careerist and politicised commanders. As Vichy's collaboration unfolded, the French soldier's experience of war in 1939–40 came to be cloaked in shame and embarrassment. In

1953 one little-known general published an early effort to set the record straight, entitled *Light on the Ruins: the Combatants of 1940 Rehabilitated.*[73] How far this rehabilitation can be taken remains one of the bits of unfinished business for military – and French – historians even after the fiftieth anniversary of the Second World War's end.

The Italian Soldier in Combat, June 1940– September 1943: Myths, Realities and Explanations

Brian R. Sullivan

Myths about incompetent or cowardly Italian performance in battle are commonplace in the English-speaking world. These attitudes appear even more widespread throughout Britain and the Commonwealth than in the United States. American experience with the Italian army, either as a wartime ally or adversary has been rather limited. But the British, Commonwealth and Indian armies had extensive contacts with their Italian counterpart in the two world wars.[1] Tales based on these encounters seem to have reinforced, rather than mitigated, negative stereotypical opinions about the Italian fighting man, particularly among those not present at the time.

Like many clichés, however, there are elements of truth in some of the anecdotes about Italian military ineffectiveness. The Italian soldier often was the victim of poor training, bad equipment and incompetent leadership. These handicaps could not but diminish his military efficiency, as they would have that of any other fighting man. But the myth of the Italian soldier as a quivering wretch forever succumbing to fright or panic is only that. In fact, in all of Italy's wars, as in the Second World War, the Italian soldier was as brave as any other. In some ways, he displayed more valour than most. Thus, both the myth and the reality of Italian combat performance in the Second World War are simple to relate. Explanations, however, are more complicated.

*

To understand the battlefield conduct of the Italian soldier in the Second World War, certain basic factors affecting his performance must be considered. One was Italian national poverty. Half a century later, when Italians enjoy roughly the same standard of living as Britons, French and Germans, it is easy to forget that in 1940 Italy had only a quarter the national income of Britain and less than half that of France. This meagre wealth was distributed among an Italian population roughly equal to the British and French. The majority of the Italian population, whether peasants or workers, struggled to survive at little more than subsistence level in the 1920s and 1930s. As a result, on average, Italian soldiers in the Second World War were less healthy, less educated and far less familiar with modern technology than British, French, German and even Soviet soldiers.[2]

Fascist Italy spent a great deal on its army before and during the Second World War. But much of this expenditure had been devoured by the wars in Ethiopia and Spain and other portions were consumed by financial corruption, profiteering, the high price of imported raw materials, industrial inefficiency, administrative incompetence and expensive autarky policies. Remaining military funds were spread out thinly on an army overly large for available resources. As a result, in all its material aspects, the Italian army displayed the poverty of the nation that had produced it.[3]

Mussolini and the responsible members of the army High Command should be cited for their criminal folly in linking fascist Italy to Nazi Germany. But the fact remains that Hitler initiated the Second World War after Mussolini and his generals had made clear that the Italian army would not be ready for a war with the West until at least 1943. In the spring of 1940, however, history presented Mussolini with an unexpected opportunity. While French resistance was collapsing that May, Mussolini decided that he must seize the opportunity this gave him to realise his imperial dreams. Much of the army High Command agreed. Mussolini and his generals gambled on a short victorious war – and lost. As a result, the average Italian soldier paid for that grave miscalculation by fighting for thirty-nine months with arms and equipment generally inferior in quantity and quality to those of his opponents.[4]

Still it would be incorrect to assume that only a few Italian soldiers entered the war with enthusiasm and that most fought only because they were coerced by a brutal régime and a harshly authoritarian officer corps. In the Second World War, the Italian army never had to resort to the draconian discipline routinely imposed under General Luigi Cadorna, the iron-willed disciplinarian who commanded the army throughout most of the First World War. During the First World War, Italian courts martial passed 4,000 death sentences (750 carried out); many other soldiers were shot without trial. In the Second World War only ninety-two Italian soldiers were condemned to death for military offences. True, the operational conditions of trench warfare in the Alps and their foot hills in the First World War varied enormously from combat in the North African desert, the Russian steppes or the *mesas* of East Africa in the Second World War. It would have been impractical to apply disciplinary methods possible in static warfare to the mobile warfare of the Italian army in 1939–45. But combat in the French Alps in June 1940 or in Albania in the winter of 1940–41 closely approximated conditions on the Italian Front in 1915–17. Neither on those fronts nor elsewhere in 1940–43 did the Italian High Command ever resort to the frequent executions by firing squad it had ordered in the First World War. This alone indicates a much greater Italian willingness to fight in the second war than in the first.[5]

True, in 1940–43, many Italian troops soldiered on with only a stoic resignation. But it would seem that twenty years of fascist propaganda had created at least some support in the ranks for Italian war aims in 1940–42. In the period November 1942–September 1943, when the average Italian soldier surely had lost hope of victory, he still fought very well, probably motivated by the desire to defend his country. One of the few positive achievements of the fascist régime was a deepening of national sentiment throughout Italian society. Most Italian soldiers do seem to have considered British presence in the Mediterranean as illegitimate interference in Italy's sphere of influence. It is difficult to be precise about the influence of these attitudes on Italian soldiers in the Second World War, but when one compares individual Italian combat performance in the two world wars, it does seem that the average Italian soldier was better motivated and fought more skilfully in the second conflict, particularly after the spring of 1941.[6]

However, when one considers the performance of the Italian army as whole in the two world wars a very different picture emerges. The Italian soldier fought very hard between 1915 and 1917. In their great August 1917 offensive the Italians came close to breaking their Austro-Hungarian opponents. After its terrible setback at Caporetto in October 1917, the Italian army recovered and ended the First World War in victory, the only Allied army to have defeated its opponent in the field. In contrast, the Italian army fought very badly in its first year of the Second World War, then improved considerably. It performed quite well for the next two years, only to collapse totally in the summer of 1943. That final débâcle can be blamed on the irresponsible behaviour of the government and the High Command, rather than on the army as a whole. Still, not even the French army in June 1940 disintegrated as did the Italian army thirty-nine months later. What explains the disasters of 1940–41, the recovery of 1941–42, the dogged resistance of 1942–43 and the disgrace of September 1943? To answer these questions requires an understanding of how the Italian soldier was trained, equipped, organised and led both in the immediate pre-war period and during the Second World War.

*

Poor training was probably the single most serious weakness of the Italian army in 1940–41. The military training law of 1925, amplified by another in 1934, had established on paper an elaborate system of pre-military training for men eighteen to twenty-one years of age. In theory, conscripts should have reported for active service well instructed in drill, basic military knowledge and the use of small arms. In practice, however, such training accomplished little or nothing because it was assigned to the Fascist Militia. The Blackshirts lacked the leadership, the discipline or the experience to carry out such instruction.[7]

Physically fit young men (about three out of four recruits in the late 1930s) reported for active duty the year they turned twenty-one. Generally, men from two different regions were assigned to a regiment[8] headquartered in a third part of Italy. Each regiment instructed new conscripts at its depot (*deposito*) under the distant supervision of the unit's senior lieutenant-colonel. Each regiment had its own individual training plan. The Italian army entered the Second World War with 75 divisions, of which 3 were Blackshirts, with a total of 75 artillery and 150 infantry, armour and cavalry regiments. To provide a comprehensive description of Italian army training at the time, one would have to discuss hundreds of separate training programmes. But with a number of notable exceptions, regimental training generally ranged from poor to truly wretched.[9]

Infantry training was particularly bad. One of the Italian army's finest officers was astounded when he observed German army infantry training in 1937. He noted the inner-directed discipline, the enthusiastic obedience and universal sense of comradeship that pervaded enlisted/non-commissioned officer/officer relations. He was particularly impressed by the intensity and realism of training, the stress on combined arms tactics, the excellent quality of weapons and equipment, and the high standards of food, barracks and uniforms which moulded the attitudes of German infantrymen.[10] All were absent from standard Italian infantry training.

Instead, Italian enlisted–officer relations were stunted by the kind of distance and formality that had characterised the British army in the nineteenth century. This influenced all aspects of instruction. Italian infantry training largely consisted of close-order drill, callisthenics and rigorous marching with full pack. Live-fire practice was minimal since the army could not afford the expenditure of much ammunition. The limited combat training the men actually received was directed by ill-educated junior reserve officers with little understanding of effective techniques and even less of current tactical doctrine. Combined arms exercises for the infantry were virtually unknown, but those few that occurred were confined to infantry–artillery co-operation. Most infantry went into combat in 1940–41 without ever having seen a tank, let alone trained with one.[11]

Pre-war infantry combat doctrine and training emphasised fire at the expense of manœuvre, as well as subordination to the original plan instead of initiative. Offensive tactics stressed attack along a single pre-selected axis of advance, backed by heavy crew-served weapons and artillery fire. If a unit faltered, it would be replaced by another unit, rather than ordered to attack in a different direction. This doctrine effectively ignored the possibility of infiltration tactics and virtually ensured plodding advances accompanied by heavy casualties. Defensive tactics were also based on fire with little use of reserves or counter-attacks. The result was brittle defences prone to collapse when ammunition ran low. This proved especially true because the Italian army lacked sufficient artillery and motor transport. Shortages of the former led to heavy reliance on mortars, but they could hardly replace the greater range and

firepower of artillery. Furthermore, when mortar crews exhausted their ammunition – as they necessarily would in trying to substitute their weapons for
artillery – the insufficient numbers of trucks meant that resupply would be
slow. This would leave the infantry with little or no fire support at all.[12]

After completing training, conscripts served the remainder of their military
obligation preparing for the annual August manœuvres. Depending on the
period in which they were called to active duty, conscript service ran from
twelve to eighteen months in the period from 1925 to 1940. After release from
active duty at the age of twenty-two or twenty-three, an Italian citizen-soldier
remained in the active reserves until aged thirty-two. Under the law, he was
expected to attend post-military training under the tutelage of the militia in
order to retain his military skills. But this programme was run as badly as pre-
military training. Reservists recalled in 1939–40 required another period of
basic training to return to their former level of proficiency.[13]

The food, shelter and clothing of Italian soldiers in the immediate pre-war
period and continuing into 1941 were abominable. Rations tended to improve
somewhat as the war continued but were never very good. The army always
lacked enough barracks and tents to shelter its mobilised forces. Existing
barracks tended to be unheated, badly ventilated, decayed and often filthy.
Standard Italian enlisted wool uniforms of the 1930s were basically adequate.
None the less, they tended to be too heavy for comfortable summer wear yet
too light for sufficient warmth in the depth of winter. Tropical uniforms for
service in Africa were better designed. But there was a serious shortage of
cotton uniforms in 1939–41, leaving many troops in Libya and East Africa to
swelter in wool uniforms during the hotter months. The uniforms hastily
manufactured in 1940–41 tended to be stitched badly and fell apart easily.

The model 1912 boot was a good design, although some batches were not
well made and tended to wear out rather quickly. Stories about Italian army
boots of synthetic material dissolving in the snow are without foundation. But
there were insufficient numbers of boots in 1940–41, which left many soldiers
serving in the Albanian campaign with worn-out footwear. Worse, Italian army
socks were open in front. Since the supply corps realised too late that boots
for winter wear needed to be issued in sizes larger than a soldier's foot, to
allow for thermal packing with rags or straw, many cases of frostbite occurred.
Alpini footwear was superior for cold weather warfare but even that proved
inadequate in Russia. That an army with extensive experience with mountain
warfare in the First World War would suffer from such problems a quarter-
century later offers a damning commentary on the military effectiveness of the
fascist régime.[14]

Recalled reservists generally were equipped, housed and fed even worse
than conscripts, since the reserves received the leavings of an already inadequate system. Beginning with the Czech Crisis in September 1938, the army
carried out a number of partial mobilisations, as well as full mobilisations in
the autumn of 1939 and the spring of 1940. Observers noted reservists, lacking
uniforms and any other military equipment, forced to sleep in the streets

outside already crammed barracks. The army demobilised partially in July and more completely in October 1940. These call-ups and stand-downs further lowered the living standards of the troops remaining on active duty and seriously disorganised training for recruits throughout 1939 and 1940.

Soldiers serving in Libya had been subjected to particularly harsh living conditions since the crisis of 1935–36 when the colony had been heavily reinforced. During the renewed build-up of Italian forces in North Africa from mid-1939 to early 1941, the troops experienced severe shortages of water and firewood. This left them constantly hungry, thirsty, without warm food and highly susceptible to infection and disease. Enlisted awareness that staff and rear-area officers enjoyed a far superior diet provoked considerable resentment. As a form of rough justice, dysentery afflicted the entire army. This increased the clouds of insects which normally tormented the troops.[15]

The weaknesses of army regular infantry training were even more pro-nounced for the combat units of the Fascist Militia because of notoriously lax Blackshirt discipline. On the other hand, training in Alpini, Bersaglieri, artillery, armoured, combat engineer and cavalry units was better than average because of the technical demands and the generally superior quality of the officers and non-commissioned officers (NCOs) of those branches. But even these regiments suffered from inadequate quantity and quality of equipment which adversely influenced training. The quality of Italian tanks was particu-larly poor. Still, the training of armoured division personnel was rather good because the lessons learned in the Spanish Civil War had been absorbed and applied. By 1940 Italian armour doctrine stressed combined arms warfare, even if the actual arms available were distinctly inferior to German or British.[16]

Shortages of NCOs and the inexperience of the inadequate numbers the army actually possessed also retarded both enlisted and NCO training. Junior NCOs were chosen from each year's intake of conscripts and were rapidly and superficially instructed. They served a year or so, then were released from active duty with their fellow conscripts. Few chose to make the army a career because of low pay, poor benefits and lack of government-guaranteed employ-ment after retirement. The small number of career NCOs and warrant officers (*marescialli*) were employed in administration or provided basic training to officer candidates. Their officers discouraged displays of initiative and most NCOs complied. The Italian army had very few of the mature, self-reliant, often combat-tested NCOs who provided the backbone of the British and German armies in the 1930s. Worse, the Italian army had too few NCOs of any kind. The contrast with the American army – admittedly a case of extremes – was striking. In June 1940 the ratio of NCOs to privates in the Italian army was 1:33. In the US army in 1945 the ratio was 1:2.[17]

Regular junior Italian army officers generally were well educated, although they also suffered from insufficient combat training. All spent two years in a military academy, then one more year in a school of application, if they were in the infantry or cavalry, or two years, if they were in the artillery or engineers.

But the Italian army was notably deficient in regular junior officers. A budget crisis in the mid-1920s had forced the army to dismiss half of its regular lieutenants. Thereafter, academy admissions were limited to guarantee a full career to all second lieutenants who were commissioned. To make up for its lack of regular junior officers, the Italian army relied on reservists to command and staff its platoons, companies and batteries. This system proved disastrous in 1939–41, particularly for the infantry.

In those years, reserve officers on active duty fell into three groups: (1) men who had served as lieutenants in 1917–19 and mostly were recalled to active duty as majors or lieutenant-colonels; (2) reserve officers who had last served on active duty in the 1920s or early 1930s and were recalled to active duty as captains; (3) men who had been on active duty in the mid or late 1930s and were recalled as first lieutenants. In 1940, only a small minority of reserve officers had fought in Ethiopia or Spain. The majority had no combat experience, even in the First World War. Officially, no reserve officer could be promoted to captain without training additional to that which he had received as an officer candidate. But such courses consisted of no more than two weeks of very rudimentary instruction. Some captains did not even get that training. Furthermore, many reserve officers were recalled in 1939 as first lieutenants, having had neither military service nor training for fifteen to twenty years.

Intense training courses for reserve officers were initiated in the autumn of 1939 but then interrupted by the demobilisation of October–November 1939. When the army was mobilised again in May 1940 these courses were not resumed. Making matters worse, the best reserve officers and NCOs were pulled out of their units to serve as basic training instructors for new recruits in the spring of 1940. This disrupted unit training and left many divisions unfit for combat. Perhaps only one-third of the reserve officers and enlisted men thrown into battle against the French in June 1940 could be considered properly trained.

Generally, battalions had only two or three regular officers in 1939–40, which meant that small units, and even a certain number of battalions, were commanded by reserve officers. The insufficient numbers of regular officers and the inadequate training of reserve officers resulted in oversupervision by superiors. This tended to stifle any display of initiative. The problem was aggravated by fact that in 1940–41 most reserve captains and majors ranged in age from their late thirties to mid-forties and did not possess the requisite energy or health to command in battle. As a result, below the regimental level, unit commanders tended to wait for instructions before acting and rarely exercised their own judgement.[18]

*

Poor training was only one of the problems afflicting the Italian army in 1940. Well-trained troops can fight effectively, even when poorly armed and equipped. In contrast, superb weapons are of little use in the hands of badly instructed soldiers. However, in the year following Italy's entry into the war, the men of the Italian army were not only badly trained but also supplied with generally substandard and insufficient numbers of weapons.

While false stories have circulated about its unreliability, the standard Italian Mannlicher-Carcano model 1891 bolt-action rifle was a fine weapon. (Lee Harvey Oswald used such a rifle in one of this century's most impressive but infamous displays of marksmanship.) In 1938, however, the army had begun issuing rifles of 7.35-mm calibre, rather than 6.5-mm as before. This was in order to use a round of increased lethality, as a result of disappointing experiences with the smaller round in recent fighting in East Africa. The army was preparing to introduce a new semi-automatic rifle as well. The unexpected collapse of France hastened Italian entry into the Second World War with the army's new rifle programmes incomplete. This forced the army to return to the use of the 6.5-mm rifle. Both calibre rifles were issued to the troops. The resulting need to manufacture and distribute two different rifle rounds created serious ammunition supply problems. Suspension of plans for the manufacture of the semi-automatic rifle and the cessation of production of the 7.35-mm rifle disrupted output and reduced the overall manufacture of rifles. Shortages led the army to issue 8-mm Austrian rifles captured at the end of the First World War and even some of its original nineteenth-century bolt-action 10.35-mm rifles, creating additional problems with ammunition supply.

Beginning in 1938, the army began receiving a small number of the excellent Beretta submachine-gun, a weapon exceedingly popular with those few troops lucky enough to be issued one. Still, the Beretta required a different 9-mm round from the standard pistol round, adding to difficulties in ammunition resupply. Furthermore, the use of moulded parts greatly limited the numbers of Beretta submachine-guns produced until a simpler model using sheet steel was introduced in 1942.[19]

Larger Italian automatic weapons were plagued by deficiencies in design as well as in supply. The standard Breda 6.5-mm light machine-gun was fed from twenty-round chargers which were introduced to a permanently attached oil-lubricated magazine. These arrangements greatly reduced the rate of fire and required the frequent cleaning of the weapon to prevent jams, especially in the desert. Damage to the permanent magazine rendered the weapon useless until repaired by an armourer. No grip or handle was provided to change hot barrels. The FIAT 8-mm heavy machine-gun also required lubricated ammunition, overheated with frequent 'cookoffs' and often broke the fingers of those attempting to clear its jammed action since it fired from a closed bolt. The only Italian machine-guns considered successful were the five heavy Breda models, ranging from 7.7-mm to 13.2-mm. But these weapons were fed from magazines which retained the cartridge cases of fired rounds. Gunners had to empty the magazines and reload them before using them again. Once loaded magazines were exhausted, this greatly reduced rates of fire. Adding to the problems of machine-gun crews were the army's seven different calibres of machine-gun, requiring the supply of seven different rounds.

Italian hand grenades, meant for offensive use only, were not very effective due to their unreliability, small explosive charges and poor fragmentation. Three nearly identical models were produced. This redundancy retarded production, although it preserved the profits of the various manufacturers. Italian-anti-tank grenades proved useless against the thick armour of British tanks, let alone Soviet T-34s.

The light 45-mm mortar functioned well. But, the puny round led many mortar units to abandon the weapon entirely and rely exclusively on the reliable 81-mm medium mortar. However, the 81-mm rounds also failed to fragment with a lethality equal to similar shells used by other armies.[20]

In contrast, the Italian army was better served by its 20-mm dual-purpose anti-aircraft/anti-tank gun and its 47-mm anti-tank gun. The latter lacked a shield, resulting in unnecessary shell-fragment wounds to its crews in combat. None the less, the gun was so effective (at least against the tank armour of 1939–40) that the British, Argentines and Romanians attempted to buy large numbers in 1939. Italian armaments firms could not manufacture enough to meet even its own army's needs. But the Italian government's desperate need for foreign exchange led it to sell the guns to the Argentines and Romanians, although not to the British. Nor could the Italians produce sufficient numbers of the dual-purpose 20-mm gun. The government was forced to dip into precious hard currency reserves to purchase 20-mm anti-aircraft and anti-tank guns from the Swiss. The army also supplied some anti-tank units with First World War-era 65-mm mountain guns adapted to an anti-tank role, but with rather unsatisfactory results. As enemy tank armour and aircraft performance improved, however, all these weapons became progressively less effective.[21]

The artillery was the finest branch of the Italian army in the Second World War. But the high quality of its training, discipline and morale could only compensate partially for the obsolescence of the artillery's pieces. While the High Command had drawn up plans in 1929 for completely re-equipping the army with 15,000 new guns, lack of funds had prevented implementation of most of the programme. Starting in 1933, some new artillery had been produced. But the vast majority had been 20-mm and 47-mm guns. Over the next seven years, the major Italian artillery manufacturer, Ansaldo, built only 114 75-mm field pieces, 76 75-mm anti-aircraft guns, 40 149-mm guns and 24 210-mm howitzers. To these 250-odd pieces, from June 1940 to June 1943, Ansaldo could add only another 2,300 tubes of 75-mm or greater. However, the majority of these were not produced for artillery pieces but for self-propelled guns and tanks. Italian artillery production throughout the Second World War was characterised by extraordinary inefficiency.

In June 1940, the vast majority of the Italian army's 10,000-odd artillery pieces were of pre-First World War design, many captured from the Austro-Hungarians in 1918. These had been fine guns and howitzers for their time, particularly those manufactured by Skoda. But by 1940–41, they had been outclassed by the modern weapons of other armies. For example, the British 25-pounder gun/howitzer could fire its high explosive shell 13,400 yards; the Italian 100/17 howitzer could send its 29-pound shell a maximum of 10,000 yards. The British 4.5-inch gun could fire its 55-pound projectile to a range of 20,500 yards; the Italian 105/28 gun could shoot its 35-pound shell a maximum of 14,800 yards; the Italian 149/13 howitzer could fire its 94-pound high explosive round only 9,600 yards. Adding to the weaknesses of the Italian artillery was yet another ammunition resupply problem, created by the multiplicity of Italian and ex-Austro-Hungarian artillery pieces in Italian service.[22]

Italian armoured units suffered from the particularly poor quality of their fighting vehicles throughout the first year of the Italian war. The army entered the conflict with about 1,500 tanks, some 1,300 of which were the tiny turretless three-and-a-half ton, two-man L3/35, based on the Carden Lloyd Bren machine-gun carrier. Most were armed with two 8-mm Breda machine-guns, although some carried heavier machine-guns, 20-mm cannon or flame-throwers. These vehicles hardly deserved the name 'tank'. While fairly speedy and mechanically reliable, the L3/35 was starkly inferior both defensively and offensively to any other armoured vehicle it encountered on the battlefield. The Italian army also possessed about 100 FIAT 3000 light tanks, a copy of the French Renault of the First World War and 100 new M11/39 tanks with a hull-mounted 37-mm cannon, plus two 8-mm machine-guns in a turret. The FIAT 3000 obviously was obsolete and the M11/39 has been described as the worst design of its era. Its frontal armour was easily pierced by any anti-tank gun of the day. Any round penetrating the glacis armour would pass through the transmission, drawing molten metal into the combustible transmission fluid. It required brave men to crew these tanks. But none of them carried radios. Thus, the courage of Italian tank crews often was rendered futile by their inability to co-ordinate armoured formations in battle.[23]

The Italian army never possessed sufficient motor vehicles for its needs in the Second World War. Many vehicles had been worn out or left behind in Ethiopia or Spain. The Italian automobile industry lacked the productive capacity to make good these losses by 1940 and could never meet the army's needs during the war. The army began the war with about 49,000 motor vehicles, filling about two-thirds of its needs and the most it would ever have throughout the conflict. By 1941, the army was down to less than 39,000 vehicles and had only some 30,000 by 1942. The army had no tracked or half-tracked transport vehicles, and its wheeled trucks were not designed for off-road movement. Truck quality was patchy, although the army did possess some good models. But again the multiplicity of types – a phenomenon that afflicted all classes of Italian arms and equipment – created great difficulties for supplying spare parts and making repairs.

Of course, armoured and transport vehicles of whatever quality are of little use without fuel. All the Italian armed forces were hobbled throughout the war by severe petroleum shortages, since Italy possessed only a few small oil-fields and depended on Germany to supply most of its needs. Since the Germans themselves had insufficient petroleum, the Italian army received far less than its operational requirements.[24]

Adding to the army's weaknesses caused by poor training and equipment was the misconceived reorganisation of its divisions in 1938–40. Army chief of staff Alberto Pariani transformed the standard three-infantry-regiment infantry division into the two-infantry-regiment *divisione binaria* (binary division). Pariani believed such a transformation would provide a more agile and easily transported unit. But he also reorganised the army with the stated goal of making Italy seem more powerful by the creation of 50 per cent more divisions from the same quantity of men and arms. The result was a division lacking sufficient offensive power, defensive depth and manœuvrability.

Pariani intended to address some of these weaknesses by adding more artillery at the corps level and providing each of the new divisions with the modern arms and equipment he expected the army to acquire in the early 1940s. But Italy's premature entry into war made the fulfilment of such plans impossible. The actual binary division of 1940 lacked sufficient firepower. Each of its infantry regiments contained only 32 machine-guns, 6 81-mm mortars and 8 47-mm and 4 65-mm anti-tank guns. The artillery regiments of these divisions had only 24 75-mm and 12 100-mm artillery pieces: roughly half the artillery of foreign infantry divisions. After Mussolini dismissed Pariani in November 1939 as a scapegoat for the bad condition of the Italian army, certain hasty attempts were made to improve the *divisione binaria*. To increase the division's infantry strength, each received the reinforcement of two weak battalions of Fascist Militia and a Blackshirt machine-gun company. But this merely added particularly ill-trained infantry to the division and actually aggravated the problems of lack of artillery and motor transport. After it entered combat in the spring of 1940, the binary division was exposed as a failure, particularly in the Greek-Albanian Campaign. But the army retained this division structure throughout the Second World War.[25]

*

Having examined aspects of how the Italian soldier was trained, equipped and organised, we can better understand how he fought in the Second World War. If distinctions are made among the successive phases of Italian ground operations in the Second World War, one can be more precise in describing the competence of Italian troops. The 1940–43 operations can be divided into three phases. The first extended from Italian entry into the war in June 1940 until the spring of 1941, encompassing about ten months. The French Alpine campaign of June 1940, the Greek-Albanian campaign of October 1940–April 1941 and the first North African campaign of September 1940–February 1941 fell within this period. The second period ran from March–April 1941 until July 1943, nearly two-and-a-half years. The remainder of the North African campaigns, the invasion of Yugoslavia, the counter-insurgency campaigns in Greece, Albania and Yugoslavia and the Italian expedition to Russia took place in the second time period. The brief third phase encompassed the defence of Sicily and the Italian–German clashes in September 1943. Operations in East Africa in 1940–41 formed a special case. This was due both to the ethnic mixture of European and African troops in the Italian forces in the region and their near-total geographical isolation. These factors left them largely unaffected by the influences that transformed the rest of the army in the second phase of the Italians' war.

A number of ground units performed well in combat in 1940–41. Notable examples include the Alpini divisions – especially the Julia – in Albania, the 11th Regiment of the Granatieri di Savoia at Keren in Eritrea and the Centauro armoured division in the brief Yugoslav campaign.[26] However, most army and Fascist Militia combat units fought poorly until the spring of 1941. The five months from October 1940 to March 1941 marked the nadir of Italian military effectiveness. During that period two events occurred that have come to epitomise the conduct of the Italian army in the Second World War: the retreat from Egypt and Cyrenaica of the Italian Tenth Army and the capture of 130,000 of its men; the repulse of the Italian invasion of Greece, the successful Greek counter-offensive and the inability of the Italian Ninth and Eleventh Armies to dislodge their opponents until the Germans arrived and forced Greek surrender.

The war already had begun badly for the Italian army in June 1940 when it attempted to drive back the six French divisions guarding the Riviera and the Alps. Twenty-two Italian divisions backed by fourteen more in reserve barely budged their opponents in a short four-day campaign. Lack of artillery and poor training help explain the Italian failure. But losses of 1,250 dead and missing indicate Italian heroism, however futile. The 2,100 cases of frostbite suffered in summer weather foreshadowed the far worse tragedies that awaited the ill-shod Italians in Albania.[27]

The Italian disaster in North Africa that winter arose primarily from sending a huge army of infantry on a slow foot march into Egypt. The trucks that might have carried these troops remained in the Po Valley for a projected invasion of Yugoslavia, which finally took place in March 1941. Furthermore, the first Italian armoured division arrived in Libya only in April 1941. The Commonwealth victories of December 1940–February 1941 can be credited to superb training – particularly that of the 7th Armoured Division – superior British tanks against which the Italians were virtually helpless at first, the motorised mobility of Commonwealth infantry and excellent British generalship. Time and again, after masses of Italian infantry without motor transport were cut off, they had little choice but to surrender or die of hunger and thirst. Even the reinforcement in January 1941 of the sole Italian armoured brigade in Libya with 100 of the new model medium tank, the M13/40, could not right the balance. The armoured brigade commander failed to co-ordinate his tank attacks with artillery and threw his armour into battle in penny-packets at the climactic encounter at Beda Fomm in February 1941. Still, there is no doubt that some Italians had fought bravely, particularly the artillery and armour, as the British official history attests.

By February 1941, of the 12 Italian and 2 Libyan divisions in North Africa three months earlier, only 5 Italian units remained. All 3 Blackshirt divisions had been destroyed and none ever were recreated.[28]

In terms of killed and wounded, however, the abortive invasion of Greece and the Greek counter-attack into Albania mark the greatest Italian catastrophe of 1940–41. By the time the campaign had ended, the Italian army had lost nearly 39,000 dead, 51,000 wounded, 12,000 with serious cases of frostbite and another 52,000 hospitalised with major illnesses. In contrast, the Greek army had lost 13,800 killed and about 42,000 wounded.[29]

Many factors contributed to this Italian defeat, including poor training, inadequate boots and uniforms, lack of artillery, the weaknesses of the *divisione binaria* and the extraordinary fact that the Greek defenders initially outnumbered the Italian invaders. The Italian failure in the Greek-Albanian campaign was multi-faceted. However, Mussolini's decision to demobilise one-third of the army for economic and political reasons three weeks before the invasion of Greece was probably the greatest contributor to the disaster. Two days after he had issued this order, Mussolini decided to invade Greece. But to avoid appearing foolish and to prevent the inevitable disgruntlement that would follow from a reversal of his directive, Mussolini allowed the demobilisation to continue. None of the reinforcing divisions hastily rushed to Albania after the Italian retreat from Greece had any unit cohesion.

The gaps in enlisted ranks caused by the October 1940 demobilisation were filled with men with little or no training, hastily pulled into the army from civilian life. These raw recruits arrived in the front lines with no notion of how to handle their weapons. In fact, they were lucky to receive arms; in some cases none were available. Under the circumstances, discipline evaporated. Soldiers withdrew immediately at the first sign of combat. Three or four might use the excuse that they must carry a wounded comrade to the rear, one who in fact had suffered only superficial lacerations. Standing in the frigid trenches at night, troops cried out to be shot rather than endure any longer the torture of freezing. Depression spread through the army; entire companies refused to eat despite the terrible cold. Lack of firewood and shelter from the incessant wind left others unable to cook or warm themselves.

Relatively well-trained junior NCOs, and platoon, company and battery commanders had been returned to civilian life, while hastily trained reservists took their places in the divisions sent to Greece. The lack of initiative shown by these ill-trained reserve officers and NCOs was a major factor in the defeats of November 1940–January 1941, when the Greeks counter-attacked. Considering these monumental blunders, it is surprising that the Italian soldier fought as well as he did in that bitter campaign. Yet the great majority stoically bled and died in two useless offensives in March and April 1941 after Italian troop strength had been built up. However, the Italians achieved no breakthroughs and it was the Germans who overran Greece.[30]

Meanwhile, in January 1941, Commonwealth forces had begun a multi-pronged invasion of Italian East Africa. The Italians opposed the offensives with large ground forces: 47,000 Italian army, 27,000 Fascist Militia, 182,000 white-officered colonial troops and 17,000 militarised police, a total of 273,000. But the Italian colony covered some 660,000 square miles and had a large Italian civilian population requiring protection from guerrilla attacks, while the Italian forces lacked sufficient motor transport and fuel. Due to poor general-ship, Italian colonial forces in the south offered little effective resistance to a Commonwealth offensive from Kenya. Addis Ababa fell in early April 1941. The protracted Italian defence of Keren collapsed in late March. Asmara and Massawa were captured by Commonwealth forces less than two weeks later. The viceroy, Amedeo of Savoy-Aosta, withdrew to a central mountain redoubt and fought on until surrendering in late May. Italian forces in the far south resisted until July and the last Italian units, surrounded in the Gondar area, held on into late November.

After he had received inaccurate figures indicating light Italian casualties suffered in the defence of Gondar, Mussolini complained that the troops had lacked courage. In fact, he slandered men who had fought with tenacious loyalty for eighteen months, despite little hope of victory or relief. Precise numbers are lacking but roughly 6,300 white troops had fallen and another 8,700 had been seriously wounded – some 19 per cent of the 80,000 Italian ground forces in Italian East Africa. The colonial troops suffered even heavier losses: about 22–25,000 dead and 28,000 badly wounded. This represents casualties of 26–7 per cent. The 4th Eritrean Battalion 'Toselli', the most highly decorated *askari* combat unit, had lost twelve officers and 500 men killed in action at Keren. Nearly the entire unit had fought to the death.[31]

*

After the defeats in North Africa and the Balkans, the Italian army began a startling recovery. Heavy losses necessitated the recall of the trained reservists demobilised in October 1940. These troops were reintegrated into the army over the winter of 1940–41. By late April 1941 the quality of Italian fighting units was far higher than six months earlier. Furthermore, however painful and humiliating the price, much of the army had gained combat experience. The new army chief of staff, General Mario Roatta, finally imposed uniform training procedures on each of the regimental depots and issued guidelines to bring that training into line with the realities of modern warfare. In early June, acting on the orders of General Ugo Cavallero, the new chief of the supreme general staff, Roatta created training battalions. These units would provide intensive instruction for those divisions selected for imminent deploy-ment to war zones. Special attention was given to preparing young reserve officers for combat leadership at the platoon, company and battery level. From mid-1941, officer candidates were required to take a demanding sergeants' course, serve as NCOs for two to three months with a unit stationed in Italy and then complete a realistic officer training course. Junior officer training henceforth involved realistic exercises in which candidates were presented with unexpected tactical problems and required to develop practical battlefield solutions quickly. In place of the rigidly choreographed training scenarios of the 1930s, this represented a true revolution in Italian military instruction.[32]

To a lesser extent, the effectiveness of the combat units of the Fascist Militia also rose in the second half of 1941. The ease with which the three Blackshirt divisions in North Africa had disintegrated in 1940–41 convinced both the army and militia leadership that it was temporarily impractical to raise more such units. For the moment, the army did not possess the arms, equipment and specialists necessary, nor could the militia effectively organise and train such formations on its own. Instead, some twenty units of battalion size and two of brigade size (*M* battalions and *raggruppamenti*) were formed to serve as assault troops attached to army corps. Those pairs of militia battalions previously assigned to army divisions were dissolved and their best men integrated into the new Blackshirt units. A number of capable militia officers, NCOs and Blackshirts had survived the combat of 1940–41. These veterans formed the cadre of the new units or were used to train additional recruits. While they never reached the same standards of training or equipment, the militia general staff attempted to emulate the quality of the Waffen SS with these new units. Eventually, when resources were available, a mechanised Blackshirt division (the M division) would be formed with the co-operation of the army and Heinrich Himmler to serve as a bodyguard unit for Mussolini. Very few of the new militia units were shipped to North Africa; most would be employed in Europe, especially in the Balkans.[33]

More Italian troops were deployed in the Balkans than in any other theatre from October 1940 to September 1943. After the conquest of Yugoslavia and Greece, some thirty Italian divisions remained in the region on occupation or counter-insurgency duty. The Italian Second Army maintained forces in the part of Slovenia and the Dalmatian coast annexed by Italy, the western half of the *Ustasha* state of Croatia and the Italian puppet state of Montenegro. Albania, including Kosovo and the part of Macedonia attached to the Italian protectorate, was garrisoned by the Italian Ninth Army. The Italian Eleventh Army occupied the majority of Greece and the Italian Aegean islands.

The first resistance the Italians faced was from the monarchist Serb forces, starting in the late spring of 1941. These guerrillas were mostly survivors of the Yugoslav army, known as the *Cetniks*. After the German invasion of the Soviet Union, Tito's communist partisans, drawn from all ethnic groups in former Yugoslavia, also began attacks on the Italian forces. The Italians soon faced other difficulties brought about by the murderous racist policies of the *Ustasha*, which plunged all Croatia into bloody chaos. Further to the south, the situation remained tranquil for some time. Smaller communist guerrilla forces in Albania and Greece took up arms only in mid-1942.

A short time after Operation Barbarossa began, the *Cetnik* leader, Colonel Draja Mihailovic, began an unofficial truce with the Italians. Thereafter, the *Cetniks* concentrated on war with the communists and the *Ustasha*. While the

Italians could not fight their Croatian allies, they could protect Jews, Serbs and Bosnians from the murderous rule of the *Ustasha*. In Greece, the Eleventh Army also provided protection to the Jews in its area of control. Unfortunately, that did not extend to the large Jewish population of Salonika, occupied by the Germans. Thus, throughout the two years from September 1941 to September 1943, the Italian army concentrated on keeping order in the Balkan areas it controlled and fighting the communist insurgents in Greece, Albania and, above all, Montenegro, Croatia and Slovenia.

The losses suffered by the Italians in the Balkans were divided among some thirty divisions and most operations were on a small scale. Still, the Italians experienced rather heavy casualties: 15,000 Italians died fighting in former Yugoslavia and an unknown number, probably several thousand, in Albania and Greece. Yet despite later Yugoslav claims to the contrary, the Italians held their own against the partisans. In this struggle the Italian soldier often displayed extreme brutality, as Mussolini himself had ordered. On the other hand, the average Italian soldier generally treated the non-hostile civilian population with respect – behaviour in striking contrast to that of German soldiers and members of the *Ustasha* forces.[34]

In anticipation of the invasion of Yugoslavia, Mussolini had held the Sixth Army in the Po Valley from June 1940 to March 1941. It contained the best units in the army, including all Italian armoured and motorised divisions. The March–April 1941 Axis conquest of Yugoslavia, in which the Italians played a rather small part, freed the Sixth Army divisions for service elsewhere. Following the arrival of Rommel and the first units of the Afrika Korps in Tripoli in mid-February 1941, the Ariete (Ram) Armoured Division disembarked in March. It was joined by the Trento Motorised Division in April. These units had been trained by veterans of the Spanish Civil War under the overall guidance of Sixth Army commander General Ettore Bastico. In mid-1937, Bastico had reformed the Italian forces in Spain after Guadalajara, and was one of the few Italian generals who understood the necessity for vigorous and effective instruction. The men of the Ariete and the Trento had benefited greatly from such hard schooling. In mid-July the incompetent and overly compliant Italian commander in North Africa, General Italo Gariboldi, was replaced by Bastico. Mussolini and Cavallero expected Bastico to hold the impetuous and highly independent-minded Rommel in check.[35]

While they never amounted to anything like the numbers needed, new and more effective tanks, self-propelled guns and armoured cars began leaving Italian factories and reaching Italian mechanised units in 1941. Selfish Italian corporate interests prevented the manufacture of German tanks in Italy under licence and retarded production of a number of promising Italian designs. Still, the new M13/40 tank did possess a turret-mounted 47-mm gun and thicker armour than previous Italian tank designs. It was severely underpowered, lacking sand filters and mechanically unreliable. But it was still a far more effective fighting machine than the flimsy L3/35s with which the Ariete had been equipped when it landed at Tripoli.

The arrival in Libya of new mechanised units, soon to be equipped with better Italian fighting vehicles, helps to explain the improved battle performance of the Italian soldier in North Africa during the spring and summer of 1941. The new units and arms, combined with better training and the harsh lessons of combat, allowed the Italians to face Commonwealth forces on something like equal terms for the first time. Throughout the remainder of 1941 and, more so, in 1942 the general quality of Italian tanks, armoured cars and the new self-propelled guns continued to rise. Numbers, however, were never sufficient to the need.

The Ariete and part of the Trento entered combat around Tobruk in April, revealing a much higher level of fighting skills than that of the Italian divisions which had invaded Egypt the previous year. Between June and August, the Ariete received enough M13/40 tanks to equip three armoured battalions (although it retained its surviving L3/35s). The sister to the Trento, the Trieste Motorised Division, arrived in Libya that August and September. Italian confidence rose. But the signal event that reversed the Italian sense of inferiority was the victory by the Ariete over the British 22nd Armoured Brigade at Bir el Gobi in late November 1941. Additional successes by other Italian divisions during Operation Crusader and in the Axis counter-offensive that followed cemented new-found Italian military pride. Despite the Commonwealth relief of Tobruk and the Axis abandonment of Cyrenaica in December, Italian morale remained reasonably high despite heavy losses. After all, the Italians had made a fighting withdrawal from Cyrenaica, not a disorganised retreat as in the previous January–February.[36]

Other important reinforcements reaching Libya in July 1941 were two battalions of highly select college volunteers from the Fascist Party youth organisation: the Giovani Fascisti (Young Fascists). The dislike by older regular army officers of initiative, political enthusiasm and inner-directed discipline in the ranks had resulted in official efforts to prevent the formation of these battalions. A near conspiracy by Fascist Party youth organisation officials, the volunteers themselves and a small number of junior officers were required to overcome such obstacles. The volunteers even had to arrange their own transport to Libya. However, the delays involved allowed the 1,600 young men to receive intense year-long training. In North Africa, the two battalions were joined to small artillery, armoured, anti-tank and motorcycle units to form a 3,000-man mechanised reconnaissance group. The Giovani Fascisti quickly gained a reputation for skill, daring and extreme courage, beginning with their victory over the 4th Indian Brigade at the second battle of Bir el Gobi in early December 1941. Thereafter, the reconnaissance group served the Italian mechanised divisions as a scouting force.[37]

However, the majority of the Italian combat forces in North Africa neither matched the proficiency nor possessed the equipment of these mechanised units. Four infantry *divisioni binarie* had survived intact the débâcle of December 1940–February 1941. Another such division, shattered in the recent fighting, was reconstituted in the spring of 1941. These five units were reorganised from mid to late 1941 with the addition of anti-tank weapons, a company of light tanks and some motor vehicles, along with a major reduction of their service units. The reordered divisions, as well as many autonomous smaller infantry units, were employed in manning fixed positions, guarding the long lines of communication back to the port of Tripoli or maintaining garrisons. The inability of these units either to advance or retreat faster than a man could walk rendered them of little other use in the North African desert. None the less, even Rommel depended on them for the sieges of Tobruk and other defensive missions. These five divisions constituted the bulk of the Italian North African forces until the middle of 1942.[38]

At this time, when the tactical and operational situation facing the Italian soldier in Libya appeared to improve, German strategy was undermining the very premise of the entire Italian campaign in Africa. At first, prospects for a huge Italian offensive to overrun north-east Africa and the Middle East seemed promising to Mussolini. In mid-April 1941, following the collapse of Greek and Yugoslav resistance, he proposed creating an army group of some twenty divisions in Libya. In addition to the 1 armoured, 1 motorised and 5 infantry divisions already there, the Duce decided to send 3 more armoured, 5 more motorised and 6 more infantry divisions to North Africa, equipped with thousands of new motor vehicles. Five divisions would guard the frontier with Tunisia, while the remainder of the army group would invade Egypt to seize the Delta and the Suez Canal. After leaving behind an occupation force in Egypt, one Italian army would advance through Palestine and Transjordan toward the Persian Gulf oil-fields. The second army would move down the Nile and reconquer Italian East Africa. That campaign completed, it would drive across central Africa to the Atlantic coast of Cameroon. Meanwhile, the Italian forces in Tripolitania would attack Tunisia from the south, simultaneously with an amphibious landing near Bizerte. The Italians would then conquer Algeria, while leaving French Morocco for partition between the Spanish and the Germans.

Even the most optimistic of Mussolini's military advisers calculated that the low production rates of the Italian automotive industry would delay the Duce's proposed invasion of Egypt until December 1941. The remainder of the grandiose series of campaigns could not begin until the autumn of 1942. However, additional planning demonstrated that the limited capacity of the Libyan ports, as well as probable shipping losses to British attacks, would delay the initiation of the Egyptian operation until late 1942. In truth it seems unlikely that the 700–800 medium tanks, 850 artillery pieces and 16,000 motor vehicles needed to accomplish Mussolini's fantastic vision ever would have been manufactured – even over a period of eighteen months. Meanwhile, the Commonwealth forces would hardly have sat quietly waiting to be attacked. Nor was Italian industrial capacity equal to sustaining ongoing operations in North Africa while building such an arsenal at home.[39]

That Mussolini could contemplate shipping two entire armies to Libya in 1941–42 revealed that he had not grasped the lessons of the Italian defeats in

Greece, Albania, Egypt and Cyrenaica. Nor had he understood the reasons for his forces' partial recovery in the spring of 1941. Given the limitations of Italian resources and the demands of mechanised warfare, Italy could have either large ineffective ground forces or a small capable army. The latter was obviously preferable. With good training, leadership and a modicum of modern equipment, the Italian soldier had shown himself to be the equal of his opponents on the battlefield. But if he was to serve in a army rebuilt in 1941–42 largely along the lines of the army of 1939–40, he was likely to be defeated once again. Ironically, Mussolini was prevented from attempting the impossible by an even more overreaching action by Hitler.

In late May 1941, Mussolini received highly confidential information – presumably from Hitler himself – that a German–Soviet war would begin in the immediate future. Mussolini ordered Cavallero to assemble a mechanised corps of three divisions in the Balkans and ready it to take part in the invasion of the Soviet Union. Such a move did not rule out an eventual Italian campaign to conquer Africa and the Middle East. Hitler confidently expected the Red Army to collapse under the blows of the Wehrmacht in a matter of weeks. Thereafter, the Germans could advance south down both shores of the Caspian Sea into Turkey and Iran, while the Italians (and the Afrika Korps) drove into Egypt, then linked up with their allies in the Syria–Iraq area. For the time being, however, Mussolini chose to divide his limited mechanised forces between the North African and Russian campaigns. He believed that he had to associate himself with the destruction of communism. The conquest of Africa would have to wait.[40]

Thus, Hitler's launching of Operation Barbarossa and Mussolini's decision that Italian forces must participate created a situation which placed insupportable demands on the Italian soldier. Mussolini had initiated a strategy that required neither an army of mass nor of quality, but one demanding both. Sizeable mechanised forces would be deployed both in Russia and in North Africa. At the same time, a huge occupation force of infantry would be maintained in the Balkans while smaller infantry armies would keep watch on the French, and guard the Italian peninsula and islands against invasion. The Italian army numbered 63 divisions in mid-1941, including 3 armoured and 2 motorised. Mussolini proposed raising another 26 divisions over the following year, including 3 more armoured and 5 to 6 more motorised divisions. While less grandiose than Mussolini's plans of April–May 1941, this augmentation of the Italian army was still wildly impracticable.

Given the balance of power within the Axis in mid-1941, Mussolini had no choice but to follow Hitler's strategic lead. However, Mussolini decided to commit significant forces not only to help the Germans pursue their goals in the east but also to pursue Italian aims in the Mediterranean region. (In late July Mussolini decided to despatch a second motorised corps to the Russian Front. In August he offered Hitler an army of nine Italian divisions for the Eastern Front.) Hitler may have enjoyed the military resources to mount a massive invasion of the Soviet Union as well as to maintain the

Afrika Korps in Libya. Mussolini did not command the forces to carry out a parallel set of operations – at least with any reasonable chance of success. But Mussolini's megalomania drove him forward and the Italian soldier paid the price.[41]

In reality, Mussolini could send only three divisions to Russia in 1941, none of them fully motorised. The Corpo di Spedizione in Russia (CSIR) numbered 62,000 men divided among two *autotrasportabili* ('truckborne') infantry divisions, one *celere* ('mobile') division, a Blackshirt *raggruppamento* and a corps artillery group. While not motorised, the CSIR possessed enough trucks to move its infantry units forward in shuttles. The *celere* division was one of three such hybrid formations in the Italian army. It consisted of two horse cavalry regiments, one truck and motorcycle-equipped Bersaglieri regiment, a battalion of light tanks and an artillery regiment composed of both horse-drawn and motorised battalions.[42]

Untrained and unequipped for mechanised warfare, the CSIR crossed the Balkans into Ukraine in July, trailing behind the lightning German advances. The German High Command broke its promises to Mussolini not to employ the Italian units separately. Instead, the Germans committed the three Italian divisions piecemeal in rearguard actions between the Bug and the Dneister. While the Germans refused to supply the Italians with food, fuel or transport, they confiscated all booty and prisoners captured by the CSIR. Constantly strung out for hundreds of miles, the CSIR finally entered combat as a united corps during the encirclement of Kiev in September. Following that action, in which it captured over 10,000 Red Army prisoners, the Italians moved into the Donetz Basin in October–November. They encountered only light resistance, but continued to suffer from transport problems. Finally, the CSIR commander, Lieutenant-General Giovanni Messe, halted his force about twenty-five miles north-east of Donetz. The Italians dug in at Gorlovka along the Donetz–Donbas canal in early December 1941. After the forward Italian positions were probed by enemy patrols, Messe decided a Soviet attack was imminent. He rejected German orders to continue the advance and had the CSIR site its artillery and perfect its defences. As a result, the Italians were able to hold off a series of ferocious Red Army attacks in the 'Christmas Battle', until rescued by the intervention of a Panzer battalion.

Messe had done his best with available resources to prepare his men for winter warfare. But many of the clothing and footwear shortcomings that had bedevilled the Italians in Albania returned to torment the men of the CSIR. Mussolini and Cavallero had accepted Hitler's assurances that the Soviet Union would collapse before the snow fell. As a result, the troops of the CSIR were not issued proper winter gear. In frigid weather from late December 1941 to late March 1942, the Italians alternated defence against Russian assaults with their own local counter-attacks to regain positions overrun. During this three months of almost incessant combat, Italian casualties from frostbite (over 3,600) rivalled combat losses. Given its insufficient number of anti-tank guns and the inadequacies of its light tanks, the CSIR repulse of repeated attacks proved the mettle of its troops and the skill of its commander. The spring thaw ended the ordeal of Messe's men.[43]

The immensity of the German–Soviet struggle meant that the heroic ordeal of the CSIR counted for very little in the outcome of the Barbarossa campaign. But in the context of the Italian war effort, the 5,500 motor vehicles and the 16 modern 75-mm dual-purpose anti-aircraft/anti-tank guns (the army had only 115) assigned to the Russian expedition represented a precious resource. In all North Africa at the time, the Italian army had just 6,500 functioning trucks. Italian industry produced only 27,000 in 1941. By August 1941 the army general staff was forced to beg civilian transport agencies for vehicles to send to Libya. It managed to acquire 280. By early December 1941, after Hitler and Mussolini had declared war on the United States, the Italians faced a future in which their industrial disadvantages *vis-à-vis* their enemies could only grow worse. In November, however, Mussolini had finally persuaded Hitler to accept another seven Italian divisions for service on the Eastern Front in 1942. At a time in which any fleeting chances for an Italian victory required a concentration of effort – and the North African theatre was the only choice that made strategic sense – Mussolini had decided on an additional dispersal of his few mechanised units.[44]

The diversion of vehicles and artillery to the CSIR helps to explain the Axis defeat in Cyrenaica in late 1941. In January 1942, however, Rommel began an unauthorised offensive toward the Egyptian frontier without informing the Italian High Command. His success in reconquering two-thirds of Cyrenaica gained grudging Italian acquiescence after the fact. But Cavallero, cognisant of the need to send even larger Italian mechanised forces to the east in 1942, extracted a promise from the Afrika Korps commander to limit any Axis operations in North Africa to local attacks. The chief of the supreme general staff also refused Rommel reinforcements to ensure compliance. Meanwhile, aided by their new American allies, the British command was able to rush enough arms and equipment to the battle front to establish a solid defence. Neither side moved from early February until late May 1942.

In the late winter Italian operations on all fronts came to a halt due to the weather, lack of supplies and strategic indecision on the part of the High Command. The Italian army had grown to unprecedented size, numbering eighty-six divisions, three short of the goal set by Mussolini in mid-1941. Most of these formations lacked sufficient vehicles and arms, however. Cavallero decided that the army must remain largely on the defensive in 1942, waiting until it received enough equipment to resume major offensive operations in 1943. He expected the decisive campaigns of the war to take place in 1944. In early 1942 most of the Italian army consisted of ill-equipped infantry divisions on occupation or counter-insurgency duty in the Balkans, guarding the Italian homeland, keeping watch on the French or undergoing training and refit. The army was deployed as follows: in Croatia, 12 divisions; in Albania, 6 divisions; in Greece, 8 divisions; in the Dodecanese islands, 3 divisions; for a total of 29 divisions in the Balkans. Along the French Alps, the Italians deployed 6 divisions; on Sicily and Sardinia, 3 divisions each; in the rest of Italy, 30 divisions, for a total of 42 in the peninsula or the islands. In Libya the Italians had 7 divisions with one more on its way; in Ukraine, 3 divisions. Very few of these formations were at a high level of readiness, however.

Leaving aside the 11 divisions deployed to Libya or Ukraine, 7 divisions in Italy and the Balkans were being readied for movement to the Eastern Front; the Littorio (Lictor) Armoured Division was on its way to North Africa; 2 motorised divisions were preparing for shipment to Libya. In addition, a new paratroop division, the Folgore (Lightning), and a new 'air-transportable' division, the Spezia, were fit for operations. The Centauro Armoured Division was nearly up to strength after re-equipping with the new M14/41 tanks, a supposedly improved version of the M13/40. Both were stationed in Italy. These 24 divisions were at or close to full manpower levels, reasonably equipped and trained for war. The other 62 divisions ranged from good to fair in terms of morale, combat effectiveness and force levels. Some, including a number stationed in the Balkans, were effective fighting formations in terms of light infantry counter-guerrilla operations. But none of these three score divisions was capable of operations against mechanised forces in Libya or on the Eastern Front. In 1942 that would be all that really mattered.[45]

*

In the spring and summer of 1942 the Italian army was at the peak of its Second World War effectiveness in regard to training, equipment, morale and tactical-operational leadership. But the strategy pursued by Mussolini and implemented by Cavallero would render the efforts of the Italian fighting man futile. Mussolini sent two more corps to be swallowed up on the Eastern Front. To the great detriment of the forces in North Africa, Cavallero lavished upon the corps destined for Russia much of the modern artillery and motor transport painfully acquired from Italian industry or the Germans over the previous year.

The Armata Italiana in Russia (ARMIR), formally the Italian Eighth Army, was established in April 1942. When joined to the CSIR, the army would consist of 229,000 men, 22,300 motor vehicles and 1,100 artillery pieces of all types. Cavallero sent only 50 new light tanks and self-propelled guns to Russia. But he bestowed the best artillery the army possessed on the ARMIR: the unique regiment of 210-mm howitzers, 36 of the 51 modern 149-mm guns available, 24 of the 38 149-mm Krupp howitzers the Germans had given the Italians, 72 of the new 75-mm howitzers and the only 75-mm anti-tank guns in the army (36 in all).

In contrast, the North African forces had nothing approaching the powerful 210-mm howitzer – not even artillery of First World War vintage – and no modern 75-mm howitzers or anti-tank guns. They possessed only 12 Italian and 14 Krupp 149-mm pieces. Given Axis control of the central Mediterranean in the first half of 1942, most of the vehicles and artillery destined for the Eastern Front could have been shipped to North Africa. Only within the logic of Mussolini's strategic fantasies did the actual allocation of motor vehicle and artillery resources make sense. Adding to his mistakes, Mussolini gave command of the well-endowed Eighth Army to none other than Italo Gariboldi. He had been selected purely on the basis of seniority.

The ARMIR sped across the Balkans by rail. Once in the Soviet Union, however, it had to move by road. Despite its immense vehicle train, it could only move forward one corps at a time. One Alpini corps and one corps of infantry, each supplemented by a Blackshirt *raggruppamento*, plus an additional division to guard lines of communication, departed from Italy between early June and late July. The CSIR struggled forward to the Don near Stalingrad, its 3rd and 6th Bersaglieri regiments, backed by artillery, driving off a heavy Soviet tank attack at Serafimovich in the process. Joined by II Corps in early August and the 6th Alpini Regiment at the end of the month, Messe's expanded command managed to stave off another series of Soviet counter-attacks between 20 August and 1 September. The men of the Sforzesca Division distinguished themselves in this action.

The bulk of the Alpini Corps arrived in late September. With that, Gariboldi dissolved the CSIR, created a new corps structure and accepted three German divisions, enough to create a fourth corps for his army. Stiffening Red Army resistance convinced Messe that the Axis forces could push no further in Russia. Since the ARMIR had no medium tanks save fifty-odd in one of its German divisions and held the northern flank of the Stalingrad salient, Messe expected another Soviet counter-offensive aimed at the Italians once the weather froze. He urged Gariboldi to create a deep zone of field fortifications and prepare a mobile defence plan. Gariboldi refused. He remained optimistic that the Germans soon would make new advances. After all, he had been given the Alpini Corps (with its 20,000 mules) to join the coming drive through the Caucasus: this was part of the Axis strategic design to overrun the Middle East in 1943. These differences between Gariboldi and Messe led to Messe's repatriation in early November. His requests to Rome for proper winter clothing, lubricants, sledges and rations had been rejected. Gariboldi ignored these matters.[46]

While the Eighth Army had been assembling in Italy, Cavallero had decided on one offensive operation in the Mediterranean theatre: the seizure of Malta. Italian possession of the island would greatly facilitate the transport to Libya of the army group which he and Mussolini expected would invade the Middle East in 1943. Hitler, however, felt uneasy about leaving the initiative in North Africa to the British for the rest of 1942. Furthermore, the painful losses inflicted on German airborne units during their assault on Crete in May 1941 made Hitler doubt Italian ability to seize Malta using the Folgore and Spezia Divisions. A compromise was reached. The Axis forces in Cyrenaica would drive forward to the Egyptian frontier in late May–early June and set back any Commonwealth plans for a new offensive. A combined German–Italian airborne assault would take Malta in July. Then Libya could receive the huge reinforcements planned by Mussolini. While one Axis army invaded Egypt, other Axis armies would sweep through the Caucasus. A link-up in the Middle East would take place sometime in early 1943.

In the attack launched by Rommel in late May the Italian forces performed superbly. Their fighting qualities, especially those of the Ariete and Trieste Divisions, played a decisive role in the unexpected capture of Tobruk in late June and the rout of the Commonwealth forces which followed. When Rommel argued that the supplies captured would allow him to drive to the Suez Canal, the Axis leaders agreed to cancel the Malta operation and commit the forces designated to the Egyptian offensive instead. The promising start of the German summer offensive on the Eastern Front seemed to presage the Axis envelopment of the Middle East envisioned by Hitler and Mussolini. Rommel's forces did reach El Alamein in late June. But repeated assaults, in which the newly arrived Littorio Division fought with skill and courage, and Commonwealth counter-attacks, led only to a stalemate in July. Both armies waited for reinforcements.

All Rommel received, however, was the Folgore Paratroop Division and the Ramke Paratroop Brigade, previously destined for the Malta operation. The Folgore had outstanding qualities, being composed of picked men, trained and motivated beyond any other large unit in the Italian army. Still, it was only a light infantry division of 5,200 men. Rommel needed more tanks, trucks and guns. The Italian arms and equipment on their way to the Don might have given him the power to punch through the Commonwealth defences. As it was, despite an all-out effort, the Axis army was repulsed at the Battle of Alam el Halfa, 30 August–5 September 1942. The Trieste, Folgore and Brescia Divisions fought with great skill and courage. The fine performance of the Brescia, just an ordinary infantry division, indicated the great improvement in Italian fighting capabilities since early 1941. But such effectiveness and determination were not enough to overcome the material advantages of the enemy. Thereafter, the initiative passed to the British Eighth Army.[47]

On 23 October the Commonwealth forces began a massive attack on the Axis lines at El Alamein. Both Germans and Italians resisted beyond any reasonable expectations. By the time the Axis began its retreat on 5 November, the extraordinary fact was not that the Commonwealth forces had broken through but that the Axis forces had come so close to stopping them. Of the eight Italian divisions, only the Trieste managed to escape more or less intact. A few scattered individuals and sub-units from the other Italian divisions avoided death or capture. Even some paratroopers from the Folgore, which had been effectively annihilated, managed to make their way back to join the tail end of the Axis retreat. However, had Rommel and the Italian High Command agreed, the provision of motor transport might have saved the Folgore, Brescia and Pavia divisions. As it was, abandoned without water or vehicles in the desert on the southern flank of the Axis lines, the survivors of the three units had little choice but to surrender.

Commonwealth forces' exhaustion, lack of fuel, rain and Montgomery's cautious pursuit allowed the Axis forces to make a relatively slow retreat back into Cyrenaica. There the Spezia, Pistoia and newly formed Giovani Fascisti Divisions had dug in. Thus reinforced, the Axis army attempted to halt the Commonwealth advance. This proved impossible. The Italians lacked sufficient vehicles for a mobile defence and the Commonwealth forces could always outflank them. The Axis army had no choice but to fall back across Libya. Tripoli fell to the Commonwealth forces at the end of January 1943. But German troops had taken over Vichy France and the eastern half of Tunisia in early November 1942. (At the same time, Italian forces had occupied Corsica and southern France as far as the Rhone.) The retreating Axis forces, joined by the Centaro Armoured Division, were able to escape into the neighbouring French protectorate. There, using terrain advantageous for the defence, the Italians would make their last stand in Africa.[48]

During the same period an even worse catastrophe had enveloped the ARMIR. The initial phases of the Soviet Stalingrad offensive spared the Italians. But Gariboldi had neither made plans for a withdrawal in stages nor stationed his army's motor vehicles for easy access in case of need for a speedy retreat. The Red Army struck at the IInd Corps (Cosseria, Ravenna and Torino Divisions; 3 Gennaio Blackshirt *raggruppamento*) on 11 December 1942. The Italians held off the attack for a week, despite repeated assaults from the greatest concentration of Soviet tank forces yet seen on so narrow a front, some 750, including 400 KVs and T-34s. When II Corps finally was forced to fall back, however, the gap in its lines required the XXXVth and XXIXth Corps to withdraw as well. Despite the large motor park with which the ARMIR had been endowed, its troops were forced to retreat on foot. The march towards the rear soon degenerated into a frozen nightmare, punctuated by continuous Soviet bombardments and armoured assaults.

The Alpini Corps had not been subjected to the same intense pressures and managed to hold in place against Red Army attacks until 17 January. Then, after being ordered to pull back, it too came under ferocious assault by the weather and the enemy. Thanks to their superior leadership, discipline and cold-weather training, the Alpini held together better than the regular infantry of the ARMIR. For all practical purposes, the other corps disintegrated. But only the alpine Tridentina division survived the retreat as an integral formation. The heroic Julia had been reduced to a few half-frozen squads and platoons. Virtually all the equipment of the ARMIR had been lost.

By early February 1943, 117,000 Italian troops had reached safety behind the new German lines. Several thousand straggled in later. Nearly 50,000 Italian soldiers had been evacuated home over the previous eighteen months for wounds, sickness or severe frostbite. The rest of the Eighth Army had died of exposure, fallen in battle or been taken prisoner. 75,000 Italian soldiers died in Ukraine and Russia, the great majority in the seven weeks between mid-December 1942 and early February 1943. These horrendous losses were over one-fifth of all Italian combat deaths between June 1940 and September 1943. In March the shattered remnants of the ARMIR returned to Italy under official silence. Only some 10,000 Italian captives ever returned from Soviet prison camps.[49]

The remainder of 1943 brought three more great defeats to the Italian army, the final one a catastrophe of unprecedented proportions. Between December 1942 and May 1943 the army conducted a skilful defence of Tunisia. During this five-month campaign, the Italian soldier gave the best account of himself in the entire war. One reason was the high quality of the troops employed. To the hardened veterans of the battles in Egypt and Libya were joined a small number of highly trained and well-equipped units brought over to defend the North African redoubt. Formed into the First Army, these included the Giovani Fascisti, Trieste, Spezia, Pistoia and Centauro Divisions, the army-division-size Raggruppamento Sahariano, as well as the Monferrato and Nizza armoured car groups and a sizeable number of autonomous Blackshirt, artillery and combat engineer units. To this formidable formation the Germans added three divisions of their own and a powerful Flak group. Mussolini gave the command to the best of his generals, Giovanni Messe. The Italian First Army defended southern Tunisia along the French-built Mareth Line against the advance of the British Eighth Army.

The Italians fought extremely well against very heavy odds. The Centauro Armoured Division took part in the bloody repulse of the American II Corps at the Kasserine Pass in late February. Messe inflicted a series of sharp reverses on Montgomery's forces, first at the Mareth Line in March, then in the Enfidaville area in April. Yet it was all in vain.

The strategic premise of the Axis defence of Tunisia, a bridgehead from which to launch the reconquest of North Africa, was doomed from the start by the material superiority of the American and Commonwealth forces. Once that became clear, the only sensible option was the withdrawal of the Axis forces to provide precious equipment and irreplaceable combat forces for the defence of the Mediterranean shore of *Festung Europa*. But the German High Command refused even the aerial evacuation of a few hundred Italian tank crews, specialists who would have been of enormous value to the two new armoured divisions then forming in Italy. Instead, when the last of Messe's men surrendered on 13 May, the Tunisian campaign had consumed over 100,000 Italian killed, wounded and prisoners. The capitulation ended a 35-month campaign which had cost the Italian army some 38,000 dead in the field, another 17,000 lost at sea from North African convoys and hundreds of thousands of prisoners.[50]

*

The loss of so many men and arms in Tunisia made it difficult for the Italian army to contribute much to the defence of Sicily two months later. The island was garrisoned by the Sixth Army, only a shell of its former self with 9 weak divisions, 2 brigades, 1 independent infantry regiment and some 20 Bersaglieri, Blackshirt, tank and self-propelled gun battalions: in all, some 192,000 Italian combat troops and 60,000 service personnel. Of the larger combat units all save 4 divisions were immobile 'coast defence' units. Even the 4 so-called manœuvre divisions were largely without motor transport. Three had trucks sufficient to move only one battalion at a time. One division, the Livorno, could be considered semi-motorised.

Considering that the Allied landings on Sicily were larger than those that would take place at Normandy eleven months later, the Italians and their German allies (three divisions) resisted well at first. The Livorno and Hermann Goering Divisions made a fierce counter-attack against the American beach-head at Gela. In this they were joined by the Italian CI tank battalion with its obsolete French Renault-35 tanks. However, American warships offshore responded with a highly accurate naval gunfire barrage. It annihilated the CI battalion and smashed the Livorno, which lacked the means to pull out of range quickly. (The mechanised Hermann Goering division was able to withdraw.) Less significant Italian resistance to other Allied landings was disposed of in much the same fashion. Thereafter, while the Germans conducted a skilful withdrawal across Sicily and the Straits of Messina, much of the Italian Sixth Army collapsed or surrendered. By the end of the campaign on 17 August the Italian army had lost 4,500–5,000 dead and 117,000 prisoners. About 95,000 troops, wounded or well, had escaped to the mainland. The rest of the Sixth Army, perhaps 35,000 men, seems to have deserted. This was a sign of worse to come.[51]

On the orders of King Vittorio Emanuele III, the army had overthrown Mussolini on 25 July. Neither the M Division, then forming on the outskirts of Rome, nor any other Blackshirt unit intervened to rescue the Duce. The men of the militia were just as weary of the war as every other Italian. The new government, headed by Marshal Pietro Badoglio, began secret negotiations with the Allies. The Italians hoped to withdraw from the war. The British and Americans insisted that Italian neutrality was impossible. The best that the Italians could do would be to surrender to the Allies and allow them to come ashore to fight the Germans. Reluctantly and hesitantly, the Badoglio government began to prepare to switch sides. Meanwhile the Germans had deduced that their ally was about to abandon them. Using a mixture of threats and trickery, the Germans occupied the Alpine passes, seized commanding positions in Italian-occupied Europe and moved large forces into Italy.

The ambivalent manner in which the military representatives of the Badoglio government had negotiated the covert armistice caused great suspicion among the Allied leadership. As a result, the British and Americans would not reveal the date of their landings in Italy. This limited what the Italian High Command could do to prepare to welcome the Allies and resist the Germans. Yet it did far less than it could have done, even without knowledge of the Italian D-Day, leaving the officers and men of the army in ignorant helplessness.

Without warning, the British Eighth Army began landing in Calabria on 3 September. Four immobile Italian 'coastal' divisions defended the beaches; the

Mantova Infantry Division and the German 29th *Panzergrenadier* Division
waited further inland. Caught by surprise and pummelled by naval gunfire,
the defenders offered scant resistance and fell back up the toe of the Italian
boot. By 8 September the Commonwealth forces had advanced to the borders
of Basilicata with very few casualties.[52]

A great many Italian soldiers, from generals to privates, had come to hate
and fear the German army by September 1943. They had seen how the Germans had mistreated and murdered Jews in France, the Balkans and the Soviet
Union. They had witnessed similar German atrocities against civilians and
military prisoners throughout Eastern Europe. At the same time, the Italians
recognised the superiority of German weapons, fighting skills and generalship.
They knew well how the German army would treat those it considered traitors.
These attitudes strongly influenced the king and the High Command. The
thought of capture by the Germans terrified them. They had decided to flee
Rome in secret at the public proclamation of the Armistice.

Late on 8 September the Allies announced the agreement with the Italian
government. Early on 9 September American and British forces began landing
in the Salerno area. In the meantime, the king, Badoglio and the entire general
staff had slipped out of the capital, leaving the Italian army without orders save
the cryptic advice: 'Resist the enemy.'

In the Rome area Italian forces roughly equalled the Germans. Since all roads
did lead to Rome, had the Italians been able to hold the city, German forces to
the south would have been trapped. In that case, Italy south of Rome might
have been cleared of the Wehrmacht by October 1943. But lacking high-level
direction and any plans, the Italian troops defending the capital were able
to resist for only one day. Elsewhere across Italy and southern Europe most
of the army collapsed from fear of the Germans or a desire to end what seemed
a hopeless struggle. Only in places empty of Germans were major Italian units
able to maintain discipline and retreat to the south. Sixteen divisions, of which
seven were 'coastal' formations, managed to escape. Six hundred thousand
Italian soldiers were captured by the Germans; another 19,000 were killed or
wounded fighting them. Hundreds of thousands of other Italian troops fled to
their homes in Italy or took to the mountains in the Balkans. Tens of
thousands would join the Resistance. About 30,000 Italian troops took refuge
in Switzerland. A few Italian units resisted for some time before surrendering to
the Germans. The men of these formations, such as the Acqui Division on
Cephalonia and the garrison of Leros, were shot without mercy.[53]

The shame resulting from its September 1943 disintegration continues to
hang invisibly over the Italian army even today. The bravery of the reconstituted Italian units which fought alongside the Allies in the later stages of the
Italian campaign and the valour of those Italian soldiers who became *partigiani*
did not erase the national sense of dishonour arising from the failure to defend
Rome and smite the German invader. But to blame the Italian soldier for
those lost opportunities is a grave injustice. Vittorio Emanuele III and the
general staff share that opprobrium.

Throughout the war, whether under Mussolini or Badoglio, the Italian soldier had done his best, fighting for what he was told were his country's interests and honour. Three hundred and fifty-five thousand died in battle, succumbed to wounds or perished in prison camps. The incompetent leaders at the summit of the fascist régime and its military-monarchist successor failed to match such selflessness and dedication. The Italian soldier deserved better, even considering the purpose of his war. That he fought for an evil cause in 1940–43 is incontestable. That he fought heroically is undeniable.[54]

The Italian Job: Five Armies in Italy, 1943–45

Richard Holmes

My story is an old one, at least as old as the twentieth century and, in some key aspects, as old as war itself. We must disabuse ourselves of the notion that there was a sharp gulf between the soldier's experience of the First World War and that of the Second – that the former was a soldier's war and the latter, somehow or other, was not.

Of course, a comparison at this very general level misses the important point that each of the world wars was not one conflict but many. In the First not only was the Western Front different from, say, East Africa, Mesopotamia and Gallipoli, but sectors even within the Western Front were themselves so dissimilar that they might almost have been separate wars. This was even more true of the Second World War: it is not altogether fanciful to suggest that an unlucky individual might manage to serve in France in 1940, in the desert in 1942, in Italy in 1943 and in Burma in 1945, getting, in terms of experience of different theatres, four wars for the price of one.

My broad assertion is that the campaign in Italy in 1943–45 was as much ancient as modern. Of course most of the essential ingredients of modern war were present: strategic bombers; ground-attack aircraft; tanks; and even, in the shape of the FX-100 *Fritz* glider-bomb which sank the Italian flagship *Roma* and damaged HMS *Warspite*, precision-guided munitions. But when many of the combatants strove to find yardsticks to measure their experience in Italy, they looked back to the First World War. General Fridolin von Senger und Etterlin, commanding XIV Panzer Corps on the Cassino front, wrote that:

> In contrast to the wide-ranging mobile battles in Russia, the conflict here resembled the static fighting of the First World War...What I saw took me back across twenty-eight years, when I experienced the same loneliness crossing the battlefield of the Somme. Hitler was right when he later told me that here was the only battlefield of this war that resembled those of the first.[1]

As John Ellis has written in his masterly study of Cassino: 'Comparison with the charnel houses of the Western Front, between 1914 and 1918, is in no way misplaced. The battles for Cassino should be borne in mind by anyone who might think that World War Two was any sort of soft option for the men at the sharp end.'[2]

As G. D. Sheffield has already told us (pp. 29–39), analogies with the First World War went well beyond this level of connective personal experience, as frustrated officers and men found, in the failures of 1914–18, ugly parallels with their own miseries. When the *Daily Telegraph*'s war correspondent Christopher Buckley reached Sicily he experienced 'a sinking of my heart. It seemed so terribly reminiscent of 1916, of a war fought in terms of advance of a few hundred yards over shell-torn ground, every yard purchased with a man's life. There had been so little of this in Africa.'[3] Major-General John P. Lucas, the corps commander at Anzio, wrote: 'This whole affair had a strong odour of Gallipoli: and apparently the same amateur [Churchill] was still on the coach's bench.'[4] He might have gained an ironic satisfaction from the fact that Churchill was complaining that the stalemate at Anzio was really 'Suvla Bay all over again'.[5] And in March 1944, when Lieutenant-General Bernard Freyberg was grappling with the knotty problem of attacking Monte Cassino, he declared: 'Reminds you of Passchendaele, doesn't it? But we'll have no more Passchendaeles.'[6] His confidence was not, alas, justified by events.

What follows is an assessment of the soldier's experience of the battle for Italy. It was a battle which began when Allied forces landed at Reggio Calabria, at the tip of Italy's toe, on 3 September 1943, and ended with a cease-fire on 2 May 1944 which followed German unconditional surrender. In the process Allied forces mounted two major amphibious landings, at Salerno on 7 September 1943 and Anzio on 22 January 1944, and forced a series of German defensive lines running east–west across the leg of the boot. The assault on the Gustav Line in the first third of 1944, including four distinct battles for Monte Cassino, had a ghastly quality all of its own. Over a year and a half the campaign cost the Allies 312,000 killed, wounded and prisoners and the Germans some 435,000. This is not the place for an analysis of the campaign from the strategic or operational point of view, but it is hard to disagree with Dominick Graham and Shelford Bidwell that 'it was a war of attrition designed to wear down the German strategic reserves', and in its essentially attritionalist nature we have yet another link with Allied conduct of the First World War on the Western Front.[7]

Graham and Bidwell point out that 'in simply the grim figures of attrition the Allied effort in Italy was profitable and not unduly severe'.[8] On the other hand, an American official history was decidedly sombre in its assessment of the campaign, declaring:

> It is difficult to justify the heavy investment of Allied troops and material into the Mediterranean theatre during 1944...While the Allies did tie down a significant number of enemy divisions in Italy, it was often not apparent in 1944 whether it was the Allies or the Germans who were actually doing the tying down.[9]

Cynics might see this as a classical case of a post-war assessment reflecting a wartime division of Allied opinion, for, in sharp contrast to the British, the Americans had deep-seated objections to the Mediterranean theatre. I will return to the question of casualties later on, but it is important to note, even at this early stage, that what might seem like relatively profitable attrition from a strategist's point of view looked markedly different to the officers and men – by and large concentrated in the rifle companies that made up the cutting edge of armies in the campaign – that actually did the attriting.

The five armies of my title are the British, American, German, French and Polish, although I confess to wandering beyond these self-imposed constraints to consider, at least in passing, the Canadians, New Zealanders and Indians. Some years ago, in my book *Firing Line*, I inveighed against 'the historian as copy-typist' and one of the problems with this sort of talk is the temptation simply to stack quotes, cemented by a little judicious comment. Although I hope to keep well away from that, I really must give you just three quotes to leave you in no doubt as to the sort of human existence that forms the basis for our reflections.

Two members of that robust Midland regiment, the Sherwood Foresters, described Anzio:

> We used to pick up the rations at night and then bring out any dead and then they were buried next day in the cemetery... You could smell them a mile away... After three months it was demoralising... It was every night, every night everybody was hunting Germans, everybody was out to kill anybody... we was insane... We did become like animals in the end... Yes, just like rats... It was far worse than the desert. You were stuck in the same place. You had nowhere to go. You didn't get no rest, like in the desert. No sleep... You never expected to see the end of it. You just forgot why you were there.[10]

And now the French commander of an Algerian *tirailleur* company in the epic struggle for the Colle Belvedere in late January 1944:

> Visibility less than two yards. Pause every 150 feet of the climb. Arduous ascent. Muscles and nerves at breaking point. The rumble of enemy artillery and mortars... Explosions 60 feet above us. Men slip, fall six or seven feet, risk breaking their backs... Such human suffering! Packs, weapons. I joke with the *tirailleurs*. Night black, visibility zero, we trample over corpses; they're ours, one with no head, his guts spilling out.[11]

And finally, a British surgeon describes his casualty clearing station at Anzio:

> some (too many, far too many) were carried in dying, with gross combinations of shattered limbs, protrusions of intestines and brain from great holes in their poor frames torn by 88-millimetre shells, mortars and anti-personnel bombs. Some lay quiet and still, with legs drawn up – penetrating wounds of the abdomen. Some were carried in sitting up on the stretchers, gasping and coughing, shot through the lungs... All were exhausted after being under continuous fire, and after lying in the mud for hours or days.[12]

The title 'The Italian Job' is not simply an attempt at an amusing aside: it goes straight to the heart of the question of motivation in the campaign. There were many combatants for whom the struggle was a job – a cold, dirty, wet and dangerous one, but a job none the less. Their aim was to get it done and get home safely. Sapper Richard Eke, of 754 Army Field Company Royal Engineers touched one of the oldest truths of combat when he described the relationship between his decidedly non-martial section and its commander, a veteran of the Spanish Civil War: 'Slightly mad and brave [what an interesting juxtaposition of words], and a little out of place in a section that hoped to avoid being heroic at all cost. He volunteered for everything, and we had never seen him show any fear.'[13] On one occasion the corporal's enthusiasm infected even Sapper Eke, who became involved in a fire fight. 'The experience was terrifying yet exhilarating,' he wrote. 'Against all my instincts I had put my life at risk, and told myself, never again would I be infected with this front line madness.'[14] When a group of Special Forces men passed by, Eke mused that: 'we had seen and almost envied the opposite type of soldier to ourselves, dedicated mercenaries that turn up in every war and fight for the sheer danger and love of fighting'.[15]

Almost any unit, in almost any army, contains a blend of Dick Ekes and Corporal Pearsons. A military sociologist would, I suppose, describe it as a relationship between those whose ideal organisational model is occupational (hence my emphasis on the job) and those who are attracted by the institutional model. Captain J. B. Tomlinson, a wartime soldier commissioned from the ranks, had a different view from his fellow Royal Engineer, Sapper Eke. 'The army is not, never has been and never can be, a democracy,' he wrote. 'Democratic methods are anathema, a curse, a bar to efficiency.'[16] This view was not typical even in the more traditionalist British army, while for the Americans the questionnaires administered by Samuel Stouffer and his colleagues point to 'a swelling chorus of discontent especially amongst the better educated enlisted men', who were especially hostile to training methods which seemed to them to be geared to the most stupid.[17] 'I thought the caste system was restricted to India,' moaned one American private. 'These officers think they are tin gods or the next thing to it.'[18]

The causes and nature of this mix and the effect of its chemistry on combat performance are almost infinitely changeable. National preferences, military culture and the combat variables – terrain, climate, weather, logistics, leadership, training, the type and intensity of enemy resistance and the duration of hostilities – all played their part. The chemistry was different in each of my five armies. We must resist any attempt to place them on a sort of linear graph, with, say, apparently totalitarian and militaristic Germans at one extreme and democratic American citizen-soldiers at the other. This sort of chemistry is not amenable to such relatively simple classification. On the one hand it was possible to find Germans who deserted enthusiastically at the first opportunity, and laughed heartily, as did a jovial Austrian confronted by Lieutenant David Cole of the Inniskilling Fusiliers, when asked if they were Nazis.[19] On the other there were American units whose freebooting spirit would not have been out of place in the Thirty Years War. Lieutenant Gus Heilman's platoon of the American-Canadian Special Service Brigade took over the village of Borgo Sabatino, renaming it Gusville and sallying forth from it to wreak mayhem. 'We always got a kick out of driving jerries upstairs,' recalled one of what the Hermann Goering Division called 'The Black Devils', 'and then blowing the building out beneath them.'[20]

Yet there were important national differences in the chemistry of combat. Stouffer and his colleagues found that little attention was paid to 'idealistic motives – patriotism and concern about war aims'. The principal desire was to get the job done and go home. 'What are we fighting for?' asked one veteran,

> Ask any dogface in the line. You're fighting for your skin in the line. When I enlisted I was patriotic as all hell. There's no patriotism in the line. A boy up there 60 days in the line is in danger every minute. He ain't fighting for patriotism.[21]

In contrast, heroic leadership by officers and NCOs was regarded as fundamentally important. 'In combat', reported a US infantry platoon commander, 'you have to be out in front leading men, not directing them from the rear. The men say, "If the officer's going to stay back a hundred yards, then I'm going to stay with him." You can't direct them – you have to lead them.'[22] And even in the US army, whose regimental structure seemed so ephemeral to the British, the bonds of mateship woven within good units tied men willingly to the altar of battle.

> The fellows don't want to leave when they're sick [reported an American soldier recovering in hospital]. They're afraid to leave their own men – the men they know. They don't want to get put in a different outfit. Your own outfit – they are the men you have confidence in. It gives you more guts to be with them.[23]

Much the same is true, albeit in a subtly different way, for the British army. We have a hint that all was not well in Dick Eke's unit when he tells us: 'the whole campaign from Sicily had been conducted without the visible intervention of officers and company sergeants. The Platoon Officer was alright, but we did not place much importance on him.'[24] Rifleman Alex Bowlby's battalion, burnt out by the combined effects of triumphs in the desert and subsequent restructuring, had officer leadership of mixed quality, but when it was good it was very good indeed. His company commander, Captain Kendall, turned a shaky company into a passably good one by public displays of sheer guts. 'Look at me,' he said quietly, walking from man to man under fire. 'They can't hit me. Look at me.'[25]

The regimental system unquestionably played its part in promoting cohesion. At its best it worked very well indeed, as it did with the sister battalions of the Queen's Royal Regiment in 131 and 169 Infantry Brigades. These units were very close-knit, and Brigadier Geoffrey Curtis remembered that when the two brigades met there were 'emotional scenes as brother met brother and sons and fathers were briefly united'. He pointed out that for the Queensmen: 'It was a family business and once you were accepted into that family you had an inner feeling of confidence and belonging.'[26] Alex Bowlby's battalion was admittedly past its best, but he clearly remembered one morning when it had to march up a road lined by tanks of the King's Dragoon Guards, who could not get forward until the way had been opened by infantry. The tank crews applauded as the riflemen went on to do their dangerous job. Bowlby knew that he might have been going up to face his death, but wrote: 'I wouldn't have changed placed with anyone.'[27]

We must not get too misty-eyed about the pulling-power of the regiment. In a sense the system had always worked best when it was needed least: in peacetime or small wars. By 1944 it was in tatters, as it had been a generation before in 1917, and scarce replacements arriving in theatre might be posted haphazardly to whichever regiment had the greatest need of them. One battalion in 56th Division received reinforcements from fourteen different regiments just before it went into the attack. Its commanding officer urgently requested a supply of cap badges, 'so that', as he put it, 'at least the men could have with them the badge of the regiment in which they had to be prepared to die'.[28]

Often, too, what may seem the triumph of the regimental system reflects the deeper comradeship of the rifle section or tank crew. As C. E. Montagu had written of an earlier war:

> whatever its size a man's world was his section – at most, his platoon; all that mattered to him was the one little boatload of castaways with whom he was marooned on a desert island making shift to keep off the weather and any sudden attack by wild beasts.[29]

When Alex Bowlby rejoined his unit after a spell in hospital, his section commander said, 'It's good to have you back.' 'It was the greatest compliment I've ever had,' said Bowlby, 'and it still is.'[30]

Tribal markings made their own contribution. The Inniskilling Fusiliers, for example, took to wearing their traditional headdress, the caubeen, when the campaign was well under way and caubeens had to be manufactured locally. Some battalions maintained splendid pipe bands. Regimental days were commemorated whenever possible. In 1944 the Hampshires celebrated Minden Day, with their usual vinous enthusiasm, as Captain A. G. Oakley noted in his diary:

> *1st August*...Hell of a sports programme today for Minden day...What a day. I am half gone as I write this.
> *2nd August.* Well, we have recovered from Minden Day apart from a slight headache.[31]

As an aside, we might note that headaches were almost universal in a theatre of war where wine and brandy were readily available. General von Senger acknowledged gentlemanly tussles with the occasional bottle; Captain Oakley, not long recovered from his Minden Day hangover, admitted to 'a helluva piss up last night; great quantities of vino drunk'; and Lieutenant Philip Brutton admitted that he 'got very tight and fell down in a slit trench'.[32] Some of the more inventive defenders of the Anzio beach-head collected copper wire from crashed aircraft to make stills which produced a savage raisin jack, a welcome alternative to the 'swipe' made from after-shave mixed with powdered orange juice.

Sometimes it was a divisional identity that focused comradeship. Lieutenant Birdie Smith of 7th Gurkha Rifles, hitching a lift without his badges of rank, was picked up by New Zealanders who spotted the red 'shite hawk' of 4th Indian Division on his battle-dress blouse – 4th Indian and 2nd New Zealand divisions were in the same corps – and 'treated me as if I was a fellow countryman'.[33]

There was often a downside to all this. Units could be stand-offish, and failures, real or imagined, might be too freshly remembered. In Italy there was, recalled Geoffrey Curtis, 'a typical love–hate relationship between infantry and armour'.[34] Armour was frequently helpless off roads – which were so easily covered by the ferocious 88-mm guns – but hard-pressed infantry were rarely disposed to give tanks the benefit of the doubt. New arrivals often felt excluded: Major T. de F. Jago was upstaged by the loud, confident and bemedalled Yeomanry officers he met in Naples.[35] Dick Eke and his comrades thought rather meanly of themselves until a patrol of Highlanders met them at their stock-in-trade, clearing mines: 'The unexpected regard that these crack infantry men had for us and our work gave us for the first time the proud feeling of being recognised as part of the élite 8th Army.'[36]

The 'Salerno Mutiny' of September 1943 shows how a system which bonded men firmly into tribal groups might work to the disadvantage of the army as a whole. Experienced soldiers from the 50th Northumbrian and 51st Highland Divisions, some of them still recovering from wounds or illness, were landed at Salerno and told that they would be sent up the line as reinforcements, not back to their own units. Despite the efforts of several officers – including the corps commander, Lieutenant General Sir Richard McCreery – almost 200 refused to go up the line and were eventually court-martialled. After a legal process which the adjutant-general described as 'one of the worst things we have ever done', most of the defendants were sentenced to terms of imprisonment. Three were even sentenced to death, though these sentences were in fact commuted. The NCOs amongst them were also reduced to the ranks, and holders of the Military Medal had their decorations withdrawn.

Major-General Douglas Wimberley, a former commander of 51st Highland who had done so much to give that marvellous division its own distinctive character, felt that the whole business had been tragically mishandled. One of the division's former brigadiers was nearby, and, suggests Wimberley:

Had only someone thought of getting him to talk to our Jocks for fifteen minutes I feel quite certain the whole affair would never have occurred. To my mind it was quite obvious, to someone who understood our Jocks, what should have been done...I remain convinced, that if the Jocks had really believed, from an officer they knew and trusted, that they would be kept together, and merely attached to the English regiments, still wearing their HD flashes and tartans, there would have been few, if any, Courts Martial.[37]

This unhappy affair reflected, at least in part, the British army's failure to cushion soldiers at one of those periods of transition which most soldiers find hard to cope with. It was not easy even for middle-piece officers. Major Jago found life in No. I Corps Reception Centre at Naples decidedly tedious, and was dismayed to discover that although the war seemed to be going at full pitch there were no jobs for gunner majors. He set off for the massive Allied headquarters, comfortably ensconced in the king of Naples's palace at Caserta, and by dint of manipulating the old-boy network so dear to regular officers, eventually wangled a posting to the Anzio beach-head.[38]

The German army's approach was altogether different. In the first place, it was the beneficiary of a culture which had long emphasised the importance of soldierly virtues. Next, although some of the literature – notably the pioneering work by Edward Shils and Morris Janowitz – tends to de-emphasise the importance of politics as a motivator, it is clear that the German army in Italy was bound together if not by national socialism then certainly by nationalism.[39] Iris Origo, an Englishwoman married to an Italian marquis, was impressed by the two 'correct and polite' German officers who visited her house late in 1943. One radiated 'pride in his country and his men, and above all his unshakable certainty, even now, of victory. He was not being a propagandist, simply stating a creed.' Later, when the Germans were in full retreat through her estate near Montepulciano, she noted that:

> they are all quite frankly tired of the war and of five years away from their homes and families, appalled by the bombing of Germany, and depressed by the turn of events here and in France. But there is not one of them who does not express his blind conviction that Germany *cannot* be beaten.[40]

General von Senger, a Bavarian aristocrat, former Rhodes scholar and devout Roman Catholic, would not have agreed. When posted to Italy in 1943 he had not liked 'having to act as an agent of its criminal alliance with Hitler', and a year later he maintained, 'we...saw clearly that from every point of view the war was lost, that wars cannot be decided with uncompleted secret weapons, that it was immaterial whether the fight was continued on this or that battle line...'[41] But as the Battle of Cassino so amply demonstrates, his soldierly convictions kept him at a task in whose outcome he had no confidence and whose moral validity he doubted.

At a practical level, the Germans took pains to organise their army so as to generate the maximum fighting power and to avoid any 'increase in the burden resting on the troops'. The division or the arm of service, rather than the regiment, was the focus of a man's loyalty, and the system of march battalions, which brought replacements from the interior of Germany to the theatre of war, and field replacement battalions, which eased their integration into divisions, generally worked well until the manpower crises of mid-1944 threw an almost intolerable strain on the entire system. In contrasting the German practice with the American preference for replacement depots – the bitterly unpopular 'repple depples', which were in fact huge peripatetic transit camps, Martin van Creveld maintains that 'perhaps more than any other single factor, it was this system which was responsible for the weaknesses displayed by the US Army during World War II'.[42]

The French and Poles were different yet again. As far as the French were concerned, it is, I fear, true that most anglophone military historians (with John Ellis as a very creditable exception) have failed to do justice to the considerable achievement of General Alphonse Juin's French Expeditionary Corps. When von Senger was contemplating the prospects of a renewed frontal assault on Cassino, he admitted that 'what I feared even more was an attack by Juin's corps with its superb Moroccan and Algerian divisions'.[43] Although there were tensions between Gaullists and long-standing members of the Armée d'Afrique, all military participants in the French political debate were convinced that the Italian campaign was more than a military adventure. It was quite literally a matter of honour, an opportunity to demonstrate that the stains of 1940 could be blotted out. And for many of the native Algerian and Moroccan combatants it was something even more: the chance of proving that they had earned, in combat, a Frenchman's rights. Lieutenant el Hadi of the Tunisian *tirailleurs* took over the assault on one of the peaks on the Colle Belvedere when his captain was wounded. He lost an arm, but by superhuman efforts took the crest despite a fierce counter-attack. In the very moment of victory:

> he was hit across the body by a burst of machine-gun fire. He shouted to *Tirailleur* Barelli who was alongside him, 'You, fire the flare!' Then he stood up, shouted to the heavens, 'Vive la France' and fell dead on the conquered peak.[44]

The political mainsprings which helped power the achievement of General Anders's II Polish Corps were no less potent. In March 1944 Anders was summoned to see General Sir Oliver Leese, who had succeeded Montgomery at the head of Eighth Army, and asked if he would consider attacking Cassino. If he declined, the corps would be used elsewhere and another formation would do the job. Given the fact that this formidable position had already resisted American, British, Indian and New Zealand attacks, Anders and his generals took advantage of the decision time offered by Leese. But they soon concluded that a mixture of motives – not least their desire to get to grips with the Germans and the obvious international impact of a Polish victory at Cassino – meant that they ought to undertake the mission. Anders was back, with a favourable response, well inside the allotted ten minutes. His opening Order of the Day caught his countrymen's mood perfectly. 'Soldiers! The moment for battle has arrived. We have long awaited the moment for revenge and retribution over our hereditary enemy.'[45]

To their Allies the Poles were men apart. Major Fred Majdalany of the Lancashire Fusiliers recalled:

We got along very well together, though they could never wholly conceal their slight impatience with our attitude. They hated the Germans, and their military outlook was dominated by their hate. Their one idea was to find out where the nearest Germans were and go after them...they thought we were far too casual because we didn't breathe blind hate all the time.[46]

Major John Horsfall of the Irish Brigade found the Poles

very remarkable people...their motives were as clear as they were simple. They wished only to kill Germans, and they did not bother at all about the usual refinements when they took over our posts. They just walked in with their weapons and that was that.[47]

Captain Tomlinson, for his part, was mightily impressed by the Poles, 'Their cheerful, uncomplicated methods of making war, notably their contempt for secrecy and silence'.[48] Dick Eke, as we might expect by now, found them 'reckless and brave'.[49]

If hatred of the enemy was a compelling force as far as the Poles were concerned it was less evident amongst other combatants. Captain Tomlinson felt psychologically as well as physically close to the Germans when he took over a dugout still permeated with the smell of its former occupants. 'The German soldier in many respects differs little from the British,' he wrote:

The tastes and morals of individual fighting men, on whatever side they fought, were not solely dependent on nationality. Gazing at the rubbish, the clutter, the signs of a hurried departure, one could even find it in one's heart to pity.[50]

Dick Eke and his mates saw a Heinkel shot down into the sea and looked hard to be sure that there were no survivors. 'It was strange that we would not have hesitated going into the sea to rescue them,' he reflected, 'even though their last desperate bid had been to kill a few more English soldiers.'[51] When Lieutenant David Cole saw a group of German prisoners he admitted that 'looking at them for a moment as fellow infantrymen, I felt a sneaking sympathy'.[52] Birdie Smith was peering through his binoculars and was shocked to see a German, who seemed so remarkably close, doing just the same. The German saluted ironically and disappeared, and although Smith called artillery fire down on to the spot, 'secretly I hoped that the brave impudent soul escaped'.[53]

The Germans often reciprocated. One decorated German officer who had just led a determined and successful counter-attack took pains to assure a wounded Inniskilling Fusilier that he would be well looked after. And while

Major Dennis Beckett's company of the Essex Regiment was repulsing a ferocious attack on the castle at Cassino, a German warrant officer, captured shortly before, stalked about attentively, for all the world like an umpire on an exercise. When the attack was at last beaten off the German formally reported to Beckett, congratulated him on a well-handled defensive action, and presented him with his leather gloves, for which he knew he would have no use in a prisoner-of-war camp. The determination of German parachutists, fostered by a powerful sense of solidarity as a new arm with young, rule-breaking officers and the distinctive dress of rimless helmets and jump smocks, was widely admired. 'Unfortunately,' as General Sir Harold Alexander told General Sir Alan Brooke, 'we are fighting the best soldiers in the world – what men! I do not think any other troops could have stood up to it except perhaps these para boys.'[54]

So many of the miseries of Italy sprang from the terrain and climate against which all combatants, regardless of nationality, were compelled to struggle, that we ought not to be surprised that, despite all these differences amongst and between Allies and opponents, the war was a shared experience. Men might feel curiously closer to their opponents than to their own commanders: David Cole spoke of 'the inarticulate communion between the infantry of both sides'.[55] A surprising number of survivors recalled Montgomery as an egotistical show-off, and General Leese, who replaced him at the head of Eighth Army, had a less than assured manner with soldiers. Birdie Smith's battalion received a pep talk from which 'the worthy man emerged as a figure of fun rather than a leader inspiring us to deeds of valour'.[56] A British officer complained of US Lieutenant-General Mark Clark's 'inability to sacrifice personal glory to the common cause', and if Allied troops often got on well enough at a personal level – though one Brit admitted that the Yanks found his countrymen 'snobbish and unfriendly' – the dissension amongst their commanders is worthy of a paper of its own.[57] Troops whose duties kept them in the rear were reviled by front-line men. The American soldier-cartoonist Bill Mauldin kept a hot spot in hell for those he called 'Garritroopers' and lived 'too far forward to wear ties and too far back to get shot', and tried to dress 'like the combat men they saw in the magazines'.[58]

There were often times when the only way that wounded could be evacuated in daylight was in the open, under cover of a Red Cross flag, and the Red Cross was generally respected by both sides. On one occasion the Germans borrowed stretchers off the British to collect wounded strewn between the lines, and returned them when the job was done. On another a British ambulance was captured by the Germans but allowed to go on its way after they had checked that it was not carrying ammunition.

Artillery fire was the chief cause of Allied and German battle casualties throughout the campaign. I deliberately specify battle casualties, for the inhospitable terrain and climate of Italy imposed a steady drain of non-battle casualties. At Anzio the Allies lost 7,000 killed and 36,000 wounded but another 44,000 sick or injured. Captain H. Morus Jones, medical officer of a Durham Light Infantry battalion from March 1943 till November 1944, analysed the casualties he treated over a seven-month period. He reported 960 hospital admissions, 285 of these for wounds caused by fire. The majority – 181 – were the results of shells or mortar bombs. There were 63 gunshot wounds, 26 wounds caused by mines – the German Schuh mine, sown by the thousand, was a constant source of worry to Allied infantry – and 16 grenade wounds.[59]

Prodigious quantities of shells and mortar bombs were fired to produce results like this. In the attack on Monte Camino on 2–3 December 1944 the Allies used 206,929 rounds weighing a total of over 4,000 tons: this was a third as many shells, and rather over half the weight, of the British bombardment fired on the Somme on 1 July 1916. In a single hour of the Monte Camino battle, 346 guns fired 22,000 rounds.[60] One British 4.2-inch mortar platoon fired 2,600 rounds in two hours at Anzio, and British artillery in the beach-head regularly fired 200–350 rounds per gun per day in busy periods.[61]

Given that the Allies enjoyed command of the skies, their ability to stockpile ammunition was rarely disputed, but although German lines of communication were constantly harried by Allied air power the effects of this interdiction were irritating rather than decisive. German gunners were rarely as well provided with ammunition as their opponents, but their contribution was none the less dramatic. The dual-purpose 88-mm could knock out tanks or deliver a deadly air-burst shell with little warning; Nebelwerfers fired with a banshee wail that ate away at the nerves, and the 82-mm mortar, the arrival of whose bombs was heralded by a peculiar feathery shuffle, took a steady toll of Allied infantry.

The combined effect of all these razors in the air was to make infantrymen feel like passive targets of the enemy gunners. 'I am done,' confided a German NCO to his diary on 22 January 1944. 'The artillery fire is driving me crazy. I am frightened as never before...During the night one cannot leave one's hole. The last days have finished me off altogether. I am in need of someone to hold on to.'[62] As Birdie Smith observed, even when the infantry were at close range up on Monte Cassino, 'more casualties were caused by bomb and shell than by bullet and grenade'.[63]

There could be single devastating blows: David Cole's battalion headquarters was hit by a shell which reduced it to a 'pile of steaming mutilated bodies'.[64] Lucky – or unlucky – shells reached out to supposedly quiet corners to underline the sheer capriciousness of war. 'Jenkins died at twenty to five on his way to the latrine as they were shouting out that tea was ready,' remembered Fred Majdalany. 'Two haphazard rounds of mortar whistled in from nowhere, and the second one landed at the feet of Jenkins.' Yet it was always remarkable, as Majdalany noted, that shelling killed so few, given the weight of shells that fell.[65]

Shelling helped break minds even when it did not harm bodies. There was its constant noise, so often magnified by the backdrop of the mountains. The dehumanising nature of the death it brought – 'what a way to meet one's maker', mused David Cole, 'being lifted on a shovel in pieces into an old blanket and dumped into a shallow earthy ditch' – also preyed on men's minds.[66] There was a constant and heavy drain of psychiatric casualties: in the US 2nd Armoured Division they ran at the rate of 54 per cent of all casualties during a period of intensive fighting in 1944.

Captain Morus Jones reported seventy-five cases of what were then termed 'physical exhaustion' during the seven months for which he kept a record in his battalion. These formed three groups. There were good men, who recovered after sleep and went back to their companies; the 'bomb-happy', who could be given safe jobs in the rear, and finally the victims of 'nerves', for whom the medical officer was the last chance of an honourable escape.[67] Lieutenant Brutton of the Welsh Guards was standing in his trench when a corporal came to its edge and saluted. 'My nerves have gone, Sir,' he reported. 'I ask leave to report to the medical officer.'[68]

There were many who asked no leave, and the personal accounts of British and American combatants are speckled with mentions of soldiers who had had enough of the war and simply deserted. At one time it was rumoured that there were 20,000 Allied deserters in Italy, and although this figure exceeds the official total by almost ten to one, it is clear that desertion was a major problem. When five of Alex Bowlby's comrades (including his section commander) deserted early on in the campaign there was a measure of sympathy for them: 'If they've 'ad enough, now's the time to pack it in.' Later, Bowlby's comrades warned a deserter that he would receive three years' imprisonment. 'And I'll be 'ere when you're pushing up daisies', he replied with perfect equanimity.[69]

Whatever the rights or wrongs of his action, the deserter was quite correct in assessing that his chances of survival in a rifle company were decidedly poor. Battalions in all the armies I am considering regularly suffered losses amounting to some 40 per cent of their rifle companies, with the burden usually falling disproportionately upon officers. Lieutenant J. D. A. Stainton of 2nd Scots Guards observed that 'the distressingly high rate of casualties amongst the junior officers was a sure sign that too much was being asked of them by a battalion increasingly short of experienced NCOs'.[70] In a single day at Anzio 5th Grenadier Guards lost 9 out of 13 company officers, and Phillip Brutton later found himself 'the only subaltern in battle unwounded and alive' in 2nd Welsh Guards.[71] Casualties in battalions at constant grips with the enemy were dreadfully corrosive. 1st Scots Guards had lost 137 killed, 312 wounded and 217 missing after a month at Anzio: there were only 120 men left in the battalion. On 16–17 February 1944 the US 179th Regiment lost 55 per cent of its effectives: 142 killed, 367 wounded and 728 missing. During the same period the attacking Germans suffered 2,569 casualties. The 2nd Sherwood Foresters went into battle on 31 January 1944 with 35 officers and 786 men: it emerged with 8 and 250 respectively, a toll not out of place on the Somme or at Waterloo.[72] The Polish Corps lost 9.1 per cent of its strength at Cassino, a burden which fell, as such burdens always do, on the forward rifle companies whose loss was proportionately far heavier, and by the end of the campaign no less than 18.2 per cent of the Poles who fought in Italy had become casualties.[73] Operation Olive in September 1944 cost the Canadian

Corps 4,511 men, and a single Canadian Brigade lost 1,372 men in the grim street-fighting in Ortona.[74]

As Raleigh Trevelyan discovered when he revisited Anzio cemetery long after the war, most of the dead were so very, very young: 'Everybody, I noticed, had been between eighteen and twenty-two.'[75] It was indeed a young man's war. At one extreme Birdie Smith's Gurkhas were 'sixteen year olds pretending to be men'.[76] At the other, as Frido von Senger noted, 'The severe fighting…was the business of twenty-five-year-old battalion commanders at the head of assault parties, each consisting of a handful of first-class fighting men.'[77] Birdie Smith would have agreed. He watched a once cheerful commanding officer, a veteran of the First World War, reach the end of his tether, demonstrating yet again that 'even early middle age was too old for an infantry commanding officer in Italy'.[78]

As John Terraine suggested over thirty years ago, 'Casualties – even very great casualties – can be made bearable if they are accompanied by striking achievements, best of all if they lead to swift and decisive results.'[79] In Italy that huge butcher's bill all too often bought nothing, and the accounts of British and American combatants are often critical of a High Command which seemed bereft of novel ideas or sometimes simply of common sense. Even Captain Tomlinson, avowedly patriotic and old-fashioned, seethed at what he saw as the callousness of his own generals. 'Had they the faintest conception', he wrote, 'of what they were asking flesh and blood to do? Had they never heard of the winter climate in the Apennines? If they had, they must have had maps without contours.'[80]

Italy left a mosaic of memories in the minds of survivors. For some it buffed up classical history left dusty since Oxford or Yale: Ravenna with its Byzantine churches, the ruins of Pompeii, and the temple of Neptune at Paesto in the Salerno beach-head. Others were struck by the simple resignation of the civilian population: a father saying gently, 'That's war' when explaining the loss of his three soldier sons in North Africa, a peasant woman casually offering herself to reward a section for letting her travel in its truck (none of them, virgin soldiers as they were, could quite rise to the challenge).[81] Snow on the Apennines: sunlight on the Arno. Tea, Chianti, grappa, raisin jack. The privilege of brave and generous comrades; the emptiness of bereavement.

Even for those apparently on the winning side, victory was sometimes bitter. For Americans who wondered why they were there in the first place and found no comfort in their own official history; for British who deeply resented the gibe that they were D-Day dodgers on a sunny Italian spree. General Juin's *africains* never received the acknowledgement they deserved in France, where Leclerc's Deuxième Division Blindée somehow scooped the pool of national gratitude. And for many indigenous officers and NCOs French post-war attitudes were a betrayal of all they had fought for.

Perhaps most cruelly of all, for the Poles, victory in Italy was followed by betrayal in eastern Europe, and too many of the Carpathian riflemen and Podolski Lancers who had risked their all at Cassino ended their days in

England as kitchen porters or wardens on rifle ranges. Yet for me it is their memorial at Cassino which comes closest to catching the spiritual dimension of the whole rotten business.

> We Polish soldiers
> For our freedom and for yours
> Have given our souls to God
> Our bodies to the soil of Italy
> And our hearts to Poland.

Peasant Scapegoat to Industrial Slaughter: the Romanian Soldier at the Siege of Odessa

Mark Axworthy

There is no question that Italy's armed forces as a whole were much more important than Romania's. However, there is a strong case to be made that the front-line commitment and effectiveness of the Romanian army compares favourably with that of the Italian army.[1] To take the crude measure of division-months spent on an active battlefront, the Romanian army was as heavily committed as the Italian army between its own entry into the war in June 1941 and Italy's defection in September 1943. Indeed, every Romanian division served at the front during this time, whereas most of the Italian army was on occupation duties. During the key months of October–November 1942, when the battles of El Alamein and Stalingrad were fought, the Romanians had more men and divisions engaged at the front on the Don and in the Caucasus than the Italians had in Egypt and Russia combined.[2] Furthermore, the manpower and weapons establishments of Romanian divisions were generally larger than those of their Italian equivalents.

On 23 August 1944 Romania successfully changed sides, briefly providing the Allies in Europe with their fourth largest, if not fourth most powerful, army. By May 1945 Romania had lost more dead, wounded and missing in Allied ranks than Britain had lost in north western Europe over the longer period of June 1944 to May 1945 – telling evidence of the enormously greater human toll extracted by combat on the Eastern Front, even on secondary fronts.

Unlike such armies as the Polish, Yugoslav or French, whose character, organisation and equipment are frozen in time by the brevity of their campaigns in 1939, 1940 and 1941, the Romanian army displayed a continuous evolution over 1940–45. Three waves of divisions were fielded. Every division of the original army was rendered *hors de combat* at Odessa and Stalingrad or in the Crimea over 1941–4. The equipment and much of the manpower of a second complete army, which was raised in 1943–4 to replace them, was lost at the time of Romania's defection in August 1944. A third wave of much weaker

divisions was then raised from training formations in order to campaign with
the Allies in 1944–5. This contribution deals with the formative experience of
the first wave of divisions at the bloody siege of Odessa, the heavy casualties
of which cast a pall over all subsequent Romanian operations.

*

Romania's war was not Germany's war. Thus the motivation of their troops
was different. Romania went to war with the limited objective of recovering
the territories lost in 1940 as a result of the Molotov–Ribbentrop Pact of 1939.
The most important of these were Bessarabia and northern Bucovina, which
had been occupied by the Soviet Union, and northern Transylvania, which an
unsolicited German-Italian arbitration had awarded to Hungary. The problem
for Romania was that these territories were lost to both Axis and Allied
powers. Romania attacked the Soviet Union in 1941 primarily because the
opportunity to recover Bessarabia and northern Bucovina was the first to
present itself, but Transylvania was closest to Romanian hearts and always
remained the ultimate goal. Thus, once the retention of Bessarabia and northern Bucovina was no longer possible, Romania changed sides and fought for
the recovery of northern Transylvania in 1944–5. In the Second World War
Romania was not consistent in its alliances, but it was consistent in its war aim:
the recovery of its interwar frontiers. Romanian troops engaged in reconquering the lost territories in both east and west could be highly motivated.
However, once these limited goals were exceeded, a debate was automatically
triggered which not only detracted from the determination of Romania's
pursuit of campaigns deep in the Soviet Union in 1942–4, but similarly
inhibited enthusiasm for the 1944–5 campaigns in Hungary and Slovakia.

*

During the Second World War Romania was overwhelmingly an agrarian
country and about 75 per cent of the Romanian population were peasants.
They provided a similar proportion of conscripts. Perhaps the only beneficial
by-product of the loss of Bessarabia, northern Bucovina and northern Transylvania had been to strip the country of most of its non-Romanian conscripts,
thereby making the Romanian army of June 1941 a much more homogeneous
and reliable instrument than it had been the year before, and thus minimising
one of the major problems that had weakened both the Poles and Yugoslavs.

The Romanian peasant was not well educated, nor was he familiar with the
modern industrial world. A German estimate was that half the Romanian
peasant conscripts were illiterate. The proportion was probably even higher
in the line infantry. In 1938 one junior infantry officer found that twenty-five
of the thirty-two conscript soldiers in his platoon were functionally illiterate. In
the previous year's intake only three men in the same platoon had passed the
reading test.[3] By definition, none of the illiterates could be developed far as

NCOs or in the technical arms of service. There is no doubt that in terms of education, industrial morale and technical adaptability, the typical Romanian peasant soldier was not well suited either to conduct or face modern mechanised warfare.

However, this does not mean that he did not possess a number of military virtues. French, British and German observers independently agreed that the Romanian peasant soldier was promising raw material; hardy, willing and uncomplaining, with good marching powers and the capacity to subsist on lower scales of rations in worse conditions than most of their own troops. The problem was that the standards of training, equipment and leadership in the Romanian army were seldom adequate to develop these attributes to their full potential.

The corps of NCOs was drawn from the same manpower pool as the soldiers, which restricted their development potential. According to one memoir: 'No special qualifications were required, though it was implied that they should be able to read and write.'[4] While internal training could make them competent within their own restricted field of expertise, its narrowness and their limited education did not allow NCOs to display the full range of initiative expected in a modern army. Nevertheless, under German influence, their numbers were expanded rapidly during the war. Between 1940 and 1944 the number of regular NCOs grew by 130 per cent, whereas the number of regular officers grew by only 23 per cent. An enormous social and educational gap existed between the officer corps and NCOs. As a result, promotion of an NCO to officer rank was virtually impossible.

The Romanian junior officer corps differed markedly from its German and Soviet equivalents, which achieved comparatively high levels of political penetration by the ruling parties, both in terms of party membership and the provision of political officers. This political penetration by all-pervasive totalitarian states stifled diversity of opinion within their armed forces, and incessant ideological propaganda could instil a degree of single-minded fanaticism and consequent combat endurance rarely present in the Romanian army. Romania had had a ruling fascist political party in the Iron Guard, but General Antonescu had suppressed it in January 1941, with the virtually unanimous support of the regular officer corps, and with Hitler's backing. Although the regular officer corps shared much of the Iron Guard's nationalism and prejudices, it viewed the militant party's green-shirted legionaries as rivals in much the same way that the German army had regarded the Brownshirts in 1934. The Iron Guard had a greater impact on reservist officers, some of whom were more inclined to see the party as a vehicle for their own civilian ambitions.

Although almost uniformly anti-Bolshevik, the Romanian officer corps was primarily motivated by limited nationalist, not universal ideological goals. In its tolerance of a diversity of political opinion, which most often surfaced in open advocacy of suspending operations against the Soviet Union in favour of attacking the rival Hungarians in northern Transylvania, the character of the Romanian officer corps differed from that of either of the main protagonists in the east. As a result, outside their own national borders, Romanian troops seldom matched the single-minded determination of the Ostheer or Red Army. Where they did, this was usually attributable to professional pride in élite formations such as the cavalry or mountain rifles.

In 1937–8 only 2.7 per cent of Romanian children were in secondary school and only a small proportion of these received significant technical education.[5] Although this highly selective education system produced a small number of excellent engineers, by and large the Romanians faced severe problems in finding sufficient numbers of technically qualified personnel to construct, maintain and operate advanced weaponry. In conflict with other Balkan states none of this would have relatively disadvantaged the Romanian armed forces, but to counter the large numbers of tanks, aircraft and ships fielded by the Soviet Union required a serviceability rate and qualitative advantage unattainable by even the far better resourced Germans.

*

In June 1940 the Romanian army had mobilised nearly 1,200,000 men in 36 divisions and 9 independent brigades; a total of 45 major formations. However, this was a hollow army. It bulged with raw manpower but was desperately low on firepower and mechanisation. On the modern battlefield it had no more life expectancy than had the Polish army in 1939. The following months saw the loss of a third of Romanian territory and its population, and the rise to power of Ion Antonescu, a professional soldier and accomplished staff officer. These factors simultaneously forced a contraction of the army and placed the remainder in more competent hands.

Antonescu initiated a pragmatic policy of contracting the army until it became a leaner, more potent and readily sustainable instrument. Over the winter of 1940–41 it was therefore quickly reduced to thirty-nine major formations capable of front-line service. Simultaneously Antonescu invited in a German army mission in order to supersede the failed French military doctrines which had hitherto formed the basis of Romanian military practice.

Romania had been rearming since the mid-1930s, but this process was far from complete by the time it entered the war. Its weak industry had put a Czech LMG, French 60-mm and 81-mm mortars and a French 47-mm anti-tank gun into production for the army, but imports remained the main source of new weaponry. The army artillery lacked any of the very heavy guns used by the Germans and Russians. Although a modern motorised corps artillery of Czech 150-mm howitzers and French 105-mm guns had been built up, the divisional artillery remained entirely horsedrawn, and new Czech 100-mm howitzers had not yet superseded 75-mm guns of pre-First World War vintage as the main pieces in its service. Only half the planned mechanisation of six cavalry regiments on Czech all-wheel-drive trucks had been achieved, and the armoured division, using Czech and French tanks and trucks, was not fully formed.

Since late 1939 a major effort had been under way to buy additional light anti-tank guns and light and medium mortars from captured German stocks of Polish, Austrian and Czech origin, but their issue remained well below establishment levels in June 1941. The issue of modern Czech rifles, light machine-guns and heavy machine-guns to the infantry was well advanced, but significant gaps in its inventory remained, most notably lack of a submachine-gun, anti-tank rifle or sniper's rifle; all weapons that later did much to bolster the firepower and confidence of opposing Soviet infantry. It is notable that although a major rearmament was under way in Romania, Germany was able to pitch the Romanian army into the war without having to supply it with any weaponry from the core Reich armament industry beyond some light 20-mm anti-aircraft guns, very few of which had been delivered by June 1940.

Although the ratio of firepower to manpower had been raised since 1940, it was still very low by German or Soviet standards at the outbreak of war. Thus a Romanian infantry division at establishment strength could fire barely a third of the weight of shell of a German or Soviet infantry division at full establishment at the outbreak of war, although it should be noted that official Soviet artillery establishments had fallen to Romanian levels by late 1941. Furthermore, it was consistently outranged by both. Given the fact that nearly half of all battle casualties on the Eastern Front were from artillery fire, the Romanian army found itself at a great disadvantage throughout.

Even given the generally lower level of mechanisation on the Eastern Front, the problem facing the Romanian army was of an entirely different order of magnitude to that facing the Wehrmacht or the Red Army and was to be reflected in its limited operational results. Lack of armour and other motor vehicles, meant that the Romanian army was constitutionally incapable of conducting or exploiting the major breakthrough and encirclement battles which were the decisive feature of both German and Soviet operations. Indeed, the same deficiencies made the Romanian army particularly vulnerable to them.

By June 1941 only three infantry divisions and the embryonic armoured division had benefited significantly from German instruction, and even these lacked German-trained reserves capable of perpetuating the new standards. Otherwise the penetration of German training was very limited, and often widely resisted by conservative senior officers. In short, the bulk of the Romanian army entered the war still shackled by twenty years of adherence to a now discredited, French-inspired, static, defensive operational philosophy to which the bulk of its organisation, tactics, training, logistics and equipment were still attuned. It was thus quite unprepared for the deep-ranging offensive operations now likely to be required of it.

*

The Romanian army's first campaign was the liberation of Bessarabia and northern Bucovina. There was an overwhelming consensus in Romanian society that their recovery was a legitimate and necessary war aim, and this provided a powerful motivation for the army. In the event, the Red Army only conducted a stubborn delaying action in the two provinces, as by the end of July 1941 it was obliged to withdraw because of defeats elsewhere. In the general euphoria of victory it was accepted that Romanian casualties of nearly 22,000 were an acceptable price to pay for the liberation of this national territory. However, a closer analysis would already have betrayed some worrying portents for the future. In gaining a bridgehead across the Prut at Falciu in early July, the 21st Infantry Division had displayed a dogged determination, but it had suffered 6,222 casualties, which had wiped out over two-thirds of its infantry. The 35th Reserve Division had displayed less sterling qualities during a local Soviet counter-attack at Cornesti. This incident revealed that the six particularly ill-equipped and ill-trained reserve divisions were not up to front-line combat and none were used again. Although the embryonic 1st Armoured Division had had a leading role in bundling the Red Army out of Bessarabia, it was deprived of the chance to take a full part in that year's successful *blitzkrieg* operations by twice being ordered to halt on the Dnestr.

In early August national consensus, and with it consensus in the army, began to break down when Antonescu ordered the Fourth Army to cross the Dnestr into unquestionably Soviet territory. Both Romania and Germany had recognised that the offensive potential of Romanian infantry divisions was very limited, and neither had planned for independent offensive operations by them beyond the Dnestr. However, the German Eleventh Army on the Dnepr was already badly overstretched and needed its lines of communications through the port of Odessa cleared quickly. The Germans therefore 'urgently desired' Antonescu to undertake this task.

At this time, due to equipment shortages, little more than half the Romanian line infantry divisions had been fully mobilised, all of which were now allocated to Fourth Army. They included the only five infantry divisions which may be considered to have had significant offensive potential: the three trained by the Germans and the Guard and Frontier Divisions. The incomplete armoured division and three horsed cavalry brigades were also present, but, for the most part, the best troops (the mechanised cavalry and mountain rifles) were elsewhere with Third Army. Thus Fourth Army was not composed of the best Romanian troops, only the best available.

After using its cavalry to cut off the port from the main Soviet front on 14 August, Fourth Army launched a series of attacks, which had involved twelve infantry divisions and the armoured division by 28 August. The Soviet perimeter was reduced and came close to collapse in the east, but just held. Romanian losses were already over 29,000 men and most of the armoured division's tanks. By this early stage the always limited offensive potential of the Romanian line infantry may be considered to have already been seriously degraded. Nine of the attacking infantry divisions had suffered between 2,000 and 3,000 casualties, whilst 3rd Infantry Division had suffered over 4,000. The last two divisions had suffered over 1,000 casualties. However, in

1941 Antonescu would not accept such casualty levels as evidence of the exhaustion of offensive potential and, largely for reasons of national prestige, he mounted two more, equally expensive assaults on Odessa.[6]

The second assault from the west, between 28 August and 5 September, again brought the Soviet garrison to the verge of collapse, and it was only saved by 10,000 reinforcements from the Caucasus. By 11 September an additional 30,000 Romanian casualties had been suffered. At this point Fourth Army's commander reported that 'nearly all our divisions have exhausted their offensive potential, both physically and morally'. Antonescu promptly dismissed him, 'because he lacked offensive spirit and confidence in the battle capacity of the Romanian army'.

A third assault was made from the west between 12 and 21 September. The Soviet perimeter was halved, and only held after a fresh rifle division was brought in by sea from the Caucasus. Then, on 23 September, a surprise Soviet counter-attack on the eastern perimeter threw the most advanced Romanians back, and Antonescu finally accepted that Fourth Army could not succeed without German support. The front settled down to local actions thereafter. In early October, pressure elsewhere decided the Russians to withdraw from Odessa; a task they skilfully accomplished on the night of 15–16 October. The Romanians entered the following day, having suffered an additional 39,000 casualties since 9 September.

This dry description reveals that Odessa was a major battle by any standards. About 25 per cent of all Axis casualties on the Eastern Front in August and September 1941 were suffered by the Romanians in front of Odessa.[7] In only two months their casualties comfortably exceeded the total battle casualties of the Italians against Greece in six months over the winter of 1940–41. Each of its three major assaults on Odessa had cost Fourth Army more men than the final German assault on Sevastopol, which the Germans regarded as an epic of endurance by their infantry. In all, some 98,000 Romanian casualties were suffered at Odessa. The overwhelming majority of these were borne by the twelve infantry divisions initially employed. As their establishment infantry strength was 9,249 men, it is clear that the average loss in their infantry regiments was in the region of 80 per cent in only two months. Had it occurred in the Mediterranean, the siege of Odessa would have been the bloodiest battle in that theatre throughout the war. In short, Odessa was the most determinedly pressed of all offensive operations by a minor Axis power in Europe.[8] Indeed, it is not immediately apparent if there were any German offensives that involved such a proportionally heavy loss over such a short timescale.

Opinion amongst Romanian veterans today is divided. Some consider themselves to have been uselessly sacrificed, whilst others prefer to take pride in the eventual capture of Odessa, a city comparable in size with contemporary Bucharest. From a purely Romanian perspective it is difficult to see Odessa as a clear-cut victory, but from a wider Axis point of view the Romanian siege had some positive repercussions. The main beneficiaries were Manstein's

German Eleventh Army trying to break into the Crimea, for the presence there of Odessa's garrison might well have made the very hard-won break-through of the Perekop isthmus impossible.

*

The trauma of Odessa provoked some remarkably candid self-analysis in the Romanian army.[9] A post-battle morale report on Fourth Army, distilled from individual unit returns, gives some revealing insights into the condition of the troops. Its criticisms ranged from the broadly historical to the immediate and specific. In the former category it noted that the entire armed forces were still infected by a defensive mentality induced by the fact that Romania had so completely fulfilled its territorial ambitions at the end of the First World War. The dissolute character of King Carol's interwar régime was held to have had an injurious effect on the moral health of the nation and armed forces.

The report found that initially there was a strong common purpose that bound together all ranks. Part of this was attributable to the moral momentum gained in liberating Bessarabia. However, more fundamentally, this mutual confidence was attributed to the long months of pre-war mobilisation of the infantry divisions first deployed, which had increased familiarity and trust between troops of all ranks and between men of different call-up years. However, it was also felt that the period of pre-war concentration had not been adequately used to prepare most of the troops. As the siege progressed, officer and troop losses led to their replacement by drafts of unfamiliar, older and less-well-instructed men, who were often not trained on the new weap-onry introduced since their conscript service. As a result there was a decline in cohesion, solidarity and competence on the battlefield amongst even the best and longest-serving infantry divisions.

Especially initially, there were many reports of acts of courage by troops in all arms, particularly the young and well instructed. However, all ranks were told that demobilisation would follow the fall of Odessa, and this tended to discourage initiative and aggression the nearer they approached the city, because no one wanted to be the last man to die. Consequently later, after heavy losses and poor replacements, acts of exceptional courage were rarer.

Despite the huge casualties suffered, there were no recorded cases of direct refusal to obey orders. However, there were numerous cases in virtually every division where troops failed to follow their officers, leaving them exposed to fire alone. This resulted in disproportionately heavy officer casualties. On 22 June Fourth Army had had 4,821 officers. By 15 October it had lost 4,599. Generally officers gave their troops good examples in terms of personal bravery and sacrifice, and losses of regimental infantry officers in the original divisions fielded often exceeded establishment. There were some acts of cowardice by both officers and men, but cases of officers deserting their troops were isolated and exceptional.

If officer casualties were higher than those of their men, those of regular officers were higher than those of reservists. All regular officers were volunteers with a minimum of secondary education and had undergone two years at military schools before joining their units. Although their casualties indicate that they were not deficient in personal courage, Romanian officers were certainly lacking in modern man-management techniques. The wide social gulf between officers and men was exacerbated by an archaic and brutal disciplinary system and institutionalised corruption. For example, flogging was still an official punishment, and food parcels seldom reached the officers and men at the front. Moreover, personal courage aside, officers generally failed to display the sort of behaviour designed to raise the dignity, self-confidence, courage, aggression and initiative of their men. It was felt that the habit of officers and NCOs of verbally browbeating their men, and often physically beating them, created a mind-set in the troops that prepared them also to be beaten by the enemy. There was also an observable lack of well-directed, constructive energy on the part of officers and NCOs. In its stead, nervous tension was widely exhibited in loss of temper, shouting, insults and swearing. Officers also tended to live very comfortably compared with their troops, who eased their own lot through plunder. This all tended to weaken the moral bonding of officers and men, and on occasion troops looted their own dead officers.

The training of the reservist officers who made up the bulk of the junior officer corps also left much to be desired. In 1941 they were still all peacetime volunteers, who had had a single year at military school. However, this had been followed by only two months serving with their units. Thus they were not familiar with their troops. Modern professional and technical skills were often deficient and few were confident enough to display personal initiative and tactical subtlety, preferring a crude and predictable adherence to textbook tactics. As a result, uncomplicated mass frontal assaults were usually favoured. The Soviet defenders reported that some Romanian attacks were delivered almost literally 'off the march', with enormously high casualties the inevitable result. Peacetime training had failed to teach officers and men either to expect or employ invention, cunning, camouflage or stratagems.

It was felt that inadequate training meant that in their first attack many divisions suffered excessive casualties, which discouraged them later, especially as replacements were even less well trained. These replacements were found not to have been properly motivated before leaving depot, and no care was taken to avoid them being exposed to the demoralising sight of Romanian wounded being repatriated. Replacements were rushed too quickly into the line, before they were assimilated into their units or battle-inoculated. This also gave no time for new officers to familiarise themselves with their men, and it was found that it was mostly in these circumstances that men failed to follow their officers. Badly instructed replacement troops were especially unnerved by mortars and tanks, and were particularly prone to self-inflicted wounds.

The influx of replacements meant that most divisions were mistakenly never rested, thereby completely exhausting their officers and men. Repeated attacks in this condition bore diminishing returns and led to a growing conviction of wasted effort. The fact that nearly half the infantry divisions initially remained unmobilised, and only began to arrive at Odessa after the three main assaults had subsided, also undermined the morale of those committed early, who felt that they were carrying a disproportionate burden.

Although replacements did not raise morale due to their poor training and unfamiliarity, the arrival of new divisions lifted spirits considerably, as it raised the prospect of unit rotation, and with it real rest and recuperation. However, the potential of these new divisions, which had only been mobilised since the outbreak of war, was limited, and the original divisions often had to remain in the line. Only in late September did divisional rotation become systematic, by which time the Romanian assault had been suspended pending the arrival of substantial German technical support. Essentially, the entire burden of the three assaults on Odessa was borne by the first twelve infantry divisions committed from the beginning of the siege, whose manpower losses were partially topped up by replacements of ever-declining quality.

Romanian troops also had a strong and justified sense that their equipment was inferior, especially in terms of mortars, automatic weapons, snipers' rifles and anti-tank weapons. They were particularly sensitive to Soviet armour, and the mere cry of 'Tanks!' could start a panic. Even improvised tanks made from Soviet agricultural tractors had a disproportionally damaging psychological effect. Therefore, Soviet counter-attacks had a very big moral impact. Heavy artillery and mortar fire, in which the Russians had superiority, would automatically set up the cry 'Counter-attack!' and, without prompt intervention by officers, positions could sometimes be abandoned without the Russians ever intending to leave their trenches. The widely reported Russian practice of shooting prisoners, sometimes after torture, contributed to some units falling back prematurely before counter-attacks. On the other hand, it also discouraged defections. Later, rescued prisoners revealed that execution was not an automatic fate and nerves steadied. With good leadership, which was at an increasing premium, Romanian troops could be convinced that Soviet armoured counter-attacks could be supressed, but this was an acquired characteristic often reached through bitter experience rather than prior training.

All was not altogether well in the technical arms either. The artillery had been the beneficiary of most interwar technical training and its morale remained high. Morale was helped by the fact that the artillery's exposure and casualties were much lower than the infantry's, and that it contained a higher ratio of officers to keep the men under a direct eye. The artillery's main failing was inadequate co-ordination of its fire with infantry movement, frequently leaving the infantry exposed.

During the siege, all Romanian tanks were subordinated to a series of infantry corps, which frittered away both armoured regiments in poorly co-ordinated assaults, ill-supported by unprepared infantry and artillery. The 1st Armoured Division's tank losses were already so heavy by 21 August that Antonescu had to refuse German requests that it be sent to Third Army for mobile operations. The German military mission was almost gushing in its praise of the persistent bravery of the tank crews, but this was no compensation for the inadequacies of their ponderous French infantry tanks, which were unable to sustain genuine mobile operations, or its Czech tanks, which were vulnerable even to the anti-tank rifles of Soviet infantry. As a result, they failed to have a decisive impact at Odessa.

The Romanian army was also logistically unprepared for campaigning beyond the Dnestr, and supplies sufficient to sustain a simultaneous assault along the whole Soviet perimeter at Odessa could not be brought up. As a result, the Russians were able to switch their meagre reserves to meet each successive Romanian attack in turn. In the case of some older heavy artillery, shell supplies ran out completely and the guns had to be repatriated to Romania.

Logistical difficulties also resulted in food shortages, which lowered morale and led to early physical exhaustion. At times troops had to resort to plundering rear areas in order to sustain themselves. However, in the summer heat, dehydration was an even worse scourge and led to cases of men leaving trenches under fire to look for potable water. In the last weeks of the siege increasing cold became an added burden. Hygiene conditions were appalling. Lack of relief meant that the infantry could never wash or change their clothes. Underwear became torn and filthy. Unwashed bodies became infested with lice. 'In what other army', asked one doctor, 'do even senior officers have to disinfect themselves of lice by hand, like any tramp on the side of the road?'

In most cases Soviet propaganda had no influence, especially on trained troops, although later replacements, brought in to reinforce failure, were observed to be more susceptible. British propaganda had attempted to undermine Romanian home morale by characterising the crossing of the Dnestr as a diplomatic Rubicon from beyond which there was no return to the Western camp. This was the majority view of the middle and upper classes in Romania, and therefore of the officer corps, and it was certainly beginning to infect the attitudes of reserve officers sent to the front in the later stages of the siege.

A major weakness was that no effort had been made to explain to the troops precisely why Romania had to cross the Dnestr and confront Bolshevism in its own home, and, more specifically, why Odessa was worth such a butcher's bill. It had been a reasonable assumption that patriotism was sufficient to motivate Fourth Army in liberating Bessarabia, but little consideration had been given to the wider political orientation necessary to impel it to mount a sustained attack into clearly Soviet territory with real enthusiasm, let alone fanaticism. The Soviet and German armies were able to churn out ideological justifications for aggressive campaigning beyond their borders, but such an ideological crutch was never available to the Romanians, who had narrower, national goals. There was also a lack of information for officers and men in the front line on anything between their own local situation and the global war situation, leaving them totally disorientated.

The Romanian troops had an expectation that significant German assistance would arrive. Its repeated failure to do so in strength lowered morale, especially when compared to the regular appearance of Soviet reinforcements from the sea. Morale reports indicated that Romanian troops had an in-built assumption about German superiority in almost everything – training, armament, mechanisation, equipment, food, cultural level, and so on – and the presence of German troops generally stimulated them to greater efforts. However, such effects were temporary, and in the long term exposure to German technical efficiency could have a depressive effect. For example, unflattering comparisons between basic Romanian medical services and advanced German veterinary provision gave rise to aphorisms such as: 'Better a horse in the German army than a soldier in the Romanian army.' During the siege of Odessa, the staff of Fourth Army had to endure weeks of well-intentioned but patronising German criticism of their shortcomings, and there was an ill-concealed sense of *schadenfreude* when the first intervention by German troops miscarried.

*

Odessa was a decisive battle in Romania's war, because it prevented the bulk of the Romanian army, the line infantry, from gaining a moral ascendancy over its Soviet equivalent at the most opportune moment, and the awful casualties suffered there continued to overshadow Romanian operations to the end of the war. Perhaps the most relevant comment on the conduct of the siege was made in bold print in a Romanian military journal in 1944: 'Commanders must remember that their men are only flesh and blood.'[10]

As a consequence of Odessa, the Romanians made significant improvements in their organisation, training, equipment, leadership, motivation and firepower, which resulted in greater combat effectiveness in later years. However, these were more than offset by a much greater strengthening of the Red Army, which even the better resourced Germans could not match. As a result, improvements in the Romanian army did little to lessen the decisiveness of defeats at Stalingrad in November 1942, or Iaşi-Chişinau in August 1944, which left the Romanian peasant soldier a convenient scapegoat for the Wehrmacht's own failings.

Red Army Battlefield Performance, 1941–45: the System and the Soldier

John Erickson

Whether in imperial, Soviet or post-Soviet guise, how Russia wages war has long been the subject of intensive investigation and close scrutiny by friend and foe alike. The constituents of an effective army and the requirements of proficiency in the individual soldier had preoccupied Russian generals as far back as the mid-eighteenth century, not to mention Peter the Great's own intimations on the vital importance of these matters. Catherine the Great's military commission convened between 1763 and 1764 and presided over by Field-Marshal Petr Saltykov, affirmed that 'the strength of an army consists not in great numbers' but rather 'in the upholding of good discipline, the quality of training, sound maintenance, but *above all else, the existence of a common language, religion, customs and blood ties*'.[1]

These were precepts and practices which the Red Army had to learn afresh and under fire during the 'Great Patriotic War of the Soviet Union 1941–45', suffering a calamitous transition from a 'class-based army' to one conforming more closely to those principles elaborated by Catherine's generals. Lieutenant-General Nikolai Pavlenko identified the seeds of inevitable disaster in the 'paranoidal absurdity' of Stalin's obsession with building an army on 'class principles', obliterating continuities with the 'old cadres', finally consigning thousands to the firing squad. This army, nurtured on ideas of promoting world revolution, learned nothing of defensive strategy, of how to fall back, launch counter-blows and finally turn to the counter-offensive. It was from such febrile stuff that the Stalinist military system was fashioned, leaving the crucial lessons of war to be learned at horrendous cost in blood and treasure.[2]

General Pavlenko was simply affirming what Catherine's generals understood instinctively and what Stalin's obsession fatally demonstrated. The soldier cannot be sensibly divorced from society and its system. It is against this background that the military qualities of the Russian soldier have been variously debated and dissected over time. In his study of the First World War General Golovin attempted to categorise the fighting qualities of the Russian soldier, ascribing the lower quality of the northern 'Great Russian contingent' to the influence of the 'communal system of agriculture' which destroyed the peasants' initiative. This was in marked contrast to the verve of soldiers from the individual farmsteads (*khutor*) of the Ukraine.[3] Trotskii in 1918, bent on raising a Red Army capable of fighting and winning, castigated that feckless-ness of 'vagabond Russia', *brodyachaya Rossiya*, so unsuited to the battlefield.

Colonel Nikolai Galay, investigating Red Army performance in the Great Patriotic War, attempted to identify the origins of the 'best' Soviet soldiers by examining those areas and locations which provided men for divisions accorded Guards status, a designation earned through battlefield excellence. Cities with a population of 50,000 and more accounted for almost two-thirds of Guards formations (rifle divisions, 120 out of 195) and three-quarters of tank and mechanised brigades. Three military districts – Moscow, northern Caucasus and the Urals – provided more than half (55 per cent) of all Guards formations. 'Workers are better fighters than peasants,' peasants from the north are 'better fighters' than those from the south, largely contradicting Golovin's earlier view.[4]

Subjective assessments of the Soviet soldier abound: in literature, Simonov, Nekrasov, Grossman, Bek, Baklanov; in memoirs, diaries, wartime German reports.[5] He was probed ideologically and psychologically in *Der sowjetische Soldat*, a prize-winning study which attracted Himmler's notice and was published by Reichsführer SS (SS Hauptamt) during the war. The section on 'Volkscharakter und Geschichte' is something akin to a counterpart in a recent Russian study on 'the Russian fighting man', 'Rossiiskii kombatant' published in the journal *Otechestvennaya istoriya*.[6]

The wartime letters of Red Army soldiers speak for themselves, many being brief reassurances for the family on the lines of 'I am well. Don't worry,' others more reflective, fiercely patriotic or touchingly valedictory.[7] Curzio Malaparte provided a singular wartime portrait of the Soviet soldier and front-line reality in *The Volga Rises in Europe*, only to be punished for its objectivity by the Axis authorities. Major-General F. W. von Mellenthin in *Panzer Battles* aimed to avoid 'individual prejudice or patriotic sentiment', though he was obviously not enamoured of 'these Asiatics'. Fighting Russians meant getting accustomed to 'a new sort of warfare', primitive, ruthless, rapid and versatile. The Soviet soldier as often as not followed his instinct rather than tactical principles, not infrequently coming off best.[8]

The great Soviet advantage which vastly enhanced battlefield effectiveness was the speedy, ruthless replacement system but what bordered on the near miraculous was the extraordinary evolution of the Russian tank arm from 'an apathetic and ignorant crowd, without training or natural aptitude' into tank crews 'endowed with brain and nerves...elevated far above their original level'. The tank armies were handled by 'daring and capable commanders', 'even the junior officers became remarkably efficient', showing 'determination and initiative...willing to shoulder responsibility'.[9] One senior German officer offered his own personal interpretation of the Red Army. The Soviet soldier began the war as a first-class fighter but ended as a first-class soldier, the product of the wartime transformation from a 'horde of riflemen'.

During the Cold War the evaluation of the Red Army's wartime effectiveness and the performance of the Soviet soldier became a matter of intense intelligence interest in an effort to assess the potential of the post-1945 Soviet army. The Soviet command used much Second World War data to modernise force structures, develop operational models and generate 'norms' adjusted to modern technology, but a reliable data base was, for obvious reasons, not available elsewhere. It is only with the demise of the Soviet Union and access to information hitherto classified or long withheld that a more reliable picture can be presented.[10] The all-embracing 'heroic myth' has been finally buried, and reports on competence at all levels, morale, crime and punishment, executions, even vodka deliveries, have recently seen the belated light of day. A melancholy disclosure revealed failures to recover Red Army dead from the battlefield and bury them properly, *'s dolzhnymi pochestyami*, 'with fitting honours'.[11] A complete operational narrative can now be assembled, together with Soviet assessments of 'effectiveness' (and its absence), displacing the previous preponderance of German material, useful though that proved to be.[12]

Organisational change and the operation of that 'speedy, utterly ruthless' replacement system emphasised by von Mellenthin is open for detailed inspection, coupled with the complete Red Army wartime order of battle, the five volumes of *Boevoi sostav Sovetskoi Armii* covering the period from June 1941 to September 1945, compiled by the Soviet General Staff using the Defence Ministry central archive.[13] But most important of all was the publication in 1993 of a volume of 'statistical research' on Soviet losses in men and war material, *Grif sekretnosti snyat* edited by Colonel-General G. F. Krivosheyev, a controversial volume but none the less indispensable. For the first time the stark figures on the wartime Red Army, manpower, mobilisation, losses, periodisation of loss, the workings of the vital replacement system, prisoners-of-war, variations of wartime strength, reserves, penal battalions (the *strafbats*), are to hand from a wide variety of sources.[14]

The Soviet soldier was at once victim and victor, survivor and sufferer in a system which combined numbing incompetence with ferocious ruthlessness. The same system, however, steadily generated effective organisation to use available manpower while constantly adapting unit and formation structures to the changing battlefield, equipping the latter with ever-increasing quantities of powerful, 'soldier-proof' weapons. That combination ultimately proved fatal to the Wehrmacht.[15]

General Krivosheyev's 'statistical research' and much else, a stupefying amplitude of vast numbers, provides testimony little short of terrifying in its own right. In a war which lasted for 1,481 days enlistments totalled 29,574,900, with 34,476,700 representing the full mobilised total if the strength of the armed forces on the eve of June 1941 is included.[16] In June 1945 army and naval strength stood at 11,390,600, with a further 1,046,000 in hospital recovering from wounds. This was final front-line strength after 21,636,900 men had passed in and out of the wartime armies. Of that number, 3,798,200 men were discharged or recovering from wounds (2,576,000 invalided out) with a further 3,614,600 transferred to industry, air defence, guard duties, and 1,425,000 assigned to the NKVD.

Within that figure of 21 million almost one million, 994,300 to be exact, were variously under punishment, imprisoned, deserted, lost or missing. In 1942 Stalin introduced penal units, the *strafbats*. No fewer than 422,700 Red Army men were sent to these punishment units, holding the most dangerous sectors with near suicidal missions, 436,600 were in local prisons, and 206,000 were 'struck off strength for various reasons'. There were 212,400 untraced deserters or simply listed missing. The total loss (according to front-line records) amounted to 11,444,100; those killed in action, in accidents or suicides amounted to 6,885,100, and prisoners-of-war to 4,559,000. In addition 500,000 were listed lost, taken prisoner after mobilisation but not yet entered on unit and formation nominal rolls. What is termed the 'demographic loss' is set at 8,668,400 (adjusted to account for returned prisoners-of-war, 1,836,000) and the missing 500,000. Medical casualties amounted in all to 18,344,148 (wounded 15,202,692, sick 3,047,675, frostbite 90,880).

The Red Army lost 973,260 officers, 95 per cent of all officer losses incurred by the entire Soviet armed forces, army, navy and air force. Infantry officers suffered the heaviest loss, 569,794; tank troops lost 47,105; artillery 94,189. Splitting up those figures for officer loss, 631,008 were killed in action, 392,085 missing or taken prisoner, more than 50 per cent of those losses being incurred in 1941–2. The Red Army lost 35 per cent of its officer corps (fourteen times more than officer loss in the tsarist army).[17] During the heavy breakthrough operations of 1943–5 rifle troops losses reached 86 per cent of the total manpower loss, tank troops 6 per cent. Out of a total of 5,200,000 men taken prisoner, 3,800,000 were captured in the first five months of the war (760,000 per month), compared with 800,000 taken prisoner between 1943 and 1945 (25,000 per month).[18]

Comparative Red Army/Wehrmacht loss ratios have generated furious, unending controversy. Soviet casualty figures were variously interpreted to indict a callous, profligate Stalinist leadership wasting human lives.[19] Indisputably losses were horrendous, costing the lives or well-being of virtually every fourth soldier mobilised. In round figures the Red Army fielded 661 divisions (490 rifle divisions) and 794 brigades (251 tank), losing 297 divisions and 85 brigades to enemy action, while the German army lost 474 divisions and 68 brigades on the Eastern Front (a computed aggregate loss of 339.5 Soviet formations against 508 German). The heaviest Red Army losses occurred in 1941–2 when 177 rifle divisions, 16 brigades, all existing mechanised corps and 37 per cent of all tank and motorised divisions were destroyed.[20]

Amidst this daily nightmare of mangled flesh and torn bodies the battlefields consumed machines. In 1941, from a tank park of 23,106 machines the Red Army lost 22,600. In 1943, when tank production had largely recovered, the Red Army lost 22,400 out of 43,500 produced or in service, in 1944 losing 16,900 tanks from a stock of 21,100 and 3,300 self-propelled guns. Out of a grand total of 131,700 tanks and self-propelled guns produced in Soviet factories, battlefield losses amounted to 96,500 (73.3 per cent). Elsewhere war cost the Red Army 15,500,000 infantry weapons, 317,500 guns and mortars, and 106,400 aircraft (40 per cent from non-combat accidents). Somehow, somewhere, the Red Army soldier had to survive day by day in this maelstrom of fire and destruction.[21]

What kind of army was this and what kind of soldiers served it, what process was involved to transform both the system and the *frontoviki*, the front-line soldiers, from the 'horde of riflemen' into a superior fighting machine?

*

In June 1941 the Soviet Union was a warfare state without a war machine, lacking a High Command, bereft of operational plans. The Red Army was in the midst of re-equipping and re-organising, poorly trained, lacking experienced commanders. The failure to bring it up to battle readiness on the eve of the German attack left the Red Army in a position in which it could neither attack nor defend.[22] With warnings of imminent German attack ignored, a cadre army was virtually destroyed on the frontiers. Between July and December 1941 new divisions were mobilised with frantic speed, grinding the Wehrmacht to a halt before Leningrad, Rostov and Moscow.

Switching to the offensive in the winter of 1941–2, the Red Army reformed and recouped in the spring of 1942, only to suffer further heavy defeats in the summer. Yet a third programme produced forces capable of defeating the Germans at Stalingrad. A build-up in 1943 followed, turning the Red Army into a well-equipped, well-trained force under skilled commanders. The Soviet High Command had learned the lessons of the cost-effective use of manpower, sufficient to field an army of six million men to destroy the Wehrmacht and to carry through the devastating offensives of 1944–5. With war industry catastrophically disrupted and with loss of territory and population, the Soviet command raised upwards of 500 new divisions, trained in the shortest possible time, never more than six months.

With only 8 per cent of the cadre army preserved, the battle for the frontiers was lost and German armour drove ever deeper into the interior. Mobilisation proceeded efficiently, though some time passed before divisions could be organised and committed to action.[23] To plug gaps torn in the line, barely trained, ill-equipped 'militia divisions',[24] the *narodnoe opolchenie*, culled from factory workers were flung in against crack German divisions. Across a huge battlefront catastrophic losses in men and machines mounted daily, consuming thousands upon thousands of aircraft, tanks and guns. The Reserve Armies Administration (Glavuproform KA), formed in July 1941, worked frantically to build up strategic reserves and by the middle of the month had sent the reserve front in the west 6 fresh armies (31 divisions).[25]

The strength of the rifle divisions fell to some 6,000–7,000 men, half of the original establishment. All mechanised corps had been disabled or disbanded, what remained of the armour was brigaded. The rifle divisions were stripped of half of their organic artillery to form a High Command Artillery Reserve, concentrating the available guns (and available ammunition) for direct fire. With the Soviet tank park virtually empty, a mere 1,954 tanks being deployed at the front in November 1941, a huge expansion of cavalry forces was rushed through in order to supply the Red Army with some form of mobile forces.[26]

In late November 1941 Red Army manpower fell to its lowest ever level, 2,300,000 men. German intelligence set Soviet order of battle at 200 rifle divisions, 35 cavalry divisions, 40 tank brigades (265 divisions 'in hand', reserves assembling in the Volga area and Siberia). At the beginning of December 1941 the new reserve Tenth Army arrived without heavy artillery or tanks.[27] Infantry weapons and lorries were also in grievously short supply. All divisions had to rely on what remained of their divisional artillery. German intelligence reasoned that the Russian defence before Moscow had reached its limits, all remaining strength consumed: 'Keine neuen Kräfte mehr verfügbar.' An estimate which was calamitously mistaken.[28]

Soviet strength slowly revealed itself in both reserve armies and divisions rebuilt after battle. Between 22 July and 1 December 1941 227 rifle divisions appeared in the Red Army order of battle to replace battlefield losses, 143 new divisions and 84 from divisions reformed or rebuilt. Corps level was eliminated, and rifle armies were increasingly made more compact and manageable by fielding 6 divisions at most. Manpower levels in rifle divisions fell drastically, averaging 6–7,000 men, barely half the peacetime establishments, with little compensation as yet in the way of greater firepower.[29] In spite of devastating losses, a military phoenix – the Red Army rebuilt and partially restored – launched its counter-blow at Moscow on 5 December 1941, confounding German intelligence and battering the Wehrmacht in a prelude to the counter-offensive in the winter of 1941–2.[30]

Reorganisation and replacement throughout 1942 brought the total of divisions formed since June 1941 to 318 (260 in 1941, 158 in 1942), with the creation of 10 Reserve Armies in the spring of 1942. These formations, duly redesignated field armies (First Reserve Army, for example, designated Sixty fourth Army) were fed into the gigantic operations in the south, where the German army drove into the Caucasus and advanced on Stalingrad. In March 1942 the tank corps reappeared in the Soviet order of battle, the result of increased tank production and the need to provide tactical control for the growing number of tank brigades (172) and regiments. The tank armies raised in May 1942 had a 'mixed' establishment which included rifle divisions, but the infantry could not keep pace with the tanks, a deficiency corrected by replacing the rifle divisions with the new mechanised corps, giving the tank army two tank corps and one mechanised corps.[31] However, Soviet armoured forces faced a further grim baptism of fire before attaining optimum organisation and maximum strength.

The Red Army of 1942 was no longer 'class-based' but 'national' in the sense of combining many nationalities and being infused with patriotism. The nationality mix – Russians, Ukrainians, Georgians, Uzbeks, Tartars, Kazakhs, Jews – inevitably produced language problems and exposed not only cultural differences but also precipitated a crisis of loyalties.[32] The requirement for men was insatiable, making huge demands on the Soviet replacement system and on reserves. Rotation and replacement played a vital part in maintaining Soviet front-line strength. When a division was 'ground down' in battle, it would be withdrawn to receive and train replacements, its place being taken by a division already replenished and so the process wound on, rotation combined with replacement, but on Stalin's express order in March 1942 divisions would not receive replacements while in action, only when pulled into reserve.[33]

Replacement training regiments in the rear carried out basic training, organising the despatch of trained recruits to field replacement regiments operating at front and army level. Several hundred of these replacement regiments duly assigned men to divisional replacement battalions for reception and further training, much of it badly needed. Officers, supplied from the 310 officer training schools in 1943 or from battlefield promotion, had their own replacement regiments, the earlier shortages being gradually overcome.[34] Well-trained, experienced NCOs were everywhere in demand and short supply, for all the efforts of the NCO courses.

The year 1943 marked the third fundamental transformation of the Red Army, the 'big switch' to a 'fire-power' army in which finally doctrine, 'armament norms' (tanks, guns) and *upravlenie*, improved command and control, and professional competence came into closer alignment. The result was to make different but no less extravagant demands on both the quantity and the quality of manpower, involving nothing less than a major redistribution of limited resources.

In January 1943, with the Soviet triumph at Stalingrad virtually complete and the entire German southern wing in great danger, Stalin was persuaded by the Armoured Forces Administration of the need to form genuine, 'homogeneous tank armies' (2 tank corps, 1 mechanised corps). From the 8,500 tanks at the front, with a further 400 in Stavka reserve and more than 4,000 in rear areas, Stalin authorised the formation of 5 tank armies, the new Soviet armoured shock force. This was accompanied by the reorganisation of the tank corps and the tank brigade and the strengthening of the mechanised corps.[35]

In the whole wartime period, 145,000 tanks passed into and through the Red Army including the tanks on hand in June 1941 and those supplied by Lend-Lease. Losses in tanks amounted to 96,000, losses in men 310,000, a figure falling only slightly below the total number of tank specialists trained in the Red Army, which was 403,272. The scale of tank production and the toll of battlefield loss demanded both huge training and replacement programmes to supply a minimum of 100,000 specialists per year. Tank crews trained in tank school regiments, a number of which were located near the tank factories themselves.

To the massive investment in mobility for offensive operations after 1943 must be added the huge accretion of firepower demonstrated by the development of Red Army artillery. With factory output reconstituted, on the eve of the Stalingrad counter-offensive artillery holdings (guns and mortars) already amounted to 72,505. In 1943 reinforced artillery divisions became 'artillery breakthrough divisions', followed by the 'artillery breakthrough corps', 5 of which were organised by June 1943, supplemented by a profusion of independent artillery regiments.[36] In 1945 the Red Army fielded 10 artillery corps, 105 artillery/mortar divisions and 97 independent brigades, consuming the services of one million men.

The days of 'hordes of riflemen' were long past. This 'third' Red Army was one constituted for large-scale offensive operations. It utilised both mobility and fire-power to the maximum and was structurally organised to support this form of warfare, producing increasingly diverse formation and unit establishments: specialised artillery, tank destroyer brigades, combat engineers, logistics.[37] The scope of this expanding variety in organisation and structures is dramatically demonstrated by the changes worked in the Red Army between June 1944 and January 1945. In the summer of 1944 Red Army manpower stood at 6,077,000, in January 1945 it had climbed to 6,289,000. The number of rifle, cavalry and airborne divisions for that same period had increased by 12, from 476 to 488. The number of tank/mechanised corps had dropped slightly from 37 to 35, but artillery divisions had increased by 6, from 83 to 89. The number of independent artillery and mortar brigades increased by more than a third within six months, rising from 93 in 1944 to 149 in January 1945 while the number of guns and mortars (excluding the Katyusha Multiple Rocket Launchers) climbed from 97,050 to 115,100 (108,000 at the front). Tank strength and self-propelled guns showed a similar increase, adding some 5,000 new units in six months, 9,985 with reserves, rising to 15,100 in January, 1945, 12,900 deployed with front-line units.[38]

Armour, artillery and the technical branches took the cream of the available manpower not only for replacements but also to create those additional units which accounted for the spurt, particularly in numbers of artillery units. Heavy breakthrough operations in 1944–5 involved heavy losses but also needed more men to serve more guns to apply ever more fire-power. The effects of the combined manpower redistribution and replacement problem were felt most heavily in the rifle formations and units in the wake of Soviet offensive operations, which were launched in the summer of 1943 following the defeat of the Wehrmacht at Kursk.[39]

Though the Red Army rifleman had come to resemble a walking arsenal, with machine-pistols, light machine-guns and yet more automatic weapons steadily increasing the fire-power of the rifle company, the weight of metal could not wholly compensate for a shrinking manpower pool. Simultaneously creating new units while providing replacements for front-line divisions finally proved to be an impossible arrangement once the Red Army was fully engaged in massive offensive operations. Inevitably the rifle division establishments shrank, falling steeply from 10,566 in 1942 to a nominal 5,400 in the spring of 1944, with a desperate struggle to maintain the strength of rifle companies. In the final months of the war infantry was in very short supply, with rifle-division strength dropping to between 4,500 to 3,600 men, little more than regiments rather than full divisions because of the reductions in the rifle companies.

This was a system which finally brought Red Army riflemen to the Reichstag, staving off annihilation late in 1941, reforming and reorganising in 1942, restructuring in 1943 to combine mobility with firepower for massive offensives and heavy breakthrough operations, and maintaining front-line strength and the replacement of losses in the final stage of the war. The horrendous initial losses in 1941 forced a policy of stripping out the extraneous (corps level) from the Red Army and simplifying organisation to match debilitated resources, whether in men, machines or competence. The experience of 1942, though involving further defeats, encouraged a degree of variation in and experimentation with structures and organisation: in artillery, tank forces and rifle units. That diversification proceeded apace after 1943 with an accompanying redistribution of manpower.[40]

Replacement and rotation proved to be the key to sustaining front-line strength in spite of the losses, while the rotation of divisions also had the advantage of furnishing a ready reserve. None of this would have prevailed without the prodigious volume of weapons of all types which poured from Soviet factories, complemented and supplemented by growing professional competence, particularly at higher command levels, which emplaced the appropriate organisation, the necessary degree of firepower and the minimum sustainability of units in the right place at the right time.

*

At all levels the Red Army, indeed the entire Soviet military machine, operated under Stalinist control. No one, whether marshal or private soldier, remained unaffected by the rigours of the Stalinist system. If the lives of Soviet soldiers were too often nasty, mean, brutish and tragically short, their civilian counterparts suffered equally from either Draconian coercion or callous abandonment by officialdom. The Red Army men overwhelmed by the massive German offensive in June 1941 were badly deployed and lacked adequate training, proper equipment and effective leadership. In spite of the evidence of imminent attack, Stalin's personal *diktat* forbade mobilisation and prohibited full readiness on the frontiers.[41] An inexperienced, half-educated, barely competent officer corps had only recently taken up positions once held by the thousands eliminated in the military purges.

'Like thunder from a clear sky' the roar of massive barrages swept over the Red Army across a vast frontage. Within hours divisions disintegrated and fronts began to crack wide open. Tank and infantry columns were lashed by devastating dive-bomber attacks. Newly mobilised men moving into disorganised units created more confusion or were simply thrown into costly 'human wave' infantry attacks carried out with primitive or stereotyped tactics, marching into machine-guns line abreast.[42]

To restore some order and stem the military haemorrhage at the front Stalin resorted to Civil War methods, imposing the 'discipline of the revolver' on the Red Army. Ahead of the Soviet soldier lay German guns and tanks, behind him the NKVD machine-gunners of the 'holding detachments' (*zagradotryadyi*), in his midst the newly installed military commissar brandishing his own revolver, and the NKVD counter-intelligence 'Special sections' (OO), seeking out 'spies, alien elements, panic-mongers'. They reported drunkenness, panic, incompetence, self-inflicted wounds and responded by arbitrary execution of 'traitors to the Motherland' in front of their regiments.

Neither high nor low were spared. In July 1941 army General D. Pavlov, Western Front commander, was shot after a summary trial together with his staff: the first of thirty senior commanders shot between 1941 and 1942.[43] Among the three million prisoners taken by the German army and destined to die horribly in captivity, were sixteen senior commanders, with nineteen generals untraced and missing in action. Huge masses of men drifted across the front; many were encircled or marooned in the rear and subsequently recovered as a manpower windfall by the Red Army, but in 1941 struggling to stay alive or forming improvised partisan units.

Stalin's Order No. 270, dated 16 August 1941, prescribed ruthless punishment for desertion, panic-mongering and surrender. The families of commanders and commissars who removed their rank badges and deserted would be arrested. The families of those taken prisoner would be deprived of state support and benefits. In divisions those regiment and battalion commanders who 'cowered in slit trenches' and did not lead from the front were to be reduced to the ranks, or if necessary shot on the spot and replaced by 'brave and steadfast' junior officers or outstanding soldiers.[44]

Everywhere cries went up bemoaning the lack of properly trained reserves and 'no tanks'. A cruel martyrdom awaited the 'militia divisions', terribly deficient in arms, equipment and training, who were scooped off the streets and plucked from factories to plug gaps in the line at fearful cost. Those who survived subsequently formed regular Red Army divisions.[45]

Under nightmarish conditions the Soviet soldier displayed a wide variety of behaviour, from fanatical resistance to sudden, even inexplicable defeatism, much depending on his immediate leadership. New divisions were created with surprising speed, but they were commanded by inexperienced officers and manned with inadequately trained men. Poor or non-existent reconnaissance, absence of information on the enemy and ignorance of the terrain led to 'blind' attacks with inevitable heavy losses. *Neorganizovanost'*, disorder and general shambles in formations and units speeded the collapse of discipline, the onset of panic, the abandonment of weapons and flight.[46] Beria's NKVD draped its own noose around the neck of the Red Army; but, not to be outdone, Mekhlis, head of the political administration, sent out his own commissions to investigate the conduct of Red Army party members, leading to more arrests or exclusion from the party.[47]

The efforts to train men as speedily as possible were hampered by lack of equipment and irrelevant instruction, and drill and more drill but not what was needed for the battlefield; the 325th Rifle Division, 10th Reserve Army had only 143 training rifles, 35 training grenades, 3 heavy machine-guns, 5 mock-ups of tanks and no signals or artillery equipment. Winter clothing had not arrived and men were deserting; 16 were apprehended and tried.[48]

If there were no rifles there was at least vodka. In December 1941 the 19th Rifle Division Fifth Army (Western Front) received 750 litres, the 82nd Motor-Rifle Division received 600 litres, and the 20th Tank Brigade two rations each of 220 litres.[49] Whether vodka, chance, innate resilience, occasional firm leadership and tactical competence, fear and coercion, personal bravery and, growing hatred of the enemy all played their several parts, the German command came to recognise in 1941 that for all the calamitous defeats inflicted on the Red Army the 'Russian will to fight had not been broken'. Resistance was frequently 'fanatic and dogged' (*fanatisch und verbissen*) and the German army was now forced to fight according to its manuals.

In December 1941 the skeletal Red Army, with divisions refurbished but desperately under-manned, struck back in the Moscow counter-stroke. Ammunition was available only for assault formations; infantry weapons, horse-shoes, lorries, tractors, guns, tanks – almost every conceivable item – was in short supply. Large tank formations needed for breakthrough operations were missing: 108th Tank Division possessed fifteen tanks, and the 38th one medium tank and thirty obsolete machines.

The defence of Moscow in late 1941 cost the Red Army 658,279 casualties, killed, wounded or missing, and the counter-offensive (5 December 1941–7 January 1942) 370,955 casualties.[50] At Stalin's insistence the Red Army now went over to a general offensive across almost 1,000 miles of icy front, a near reckless commitment with forces under-manned and over-tasked. Western Front supply dumps held food for only one day, with no reserves of fuel. Regimental and anti-tank artillery lacked even a single round. For the first time the Red Army heard from Stalin about the 'artillery offensive', while from the infantry he demanded more 'shock groups' though without giving front commanders time to regroup to form them. Yeremenko's 'shock troops' had barely a crust between them; to eat they had first to fight their way into the German dumps at Toropets.[51]

Breakthrough operations inflicted appalling casualties, but there was nothing resembling a second echelon to follow through. Incompetence bordering on the criminal brought needless loss. The 376th Rifle Division fighting on the River Volkhov, losing 50 per cent of its men in four days of attacks which began on the night of 30 December 1941. It had been flung into battle without information on the enemy. It was led by badly trained reserve officers and lacked air and artillery support. In the following three weeks the division lost

15,000 men and was reinforced no less than four times to the tune of 12,000 men.[52]

Lieutenant-General Sokolov, formerly of the NKVD, knew nothing of the units of his Second Shock Army or what they were supposed to do. Stalin finally replaced him. The profligate attitude to human life became too much. In his directive of 30 March 1942, Western Front commander General Zhukov referred to the letters directed to the Stavka (GHQ and Stalin) denouncing the 'criminally negligent attitude' (*prestupno-khalatnoe otnoshenie*) of all officers squandering the lives of Soviet infantrymen. 'Abnormal losses' must be reported, infantry attacks must be properly planned and prepared. Commanders throwing infantry against unsuppressed defences faced demotion. Reports on 'abnormal losses' must include identifying those responsible and the measures taken to ensure there would be no repetition.[53]

For many the crisis of 1942 was more traumatic than even the disasters of 1941. With the growing emphasis on a 'just and patriotic war', coupled with the onset of genuine hatred of the Germans and first-hand evidence of their atrocities, party propagandists and army newspapers and indoctrination all worked to raise motivation and stiffen morale. More weapons were coming into units and Stalin demanded improvements in training, though front-line units complained about badly trained replacements and the poor provision made for these men.

In the summer of 1942, after further huge defeats in the south, Stalin seemed to fear some great crisis of morale, even defeatism. He personally drafted and issued Order No. 227 in August 1942, accusing the Red Army of treason to the Motherland, demanding 'not a step back'. And for the first time he introduced penal units, *strafbats*, into the Red Army.[54] Younger commanders in the mould of Zhukov and Rokossovskii, thoroughly tempered by war, realised that the Red Army suffered from over-rigidity, from inefficient commissar control and the incompetence of Civil War relics, and was incapable of coping with the complexities of modern operations. Even Stalin, after the disaster at Kerch in May 1942, lashed out at Soviet commanders who 'had totally failed to comprehend the nature of contemporary war'.[55] The result of this policy of carrot and stick was the abolition of 'dual command', displacing the commissars and restoring 'unitary command' to the officer.

The executions continued. Chuikov, commanding inside fiery Stalingrad, dealt out retribution and punishment in a savage enforcement of Order No. 227, with 13,500 men reportedly felled by firing squads. Stalingrad did not fall. The Soviet soldier showed himself at his tenacious best in a battle of sustained, terrifying brutality. From a great slough of defeat, even despair, the Red Army stood on the verge of becoming a more viable and modern fighting machine, absorbing lessons on equipment, training and leadership. The days of vast encirclements snaring Soviet armies were over. In November 1942 the Red Army turned to carry out its own encircling offensive, trapping the German Sixth and the Fourth Panzer Army at Stalingrad. In its defensive and offensive phases, from 17 July 1942 to 2 February 1943 the battle of Stalingrad cost the Red Army 1,129,619 men, killed, wounded and missing.[56]

The Soviet soldier passed through the extreme of nightmarish experiences at Stalingrad, with divisions ground to pieces, unceasing gunfire day and night and hand-to-hand grappling. The wounded had to face a perilous evacuation under fire across the Volga. In a war of unexampled ferocity the military medical services faced an almost superhuman task, their resources stretched beyond breaking point.[57] Pressure on manpower caused orderlies and stretcher-bearers to serve simultaneously in an infantry role.

Ever-more women 'medics' were attached to front-line units and frequently flung at company level into hand-to-hand fighting, 'awful...not for human beings...men strike, thrust their bayonets into stomachs, eyes, strangle one another. Howling, shouts, groans.'[58] At the front 41 per cent of all doctors and 100 per cent of the nurses were women. Losses among women medics with rifle battalions were second only to those of the infantrymen themselves. Forty thousand battle-tested women medics were decorated, fifteen of them with the highest award, Hero of the Soviet Union.[59]

To the ceaseless torrent of wounded, which finally totalled some eighteen million (though a reported 72 per cent recovered to return to active service), were added the ravages of disease and epidemics, aggravated by the filth and squalor of front-line service. There were few facilities for personal hygiene, little soap, few laundries. The available anti-epidemic units had neither sufficient transport nor adequate immunisation supplies.[60]

Across this gigantic front – though the tally of prisoners taken by the German army was falling – the disaffected, displaced and despised, men from central Asia and the Caucasus, Russians also led by the captured General Vlasov, criss-crossed the lines to serve with the German army.[61] This human tragedy of the alienated, the persecuted and the dispossessed had the bizarre effect of supplying the German army in the east, with 20 per cent of its manpower drawn from former Soviet citizens. After 1943 the number of non-Slavs serving with the Red Army declined, to a degree that the widest range of Soviet nationalities was almost better represented in the Wehrmacht than in the Red Army itself. For this flagrant disaffection Stalin revenged himself after 1943, by deporting Kalmyk, Chechen, Ingush, Crimean Tartars and Balkar populations to Siberia and central Asia.[62]

For its feat at Stalingrad the Red Army was rewarded with its own gold braid, the restoration of rank shoulder-boards, the *pogon* and the designation 'officer' returned to full respectability, with special decorations assigned for officers only. The stabilisation of the command and the emphasis on professionalism were subjected to an immediate test, the giant Battle of Kursk in 1943, the Red Army's planned strategic defensive operation culminating in the massive armoured joust which mangled the German Panzer arm. Soviet troops trained intensively and prepared on a scale not previously seen, all to confound Stalin's misgiving that the Red Army could not fight effectively on the defensive.[63]

The obsession with the political reliability and moral sturdiness of the army never abated, with a new counter-intelligence organisation, SMERSH ('Death to Spies') established in 1943 to seek out 'alien elements', sabotage, subversion and anti-Sovietism. The deluge of propaganda continued, promoting professionalism in the Red Army, instilling hatred of the Germans, though in 1943–4 there was a noticeable shift away from the theme of 'Russian patriotism' towards an emphasis on 'Soviet patriotism' together with reaffirmation of the party presence in the military.[64]

The Red Army in 1943 went back to school or into school for the first time. Nineteen military academies trained officers for higher command and more than 300 military schools turned out officers, with each arm – infantry, artillery, engineers – running specialist courses, with preference given to men with secondary school education. The Voroshilov General Staff Academy prepared officers for divisional and higher command in accelerated courses lasting up to six months, the Frunze Academy trained staff officers. In front commands and in military districts junior officers were trained in short three-month courses, in addition to being given further training at the front. As the severe crisis with the supply of officers passed, the length of courses was increased and the training improved.[65]

With the emergence of the new élite arm, the tank forces, training for both crews and commanders was vastly expanded. The tank corps committed in 1942 had suffered from poor organisation and lack of training. In 1943 tank crews trained in the 50 tank schools, with battalions in the tank-school regiments organised to familiarise men with specific tank types – KV heavy tanks, T-34s, T-60s and T-70s. Among the tank crews gunners trained for four months, driver-mechanics for nine months, loaders only two months. Tank officers and sergeants were trained in their own schools, with instruction lasting one year and specialising on one type. General Mellenthin was not alone in both observing at first hand and experiencing, to his cost, the dramatic changes in both quality and quantity in the Soviet armoured forces, whom he characterised in unsavoury fashion as an 'ignorant and apathetic crowd' magically galvanised.

As the Red Army cleared German-occupied territory it scooped up manpower from civilians, soldiers marooned since 1941 and partisans, drafting all men between the ages of sixteen and fifty. These thousands of 'new' soldiers, many far from young, were given ten days' training before being posted to units in the field; they were additional manpower but far from properly trained replacements or reinforcements. The Red Army soldier in the later stages of the war could well find himself a hewer of wood and a drawer of water, building roads, dragging guns, repairing rail tracks, securing bridges, as the Red Army advanced into devastated territory, 'scorched earth' in reverse as the German army retreated and Soviet armies outran their immediate support. Ever more women appeared in front-line units: *zhenschiny-voiny*, girl machine-gunners, *devushki-pulemyotchiki*, girl snipers, *devushkisnaipery*, women mortar crews, sappers (graduates of the Moscow Military Engineers' School), driver-mechanics in tank units, women signallers and military police. Women ran the military mail and the field post offices, frequently coming under fire to deliver 'long-range shells' (the mail) to front-line units. As the Red Army advanced it was followed by a 'second front' – laundresses, bakers, cooks. Women political officers supervised the civilian laundresses in field laundry detachments, working 'in some kind of house or dug out' impregnating the clothes with 'K' soap to get rid of the lice. 'We had delousing powder but it didn't help, so we used "K" soap which really stank – it had a terrible smell. Twenty or twenty-five grammes of soap were issued per soldier for washing clothes. It was black, like earth.' By 1945 the figure for women *frontoviki* had risen to 246,530, the average complement of women soldiers serving in field armies varying between 2,000 and 3,500.[66]

For the front-line Soviet soldiers, the *frontoviki*, there was little relief and virtually no hope of escape from the fighting. Stalin is reported to have said that it took a brave man to be a coward in the Red Army. The growing scale of Soviet successes, in spite of the losses incurred in breaking into heavily fortified positions, sustained morale, but the driving force was consuming hatred of the enemy, exemplified in the propaganda but personalised in exacting vengeance by millions who had seen for themselves or suffered through their families what German rule had done. These hard soldiers lived hard, 'home' his (or her) machine-gun, mortar or tank. They were not always fully fed, making the best use of extra bread and anything else to hand, but ammunition and fuel always had first priority.[67] Alcohol did not make the soldier any braver, only more foolhardy and often dangerously reckless. Discipline was fierce, though officer–soldier relations were generally good.[68] In the absence of sufficient NCOs, junior officers, company commanders especially, had a gruelling time: with enemy artillery never entirely suppressed; their men led by but often caught in inaccurate Soviet artillery fire; fewer men in reduced rifle companies needed to move weapons forward; a rush to the objective; company strength dwindling under fire and survivors preparing to fight off the inevitable counter-attack.[69] The unit now suffered more casualties, among them very likely the company commander himself. But if tactical gains led to success at the *operational* level, engulfing entire enemy formations, then clumsiness, profligacy and sacrifice were deemed justifiable.

*

Few armies in history have undergone such devastation as the Red Army suffered in 1941 and gone on not only to survive but to win out at the end. That final triumph had long been attributed to sheer weight of endless numbers, 'accepted wisdom' which has done grave damage to a proper understanding of the operation of the Soviet system and the place of the Soviet soldier. The 'hordes of riflemen' vanished all too quickly after 1941, the manpower pool was drastically thinned not only by battlefield losses but by German occupation of great swathes of Soviet territory, which drained off sixty million people. Ruthless almost beyond belief, the Soviet method of rotation, replacement and reserves nevertheless sustained front-line order of battle. Draconian discipline was reinforced by the barrel of a gun pointed at the soldier and frequently held to the head of his commanders; the loyalty of both was questioned, their endurance tested to the limit.

None of this would have been effective, brutal though it was, without intelligent reorganisation and restructuring throughout 1942 and beyond, principally the transformation in 1943 to an offensive force maximising fire-power and mobility. Not the mass but the *organisation* of manpower, coupled with the production of effective battlefield weaponry, proved to be the war-winning formula. The evolution of the wartime rifle division was itself a miniature illustration of this process, dwindling manpower being offset by steadily augmented firepower and weight of metal.

The story of the Soviet soldier in the 'Great Patriotic War' is intimately bound up with the costly struggle to produce effective commanders, suitable organisation, superior weapons and appropriate tactics. Though Soviet official histories perforce emphasised the planned and premeditated aspects of these developments, the system and the soldier both proved hugely adept at impro-visation, which was repeatedly required in the face of acute shortages, lack of special equipment and unforeseen crises. In terms of tactical performance the Red Army progressed from the poor, often the inept, to the fair as the war drew on, but after 1943 the overriding aim was to achieve decisive operational success after the breakthrough. Soviet riflemen and tank crews paid a griev-ously high price for tactical deficiencies, which failed to accelerate penetration to any depth and thus left enemy reserves intact.

After the passage of more than fifty years the fundamental question affect-ing Soviet soldiers, sailors, airmen and civilians remains whether they individu-ally and collectively survived and won out because of or in spite of the system. In war soldiers and civilians alike astonished themselves with their achievements, attained through self-discipline, devotion to duty, immediate responses to good leadership and deep patriotic commitment: pride in and pity for Russia. The bemedalled peasant lad turned Red Army man, *krasnoarmeets*, undoubtedly took pride in being one of 'Stalin's soldiers' and had much to be proud of.[70] But the weight of the failings of the system, unpreparedness, the profligacy with life, blunders and incompetence could not easily be shrugged off, forgiven or forgotten. A Red Army officer's tart observation summed up the triumph and tragedy: 'Yes, we got to Berlin, but did we have to go via Stalingrad?'

Offensive Women: Women in Combat in the Red Army

Reina Pennington

A typical view of the historical role of women in combat was expressed by John Keegan in his 1994 book, *A History of Warfare*: 'Warfare is...the one human activity from which women, with the most insignificant exceptions, have always and everywhere stood apart...Women...do not fight...and they never, in any military sense, fight men.'[1] It's often been said that one of the first casualties of war is truth – and that seems to be the case here. Nearly a million women served with the Soviet armed forces in the Second World War. And whether the subject is bank accounts or personnel, any number over six figures is generally considered significant.

If historians have failed to note the presence of military women in their study of history, it is at best the result of tunnel vision, and at worst, a case of deliberate oversight. Keegan derides the habitual reluctance of military historians 'to call a spade a spade', but it seems that Keegan himself, among others, exhibits precisely this reluctance regarding the military history of women.[2] In the words of D'Ann Campbell, women continue to be the 'invisible combatants' of military history in general, and of the Second World War in particular.[3] A study of military women must be positioned at the intersection of military history and women's history, which is probably the reason it is so rarely undertaken. Specialists in military history hardly ever address so-called 'women's issues', while those who write the history of women seldom have an interest in military affairs. The overall neglect is so marked that the history of military women is virtually a historiographical 'no-man's land'.

In its use of large numbers of women in combat, the Soviet Union was unique in world history. During both world wars and its civil war, women fought in the front lines.[4] Soviet women engaged in combat in all branches of service in addition to their mass employment in support services.[5] By the end of 1943, more than 800,000 women were in military service, or about 8 per cent of the total personnel. Altogether a million Soviet women served (including partisans); half of them were at the front.[6] Soviet women were unique in being the only women soldiers during the war who fought outside the borders of their own country.

Many Western historians have dismissed information about Soviet women in combat as mere propaganda. George Quester typifies this point of view: 'Some few actual combat roles have occasionally been given great publicity, but this has typically been an exercise in public relations, designed to impress the outside world with the underdog position of the country in question.'[7] Military analyst Jeff Tuten also dismisses Soviet sources, claiming that 'little reliable information on the performance of the female combat formations is available'. However, he accepts German reports as reliable; he makes a point of noting that 'Germans were shocked to find [women] on the battlefield and were contemptuous of their fighting performance'.[8] Apparently Tuten is willing to accept one negative German opinion over numerous Soviet sources, and even other, more positive German evaluations.[9] For example, an officer of the 1st SS Panzer Division tells of an encounter with women Soviet soldiers at Kharkov in 1942; he says, 'I thought my time had come', but he made a 'lucky escape' because, when his unit retook their position, they found all those who had been left behind had been killed by the women.[10]

The view of Soviet women in the Second World War that is generally accepted among more serious scholars of Soviet history is summed up by Albert Seaton: women were recruited on a large scale during the war, but mainly as volunteers in auxiliary services such as signals, traffic control, medical, clerical and administrative service. While stories of women as pilots, snipers and tank crews were (supposedly) highly publicised, such women were rare, and there was no conscription of women except in technical specialties.[11] Many researchers accept this view; a PhD student specialising in Soviet military history believes that women in the Red Army were 'generally put in positions where they would work pretty much alone, for example as snipers, or in positions which were "out groups" relative to the front-line soldiers in platoons, such as medics who would come forward from battalion or regiment to retrieve the wounded'. Thus, he says, unit-cohesion issues were largely avoided, either by accident or design. He also claims to have found no references to women serving at platoon or company level.[12] However, Soviet sources reveal numerous references to women who worked at the platoon and company level, who even commanded those units, and many references to medics indicate that they did not stay in the relative safety of headquarters, but were assigned at the sub-unit level.[13]

There is evidence that the history of Soviet women in combat has been forgotten even by the Soviet public. Griesse and Stites note that 'for the most part public recognition of women's sacrifices and experiences in war was not played up very much. What was stressed in the postwar years were the new crucial roles for women, for instance, motherhood and the labor force.'[14] This phenomenon of selective amnesia is not unique to the Soviets. Social anthropologist Sharon Macdonald points out that because war is traditionally defined as masculine, women in combat disrupt the social order by their very existence. They are outside the social framework of understanding, and when history is written after the war, their experiences are usually explained away or simply forgotten.[15]

Fortunately, a better informed analysis is provided by historians like Anne Griesse, Richard Stites, Jean Cottam and John Erickson in the West, and Valentina Galagan, Yulia Ivanova and Vera Murmantseva in the former Soviet Union. In numerous recent works, they call attention to the experience of Soviet women in combat.[16] Through such efforts, the history of Soviet women in combat is finally coming to light. Erickson describes it as 'a saga which has yet to be told in all its astonishing variety and harrowing individual detail', and says that the lack of attention this 'extraordinary dimension of the Soviet war effort' has received from both Soviet and Western historians is 'not a little reprehensible'.[17]

*

When the Germans invaded the Soviet Union on 22 June 1941, only a few Soviet women were in military service. During the first weeks of the war tens of thousands volunteered for active duty; most were rejected. When the war began, there was no plan in place for the large-scale military mobilisation of women. Only those with selected technical skills, such as medical and communications, were liable for conscription, or were even accepted as volunteers.[18] In fact, the Soviets' first response was quite similar to that in the West: they urged women to replace men in industry and agriculture. Women would 'man' the home front so that men could go to war. This was a less radical shift in the Soviet Union than in the West, since Soviet women already comprised 40 per cent of the industrial labour force.[19]

As the war progressed, however, female volunteers were accepted increasingly, and there were many mobilisations of women, particularly for the rear services, communications and the air defence forces. The central committee of the Komsomol handled most of the mobilisations. One veteran described the Komsomol selection process as 'stringent and thorough'.[20] However, there were no tests of strength, endurance or physical fitness; the Soviets apparently assumed women would be able to do the job.

Some people are surprised that women could drive a tank or load an anti-aircraft gun. Yet serving as a cook or laundress with the Red Army, roles that we take for granted, also required great strength. Women cooks hauled huge cauldrons that were so heavy that many men said they feared these women would never be able to bear children. Laundresses also handled extremely heavy physical loads for long hours each day.[21]

Women military mechanics routinely lifted heavy weights, of course. Armourers with the all-female 46th Guards Night Bomber Regiment had only a few minutes to rearm aircraft between flights; they loaded four 100-kg bombs by hand.[22] Soviet women were accustomed to heavy physical labour before the war, and women in the Red Army were simply expected to carry ammunition, load bombs and big guns, change propellers, and so on. The Soviets seem to have taken for granted what many Westerners still see as an insurmountable obstacle.[23]

By 1943 Soviet women had been integrated into all services and all military roles, ranging from traditional support roles like medical service, to primarily defensive work in anti-aircraft defence, to offensive combat roles in the infantry, artillery and armour, as well as the partisan movement. Many women commented on the fact that when they were accepted for duty they were given men's clothing, right down to the underwear.[24] Oversized boots were a particular problem. No special women's uniforms were issued until half-way through the war.[25]

Why did so many Soviet women want to go to war? Their motivations ranged from following relatives to the front, to avenging the death of a friend or relative, to simple patriotism. Quite a few women whose fathers had been arrested during the purge years hoped to redeem their families by serving at the front.[26] Perhaps Sergeant Klara Tikhonovich, an anti-aircraft gunner, best expressed a common impulse that sent women to war. She writes, 'A young person recently told me that going off to fight was a masculine urge. No, it was a human urge ... That was how we were brought up, to take part in everything. A war had begun and that meant that we must help in some way.'[27]

Soviet women did manage to take part in nearly every wartime duty. In addition to the 800,000 in the army, thousands of women served in so-called 'fighting battalions', which were local security units similar to the British Home Guard; not part of the regular army, but still an armed force. Around 25,000 women served with the Soviet navy, and another 30,000 in river transportation.[28] In the Soviet air force there were three combat regiments that started the war as all-female, and numerous women pilots, mechanics, and so on, in other line units.[29] Within the Red Army women made up 75 per cent of military drivers and communications was heavily staffed by women.[30] Most of the Red Army's women served in positions that were considered non-combatant, but nearly all were trained to handle weapons. Even military drivers trained with rifles and bayonets.[31]

My focus here is on the women who experienced war most directly, beginning with medical personnel. Female medical orderlies and nurses served in the Red Army down to the company level.[32] More than 40 per cent of all Red Army doctors, surgeons, paramedics and medical orderlies, and 100 per cent of nurses, were women.[33] Many were quite young and had minimal training, but as one veteran noted, 'When a young boy ... has his arm or leg cut off before your very eyes, childishness quickly gets wiped out of your mind.'[34] The primary duty of medical personnel was, of course, to tend the sick and wounded. For front-line medics, that meant rescuing the wounded from under enemy fire, and whenever possible, retrieving precious weapons together with the wounded soldiers.[35]

Medic A. M. Strelkova describes what was required: 'I don't know how to explain this. We carried men who were twice or three times our own weight. On top of that, we carried their weapons, and the men themselves were wearing greatcoats and boots. We would hoist a man weighing 80 kilograms on our backs and carry him. Then we would throw the man off and go to get another one...And we did this five or six times during an attack.' This work was extremely hazardous; front-line medics had a casualty rate second only to the active infantry.[36]

Senior Sergeant Sofia Dubniakova, a medical orderly, told an interviewer: 'There are films about the war in which one sees a nurse at the front line. There she goes, so neat and clean, wearing a skirt, not padded trousers, and with a side-cap perched on top of an attractive hairdo. It is just not true. Could we have hauled a wounded man dressed like that?'[37] Medical orderlies dragged incredible numbers of wounded soldiers from the field of battle. Maria Smirnova, with the 333rd Division of the Fifty-sixth Army, rescued a total of 481 wounded from under fire during the war.[38] This number seems fantastic, but Soviet sources document some women medics who rescued twenty wounded soldiers, with their weapons, in a single day, others who carried the wounded as much as fourteen kilometres from where they fell to a medical station.[39]

Olga Omelchenko tells a terrible tale of her experience as a front-line medic with the First Company of the 118th Rifle Regiment, 37th Guards Division. After a heavy battle in 1943, she says: 'I crawled up to the last man, whose arm was completely smashed. The arm had to be amputated immediately and bandaged...But I didn't have a knife or scissors...What was I to do? I gnawed at the flesh with my teeth, gnawed it through and began to bandage the arm.'[40]

Soviet medical personnel were not, strictly speaking, non-combatants. Many medics, nurses and doctors took up weapons in battle.[41] For example, Katiusha Mikhailova, a medic with a naval infantry battalion, completed the same training course as the men in her battalion, and carried grenades, anti-tank grenades, and a submachine-gun in addition to her medical bag. She was expected to be fighting whenever she was not tending the wounded.[42]

In addition, many women who started as nurses ended as soldiers. Some had applied for combat and been forced into nursing against their will, but later managed transfers.[43] Others were trained nurses who discovered a taste for combat as the war progressed. Lieutenant-General Kolomiets, former commander of the Independent Maritime Army, wrote in his memoirs that after the death of a well-known woman machine-gunner in his unit,

a great many of the nurses who served in the Division kept asking to be transferred to machine-gun duty. We tried to convince them that [they] were also considered combatants and heroes. All the same, during the relatively quiet periods at the front we organized training in machine-gun firing for a group of nurses, and we made an exception for those who were especially proficient by transferring them to machine-gun duty.[44]

One last point on medical personnel is that nearly every Soviet woman who served in the Red Army, whatever her assigned position, was expected to fill in as a medic simply because she was a woman. Tens of thousands of women pulled this sort of double duty, and often they are dismissed as medics rather than combatants.[45]

Women also served as communications operators throughout the Soviet armed forces, and constituted about 12 per cent of the total.[46] The picture that might come to mind is a telephone operator at a switchboard in a headquarters, but this was not typical. Many women in communications served right on the front lines of battle. Signaller Antonina Valegzhaninova, for example, describes her experience at Stalingrad: 'One battle stands out in my memory. There were scores of dead... They were scattered over a huge field, like potatoes brought to the surface by a plough.'[47]

One of the first conscriptions of women was for communications work; in August 1941 10,000 women were drafted for front-line duty with the signals troops.[48] In 1942 another 50,000 were trained. By 1943 some communications regiments at the front were 80 per cent female.[49] In addition, more than 1,200 female radio-operators were trained for parachute drops behind the lines. N. M. Zaitseva, for example, made nineteen parachute drops behind enemy lines during her career. This work was extremely dangerous; a number of women were forced to kill themselves with hand grenades to avoid imminent capture and to destroy their equipment.[50]

Women in communications were expected to fight; Elena Stempkovskaia was awarded the Hero of the Soviet Union after she died defending her command post from attack; she reportedly killed some twenty enemy soldiers with a machine-gun. Tatiana Baramzina had been a sniper but transferred to communications when her vision deteriorated; she was still a good enough shot to kill a number of Germans in July 1944 when her battalion was sent behind enemy lines. Of course, she had to alternate between shooting the enemy and filling in as a medic, in addition to her communications duties. She was captured and then executed with an anti-tank rifle.[51]

More than a quarter of a million women served in the Air Defence Forces during the war. In fact, Stalin was personally involved in the integration of women into the Air Defence Forces. In 1942 there were two mobilisations to bring women into air defence. Specific goals were established: women were to replace 8 out of 10 men in instrument sections, 6 out of 10 men in machine-gun crews, 5 out of 6 men in air warning posts, 3 out of 11 men in searchlight crews, and all male privates and NCOs in the Air Defence Forces rear services. Women between the ages of 18 and 27 were subject to conscription.[52] There were many women officers, and women served in every position on gun crews.

By the end of the war, more than 121,000 women served on gun crews, and about 80,000 in Air Defence Forces aviation or in searchlight and observation posts.[53]

Klavdia Konovalova served with the 784th Anti-Aircraft Artillery Regiment. She was initially assigned as a gunlayer, but sought to be a loader. She says that being a loader 'was considered purely masculine work, since you had to be able to handle 16-kilogramme shells easily and maintain intensive fire at the rate of a salvo every five seconds'. Since she had worked as a blacksmith's striker before the war, she was more than capable of handling that role. She served as a loader for a year, then was appointed commander over a gun crew of two women and four men.[54]

Women also played a role in the partisan movement. By 1944 there were a minimum of 280,000 active partisans; women filled 25 per cent of the total, and 9.3 per cent of operational roles, or approximately 26,000 active women partisans.[55] There has been some controversy over the precise role of women in partisan groups, with historian Earl Ziemke arguing that they were little more than cooks and camp followers.[56] Kenneth Slepyan, who recently completed a dissertation on the Soviet partisans based on archival materials, holds a different view; he shows that women partisans made significant contributions in both support and combat roles.[57] First-hand accounts indicate that there was wide regional variation in the extent to which women were permitted to fight.[58] We know that virtually all partisan women were armed, and many were directly involved in combat and sabotage.[59] While there were no female detachment or brigade commanders in the partisans, some women did command companies and platoons, both all-female and mixed units.[60]

The most famous female partisan was Zoya Kosmodemianskaia, a teenager who was caught and executed by the Germans early in the war. Many women partisans performed extraordinary acts of resistance. For example, Elena Mazanik assassinated the German governor of occupied Belorussia by planting a bomb under his bed.[61] Partisan scout Antonina Kondrashova endured a special torture. Her mother had been captured by the Germans, and for two years she and other prisoners were made to lead the way whenever the Germans went on missions. Kondrashova says the Germans

> were afraid of partisan mines and always drove the local population in front of them...on more than one occasion we would be sitting in ambush and suddenly see women walking towards us, and behind them the Nazis. They would come closer and you would see that your mother was there. The most terrible thing was to wait for the commander to give orders to open fire. We all awaited this order with fear...[but] if the order to fire was given, you fired.

Her mother was eventually shot, along with other prisoners, by the Germans when they retreated from the area in 1943.[62]

Partisan work was extremely dangerous. Fekla Strui, a deputy of the Supreme Soviet turned partisan, was wounded in both legs during a battle. Her unit was encircled by Germans, preventing her evacuation. She recalls, 'My legs were taken off right there in the forest. The operation took place in the most primitive conditions. I was put on a table to be operated on, there wasn't even any iodine and my legs were sawn off with an ordinary saw, both legs... the operation was performed without anaesthetic, without anything.' She was eventually airlifted to Tashkent, where recurring gangrene forced a series of four reamputations.[63]

Few Westerners had the chance to observe Soviet partisan activity, but a number of Allies had first-hand experience in Yugoslavia. Lieutenant-Colonel Henry Peoples, a pilot with the Army Air Force's 332nd Fighter Group and recipient of the Distinguished Flying Cross, offers the following observation:

I was shot down over Yugoslavia and spent two months with Tito and his partisans. I went on a couple of their raids and they were exciting, to put it mildly. They were fine people and their women, such women! You talk about amazons; those Yugoslavian female partisans, I swear they could outfight their men. They were strictly f. and f.: fearless and ferocious.[64]

Women also served in combat engineering, and were regarded as non-combatants. A number of women served in military engineering, mainly as sappers. Their number is uncertain, but at least seventy-five women were trained as sapper platoon commanders in 1942 alone. One of them described the training as tightly disciplined and demanding, and says they were told that the average life expectancy of a sapper platoon commander was two months.[65] Another, Junior Lieutenant Appolina Litskevich-Bairak, described the work of sappers in the Red Army:

During the night the soldiers would dig a two-man foxhole in no-man's land and before dawn one of the section commanders and I would crawl out to this small trench and the men would camouflage us. We would lie like that the whole day, afraid to stir... all day we kept everything that happened under close observation and drew a map of the front line... we would guess [where the] German sappers had laid minefields... At night, this time with the sappers, you crawled up to the front line. We would clear a passage through our own minefield and crawl towards the German defences... All work at the front line was done on your belly.[66]

The work of the sappers did not end with the war; Litskevich-Bairak noted that her platoon continued clearing mines for an entire year after peace was declared.[67] Unfortunately little is known so far of the work of Soviet women in combat engineering.[68]

Although comparatively few women served in Soviet armour, they filled a variety of roles, including medic, radio operator, turret gunner, driver-mechanic, tank commander, and platoon commander. Most served with the T-34 medium tank, which was also the fastest Soviet tank.[69] Others served in test functions; for example, Polina Volodina commanded an all-woman crew that tested the T-40 floating tank and later served as the head of tank salvage and repair service of the Southern Front.[70]

Nina Vishnevskaia, who worked as a medic in a tank battalion, reports that they were reluctant to take women in tank units, even as medics. While male members of armour were issued canvas trousers with reinforced knees, the women were given cotton overalls. She recalled:

> Very soon we were ... wearing rags, because we did not sit in tanks, but had to crawl on the ground. Tanks often caught fire and the tankmen, if they remained alive, were all covered with burns. And we also got burns, because to get hold of the burning men, we had to rush right into the flames. It's very difficult to drag out a man, especially a turret gunner, from the hatch ... In tank units medical orderlies didn't last long. There was no place in a tank for us. We clung to the armour and thought about one thing only: how to keep our feet clear of the caterpillars so we wouldn't get dragged in. And all the time you had to watch if there were any tanks on fire. You had to run, to crawl there.

Vishnevskaia, who was at the Battle of Kursk, was the only survivor of five women medics who had enlisted together.[71]

A number of women served on tank crews in combat positions. Irina Levchenko started the war in 1941 as a medic, eventually serving with a tank battalion in the Crimea. After recovering for the second time from battle wounds, she applied to the Commander of the Armoured Troops, General Fedorenko to be sent for combat training in armour. She was allowed to attend the Stalingrad Tank School, and thereafter drove a T-34. She ended the war with the rank of lieutenant-colonel and a Hero of the Soviet Union medal.[72]

Marina Lagunova was a driver-mechanic who saw combat at Kursk and on the drive to the Dnieper River. In September 1943 her tank was destroyed and she lost both her legs. She recovered from her wounds and went on to become an instructor with a tank-training brigade.[73] Probably the most famous woman to serve in armour was Maria Oktiabrskaia. After her husband was killed, she donated her life savings to buy her own T-34. At the age of thirty-eight, she entered combat in October 1943 with the 26th Guards Tank Brigade, and fought until she was killed in battle near Vitebsk in 1944 while making a repair to her tank track.[74]

A number of women also served in the 45-ton IS-2 tank, equipped with a 122-mm gun. Ola Parshonok first drove the T-34, and later the IS-2; she went all the way to Berlin with the 231st Tank Regiment. Only one woman became commander of a heavy tank: Junior Lieutenant Aleksandra Boiko. She and her husband volunteered for duty and donated 50,000 roubles to build a tank, on the condition that they be crewed together. They graduated from the Cheliabinsk Tank Technical School in 1943. Aleksandra was commander of the four-person crew, and her husband was driver-mechanic. They fought in the Baltic States and into Poland, Czechoslovakia and Germany. Both were wounded and decorated.[75]

A unique role in the Red Army was that of sniper. A so-called 'sniper movement' was started in the autumn of 1941, sponsored by the Komsomol, and spread rapidly. More than 100,000 women went through sniper-training courses during the war.[76] In May 1943 the Central Sniper Training Centre for Women began operation under the command of N. P. Chegodaeva, a graduate of the Frunze Military Academy and veteran of the Spanish Civil War; its 1,500 graduates accounted for 11,280 enemy troops killed by the end of the war.[77]

The job of sniper was an exacting one. Snipers needed a good eye for distance and motion, the ability to handle a rifle blindfolded, the endurance to crawl long distances to get near enemy lines, then dig in and camouflage themselves, and the patience to wait nearly motionless for long hours in all kinds of weather. Snipers usually tried to get into position in the pre-dawn darkness, and only returned to their own lines after nightfall. They sought positions as close as 500 metres from the enemy lines, and might lie in snow, sit in a tree, or perch on a roof for twelve hours or more at a time. Many snipers worked in pairs in order to keep one another alert.[78]

Some women snipers operated in all-female platoons. A platoon of 50 women snipers in the 3rd Shock Army, commanded by Nina Lobkovskaia, was credited with killing 3,112 German soldiers.[79] Many women snipers were killed in action. The most famous woman sniper was Liudmila Pavlichenko, with 309 kills to her credit, including 78 enemy snipers. She fought with the 25th Chapaev Rifle Division during the defence of Odessa and the siege of Sevastopol. After she recovered from serious wounds, she was made a master sniper instructor, and later toured the United States to urge the opening of a second front.[80]

A number of women also served with the Red Army as machine-gunners. In 1942 alone the Komsomol Vseobuch, or Universal Military Training service, claims to have trained 4,500 women as heavy machine-gun operators, and another 7,800 on light machine-guns. Women machine-gunners served down to the platoon level, and many commanded platoons and companies. Just one example is that of Zoya Medvedeva, who served on machine-gun crews throughout the war, first as a crew member, then commanding machine-gun platoons with both air defence and rifle units, and later as a company commander.[81]

Medvedeva describes the experience of battle at Odessa. During the enemy artillery bombardment, she says, 'I sat, my face buried in my lap ... feeling like an ant on an anvil that a blind blacksmith was striking with all his strength; he never missed the anvil and each one of his blows narrowly avoided flattening me completely.' She graphically describes the rigours of serving on a machine-gun crew, of having repeatedly to lift the gun into place, remove it during artillery attacks, and move it from one location to another. In their first battle, only thirteen soldiers of her platoon survived; Medvedeva herself was wounded in the head. Ironically, she was not wounded while at her gun, but only after the battle, when she was helping bring in the wounded. Two months later, she returned to her division at Sevastopol. She was told there were no vacancies in the machine-gun companies and was forced to work as a medic. Only after demonstrating her skill with a gun was she allowed to return to her assigned position. Later, she was one of only seven soldiers to survive out of the entire company. They broke out of an encirclement and then fought with a naval infantry unit, where Medvedeva took over a machine-gun position. She sustained a concussion and shrapnel wounds and was evacuated. After recovering, she went on a course for machine-gun platoon commanders and returned to the front as a junior lieutenant.[82]

No comprehensive statistics are available for women who served in combat positions in the infantry.[83] Aside from the roles already mentioned, women served in mortar crews and as reconnaissance scouts. In 1942 more than 6,000 women were given training on mortars; more than 15,000 were trained with automatic weapons or submachine-guns.[84] Women are rarely categorised simply as soldiers; women who served with infantry units are generally designated as medics, snipers, scouts, or members of gun crews. However, in February 1942, the First Independent Women's Reserve Rifle Regiment, an infantry training unit, was formed; eventually nearly 10,000 women (3,900 privates, 2,500 NCOs and 3,500 officers) graduated from this group.[85] In addition, the Voroshilov Infantry School in Ryazan included three women's battalions and sent nearly 1,400 women platoon commanders to the field in 1943: 704 took over rifle sections; 382 machine-gun sections; and 302 mortar crews. Can these women be regarded as infantry soldiers? A man who commanded a rifle platoon would be called a soldier; there is no reason why women who did the same thing should not.[86]

Some people ask whether women had trouble killing. But there was no 'woman's' reaction to killing. Some say they never got used to it and it was always difficult. Others say it was a matter of self-defence.[87] Still others say their anger and hatred was motivation enough. Sergeant-Major Liubov Novik says, 'Whenever I recall the past now I am seized with terror, but at that time I could do anything – say, sleep next to a dead person, and I myself fired the rifle and saw blood; I remember only too well the especially strong smell of blood in the snow... [but] It wasn't that bad then and I could go through anything.'[88]

Partisan Antonina Kondrashova points out that she once heard the cries of a child who was thrown by the Germans into a well. 'After that,' she says, 'when you went on a mission, your whole spirit urged you to do only one thing: to kill them as soon as possible and as many as possible, destroy them in the cruellest way.'[89] On the other hand, one woman pilot sees killing as a military skill, and says that should not be equated with cruelty.[90] Women's reactions to killing varied widely, depending on the role they filled and their personality.

Another aspect of the experience of war is that of the prisoner-of-war. Many Soviet women (and men) expressed their fear of being captured during the war. As one nurse recalled, 'we always kept a cartridge for ourselves: we would rather die than be taken prisoner'.[91] In most published sources, the fear of being captured is attributed to the expected atrocities they would suffer at the hands of the Germans. Though Westerners seem to regard the possibility of rape as one of the most horrific things that might befall a woman prisoner-of-war, they overlook two things: women are not the only victims of rape, and there are things that happen to prisoners that are worse than rape. For example, one veteran reports, 'one of our nurses was taken prisoner. About a day later we liberated the village and found her – her eyes had been put out, her breasts lopped off. She had been impaled. It was frosty and she was all very white, her hair completely grey. She was a young girl of nineteen.'[92]

Partisan Liudmila Kashichkina was captured when she was twenty-three years old. She says,

I was kicked and beaten with whips [until my skin was in ribbons]. I learned what a fascist 'manicure' is. Your hands are put on a table and a device of some kind pushes needles under your nails, all your nails at the same time. The pain is indescribable. You lose consciousness immediately...and there was some kind of machine. You heard your bones crunching and dislocating.

Kashichkina was sent to a concentration camp in Germany, then transferred in 1944 to another camp in France. She escaped together with some French prisoners and joined the Maquis. She was later awarded the Croix de Guerre and sent home to Russia.[93]

The Soviets have not published precise figures on women's wartime casualties. Losses for the women's aviation regiments can be calculated based on unit histories and regimental albums; they varied from 22 per cent to 30 per cent of flying personnel, which were typical rates for aviation.[94] Losses among women in the front-line Red Army appear to have been proportionately heavier than among men, even though most were not in designated combat jobs.[95] The high casualty rate for front-line medics has been already noted. Women sappers undoubtedly had a casualty rate as high or higher than that of the infantry. Until hard statistical data becomes available, only a subjective conclusion is possible: women at the front appear to have taken the same risks and suffered casualty rates typical for their duties.

A brief look at what happened to Soviet military women when the war ended is instructive. By the autumn of 1945 a decree was issued demobilising all women from military service except for a few specialists, and thereafter very few Soviet women indeed served in the military.[96] Many in the West have assumed that the Soviets made a 'logical' decision to demobilise women based on the single criterion of military performance. The line of reasoning is that if the women had performed well, they would have been retained in service; since they were demobilised, they must have fought badly. But a deeper examination reveals that other factors overrode any consideration of wartime performance in the decision to exclude women from the post-war military; these factors include cultural ideas of gender roles, social policies of pronatalism, and psychological effects of war-weariness.[97] Political decisions were made long before the end of the war to bar women from the post-war Soviet military. A deliberate policy to downplay the role of women in combat and stress traditional female roles was implemented even before the end of the war. An article in *Pravda* in March 1945, on International Women's Day, stated, 'in the Red Army... women very energetically proved themselves as pilots, snipers, submachine gunners [etc.]... But they don't forget about their *primary duty* to nation and state, that of *motherhood*.'[98]

In July 1945 President Kalinin spoke to a gathering of recently demobilised women soldiers. He told them:

> Equality for women has existed in our country since the very first day of the October Revolution. But you have won equality for women in yet another sphere: in the defence of your country arms in hand. You have won equal rights for women in a field in which they hitherto have not taken such a direct part. But allow me, as one grown wise with years, to say to you: do not give yourself airs in your future practical work. Do not talk about the services you rendered, let others do it for you. That will be better.[99]

It seems unlikely that similar advice was ever given to male veterans.

In conclusion, women were never received as part of the Soviet military élite. Even while women were at the front, the Soviets instituted gender segregation in the educational system and barred women from the newly created Suvorov cadet schools. In 1943 the groundwork was already laid for exclusion of women from the post-war military. Not only was performance irrelevant to Soviet decision-making about whether to allow women to remain in military service; there is strong evidence that during the post-war period, the Soviet government deliberately obscured women's war-time achievements. Soviet military women were, in effect, consigned to planned obscurity.

After the war, women were demobilised and the armed services reverted to a traditional male composition. Historian Barton Hacker suggests that 'the common vision of armies as all-male institutions strongly distorts past realities' and that 'for thousands of years armies routinely comprised women as well as men'.[100] The Soviet experience appears to bear out this thesis. The rich history of the military roles played by Russian and Soviet women over the past several centuries is only beginning to be brought to light. This history should finally set to rest assertions that women's military participation has been 'marginal' or 'insignificant'.

Motivation and Indoctrination in the Wehrmacht, 1933–45

Jürgen Förster

More than fifty years have passed since the Second World War ended in Europe. When the invading armies of the Allies reached the heart of Germany, the Wehrmacht capitulated and the German national community disintegrated. By then the Third Reich had ceased to exist. The unconditional surrender of 7 May 1945 affected the Germans more than the Allies through the utter sense of military defeat, and the massive destruction, dislocation and humiliation. Although it was obvious that Hitler's prediction that 'International Jewry and its helpers' would extinguish the German people was not fulfilled, nevertheless, fear of reprisals and revenge mitigated relief that the long fighting and endless bombing had finally ceased.

After a half-century of writing and research on the Third Reich, the relationship between the Wehrmacht and National Socialism is still on the agenda. From Nuremberg onwards, the notion existed of separating the Führer from his Volksgenossen, the generals from their supreme commander, and the Wehrmacht from the Nazi crimes. For far too long did the memoirs of German generals, quickly translated into English during the Cold War, shape the public's image of the Wehrmacht's record in the Second World War. Hitler was dismissed as a bungling amateur in military matters; his illogical aims had supplanted the generals' objective professionalism. They presented themselves as having pursued a fundamentally just cause and as having become Hitler's victims. This concept was most cleverly stated in Field-Marshal Erich von Manstein's *Lost Victories* (1958) and General Siegfried Westphal's *Army in Chains* (1950). Moreover, with his book *On the Other Side of the Hill* (1951), B. H. Liddell Hart had already provided the generals with a large international audience which they could address about the campaigns in Europe and Africa and their purely professional role in them. It was not until the late 1960s that historians began to research the available documents and shed light on *this* side of the Second World War. Now anyone could see that the linkage between strategy and mass murder in the war policy of the Third Reich makes it impossible to posit a Wehrmacht which remained detached from its political leadership. However, many a veteran still doggedly perpetuates the myth that the German soldier only did his duty and fought a clean war. To state that the Wehrmacht committed many crimes in the Soviet Union, in the Balkans and in Italy is not to say that every single German soldier was a criminal or was equally guilty of the crimes perpetrated in the name of the régime; yet it is also true that not every soldier was a mere defender of his fatherland.

It is not easy to answer the question: Why did the German soldiers fight so well 'in spite of their demented Führer'?[1] The Wehrmacht was certainly 'one of finest fighting arm(ies) of the war'.[2] For its successes, one historian emphasised the Wehrmacht's internal organisation, 'which succeeded in creating and maintaining fighting power'.[3] In my view Germany's military effectiveness in the Second World War may not be reduced to 'fighting power'. Fighting organisations need professionalism *and* motivation. The Wehrmacht was not merely a superb professional organisation, it was also an integral part of state and society in the Third Reich. The armed forces intentionally formed the 'second pillar' of the Führer-state, alongside the party. Thus, the history of the Wehrmacht and the Third Reich has to be viewed as *one* history. Yet, the question is: What did the Wehrmacht fight for? Since the soldierly and patriotic rhetoric persisted side by side with ideological orientation, and since the relationship between the Wehrmacht and Nazi Germany was a dynamic one, it is difficult to agree with the interpretation of van Creveld that the average soldier 'did not as a rule fight out of a belief in Nazi ideology – indeed, the opposite may have been nearer to the truth in many cases'.[4] The question of the efficacy of ideological indoctrination is difficult to answer. What is the difference between motivation and indoctrination and where is the boundary between fighting spirit and Nazi *Weltanschauung*? And finally, the historian has to ask: How can I get close to the real thoughts and emotions of 'Men against Fire' and how do I differentiate them from ideology – that is, what German soldiers *believed* they were fighting for?

*

The conditions and assumptions under which the Wehrmacht functioned after 1933 were heavily shaped by the First World War and by a cultural tradition dating back to imperial Germany. Not only did soldiers cultivate military values, but an authoritarian attitude also predominated throughout the Weimar Republic. The National Socialist régime did not have to invent the glorification of the trenches as a corporate experience, dismantling social and educational barriers and uniting the whole nation except for those who, with the help of Bolshevism, had stabbed the victorious army in the back. The Weimar military regarded itself as a higher school for Germans, which would again teach men devotion to the national cause, as well as the virtues of sacrifice, courage and comradeship. Since the First World War had been their most significant experience, both Hitler and the military establishment wanted to transfer the system of values of the front-line soldier to the public at large. Thus,

German military tradition and National Socialism could easily be amalgamated. 'This is part of the story of how war, regardless of victory or defeat, but especially in defeat', came to be seen as teaching the Germans national solidarity and sacrifices and bestowing on them 'a seriousness and weightiness of experience others did not possess'.[5] It was a military pamphlet that in January 1944 praised Hitler as 'the first socialist of the Reich' and defined 'real socialism' as the doctrine of 'performing the hardest duty'.[6] The political arrangement between National Socialism and the military establishment after 1933 was deeper than that between Ebert and Groener after the First World War, since it had an ideological base from the very beginning. Hitler and Blomberg's special relationship was based on the common idea of a new community of blood and destiny, capable of enduring in the most difficult circumstances and thus preventing another national collapse of the type which had occurred in Germany in November 1918. The goal of a new *Volksgemeinschaft* incorporated not only fundamental transformation of the mind and will of the entire people, but also their cleansing of everything 'un-German'. If few officers recognised the ideological rather than the politically radical core of Hitler's grand design, the goals they shared with their Führer provided a sound basis for co-operation to do away with the disasters of the Weimar Republic.

The bond between Hitler and the military was determined by the common goal of German rearmament, her *Wiederwehrhaftmachung*. This goal went far beyond the mere enlargement of the Reichswehr. It meant the total militarisation of the German people, mentally and physically. Both sides wanted to retain the Reichswehr as the professional cadre for the envisaged people's army based on conscription, which would be introduced at the earliest diplomatically feasible moment. Hitler had promised the Reichswehr to support its rearmament by militarising the whole nation (on 3 February 1933); Blomberg in return avoided interference in domestic politics and concentrated on purely military matters. Yet the defence minister soon became convinced that the Reichswehr should adopt a committed attitude to the national revolution. In June 1933 after the far-reaching Enabling Act, Blomberg bluntly told the regional commanders that it would be a good thing if the NSDAP achieved its total power soon. Since the new government deserved the Reichswehr's public loyalty, Blomberg shunned the notion of staying aloof from political matters and declared: 'The Reichswehr has to serve the national movement with all devotion.'[7] The plan to triple the peacetime army in four years was accompanied by efforts on Blomberg's side to instruct the armed forces in the 'basic principles of the National Socialist State', since 'the ideas of both our corps spirit and National Socialism spring from the common experience of the Great War'.[8] This move towards active involvement in politics was clearly a deviation from General Hans von Seeckt's non-committal attitude towards the Weimar Republic, yet it did not mean that the armed forces were now indoctrinated with National Socialist ideology. The then still small professional and homogeneous Reichswehr was highly motivated, especially by its massive enlargement, which united professional motives and self-interest. The new régime not only offered the prospects of rapid expansion after years of restraint, but also provided a social climate in which the military would regain its lost prestige. Major-General Ludwig Beck, a future leader of the resistance movement welcomed the 'political transformation' of 1933 as the 'first glimpse of light since 1918'.[9] And Grand Admiral Erich Raeder declared ten years later that the navy had not needed to become National Socialist, because it already was so through its education before Hitler came to power.[10] All Blomberg did was to give the soldiers a new spirit of determination and duty in the interests of the *Volksgemeinschaft*. His directives of that time also show the significance he attached to the social dimensions of National Socialist doctrine. On the one hand, Germany's fate and military tradition were being identified with National Socialism, but on the other, genuine efforts were being made to sweep away the old class barriers. At the same time the armed forces were being cleansed of Jews, and German officers unwilling to accept National Socialism as their professional and spiritual compass were threatened with dismissal.[11]

After conscription had been reintroduced in the spring of 1935, the build-up of a massive people's army was accompanied by a systematic and unified educational programme. The Wehrmacht was again regarded as the most important step in a young German's education, after home, school and Labour Service. Blomberg's relevant directives now show a clear trend towards National Socialist ideology. The Wehrmacht's educational goal, as defined by Blomberg on 16 April 1935, was 'not only the thoroughly trained soldier and the master of his weapon, but also the man who is aware of his race (*Volkstum*) and of his general duties towards the State'.[12] Although it is difficult to say precisely what the views of those lower down the echelon were, divisional commanders of different social background, such as Generals Maximilian von Weichs and Ferdinand Schörner, issued directives similar to their supreme military commander's. Yet they emphasised the values of leadership, honour, decency, toughness, comradeship and pride, which, of course, were treated in terms of the nationalist dynamics of the National Socialist state. Both generals aimed to give new recruits and young officers feelings of *esprit de corps* and national solidarity, and ended with a strong appeal for devotion to 'Führer, Volk und Vaterland'.[13]

In 1936–7, Blomberg made an even more systematic attempt to politicise the Wehrmacht, when instruction in 'national policy' was made a compulsory part of the curriculum at all officer training schools and staff colleges. For the efficient execution of this programme, central courses were established to train the instructors. Selected participants on the first course on 'national policy' at the War Ministry in Berlin had the honour to hear lectures from high-ranking party leaders such as Hess, Rosenberg and Goebbels, as well as from specialists such as Frick, Himmler, Günther and Krieck. It was at this course in January 1937 that Himmler defined Germany as a theatre of war equal to those on land, in the air or at sea. In the coming war, the SS and the police would fight the ideological enemy on the home front, namely Bolshevism as the incarnation of Jewry and subhumanity. Himmler declared that it was necessary to prepare the whole nation spiritually for the coming war of annihilation. All the lectures were immediately published in a book which was distributed down to company level for further guidance.[14]

*

A new phase in preparing Germany for war began in 1938 when Hitler became the Wehrmacht's acting supreme commander and embarked on his risky expansionist foreign policy. He stressed openly and in selected circles that the German people should now be instructed towards the 'fanatical belief in final victory' and that Wehrmacht officers should lead their men into battle with weapon and *Weltanschauung*.[15] Thus, war policy was accompanied by new educational efforts from the top. On the one hand, the instruction shifted from stressing character values in nationalistic terms to clearly disseminating ideology. All the commanders-in-chief agreed with Hitler that the officer in the Nationalist Socialist Wehrmacht had to be a National Socialist in spirit or he could not be an officer at all. The new ideal type was the 'political soldier', in whom professionalism and politics, weapon and *Weltanschauung*, would be identical.[16] Since the Army High Command was aware that not all officers subscribed to this outlook, Colonel-General Walther von Brauchitsch, issued an order for the training of officers. On 18 December 1938 he wrote: 'The [national] revolution has been stupendous. A new German being has grown up in the Third Reich, a new unique fellowship of the nation has been created. Our loyalty to [Hitler] who is both a true soldier and National Socialist and who has worked this miracle is unshakeable.' Brauchitsch then went on to elaborate the consequences for the army officer corps: 'The officer corps must not be surpassed by anybody in the purity and genuineness of its National Socialist outlook. It is its banner bearer, and therefore unshakeable, if all else should fail. The officer shall also be the leader of his men in regard to politics. This does not mean that he shall talk a great deal with them over politics, but he must have command of the basic ideas of National Socialism', to answer the questions put to him by his soldiers.[17]

Here we have the traditional German concept of educating the troops, whereby the commander (helped by his intelligence officer at the higher levels), bore sole responsibility, conducting it on a personal, one-to-one level. It was the firm belief of the military that the officer should lead and educate his men by example rather than by boring lectures, thus moulding the morale, *ésprit de corps* and political conviction of the soldiers. The commissar system of the antagonistic Bolshevik ideology was expressedly defined as being unfit for the Wehrmacht because its officer was the tactical and political leader of his men and did not need party interference to do his job.

On the other hand, the new emphasis on 'Instruction about National Socialist *Weltanschauung* and national-political aims' was given its organisational place within the Wehrmacht. Political instruction became part of the system of *Geistige Betreuung*, which was handled by the third general staff officer along with his intelligence matters. Other means used to keep up motivation were daily newspapers, broadcasts, films, entertainment and sport. Relevant pamphlets for the systematic political instruction of the Wehrmacht in the principles of National Socialism were first issued by the OKW (*Oberkommando der Wehrmacht*) from the spring of 1939 onwards. The so-called *Schulungshefte* opened with an article about the 'political soldier of the Wehrmacht' and the officer's role in achieving this goal. Though the Wehrmacht was geared towards war with Poland and appropriate ideological articles were distributed, among them a nasty one directed against Jews, the two primary means of influencing soldiers in the desired direction were still seen to be the officer's example – his ceaseless care for his subordinates, their resulting confidence in him – and discipline. The relevant paragraph on *Geistige Betreuung* in the general staff manual rightly stated that military successes would keep up troop morale better than anything else, but, as the experience of the First World War had shown, *Geistige Betreuung* would acquire a greater importance should the war be prolonged.[18] Two different departments within the OKW handled motivation and indoctrination, *Wehrmachtpropaganda* and *Inland*. While the former was in charge of *Geistige Betreuung*, the latter handled ideological-political instruction and off-duty activities, which were also seen as a means of influencing the soldiers ideologically.[19] The predominant attitude of the Wehrmacht before the war was aptly characterised in an address by a divisional commander: 'The Führer Adolf Hitler gives us the example. We will follow him gladly into the German future – come what will.'[20]

In September 1939 Hitler plunged the nation into war with Poland. He wanted to secure Germany's future by expanding her *Lebensraum*. He was calculating that Britain and France would be deterred from defending Poland by the Wehrmacht's might and the backing of the Soviet Union. Although the strike against Poland was not unpopular in Germany, the general mood differed from that in August 1914. Germany did not go to war with the same enthusiasm as in the First World War. Hitler's fear of November 1938, that his public stress on peace policy since 1933 might have misled the German people, proved right. The psychological re-orientation towards calling for the use of force themselves should peaceful means fail had come too late. The quick victory over Poland, however, was soon to strengthen the pride and self-confidence of the nation and lead it to disregard the likelihood of another long war with France and Britain. This optimistic attitude was supported by a propagandistic effort to make the Germans firmly believe that in the end all that was necessary for the future of the National Socialist *Volksgemeinschaft* would be achieved.

The war which Hitler unleashed in Europe in September 1939 was ideological from the start, both in Poland and on the home front. Yet the Wehrmacht fought it traditionally. For his racial struggle and the purification of the acquired living space, Hitler relied on the various formations of the SS, especially the Einsatzgruppen. The attitude of the Wehrmacht to the indigenous population in Poland was a mixture of racial arrogance, insecurity, and naïve trust in the use of force. The army's official policy was to deal ruthlessly with insurgents, yet individual soldiers looted and indiscriminately executed civilians, among them many Jews. The 'barbarisation of warfare', to cite a phrase from Omer Bartov, began in Poland not in the Soviet Union. For Brauchitsch's intelligence officer, these criminal acts came as 'no surprise after the long years of education!'[21] Hitler pardoned all convicted military and SS personnel, granted the SS in the field its own 'special jurisdiction', and justified the crimes as the natural result of bitterness unleashed by Polish behaviour towards Germans. Brauchitsch and other commanders urged the officer corps to accept the SS's 'radical measures' in carrying out the necessary racial struggle in the annexed Polish territories. What they feared was the brutalisation of their soldiers, which they tried to prevent by stressing spirit and discipline to the troops. The army was happy to wash its hands of responsibility for administering Poland and was busy implementing Hitler's decision to strike in the West as quickly as possible.

The war in Europe developed into a patriotic affair as soon as the old foe France was surprisingly beaten so quickly. The bond between Hitler, the military, and the whole nation was never stronger. The commanders-in-chief unanimously praised the Führer as the top soldier of the Reich who would definitely secure its future. Nothing seemed impossible for the German soldier. It was then that Hitler decided to attack the Soviet Union in 1941. The great organisational effort, through a cadre system, to gear the armoured and motorised divisions to the envisaged *blitzkrieg* in the East was accompanied by an initiative to combine training soldiers to be determined and aggressive fighters with their education in the Nationalist sense. For the first time, *weltanschauliche Erziehung* (ideological instruction) was separated from *Geistige Betreuung* and specific topics were enumerated, such as clean race and living space, but the envisaged adversary was not mentioned, since the Soviet Union was the Reich's official ally, who was helping to fight off the British blockade.[22] In general the troops were to be trained to fight an 'equal adversary'. Yet, four months later, the army's commander-in-chief underlined the value of educating both officers and men 'to the ruthless offensive spirit, to boldness and to a determined way of acting supported by confidence in the superiority of the German soldier over every enemy and by a staunch belief in final victory'.[23] Field-Marshal Brauchitsch made it very clear that the 'training of the soldier to be a determined and aggressive fighter could not be separated from a lively education in the National Socialist sense'. Neither Hitler nor the generals were interested in a sophisticated educational programme. What was desired was an instinct for the *Volksgemeinschaft*'s needs and a unshakeable belief in the Führer. Since the company or battery commander was seen to be 'the central personality' to forge the company into a compact unit in the desired mould, it was his responsibility to 'both lead the individual man into and keep him within the battle-community'.[24]

In June–July 1941, just before and after the first shot was fired on the Eastern Front, German troops were literally bombarded with the National Socialist explanation for going to war against the Soviet Union. Although the Wehrmacht had officially been deprived of anti-Bolshevik propaganda for two years, the régime and the military commanders could easily refer to enemy stereotypes that were familiar to the troops. Since the invasion of the Soviet Union was to be executed as a war of annihilation against 'Jewish Bolshevism', the character of motivation from the top changed fundamentally. Nazi ideology dominated the official publications. Enemy number one was the Jewish commissar: 'Anyone who has ever looked at the face of a red commissar knows what the Bolsheviks are like. Here is no need for theoretical expressions. We would be insulting the animals if we were to describe these men, who are mostly Jewish, as beasts. They are the embodiment of the Satanic and insane hatred against the whole noble humanity.'[25] Nazi ideology penetrated into the vocabulary of front-line commanders. There are numerous examples of orders issued to exhort their men to special warfare in the East. The army's fight against the phantom of Jewish Bolshevism, alongside the various SS-forces, had a dialectical dimension. It was inspired by ideology, but was rational in its implementation. Time and again, Jews were identified with Bolshevism *and* insurrection. As long as the mass shootings of Jews and communists were perceived and construed as military measures to crush 'all active and passive resistance' in the rear, German soldiers did not hesitate to carry them out.

The deliberate intermingling of ideology and military necessity, which Hitler had striven for and the Army Command had consciously accepted since the beginning of Barbarossa, becomes especially evident from the well-known orders of Field-Marshals von Reichenau and von Manstein (6 October and 20 November 1941 respectively), and Colonel-General Hermann Hoth (17 November 1941). They all called for the complete destruction of the Soviet war machine, for the annihilation of the Jewish-Bolshevik system, and for their soldiers' understanding of the mass executions carried out in the zone of operations by Einsatzgruppe C. In language stronger than Reichenau and Manstein's, Hoth turned his soldiers' eyes to German history and the trauma of November 1918. 'The annihilation of those same Jews (!) who support Bolshevism and its organisation for murder, the partisans, is a measure of self-preservation.'[26] Long before the German offensive came to a halt at the gates of Leningrad, Moscow and Rostov, it dawned on Hitler and the generals that the *blitzkrieg* against the Soviet Union had failed. They knew that Germany had to fight a fierce and stubborn adversary. The illusion of a short war was gone but not their belief in victory. Even the unexpected Soviet winter offensive did not severely hamper the inner momentum of the *blitzkrieg* army. Since the Wehrmacht had stood firm, the feeling of qualitative superiority over Soviet troops was retained. There was no need for Hitler as supreme commander to doubt the morale of his soldiers. He credited himself with having upheld their will to resist the Soviet offensive and with overcoming the winter crisis, 1941.

Thus, in the spring of 1942, Field-Marshal Wilhelm Keitel, in his capacity as acting commander-in-chief of the army, could issue directives for the troops which still balanced their ideological indoctrination with military virtues, although Nazi *Weltanschauung* was defined as being the basis for the instruction of the troops.[27] Even the stress on *Wehrgeistige Führung* in 1942–3 by the Wehrmacht itself was another name rather than a new concept of motivation. Radicalisation of ideological indoctrination did not come about before the winter of 1943–4. Then, Hitler and commanders alike both saw the need for a more uniform orientation of the Wehrmacht to fight the 'battles of destiny' against Jewish Bolshevism and Western plutocracy. Stronger doses of ideology should stiffen morale and thereby keep up the military performance of the troops. The result of both professional initiatives and old party aims was a highly ideological booklet with the telling title *Wofür kämpfen wir?* ('What Do We Fight For?'). It was published in 1944 by the army personnel office with an order from Hitler which emphasised the officer's role in leading his men by personal example both in battle and Nazi ideology. There was no room for non-political officers. 'He who fights with the purest will, the staunchest belief and the most fanatical determination will be victorious in the end' (8 January 1944). Colonel-General Alfred Jodl, chief of operations within the OKW, was sure that the German soldier would fight better and die more easily than his Western counterparts because he was defending his home, which was threatened with destruction.[28] The effort from the top to instil the Wehrmacht with that very belief in victory which the reality on the battlefield seemed to contradict was supported by the establishment of the *Nationalsozialistischer Führungsoffizier* (NSFO).[29] They were to aid commanders down to divisional level in uniformly instructing soldiers in questions of ideology and strengthening their will to fight this historically significant war even in the face of setbacks. The intelligence officer (*Ic*) was now free to concentrate on intelligence matters if he was not chosen to be the new NSFO. In December 1944 the Wehrmacht had 1,074 full-time NSFOs and over 46,000 part-time.[30] Since Hitler and the Wehrmacht rejected the Soviet commissar system of political leadership within the armed forces, the new phase in indoctrination in the Wehrmacht was commanded by them and not by the party. The establishment of the NSFO was a sign of military weakness in clear recognition of the 'incontrovertible laws of attrition', not the logical end of the gradual formation of the new National Socialist people's army. Yet the long-term concept of 'political soldiers' became a tactical means to fight the over-powering enemies. Many in the military continued to share the vague hope that Hitler, who was still held in high esteem by most of the soldiers, would find some way to reverse the tide. In a way, their Führer fulfilled that hope when, in December 1944, he struck a major blow at the Allies in the West in order to be done with them before the anticipated Soviet winter offensive would commence.

When the Western Allies had already crossed the Rhine at Remagen, the Red Army launched its final strike to take Germany's capital. For Hitler, it was now all or nothing. On 15 April 1945 he issued a highly ideological proclamation to instil the troops with the necessary devotion to fight off 'Jewish

Bolshevism' and thereby turn the course of war in Germany's favour.[31] Now the former offensive concept of annihilation took on a defensive posture. When Hitler's and Goebbels's national and ideological appeals failed to have the hoped-for effect and signs of demoralisation became visible, harsh measures, both by commanders on the spot and by the régime, were taken to ensure that the Wehrmacht, Waffen-SS, and the newly raised *Volkssturm* would continue to fight despite the doom already impending. Military police were given special powers and mobile courts martial were created. The increasing terror in the Wehrmacht and at home obliged everyone to remain loyal, whatever they thought privately. However, the military refused to see that capitulation might be a preferable alternative to fighting to a more costly conclusion. A Berlin joke may best reflect the attitude of the general German public in the last year of the war, when endurance dissolved into apathy and personal survival became more important than the national interest. 'I'd rather believe in victory than run around with my head cut off.'[32]

*

To conclude, the politicisation of the Wehrmacht after 1933 was carried out on the initiative of its commander-in-chief, General von Blomberg, not by the party. The relationship between Wehrmacht and National Socialism has to be seen in the context of German history after the First World War and the envisaged general militarisation of German society by both these agencies. The spirit of National Socialism, not its crude theory, and the idea of *Volksgemeinschaft* were used both to build up a corporate identity within the new Wehrmacht and to prove the politically demanded unity of state and armed forces. The stream of Nazi ideology into the Wehrmacht gathered momentum from 1938 onwards. Yet it was still the military which exercised control over the political outlook of the soldiers within the national framework of preparing German society for war. The Wehrmacht's idea of the 'political soldier' was manifested precisely in a new breed of determined fighter, in whom weapon and *Weltanschauung* had become fused. The Wehrmacht saw no need for a cadre of party functionaries as in the Red Army. It is a historical paradox that the position of the Soviet commissars was finally changed in favour of the commanders when the road to victory became visible, whereas the Nazi party gained political control over the Wehrmacht when it was doomed to total defeat. It has been easier to find what the Wehrmacht and the Nazis wanted, what they tried to instil in the minds and hearts of the soldiers, than to determine the output of these initiatives. Even more difficult is it to explain why the soldiers fought as they did. Was it because of normal soldierly motivation or of ideological indoctrination? Was it due to the soldiers' own beliefs, acquired instruction, small-unit cohesion, discipline or their instinct to survive? The task of the historian in this respect looks like quantifying the unquantifiable. One hesitates to credit the Nazi régime with much of anything. However, the dynamics of ideological motivation for the right cause of the *Volksgemeinschaft* during the war did contribute to morale in the Wehrmacht and resilience among the people to a much larger extent than even Hitler was willing to acknowledge at the close of his life. In February 1945 the Führer confessed to his secretary, Martin Bormann, that in regard to the spiritual mobilisation of Germany the war had come twenty years too soon. 'Lacking the National Socialist élite we wanted to mould, we had to put up with the existing human material. Hence the result! The war policy of a revolutionary state like the Third Reich necessarily became the policy of petty bourgeois reactionaries.'[33] The way German society fought the ideological war Hitler himself had unleashed in September 1939 proved the Führer wrong. Furthermore, it needed the effort of a powerful anti-Hitler coalition to bring down the Third Reich.

The German Soldier in Occupied Russia

Theo J. Schulte

During the Second World War the German armed forces spread death and terror across much of Nazi-occupied Europe: from the Greek islands to the Norwegian fjords, from the Atlantic coast of France to the tiniest villages in the Ukraine. In the occupied territories of the Soviet Union in particular the Wehrmacht did not confine itself to fighting a 'normal war' against the soldiers of the Red Army. It also actively planned and pursued a 'war of extermination' directed against Soviet civilians behind the front lines, long after the main combat had passed by. An apologist myth, which has survived for over half a century since the end of hostilities in 1945, sought to suppress this fact.[1] Denial was reinforced by a sentimentalisation of the Wehrmacht's 'heroic' role in defending the Reich against 'invasion' by the Red Army from 1944 onwards, and by a related myth which portrayed the million or more German soldiers who had died as prisoners-of-war in Soviet captivity as victims of the conflict.[2]

Evidence has, in fact, been available for decades which deeply implicates the German army, as an *institution*, in the large-scale plunder of the occupied territories, the systematic murder and abuse of captured Red Army soldiers and the terrorising of the civilian population of the occupied Soviet Union.[3] The various explanations that historians have advanced for German army involvement in such a war of extermination are, differences of emphasis excepted, all in accord in placing the Wehrmacht at the heart of Nazi brutality. This applies whether one writes of *slow corruption* into extermination (Messerschmidt), the *failure to differentiate* between the military and ideological (Förster), the dominance of *anti-Bolshevism* (Streit/Pätzold), the *false notion of a Jewish/partisan link as a matter of convenience* (Hillgruber), or *anti-Semitic tendencies in* the Eastern Army (Krausnick).[4]

The latest research, drawing in part on new sources from hitherto closed archives in the former Eastern bloc, more ominously points to widespread rank-and-file-level involvement in the direct implementation of the Holocaust. The emphasis in research is thus shifted from the mass murder undertaken by Nazi special agencies, particularly in the concentration camps, to the systematic extermination of Jewish communities, men, women and children, in the rear areas by regular troops of the German army. The experience and memory of the battlefield for the common soldier, the German *Landser* (squaddie/GI/

bidasse) can no longer be separated from the implementation of the Final Solution. Disturbingly large numbers of the 'ordinary men' who formed 'Hitler's Army' had not distanced themselves from the most sinister racial aspects of the war in the East. As with their comrades in the police battalions, many had been 'willing executioners' in a war pursued for ideological rather than military objectives in the 'killing fields' of the rear areas.[5] Racially motivated acts of barbarism had been committed by the German army from the commencement of the war in the East and were an essential component of the initial successful advance, rather than a forced response to the later stalemate and retreat.[6]

War of Extermination: the Crimes of the Wehrmacht, 1941–4, an exhibition organised by the Hamburg Institute for Social Research, which opened to the general public in March 1995, has been at the centre of the controversy. The exhibition, which received extensive press attention both at home and abroad and prompted a flood of letters, many from Wehrmacht veterans who had experienced the war in Russia, embarked on a tour of the Federal Republic and Austria that is scheduled to finish in Marburg in October 1997.[7] The harrowing photographic images, frequently taken from the personal mementoes of rank-and-file soldiers, and depicting scenes of great cruelty and barbarism perpetrated against innocent civilians and captured prisoners-of-war, are reproduced in the exhibition catalogue.[8] Page after page of ordinary soldiers' 'snapshots' show scenes in which troops of the Wehrmacht ill-treat and humiliate the old, the young and the weak. Numerous pictures are of the summary executions of individuals (including women) who were conveniently labelled as 'partisans'. Alongside these are yet more graphic scenes of mass killings of huddles of frightened civilians, usually Jews. Far too often the German soldiers who appear in these photographs appear to be engaging in what has been described as a form of 'execution tourism'.[9] At best, many of the troops who are 'merely' bystanders appear indifferent to the suffering all around them. The work has something of the 'pornography of death' about it, with the soldiers of the German army playing the role of both torturer and voyeur. An edited scholarly collection of some thirty articles soon followed, which included a number of case studies clearly intended to explore the charge of criminality laid against the German army and its soldiery.[10]

Fundamental questions were raised regarding the behaviour of ordinary German soldiers – in a far from ordinary war. Debate is highly charged because it invites suggestions of mass-participation in criminality, and problems of collective and individual guilt. The emotional experiences and memories of several generations of German soldiers are involved. Some nineteen million of them born from as early as 1880 and as late as the end of the 1920s served in the armed forces of the Third Reich. At least three million men participated in the attack on the Soviet Union and perhaps as many as seven million troops were involved at one time or another in this theatre of war. Mass conscription as a forced response to military, economic and demographic crisis, particularly in the period after 1942, cut across all age and social groups, so that Hitler's vast Eastern Army became a mirror image of Third Reich society.[11] To even suggest that the ordinary troops of the Wehrmacht had been directly involved in that most heinous of war crimes, the extermination of the Jews, would have been to acknowledge that an entire generation had been Nazified. Instead, those who returned from the front underwent a form of 'collective amnesia' as far as the sinister aspects of their wartime experiences were concerned.

Even some of the historians who from the late 1960s onwards had undertaken the seminal research on the *institutional* Nazification of the Wehrmacht had, because of their own wartime experiences, been inclined towards semantic niceties that obscured the extent to which the rank and file were involved in genocide. This was due not so much to a lack of comprehension on the part of such scholars as to the perverse character of the entire war in the East but rather to an inability to transcend personal individual experiences.[12]

In keeping with the interest in the topic stimulated by the fiftieth anniversary of the end of the war in Europe, the German liberal weekly *Die Zeit* published a collection of articles in spring 1995 entitled: 'Obedience to Murder? The Secret War of the German Wehrmacht'.[13] The collection took up the debate about the extent to which the German army *as a whole* could be implicated in the criminal activities associated with the trail of blood that Nazism had left across Europe. This was not an abstract study of a period of history now more than half a century away, for it involved present-day matters of personal biography and family history. A round-table discussion of interested parties, with participants who had seen service in the Wehrmacht sitting alongside the most outspoken of recent 'demythologisers', showed a marked lack of agreement on the relationship between personal and collective experiences of the war, and notions of individual and collective guilt. A good deal of tension was created by the suggestion of the veterans that recollections of their own personal experiences as soldiers were somehow superior to the 'evidence' available to latter-day professional historians in the military files. A selection of readers' letters, many from veterans, reflected this dispute. Here, again, as in the responses to the Hamburg exhibition, were many of the familiar arguments.[14] Apology, denial and refutation (usually based on anecdotes and personal reminiscences of the war) were prominent, with many correspondents again taking issue with the 'sweeping' nature of the archival evidence, postulating instead the primacy of exonerating individual experience over incriminating historical research. Appeals were made to all manner of 'explanations': notions of the universal brutalising effects of war in general; the harsh conditions under which the war was fought in the East; the sheer scale of the war, which rendered the German forces impotent when it came to dealing with the basic needs of the massive numbers of captured Red Army soldiers; the suggestion that the partisan menace had been much more real than recent studies had suggested; the need to differentiate between the decent behaviour of most regular German army units and the fanaticism of the SS/SD forces;

and the role played by fate and fortune in placing individuals in a given place at a particular time. Many of those who felt the urge to put pen to paper denied wartime knowledge of the Final Solution, let alone any suggestion of even indirect participation. Veterans were often indignant that the latest research had distorted their memories of the war in a way that amounted to the 'theft of positive experiences' of individual heroism, comradeship, shared dangers and tribulations. Far less prominent amongst the letters were moving statements of contrition on the part of former soldiers. Such examples of personal engagement with the experiences of the past often contained hitherto long-suppressed memories of both active and passive involvement in acts of cruelty. Yet even here, contrary to the line advanced by critical historians, some veterans were inclined to stress the draconian punishments that were both threatened and enforced against the rank and file who showed slight dissent, let alone resistance. A popular opinion survey conducted in April 1995 certainly pointed to the general unwillingness of the German wartime generation to recognise the findings of the latest research; for whilst 52 per cent of all age groups questioned believed that the Wehrmacht had committed war crimes, amongst those aged over sixty-five some 63 per cent could not accept this verdict.[15]

At the time of writing (July 1996) the controversy continues to be fuelled by the publication of yet more historical studies on the criminal activities of the German army and the accompanying press coverage.[16] Most recently, the confrontational tone has been reinforced by a highly publicised and much criticised parallel study, which seeks to identify virulent anti-Semitic influences in the men of the German police battalions who carried out many of the massacres in the East.[17]

To hold out against the tone of this flood tide of new research, where the words 'Wehrmacht' and 'criminal' appear almost interchangeable, yet at the same time avoid a reputation as a quasi-apologist, may seem a foolhardy task. I am very much aware of the reluctance of significant elements of German public opinion to recognise 'the repressed burden of 1941', and have no wish to associate myself in any way with factions within the Historikerstreit (particularly the ill-judged attempts at empathy with the retreating German forces in the work of Andreas Hillgruber).[18] Yet, whilst deeply entrenched denial clearly explains the impetus to provide yet further unequivocal 'proof' of Wehrmacht complicity in war crimes, I hold the view that such an approach runs the risk of merely assembling material which is over-narrowly concerned with corroboration of the highly questionable assertion that *every* German soldier was in some measure a war criminal. To cast the historian in the role of 'hanging judge' makes for bad history. Total condemnation of the Wehrmacht is no more an aid to understanding this dark chapter in German history than is total exoneration.[19] Many open questions still remain. Not least, despite the various studies which now exist, the remark made well over a decade ago that 'quantified research to determine the true scale and degree of involvement remains outstanding', still holds true.[20]

Overall, as my case studies of two specific rear area units already attempted to show nearly a decade ago, a highly nuanced view is called for of the soldier's experience of war in Russia:[21] one that takes full account of the most unsavoury aspects of military behaviour, whilst at the same time preserving an awareness of the contingent and sometimes very different circumstances under which the war in the East was fought, as well as the many and varied generational influences on the men of the Wehrmacht. And, perhaps most importantly, much more work needs to be undertaken on the matter of personal motivation and the way in which individual experiences of war are both repressed and reconstructed long after the event.

The harsh 'demodernised' conditions under which the war was fought in the East lent this quasi-colonial conflict a peculiar savagery and moral brutalisation. Racism and xenophobia were stimulated by the almost primeval backdrop to the conflict. And, as even the most fervent advocates of the ideological indoctrination approach agree, despite the views of a select few romantics in the military hierarchy, for many ordinary troops exhausted even quite early on by the pressures of the war, the backward and primitive nature of much of the occupied territory only served to reinforce long-held negative stereotypes of Russia.[22] It should also be remembered that although some 80 per cent of the territory occupied by the military was technically behind the front lines, the entire occupied territory was often an area of intense military operations. Morale in many rear units, with their high proportions of reservist officers and over-age troops, was certainly well below the level which is suggested by literature which rather stereotypically focuses on the marked 'fighting power' of the younger front-line troops of the Wehrmacht. Whether, on the basis of this, one should accept the rather élitist argument advanced by some veterans' organisations that rear-area units behaved in a more despicable fashion than the front-line combat troops remains a moot point.[23] Indeed, a different reading of the research findings can point us in the opposite direction as the case-study work on front-line units, such as that on the advance of the Sixth Army through the Ukraine on its way to its fate at Stalingrad, all too clearly shows.[24]

The deliberate abandonment of conventional rules of warfare across all the sectors of these 'lawless territories' and their replacement by ideologically based guidelines, the so-called 'criminal orders', were clearly used to obscure principles of accepted morality in order to convince the troops that brutality was not only permissible but a formal requirement.[25] German commanders were often at pains to stage executions in an organised and disciplined fashion. However weak and flimsy this legal fiction might now appear, its significance in removing the inhibitions of ordinary soldiers at the time should not be underestimated. The catch-all application of anti-partisan directives, often directed against those not even suspected of association with the guerrilla movement let alone involvement, was very much part of this strategy. Similarly, the 'necessity of war' was a pretext for legitimising the most draconian measures;

particularly hostage-taking and collective reprisals.[26] In those documented instances where such an approach was either ignored or broke down, the disquiet often expressed by the rank and file is often quite marked.[27] Certainly, this view can be placed within the context of the phenomenon identified by 'structuralist' historians of the Third Reich whereby the Nazi régime's capacity to create a 'permanent state of emergency' allowed normal law-abiding individuals to be led astray.[28]

As far as generational influences are concerned, the Wehrmacht should not be regarded as a homogeneous entity, despite the classic institutional pressures applied to create uniformity and conformity. The armed forces of the Reich contained a very broad cross-section of male German society, with individuals ranging in age from teenagers to octogenarians. Each generation carried with it into the theatre of war in the East markedly differing attitudes to National Socialism and differing life-experiences.[29] In respect of the military comman-ders, former First World War staff officers, now Hitler's generals, commanded former junior front-line officers from the 1914–18 conflict, who in turn had become staff officers, whilst below them was a corps shaped by negative reactions to the Weimar Republic, and finally the extremely young and often inexperienced officers who were products of the Hitler Youth movement.[30] Residual ideas from the imperial period on German great-power status mixed with schemes to revise Versailles, social Darwinist concepts of biological struggle and, finally, Nazi fantasies of an army representing the racial national community (*Volksheer*). The degree to which each element was willing to participate in a war of extermination against the Soviet Union, be it as an anti-Russian, anti-Bolshevik, or anti-Jewish crusade, is an area of research that needs to be pursued.

Throughout any discussion, the focus also needs to be clearly placed on the experiences of the ordinary German *Landser*. Powerful arguments have been advanced that amongst those who would later form the army which invaded the Soviet Union in the summer of 1941 there was an 'everyday racism' that transcended all ages, classes and social groups. The Nazification of significant elements of the working class is particularly controversial, since here was a key element of society traditionally seen as resistant to the appeal of fascism.[31] Yet, despite research on certain aspects of this topic, the degree to which Nazism reached into the schools, the Hitler Youth, the factory apprenticeship schemes, and the barracks of the conscript *rank-and-file* soldiery during the pre-war 1930s remains an area that warrants much more study.[32] A view from the 'other side' of the war – the perspective of a soldier in the Red Army – lends some weight to this assertion. At the same time it points to some of the universal features of the soldier's experience of war. The critical sociologist and veteran, Vladimir Subkin commented ironically,

It was the pre-war country that entered the war. Everything in that country was taken by the people to the front. The capacity for self-sacrifice, and suspicion of others. Cruelty and spiritual weakness. Baseness and naïve romanticism. Officially demonstrated devotion to the leader, and deeply concealed doubts...Nothing was left behind, nothing was forgotten.[33]

Overall, in view of the many generations involved, the debate needs to go beyond the already complex issue of the extent to which the conditions of the battlefield or the influences of the Nazi régime both before and during the war shaped behaviour. Due regard needs to be given to long-term trends in popular attitudes by way of longitudinal studies covering the entire period from the start of the age of 'classic modernity' in the 1890s to the 1940s; particularly with regard to the links between the modern and the irrational. To downplay everything except anti-Semitic influences would be to ignore a half-century of informed research.[34]

Recent work, it must be acknowledged, does provide extensive evidence of what can be described as deeply ingrained racist attitudes on the part of disturbing numbers of German troops during the war itself. This is particularly pronounced, not only in the grim atrocity photographs taken by ordinary soldiers which form the basis of much of the Hamburg Institute exhibition, but also in both the tone and content of letters home from the front (*Feldpostbriefe*).[35] For the demythologisers this is the proof that the relentless Nazi propaganda machine had produced conscripts for Hitler's army who fully shared his views on 'inferior peoples' and that it was these ideas, not the exigencies of war, that motivated atrocities.

All the same, given the lack of an all-encompassing survey, it remains debatable how far such ideologically based thinking extended throughout all the many troops of the Wehrmacht. Moreover, although the various studies clearly show ordinary troops engaged in acts characterised by murder-lust, sadism, emotional cruelty and even sexual perversion, an ideological basis for such behaviour is not always evident.[36] In particular, whilst not denying the weight of grim evidence on the wholesale murder of Jews by ordinary soldiers, the degree to which they both initiated and participated in such massacres simply because of an ingrained anti-Semitism alone remains unclear. Analysis of *Feldpostbriefe* and individual diaries may suggest that this was a key factor; a thesis reinforced by one particular comparative study on the differing behaviour of the Italian and German armed forces in the Balkans, which argues that at least as far as the officers of the Wehrmacht were concerned they, unlike their allied counterparts, rarely saw the Jews in anything other than categorical terms. But it is less certain that all of the rank and file shared such fixed views and, accordingly, cruelty and brutality may indicate simple, albeit reprehensible, moral delinquency.[37] Even the fact that certain soldiers showed more concern for the well-being of their military dogs than they did for captured Red Army soldiers or even Soviet women and children should not be over-analysed in narrow ideological terms.[38] In the same way, the tendency of many army units to avoid the more 'messy' aspects of rear area security work and pass this on to the SS and SD instead, could be interpreted in the context of research based on *Feldpostbriefe*, which suggests that soldiers behaved as if such tasks had become, with their emphasis upon diligence, stamina, perseverance, duty, obedience and subordination, a rather unpleasant form of 'work'.[39]

Studies of the terroristic activities of the German forces, including hostage-taking and collective-reprisal measures, on the broader geographical front in Greece, Italy and France, where racially motivated actions against the non-Jewish civilian population are not usually considered to have been the norm, certainly point to the difficulties of determining motivation.[40] Recent work which explores the role of police battalions and various civilian agencies in related acts of barbarism in occupied Europe also raises all sorts of questions as to the true determinants of behaviour.[41]

Peer-group pressure and dogged conformity within the group, careerism, a general all-encompassing xenophobia unrelated to any specific racial or national group (hence the inclination to murder not just Jews but all and sundry irrespective of race, age or gender), deference to authority, draconian discipline, and the motivating role of junior officers and NCOs are just some of the many other possible variables in the equation. To pursue this matter may well take us into areas of psychology, sociology, social-anthropology, perhaps even ethnology, where many historians are reluctant for various understandable reasons to venture. But, as with Holocaust studies in general, the very irrational nature of the beast may necessitate such an eclectic approach.[42]

Non-conformism on all levels warrants more attention amongst scholars, not least because it seems to have been over-marginalised lest its acknowledgement be construed as giving succour to apologists. Certainly this is a problematic area, and carries dangers that must be avoided of reviving an almost romantic view of the war, filled with individual soldiers and sometimes entire units performing acts of generosity and kindness towards both civilians and captured enemy soldiers. All the same, provided a sense of balance is maintained, and full recognition given to the many acts of barbarism and cruelty perpetrated by the men of the German army, historians should be concerned as much with 'exceptions to the rule' as with the 'norm'.[43]

Contradictions and paradoxes on a less mundane level certainly emerge throughout official Wehrmacht reports on the conduct of the war. Despite the highly bureaucratic tendency of the Nazi state to catalogue in a detailed and often unashamed fashion its violent activities (which ironically provided the material for subsequent legal prosecutions), a good deal of the evidence that allows us insights into cruelties is actually drawn from reports of the select few officers who took exception to such policies.[44] Alongside directives urging increased violence in pursuit of elusive military goals, which were often clearly determined by Nazi racial thinking, are calls for restraint in order to curb over-zealous units which engaged in the wholesale murder of men, women and children. The majority within the corps may well have been more consistent in their ideological loyalties, but this does suggest that there did exist differences in approach to the war. Whether a more marked rift existed between officers and men remains one of the open questions. Military commanders certainly reported instances where German troops expressed serious misgivings that collective reprisal measures were both unsound and unnecessarily brutal.[45] As one writer has emphasised, between the extreme positions which emphasise a quasi-religious commitment of the common soldier to the Nazi *Weltanschauung* and the image of him stolidly resisting the flood of propaganda, lies a terrain of attitudes waiting to be explored.[46]

Many of the harrowing accounts of massacres still leave unresolved questions as to why a very select few refused to participate in murder. We obviously need to know much more about 'those who refused to obey'. The frequent references to an initial reluctance of the many to participate in murder (which soon gave way to the habituation of a majority to a regular programme of killing) and the highly limited instances of moral courage through continued refusal, also pose many questions that have not been satisfactorily answered despite much pseudo-debate on this subject.[47] A study of military chaplains in the Wehrmacht would be a useful adjunct to such research.[48]

In any such analysis, one should also set against the damning evidence of popular racism contained in *Feldpostbriefe*, material taken from similar sources which points to a marked lack of ideological enthusiasm, a sense of impotence – rather than dominance – and even, occasionally, hostility to Nazism.[49] Official ideology was often filtered and sometimes diluted by the inertia of army life and the boredom of occupation duty. Reports on troop morale present a picture in which many soldiers shunned propaganda, preferring instead to retreat into the escapist world of popular light entertainment, films and theatre. True, even soldiers lived 'thick' lives that extended well beyond the 'thin' world of military duties, and one of the most disturbing aspects of the Third Reich may have been the almost symbiotic relationship between normality and barbarity. All the same, to resist the temptation to dismiss such behaviour as no more than simply displacement activities is not to present a 'monolithic and unrealistic conception of ideology', nor is it to deny the dangers of historicising the Third Reich.[50] National Socialism clearly sought to influence and distort popular beliefs on notions such as patriotism, conceptions of law and justice and military honour in the most subtle fashion. But the actual extent to which the régime succeeded in instilling its value system into the rank and file remains a vast and complex subject in which there are no obvious or simple explanations. If the vast literature on attitudes to the Nazi regime in its broadest sense (*Resistenz*) is applied to the German army rank and file, what emerges, as has been suggested elsewhere, is not a clear dichotomy between fanatics fighting a religious war, or ideology-free and perhaps bored and apathetic soldiers; but a continuum.[51]

Strategies for survival, which included desertion, self-mutilation, feigned displays of incompetence and various attempts to sit out the war (including my 'slipper soldiers' who had abandoned not only military uniform and conventions of rank but the very war itself) whilst certainly not representative of the Wehrmacht as a whole, also offer valuable alternative insights into indoctrination and motivation.[52] Studies of desertion, in particular, suggest a complex pattern of attitudes to the war ranging from moral rejection of Nazism to 'simple' self-preservation.[53] Such research needs to be better integrated with the 'fashionable' literature which sometimes seems preoccupied, albeit for understandable and often commendable motives, with the participation of the rank and file of the German army in atrocities.

Finally, it should be recognised that despite the necessary impulse which drives historical research, intrinsic problems exist in any attempt to recreate the mental world of the Third Reich, perhaps the more so in the context of the experiences of the war in the East. Many of the men who served in the German forces regarded it as a different time and place, and the political vocabulary and values of the present are often inadequate in attempting to explain the situation in which they found themselves and the way in which they behaved.[54] It should not be forgotten by historians of 'everyday life', even in dealing with soldiers, that this was often a far from easy time for ordinary decent men caught up in the most indecent of wars, which was not necessarily always of their own making. This is not, it must be stressed in the most unequivocal terms, an argument designed to provide an easy alibi for those who lapsed into barbarism or even moral indifference. All the same, Alfred de Vigny's maxim, although from a different military era, might be paraphrased to give it equal resonance for the German soldier's experience in Russia; an army is a dumb beast which kills when it is set down but its soldiers also feel pain.

Who Fought and Why? The Assignment of American Soldiers to Combat

Theodore A. Wilson

Efforts to devise policies for the most efficient and equitable use of manpower proved the most glaring failure of America's mobilisation for the Second World War. The Selective Service Act of 1940 conferred authority to conscript males for military service and, as well, to defer those occupations which were vital to the nation's economy. Only, in late 1941, however, did the huge expansion of production fuelled by Lend-Lease and the US army's 'victory program', a blueprint for building a 215-division ground force, begin to raise concern about allocating manpower as any other scarce resource.[1] How to do so became ensnared in awkward philosophical issues (the historic American commitment in theory to universal male military obligation and its aversion in practice to comprehensive national service), thorny social questions involving the utilisation of blacks and women, and the partisan and bureaucratic political imperatives that flowed from labour shortages. From a total population of slightly over 130 million, 66 million Americans either were in uniform or performing civilian jobs during the peak year of 1944. Some 20 million men were theoretically eligible for military service. The armed forces increased from some 1.5 million in 1941 to 11.4 million at the war's end.[2] How manpower was allocated is a familiar story to students of America's role in the Second World War, for it was sketched out in the first two 'green books' to be published, those volumes of the *United States Army in World War II* that treated 'procurement' of troops and the organisation of the AGF (Army Ground Forces).[3]

The performance of the military services in making the most efficient use of these millions of men and women was hotly criticised at the time. Earning most blame was the system for selection, training and assignment of soldiers to combat roles, for the confusion, delays and waste associated with the army's uncontrolled expansion after 1941 were fully documented. However, also posing serious problems for manpower allocation were the policies controlling voluntary enlistment and 'selective' service, the effects of reliance on 'aptitude' examinations and civilian vocations upon assignment to the combat arms, the influence of personalities and doctrinal loyalties as regarded priority given infantry and combined arms training, the harmful effects of 'personnel turbulence' (especially the gutting of fully trained units to provide cadre for newly created units) and a short-sighted loss-replacement policy.[4]

Any effort to explain why this débâcle occurred is largely absent from the official history, though acknowledgment was made of the army's expansion far beyond pre-war projections and with great speed – a forty-fold increase from fewer than 200,000 in 1939 to 8.3 million in 1944.[5] Further insights regarding the United States's use of military manpower in the Second World War must come from an examination of those dominant ideas and organisational imperatives that shaped the US army's manpower policies. This suggests that such questions as efficient allocation of human and other resources, methods to modify the behaviour of raw recruits, and individual and group performance under extreme stress, be related to intellectual and cultural developments in the larger society. For example, what explains the seemingly total victory of assumptions derived from scientific/industrial management theories in the process of mobilisation planning? The metaphor of the factory, with individual soldiers (identified by and as serial numbers) perceived as specialised cogs and/or interchangeable parts, assembled on a production line and plugged into the war machine as needed, is pervasive in both the literature and the historical records of the Second World War. Was the treatment of human beings as objects to be shaped, manipulated and discarded at will adopted with any concern or understanding of the possible implications for harm to the individual psyche and group morale? Did it reflect the alleged egalitarianism of the scientists/managers who established the rules for assigning men to combat or other, darker purposes? What was known about the qualities of the raw material? Did the assumption operate that the manufacturing process (training) would remove most if not all imperfections? What level of variation from the norm was to be tolerated?

Similar questions arise with regard to the selection and allocation of military manpower. Who made the decision to install what became the Army General Classification Test and to correlate military assignment with civilian vocation? What values and beliefs about the 'nature' of the American populace did they bring to the task? Was it mere coincidence that these policies ensured that the typical combat infantryman was demonstrably less capable physically and intellectually than his counterparts behind the lines and in the Army Air Forces and navy? US army practices manifestly resulted from the conviction that the Second World War would be a war of movement and concentrated fire, a 'high tech' conflict placing far greater physical and intellectual demands on the technician behind the lines than on the GI riding a truck or tank into battle. Did the American personnel gurus foresee the implications of these practices and, in particular, the resultant correlation between likelihood of seeing combat and low socio-economic standing – 'poor white' rural origin – first or second-generation immigrant background?

From the outset, US army chief of staff General George C. Marshall and other American military leaders decided that the *system* was pivotal: setting up the appropriate categories for selection of manpower (classification), the process of refining and grinding down this raw material (training), and plugging individual soliders into the correct slots (assignment). Given dominant assumptions about the nature of the raw material available to them, their goal was effective performance within rather broad tolerances. They assumed that the human components would wear out or break sooner rather than later, but their retrospective assessment of the First World War and the implications of *blitzkrieg* warfare led military leaders to conclude that battles and campaigns would be sharp, quick affairs, featuring rapid movement and concentration of fire. The prevalence of such assumptions may explain (although not justify) why so little attention was accorded unit cohesion and the training of individuals and small units.[6]

The US army initially had adopted rigorous screening criteria for the 'inductees' swelling its ranks once the Selective Training and Service Act was approved on 16 September 1940. Such physical inadequacies as deformed limbs, missing teeth and intellectual 'inaptness' (as manifested by having less than four years' schooling) led to rejection of some 30 per cent of draftees in the period before Pearl Harbor.[7] Wedded to the development of a combat force relying primarily on advanced technology and relatively small numbers of front-line soldiers, the army strongly preferred men who possessed strong bodies and above-average intelligence. With theoretically restrictive deferment policies, the presumption was that more than enough physically and intellectually qualified men would be available.[8]

In 1942 the pendulum swung the other way. What was defined as acceptable was broadened significantly. For example, the requirement that an inductee possess all his teeth was changed to 'sufficient teeth – natural or artificial – to subsist on the army ration'.[9] After 1942 some 200,000 men with venereal diseases were accepted, along with many thousands of individuals previously categorised as 'mentally and emotionally impaired' (exhibiting behaviour – such as homosexuality or prior criminal records – defined as 'detrimental to the effective functioning of the unit').[10] Significantly, the criterion that inductees have at least four years' elementary schooling and possess sufficient command of English to comprehend orders was dropped; thereafter, illiterates and non-English-speaking individuals were accepted.

The reasons for this loosening of standards were in part political. Draft boards responded to the bitter reproaches of mothers, wives and girlfriends demanding to know why their loved ones were being taken when otherwise able-bodied men were deferred for such dubious causes as being toothless, having been convicted of armed robbery, being unable to read and write, or – as in the case of popular singer Frank Sinatra – showing up with a punctured eardrum. Also, however, the army, then endeavouring to stock the gigantic combat arms training establishment (38 divisions were activated between February and December 1942) while finding sufficient numbers of qualified men for the Army Air Forces and the rapidly proliferating Army Service Forces, presumed placidly that an effectively managed system for manpower allocation would fit each soldier, whatever his limitations, into the most appropriate slot. Since acknowledged authorities asserted that this could be done scientifically, this process was thought to be blind to social, economic and even racial distinctions.[11]

While questions of job-related deferments (especially for agricultural workers) and the appropriateness of parenthood as a justification for exclusion from military service were hotly debated, the statistical breakdown of who served supports claims that the Second World War draft was generally fair and equitable. But that judgement applies only to induction and military service. Assignment to combat posed altogether different issues of equity. The US navy and the Marine Corps drew their manpower from volunteers for most of the war. They also successfully resisted acceptance of large numbers of African-Americans. Still, the US army received 72 per cent of all Americans who served in the Second World War, logically guaranteeing that a broad cross-section of the male population would find their way into the combat arms. As it turned out, however, for enlisted ranks, assignment to combat units was predominantly the fate of the less-well-educated, vocationally unqualified, more youthful segment of the manpower pool. In the local draft boards, the induction stations and the camps in which raw recruits received basic training, a filtering process took place. That process reflected deep-seated prejudices and convictions about who should fight.

The implications of this process were clearly understood at the time and are implicit in the early official historical works that dealt with selection and assignment of ground combat troops.[12] But there has been a surprising lack of attention to these questions by historians until recently. In part that was because the creators of the army's assembly line were ambivalent about the factory metaphor. Some advocated giving priority to shaping what was being produced (i.e., the human cogs and gears to be employed as interchangeable parts in assorted military constructions); others, including Marshall, McNair, and most senior commanders, were chiefly concerned with assembling and field-testing the constructions themselves – the larger units such as divisions, corps and armies. The result was a system for assignment and training of American soldiers that lurched from one crisis to another. In 1944, Brigadier-General James G. Christiansen, Deputy Chief of Staff of Army Ground Forces, characterised the army's handling of military personnel issues as 'disastrous'. He stated bluntly that the classification system was 'one of the worst things in the Army, for which we haven't paid the toll yet but may have to.'[13] Even senior officers from the agencies chiefly responsible for setting up the system admitted after the war that 'personnel management and the functions of the War Department related thereto was probably the most troublesome – and handled the least satisfactorily – of all the major functions of the War Department during the war'.[14]

*

A vital element in explaining this dog's breakfast of a system is to be found in the assumptions and ideas current among senior officers in the US army and embraced with some enthusiasm by those in the Office of the Adjutant General and G-1 (Personnel) of the War Department General Staff (WDGS). On 5 April 1941 the War Department released *Training Circular*, No. 25. This document – which replaced *Bulletin*, No. 44 (1928) and *Bulletin*, No. 46 (1931), as amended by *Bulletin*, No. 6 (1938) and *Bulletin*, No. 4 (1939) – made known to officers who 'do not have access to extensive libraries' what the Army perceived to be 'literature of especial value in their military education'. The approved list included 'selected literature' relating to philosophical, psychological, historical, political, economic and military subjects. *Training Circular*, No. 25 stipulated that the bibliography represented a systematic approach to continuing education: while officers were encouraged to read widely, they were required to read at least six books per year from the approved list and to follow a two-from-column-A-and-one-from-column-B approach, choosing works from more than one grouping 'in order properly to diversify the reading'. The presumption was that the educated officer be thoroughly grounded not only in those works immediately relevant to his specialty (e.g., Gilchrist's *World War Casualties from Gas and Other Weapons* for chemical warfare officers) but also in 'scientific' and historical studies germane to analysis of American social, political and economic organisation and the place of the United States in world affairs.[15]

A novel feature of the 1941 compilation was the category of books dealing with philosophy and psychology. Included in this brief listing were several general works (Will Durant's *The Story of Philosophy*, John H. Randall's *Making of the Modern Mind*, Peter Odegarde and E. A. Helms's *American Politics*, and Lewis Mumford's *Technics and Civilization*), books that would repose in the sitting-room bookcases of many American middle-class households. The list emphasised applied psychology: *Social Psychology*, by Daniel Katz and Richard L. [sic] Schank; Leonard W. Doob's seminal analysis, *Propaganda*; Ordway Tead, *The Art of Leadership*; Walter Van Dyke Bingham's *Aptitude and Aptitude Testing*; and Morris S. Viteles's professionally acclaimed survey, *Industrial Psychology*.[16]

The dedicated officer who read all these works would have been extremely well educated in comparison with the typical American who purchased books.[17] The army's reading list was compiled primarily for junior officers and, thus, its chief effects presumably would be experienced in that distant time when the captains and majors of 1940–41 had succeeded to the highest ranks of their respective arms and services. The list was no doubt put together by captains and majors in the Adjutant-General's Office after soliciting the advice of their fellows elsewhere in the Munitions Building. None the less, the list did reflect the priorities of that segment of the army's senior leadership who approved it. Whether Major-General Emory Adams, the Adjutant-General, had read all (or indeed any) of the approved works, the army had taken a position with regard to what it deemed important for officers to know.

The great majority of the books had been published before 1930 and many dated to course syllabuses familiar to West Point cadets before the First World War. There were two notable exceptions to the list's fustiness in both the general reading course and an attached compilation of approved works for reserve officers of the arms and services: the category of 'philosophy and psychology', and the list of books on economics and industrial mobilisation. These two fields of enquiry – reflecting the perceived failures of waging the First World War and central to how the US army went about constructing the vast forces which fought the Second World War – served to bridge the narrowly conceived concerns of American military professionals and the organisational-cultural-intellectual revolution that had transformed over the previous thirty years those institutions – the corporation, the church, the club – with which the army had most in common. How to organise groups of human beings in the most efficient and economical way and how to extract maximum effort from those so organised were questions that had captured the attention of some career officers by the mid-1930s.

Between 1915 and 1935 vigorous ferment bubbled through a number of fields of intellectual enquiry that directly related to the army's mission of choosing and assigning manpower in some future national emergency. Officers in the Adjutant-General's office and the WDGS's Organisation and Training Branch (G-3) charged with this task obtained an extremely limited grasp of these ideas, acquiring information and insights in a haphazard manner and filtering them through institutional and personal prejudices. Nevertheless, the procedures for classifying the 'raw material' coming into the army that were installed during 1940–41 embodied to a remarkable degree assumptions about human behaviour and organisational make-up espoused by a particular subset of the American intellectual community.[18]

First and foremost was the belief that industrialisation had ushered in a new kind of civilisation, one founded on the machine, organised according to mechanistic principles, and dedicated to the glorification of unceasing growth, size and energy. 'The sway of the machine', Max Lerner has observed, 'is less disputed in America than that of any other institution, including the science which made it possible, the capitalism which has organized its use, and the democracy governing the distribution of the power that flows from it.'[19] One corollary of this celebration of technology was the near-universal recourse to the special language associated with engineers and other experts to describe social and political issues. This was hardly new, for educators had likened schools to factories for much of the nineteenth century, and the factory analogue formed the basis for the cult of educational efficiency championed by Progressive Era reformers. As JoAnne Brown has noted, this 'factory model' came in two versions: weak and strong. 'In the weak version, schools were factories and children were little workers; in the strong version, schools were factories and children were raw material, to be transformed into finished products.'[20] The applicability of the factory meta-phor to society as a whole appeared self-evident to industrial management experts such as Frederick Winslow Taylor and psychologists such as Edward Lee Thorndike.[21]

Equating national strength and precepts of efficient organisation with the limitless possibilities of 'human engineering' proved an easy next step for proponents of scientific management. They harked back to the conviction of utopian novelist Edward Bellamy 'that a man's natural endowment, mental and physical, determines what he can work at most profitably to the nation and most satisfactorily to himself'.[22] During the First World War, a group of applied psychologists, designated the Committee for the Classification of Personnel, devised common-sense screening procedures to identify and pigeonhole the skills arising from those 'endowments'. Adapting current occupational categorisations, they drafted an index that enumerated and pre-scribed standard terminology for 714 distinct occupations. The committee designed a simplified 'record of achievement' card that recorded each soldier's occupational history, education, leadership experience (Boy Scout troop leader, football captain, and so on) and performance on the trade test.[23] Next came the development of a 'Table of Occupational Needs', stipulating precisely 'how many men of a particular occupational designation were required for each of the many sorts of platoons, companies, batteries, squadrons, trains, battalions, and regiments' in the army.[24] After some deliberation, the Adjutant-General's office adopted this system for classifying military manpower.

Its advocates justified the classification system on grounds of efficiency, military necessity and the objective evidence of modern science. 'However profoundly one may be committed to the Social theory of Rousseau and Jefferson that all men are created equal,' observed the history of the army's personnel effort in the First World War, 'every employer and every officer in command of troops knows that there are enormous inequalities of skill and talent. The personnel officer knows that his office exists precisely because these inequalities are extremely important to the Army.'[25]

A second and related thrust was fixation upon a hereditarian explanation for perceived differences in intellectual capacity. Such views had flourished in the nineteenth century and featured strongly in Social Darwinism. But the devel-opment of so-called 'intelligence tests' just before and during the First World War appeared to give irrefutable scientific credence to heredity as sole deter-minant of both individual and group differences. Harvard psychologist Robert Yerkes and a band of zealous associates persuaded the Surgeon-General's Office to allow testing in the US army. In 1917–18, the famous 'Alpha Test', used for identification of superior inductees (those eligible for officer training), and a parallel 'Beta Test' to sort out those with intelligence so below the norm as to be incapable of functioning in typical military situations had been administered to 1,726,966 soldiers (including some 41,000 officers). The tests appeared completely objective and thus unchallengeable. For that reason, the results reported by Yerkes and his associates sent shock waves throughout the army and the country. According to 'scientific evidence', 24.9 per cent of those tested were illiterate, 47.3 per cent of all whites and 89 per cent of all African-Americans had mental ages below thirteen, and recent immigrants from southern and eastern Europe were far less intelligent than the 'old' immigrants.[26] The obvious conclusion was that making soldiers of men ranging from 'dumb to dumber' posed difficult, if not insuperable obstacles.

In retrospect, the bias in favour of middle-class, educated inductees would appear to be self-evident. The Alpha test comprised arithmetical problems, number sequences, synonyms and antonyms, selection of analogies, and common-sense questions. It asked such questions as whether Scrooge was a character in *Vanity Fair, A Christmas Carol, Romola* or *Henry IV*; whether the unit of electrical measurement was watt, volt, ampère or ohm; whether a Percheron was a horse, cow, sheep or goat. The Beta test asked for recognition of a tennis net not set up on a court, figures on lawn-bowling green without bocce balls, and a camel without a hump. Edward M. Coffman concluded that the Alpha test was 'more indicative of the social milieu of the university-trained men who prepared the test than of the background of the majority of soldiers who took the exam'.[27] As well, the tests devised by Yerkes and his associates reflected dubious assumptions about race and intelligence. They were taken as proving that recent immigrants were diluting the intelligence pool in American society.[28] Such aphorisms as 'the lighter the skin, the brighter the recruit' poured fuel on the bonfires of racists and proponents of immigration restriction.

Charges of bias were strongly rebutted by the psychologists. James Reed has noted:

> They believed in equality of opportunity, not equality, and they shared with other American Galtonians the assumption that 'civic worth' or 'mental ability' or 'IQ' were inherited biological capacities distributed unevenly among classes and ethnic groups... Their technology of mental measurement reconciled equality of opportunity with inequality; they provided numbers that seemed to confirm the [naturalness] of social class and racial caste.[29]

Significant differences between individuals and groups resulted from hereditary attributes not cultural influences. Thus, the positive relationship between schooling and test scores which, in turn, closely reflected contemporary status hierarchy and occupational breakdowns was predictable. Those who possessed superior educational and vocational qualifications did so because of superior intelligence. 'The men responsible for developing the tests', one scholar has stated, 'were confident... that the tests measured "native intelligence" and not something else.'[30] Lewis Terman, responsible for the famous Stanford-Binet Test in 1912, suggested the analogy that measuring human intelligence was like assaying the quality of a gold-bearing vein of quartz. One ascertained from a small sample how much gold was present in a ton. Yerkes and Terman, avowed hereditarians, were in no way surprised to find how little 'gold' showed up in their assay of the US army.[31]

A third pivotal assumption dealt with the presumed causal relationship between intelligence and an individual's social status. It reinforced class distinctions in what was claimed to be a 'classless' society. The majority of Americans from middle-class, Anglo-Saxon Protestant backgrounds (such as senior army officers in 1941) accepted without question that high social and occupational status resulted from superior intelligence.[32] Early analyses of intelligence tests, grounded in hereditarian convictions, supported this view. Themselves middle class, the researchers 'discovered a relationship between test standing and social status and announced the cause of the relationship to be innate intelligence'.[33] The most celebrated study of this genre was based upon the First World War army tests and held sway in certain circles through the 1940s. Those who equated superior intelligence with such manifestations of high socio-economic status as a secondary school diploma, club or sports team affiliation and college attendance came to identify a 'good' background with the capacity for leadership, and leadership with the qualities of initiative and flexibility that were needed in the modern army. After all, they reasoned, in a land of unlimited opportunity, anyone who was poor and uneducated must either be intellectually or morally defective. The argument's profound circularity was ignored.[34]

These ideas had percolated through various offices of the army by the end of the First World War. In 1920 Major-General Henry Jervey, head of the Army War College, was sufficiently confident that listeners were attuned to the linguistic elements of human engineering that he could observe:

A man could not be considered merely a man. He was something more. He was part of machine made up of many different parts, each a man it is true, but having to play a highly specialized part. Consequently, it became necessary to economize the specialized abilities of these various spare parts and assign them where their specialized abilities would do the most good. In other words, round pegs had to be selected to fill round holes.[35]

Jervey and other enthusiasts acknowledged that the need for a 'careful classification of men before assignment' had not been fully accepted. Indeed, neither the data resulting from 'Army Alpha' about the capabilities of soldiers nor the centralised assignment system developed by the army's Committee on Classification of Personnel had been made use of to any great extent during the First World War.[36]

The search for enlightenment (transferable models) by those for whom inclination or the roll of the military assignment die had led to formal involvement with what was gradually becoming known as 'personnel management' went in several directions. Some officers sought to keep up on their own with the mushrooming literature in the fields of industrial sociology, the social psychology of work, and management and organisation theory. Here books such as Viteles's *Industrial Psychology*, asserting a hereditarian explanation of intelligence and an outmoded view regarding individual differences, served as sacred text.

A second source of information came via the interaction with civilian experts invited to lecture at the Army War College or its younger sibling, the Army Industrial College. The final stop in the formal educational process of career officers, the Army War College, included in the 1925–6 G-3 syllabus only one exercise that in any way dealt with the evaluation and assignment of newly inducted recruits ('replacements' in the jargon then current).[37] In general, the notion held sway that an undifferentiated mass of civilians would present themselves at mobilisation centres following the declaration of war by Congress, be assigned to units for training, and over succeeding weeks and months acquire the particular skills associated with the branch of service with which they had, willy-nilly, affiliated.

By the mid-1930s, however, consideration of the 'who, what, whys' of the characteristics of those men with whom the army must perforce deal in any future emergency had crept into the War College curriculum. The G-1 segment of the 1934 course included lectures on 'Methods and Means of Determining Officer Candidates' Mental Qualifications' (Professor Carl C. Brigham, Princeton); 'Development of Personality' (Dr D. A. Laird, Colgate), 'Occupational Specialists' (LTC H. E. Stephenson); 'Major Problems of the Adjutant-General under W. D. Mobilization Plan, 1933' (MG J. F. McKinley, TAG), 'Labor in War' (Wm Green, Pres, AFL); and 'Population' (Prof H. P. Fairchild, NYU). Over the next several years, committee assignments on processing of personnel, employment of manpower, replacements, morale, classification, and assignment and promotion of officers became staples of the G-1 course, with Brigham and Fairchild being joined by Douglas Southall Freeman, whose lecture, 'Methods Employed by General Lee to Maintain Morale', was wildly popular, and Dr George Gallup lecturing on 'Influencing and Evaluating Public Opinion'. However, not until the 1939 course did the G-1 component incorporate a presentation on 'Aptitude Tests' by Dr Johnson O'Connor (Stevens Institute of Technology).[38]

Ironically, the ideas embodied in this changed perspective within the military profession were in the process of giving way to doubt, backlash amounting to an intellectual counter-revolution, at precisely the time the US army's personnel gurus were embracing the practical implications of organisational and personnel management theories.[39] Indeed, in the army itself these assumptions were not easily victorious. In the years immediately after the First World War the hostility of career officers (especially physicians in the office of the Surgeon-General) combined with embarrassing public revelations of widespread illiteracy (nearly 30 per cent of those tested demonstrated themselves unable 'to read and understand newspapers and write letters home') caused the War Department to back away from psychological testing in any form for any purpose.

Military officers criticized the psychological testing program, saying that…the tests might be testing intelligence or mental alertness [but] what did that have to do with military effectiveness? What did such military virtues as obedience, enthusiasm, patriotism, and bravery have to do with intelligence?

Indeed, they asked, might not 'higher intelligence make people less rather than more obedient?'[40] Any serious effort to answer these questions would be deferred for more than two decades.

In 1923 Robert Yerkes claimed that no more than 10 per cent of the population, on the basis of army data, 'are intellectually capable of meeting the requirements for a bachelor's degree'.[41] That both gratified and alarmed racist, élitist officers, many of whom found attractive such theories as eugenics. But concern about adverse publicity should information leak about the intelligence level of recruits (particularly one sample in which southern blacks scored higher than southern whites) slowed acceptance. Typical was the Adjutant-General's Office's reaction when a Signal Corps officer, Lieutenant Ralph K. Bowers (ORC) analysed data from several tests administered to recruits during 1927 and 1928. Bowers drafted an article that purported to compare 'the intelligence or innate mental capacity of the enlisted personnel at the present time with that of the enlisted personnel during the World War'. The wartime sample included 15 per cent with a 'mental age' of sixteen or more, 11.5 per cent of ten years or less, and a median of thirteen years and three months. Asserting that those with mental ages of ten years or less 'are either morons, imbeciles or idiots', Bowers claimed that fully 36 per cent of current recruits into the US army were 'of the moron type'. By his calculations, nearly one-fifth of the army's enlisted cadre comprised persons who would have been discharged as unfit during the First World War.[42]

In 1929 G-3 reported that the War Department was continuing to use the old 'Alpha' test 'because it was thoroughly standardized and an ample supply of copies was on hand' even though 'it is not found satisfactory for use with Army men, as it gives too much weight to results of formal training [i.e., education]'. A revision was under way to deal with problems of administering the test, though its underlying assumptions were not disputed. Also being pursued was creation of 'a very simple test of intelligence that…will serve to segregate men of low intelligence from those who will make satisfactory soldiers'. But nothing further was done for ten years.[43] Resistance from reactionaries in the Surgeon-General's Office, the politically sensitive issue of testing human capabilities in a society in which all (white) men were deemed to have been created equal, and the lack of any organised advocacy group inside the army led to updated testing instruments being developed by a professional group, the American Association for Applied Psychology. During a fourteen-month period in 1939–40 a revision was made of the Alpha test, the 'Army General Classification Test'. The AGCT was, in fact, a 1940-model packaging of 1920s ideas and assumptions.[44]

The cult of efficiency found most direct expression in the merger of aptitude testing and job classification. Note the assumptions inherent in the following portrayal of psychology's contribution to modern war in the July–August 1940 issue of *Infantry Journal*:

> The increasing emphasis upon mechanization in the armed forces, the newer and more complicated weapons…made a new demand upon psychology. Trained men for the handling of this technical equipment exist only in small and totally inadequate numbers…Training on mechanical devices succeeds best and quickest when the trainees possess the basic aptitudes required by the job. The problem therefore becomes one of identifying, prior to selection and training, the individuals possessing the required aptitudes.[45]

Given evidence from the most up-to-date research that efforts to offset specific aptitudes by 'intensive and prolonged training' yielded no benefit or minimal gain 'wholly out of proportion to the time and effort expended in training', the challenge was to identify the specific abilities – visual perception, auditory perception, two-hand co-ordination, hand-and-foot co-ordination, for example – associated with outstanding performance in each and every military function. Modern techniques of testing could then find those individuals who possessed the requisite skills and, since 'our present industrial experience' reveals that the tests 'are measuring innate, basic abilities', the waste of time and effort involved in training soldiers to perform tasks beyond their 'natural' capacities could be avoided.[46] The lesson was that every inductee who met the army's minimal IQ requirements could learn how to load and fire a rifle. A significantly smaller number would with close supervision and repeated practice be able to shoot with some speed and accuracy. But only a few individuals possessed the innate abilities to benefit from special instruction in marksmanship. Therefore, given the presumed devaluation of marksmanship in modern warfare and the push to identify men who operate airplanes, forklifts and typewriters, training had to emphasise the lowest common denominator and could be organised as a mass production, assembly-line operation.

What became the dominant narrative was set forth during a wartime conference on 'Classification and Assignment of Enlisted Men': 'When life was simpler and Army jobs were confined specifically to each arm or service, the allocation of soldiers was chiefly accomplished by personal preference, either by the soldier himself, or by some other entirely personal equation or estimate.' The speaker stated that this was a natural survival of an age

> when an individual was commissioned by the King, Emperor, or in our own case, Governor of a Colony or State, to raise a company of horse or foot soldiers as the case might be. During our greatest display of arms prior to World War I, the Civil War, the need was for bodies capable of firing the

piece or able to ride a horse and preferably those who could forage pretty
well for themselves.

But he concluded:

> A modern Army employs skills most highly complex and technical. It
> demands coordination of the efforts and abilities of all soldiers. Efficiency
> of every degree of technical complexity drops when men are placed in jobs
> which are either too difficult or too easy for them. In the one case they are
> bored and easily distracted. Delinquency rates mount. Morale suffers too, if
> ability or special qualifications go unrecognized. Certainly, when the wrath
> of destruction was loosened upon us in 1941, we had notice of the
> mechanical monsters with their intricate mechanisms that we would have
> to combat. In building our Army, almost from scratch, we would have been
> derelict had we not set ourselves to the task of utilizing the civilian skills
> that would come to the Army... Since the last war the increased specializa-
> tion in industry has led to the intensive studies of selection techniques,
> whereby the natural aptitudes of individuals can be estimated... We have
> gone far in utilizing those procedures by which we can obtain pertinent data
> concerning the enlisted man's education, intelligence, aptitude, previous
> military experience, civilian work history, interests, and hobbies.[47]

Were these ideas embraced systematically and generally throughout the
army as it undertook the task of rapid expansion in 1940–41? The simple
answer is 'No'. It may be doubted that more than a handful among the army's
senior leadership had read any of the works on psychology/philosophy. Of the
nineteen books from *Training Circular* No. 25 prescribed for lieutenant-colonels
and colonels in the combat arms, not one dealt with psychology or the
application of industrial techniques to military mobilisation.[48]

Nevertheless, as the anonymous authors of a 'Working Outline: History
of Psychological Measurement in World War II' noted in 1944, the regrettable
'lapse' of interest within the army was to some degree compensated for by
'self-education' in what was being done in industry 'of a limited number of
officers interested in personnel', by the efforts of that small number of
Reserve officers who happened to be psychologists, and by self-serving civilian
advisers to the Adjutant-General's office.[49] What was done offers strong
confirmation of the incorporation of certain of these concepts – outmoded
and/or dubious by the standards being utilised by major US corporations – by
the army.

What becomes apparent, too, when one examines the career patterns of a
sample of the interwar US army officer corps is the diversity of social and
geographical backgrounds of its members and, second, the reality that high
command was almost exclusively the prerogative of West Point 'ringknockers'
and graduates of other military-oriented institutions with a leavening of those
promoted from the ranks. Beyond that, however, a sharp distinction existed
between the 'arms' and 'services' as regards assignment to staff positions,

especially time spent with the War Department General Staff.[50] By the late 1930s, the stepchildren of the General Staff were the Office of the Adjutant-General and the Personnel Division (G-1). Notably, these two agencies championed the cause of testing and classification and, lacking sustained interest in these issues by the chief of staff, gained effective control over the personnel aspects of America's military mobilisation. The system they created was in a sense the revenge of the 'managers' over the hotshots in the combat arms. It also reflected contradictory assumptions about how to organise and train, clashing priorities regarding what was important, and differing organisational ambitions within the US army.[51]

These deeply ingrained patterns of assignment and interest might have been of little consequence had the pre-Pearl Harbor timetable for expansion been followed to the letter. The circumstances that General George C. Marshall feared above all others – uncontrolled growth driven by political expediency – characterised the organisation and training of US ground combat forces during 1940 to 1942. In such an environment, two mutually reinforcing elements already present in American military thinking came to the fore. First was the 'push' resulting from a widely held view (that, like Lot's wife, metamorphosed into doctrine when the Wehrmacht blitzed Poland and France) that mobile warfare featuring highly sophisticated weapons would characterise the Second World War. The 'pull' was generated by AGO/G-1 'experts' seeking to make the field of personnel management a professional speciality. Their efforts to equate the US army with General Electric and General Motors and the preparation of individual soldiers for combat with the assembly-line process by which lightbulbs and automobiles were manufactured were accepted by default. The army's leadership had other matters with which to be concerned and, after all, any irregularities or flaws in the basic components – riflemen, artillerymen, messengers, and other assignments for which no or minimal qualifications were needed – would be compensated for by aggressive unit training and, above all, robust leadership. This approach reflected the 'strong' version of the factory model of education, treating recruits as 'raw material to be transformed into finished products'.[52] What were its effects?

*

Reflecting hereditarian and élitist assumptions, the army's leaders were concerned, first, about the relative proportion of 'dolts' and of men who possessed the requisite attributes to become specialists coming through the portal of Selective Service. Aside from the college students, first impressions were horrifying – confirming the First World War data that the 'mental age' of the typical American adult male was no higher than twelve and that many Americans were classifiable as 'moronic'. Obviously, very little could be done to 'develop' such raw material. But, fortuitously, modern warfare, it appeared, had relegated the vast majority to the sidelines and the rear areas.

That the experience of building an American army to fight the Second World War embodied 'the idea of just tossing bodies around' is generally acknowledged.[53] We can say with certainty, as well, that the procedures adopted to classify inductees proved inherently unfair. Confusion about the nature of the Army General Classification Test, viewed by its creators as identifying the aptitude of the subject to perform certain tasks but universally perceived within the army as a test of general intelligence, had fateful consequences. Designated the 'Army General Classification Test' to emphasise that the aim was to identify 'aptitudes' rather than to ascertain innate intelligence, the AGCT was nevertheless a straightforward IQ test modelled on the Stanford-Binet examination devised two decades earlier. Chief Psychologist Walter Bingham noted:

> The Army is trying to determine chiefly two things by this test: First, how difficult are the problems a man can solve, and, second, how fast he can be expected to learn the things he has to know in order to be an efficient soldier in this new, highly organized, technological Army of ours?[54]

The AGCT established five groupings, ranging from Class I (requiring a minimum score of 130), Class II (110 minimum), through Class V (69 or lower). The AGCT's cultural and educational biases today seem blatantly obvious. One question asked the 'term in lawn tennis for zero'. Another fill-in-the-blank question asked: 'The term for spatial perspective in Renaissance art is————?' Used in conjunction with information about vocational experience in civilian life, the outcomes determined one's military assignment. The system claimed to correlate identifiable skills and training potential with the demands of a given military speciality. It is notable that the category 'rifleman' designated 'Specification Serial Number' (SSN) 745 had no mandated qualifications for assignment thereto.[55] The result, predictably, was that those achieving the lowest scores and with few or no prior skills were dumped into the combat hopper. While the army personnel experts claimed that the intent was to ensure efficiency and equity, this system achieved neither aim. The American soldier's experience in the Second World War evokes an image not merely of square pegs in round holes but a constant reconfiguring of the holes and reshaping of the human pegs being pounded into them.

The realities for young Americans caught up in this impersonal machine are illustrated by what happened to three men who entered the US army induction station at Fort Leavenworth, Kansas in April 1943. First in line was Joseph A., a 20-year-old University of Kansas student, with one year in the Reserve Officer Training Corps when called up. His aim was to be admitted to OCS in the Field Artillery or Signal Corps. Next came Jack B., a 21-year-old casual labourer from the 'little Balkans' region of south-eastern Kansas. With several convictions for 'drunk and disorderly' and brawling, he was functionally illiterate. Third was James C., a 22-year-old African-American from Leavenworth with a 10th grade education. He possessed skills as a carpenter and electrician and had been employed as a handyman at the post. His grandfather had been a 'Buffalo Soldier' and Jim's ambition was to join the 2nd Cavalry Division.

Having been subjected to a battery of physical examinations, the AGCT and other tests, and a short interview by a contract psychologist, the three newly fledged soldiers reacted with varying degrees of shock and bewilderment when their assignments were posted. Joseph A. scored 122 on the AGCT, but since combat arms officer candidate schools were greatly oversubscribed, Joseph was offered a slot in the 'Army Special Training Program'. This had been set up as a result of intense lobbying by college and university presidents, who argued that their campuses, virtual ghost towns without male undergraduates now serving their country, could play an important role in training soldiers for a variety of missions. At a time when the army believed its need for officers had been met, a plan for stockpiling the most intelligent from its ranks – as the US navy was doing via its V-12 Program – was extremely appealing. Using a minimum 110 (later 115) AGCT score as the qualification, more than 150,000 soldiers were withdrawn from the training process and sent to educational institutions across the country. Since the great majority were university students or graduates, their absence contributed significantly to the existing distortion of assignment to the combat arms.[56]

Jack B.'s AGCT scores placed him in group V. Following current policy of retaining for military service all except the hopelessly 'inapt' and the psychotic, Jack's lack of vocational skills ensured assignment to the infantry. First, however, he was sent to the Fort Benjamin Harrison 'Special Training Unit' for a twelve-week course of study. The aim was to teach Jack and some 300,000 illiterates and 'marginal' soldiers to read and do simple sums in order that they could profit from basic training. Working through primers such as *MEET PRIVATE PETE* and *PRIVATE PETE EATS HIS DINNER*, Jack made steady progress. The programme inculcated patriotism as well as reading skills. For example, Lesson 17 in the 'advanced' text at Fort Benjamin Harrison read:

There are [fill in number] in our family; We must defend our families; Families should have good homes; Good homes help to make a good army; The Army defends our homes; We want to defend our homes; We must defend our country; All must work to make America safe; We must learn to read and write; I am an American; I love my country; It is our duty to defend our country; I can write the name of my country.

Jack and some 270,000 others received a 'Certificate of Completion' that affirmed ability to read, write and do basic maths at a fourth-grade level. He was then thrown into the infantry assignment pool.[57]

James C. scored 110 on the AGCT but was not informed that this qualified him for OCS. He was initially assigned to his first preference, a Signal Corps detachment with the just-reactivated 2nd Cavalry Division. However, prior to

completion of basic and advanced Signal Corps training, the 2nd Cavalry
Division was disbanded and its soldiers converted into stevedores and
labourers. Eventually, James C., who had peacetime experience driving trucks
and heavy-wheeled machinery, arranged for a transfer to the Transportation
Corps. He spent four months qualifying to be a driver and then another five
months in Britain transporting supplies from Liverpool to the US army's
camps in south-west Britain.[58]

*

These three 'case studies' and mountains of evidence documenting the experience of an additional 2.5 million Americans who served in combat units make
clear that class and race to a large degree dictated the composition of the US
army in the Second World War. Given the rationale in place at the time the
system was set up – that the Second World War was to be a war of movement
and massed fires, that level of education equated with capacity for learning and
leadership ability, and that American males (poorly motivated, passive mom-
ma's boys) constituted indifferent raw material – the traditional arms, especially
the infantry, received low priority. As a result, the best qualified in terms of
education and previous civilian vocations were skimmed off for the Army Air
Forces and Army Service Forces, officer candidate schools, and the ASTP –
leaving the 'dregs' for the infantry and other ground combat arms. By 1943 the
distribution of manpower was significantly skewed. The AAF claimed nearly
twice as many Group I and II men than the AGF, and the proportion of Is
and IIs grabbed by the ASF was 30 per cent higher than that of the AGF. At
the other end of the scale, the AGF contained more than five times as many
Group V soldiers as the AAF and almost four times as many as the ASF.[59]

By 1943 the army's senior leaders belatedly acknowledged that the 'typical'
infantry soldier, who required physical and intellectual abilities as least as high
as an airman and much greater than a stevedore or shipping clerk, was some
two inches shorter, twenty-five pounds lighter, and by the army's own
standards significantly less intelligent than his counterparts in the AAF and
ASF. The army's chief trainer, AGF Commanding General McNair, wrote a
memo for the record in December 1943, just when the AGF was completing
its job of building the ninety ground force divisions with which the United
States would fight the Second World War. Because of the misguided priorities
established when the personnel system went into operation in 1940–41,
McNair noted, the Army Ground Forces had been systematically denied their
share of the best-qualified men inducted into the armed forces. As a result,
enlisted personnel in the combat arms were demonstrably less qualified
intellectually and physically than were soldiers in the Army Air Forces and
Army Service Forces.[60] McNair observed fatalistically:

From time to time in the past, this headquarters has pointed out that certain
procedures in distributing manpower discriminate against the ground

forces. The enclosed chart...shows the accumulative effect of such measures...While the situation is viewed as unfortunate, it is realized that it is now too late for effective remedial action. This study is submitted in order to make clear the composition of our war Army in its practically complete form.[61]

Pressure to redress the situation proved too little and too late. A campaign to improve the 'image' of the infantry via such band-aid measures as the Bronze Star Medal and 'Combat Infantryman's Badge' was launched with great fanfare in the summer of 1944. By that time, AGF's demand that physical attributes be made an important determinant for assignment to combat units led to adoption of the so-called 'PULHES' system, a modified version of a mechanism for assessing physical characteristics developed by the Canadian army.[62] Wartime studies had demonstrated a strong correlation between physique and socio-economic background, given the importance of nutrition and healthy childhood environment on adult development.[63] Notably, army personnel experts adamantly opposed the introduction of physical criteria for classification purposes. They argued that giving lower priority to physical and intellectual characteristics of those in infantry and artillery had posed problems but, as one observed, 'Hannibal's troops were short and they had performed well.'[64] In any event these changes had almost no effect, for by early 1944 US ground combat forces were essentially organised.

This situation might have been of no great import had military operations developed as forecast. That did not happen. In both Europe and the Pacific expectations of mobile warfare and swift victories yielded to the reality of brutal slugfests. The US army's experience in North Africa had revealed that the infantry was taking substantially higher casualties than projected. By late 1943 infantry casualties stood at approximately 70 per cent of the total and that was to be sustained for the remainder of the war.[65] So a pivotal question is not only who served but how and when an individual was assigned determined his share of the burden and the likelihood of ultimate sacrifice.

That point is reinforced by the multiple ironies associated with the experience of the three soldiers in our sample. Joseph A. spent the summer and autumn terms of 1943–4 studying French at the University of Virginia. He had been pencilled in to serve as a civil affairs liaison officer following the invasion of France. When the ASTP was abolished in spring 1944, he was pushed through basic training, ultimately finding himself assigned as a replacement rifleman in the 106th Infantry Division. Jack B. 'graduated' from the Fort Benjamin Harrison STU, was sent to Branch Immaterial Training at Fort Meade, Maryland, and after being stockpiled for six months in a Replacement Depot in Britain, also ended up as a rifleman in the 106th Division. James C. was bitterly disappointed when the 2nd Cavalry was broken up, because of a shortage of stevedores and, manifestly, the army's hostility toward African-American combat units. He ultimately ended up as a driver for the 'Red Ball

Express'. He had just pulled into Bastogne with a load of bridging timbers when the German attack triggered what became the Battle of the Bulge. He served for a time as a rifleman with the 17th Airborne and, once he found his way back to his unit in late January, volunteered for the crash programme initiated to train African-Americans for combat roles in the infantry.

The multiple ironies of the Ardennes battle – belated changes in the Selective Service regulations regarding fathers, reassignment to the combat arms of AAF and ASF personnel who met the physical profile, and the near-collapse of the combat replacement pipeline – produced a retrospective review of the army's personnel system. But no important changes occurred, for basic assumptions were not challenged. Pivotal if always unspoken was the view that high casualties among the lowest-rated groups were, on philosophical grounds, more justifiable than losing large numbers of the nation's best and brightest. Some were convinced that 'culling' the inept, weak individuals would 'strengthen the race'.[66] Eugenics in that sense was alive and well in the US army during the Second World War. The army's almost uniform opposition to employing African-Americans in combat was not a contradiction but a reaffirmation of these views.

It is necessary to acknowledge that casualty rates among air crews and junior combat arms officers were comparable to those for infantrymen.[67] However, the former were volunteers and the numbers involved – that lowest common denominator of sacrifice – were many fewer. For those who did not volunteer, class was a critical determinant of who fought. GIs, who cynically referred to themselves as 'Government Issue' but also 'Goddamned Infantry', knew this as fact. One has only to review the findings of the army's Information and Education Division, or the bitter criticisms of the 'caste system' collected by the Doolittle Board in 1946 to appreciate how deeply ran their feelings of exploitation.[68]

In sum, the stereotypical squad portrayed in all those 1940s and 1950s war films – young, uneducated 'hicks, micks and spics' leavened by a crusty careerist and an older, well-educated idealist, 'doc' or 'prof' – was not far off the mark. The system ensured that those with less education, little or no pre-war vocational experience – the depression dropouts, the slum kids, the backwoods boys from Appalachia and the deep South – ended up as '745s', riflemen. It appears that the personnel gurus planned it that way. Their intentions, of course, were noble – to ensure the proper assignment of those qualified to do the army's *important* work such as flying airplanes, fixing tanks and classifying inductees.

For the American soldiers relegated to less glamorous roles the Second World War was emphatically a 'Not-So-Good' war. One must also acknowledge that this crew of allegedly passive, scrawny under-achievers, when able to survive the initial shocks of 'on the job training', when given effective leadership, when permitted to develop strong unit cohesion at the battalion, company and platoon levels, fought with remarkable proficiency. They are owed both recognition and the labour to understand what put them in harm's way in far greater numbers than their better-educated, vocationally qualified, bigger, stronger counterparts.

The GI in Europe and the American Military Tradition

Reid Mitchell

'You Americans! The way you fight! This is not war! This is madness.'
A German officer to his American captors, Normandy, 1944[1]

There is no universal agreement about what constitutes the American military tradition and whether there is such a thing. Russell Weigley has found it in the contradiction between the American way of war based on power and an army historically structured around mobility. Others have pointed to a reliance on technology and material as somehow distinctly American. Clearly one characteristic of the military first in Anglo-America and then in the United States was the maintenance of small professional forces which would be augmented by rapid mobilisation of citizens at the time of war. This was certainly the case in the two greatest wars the United States has fought, its own Civil War and the Second World War. Both wars saw a peacetime army expanded to such an extent by the addition of officers and soldiers with little military training that it was also transformed – confirming the old Marxist saw that enough quantitative change is qualitative change.

What those of us who have worked as social historians of the American military and the American fighting men have not yet done is to explore whether or not there may be significant continuities in the experience of the American soldier that stretch from Anglo-America to the present. There are now better or worse social histories of the Continental soldier, the Mexican War soldier, the frontier soldiers before and after the war, the Civil War soldier (probably the best developed field) and the GI. This paper offers very tentative suggestions about the rifle infantryman in Europe and some parallels with other American soldiers.

As for which GIs we should consider combat troops, the authorities took their own survey at the time. 'More than 70 per cent of all the men, regardless of their own type of duty, agreed upon full combat credit for men in infantry rifle and heavy weapons companies, tank and tank destroyer companies, combat engineers, medical aid men, and other front-line troops.' With one exception, the further away from the front line a soldier was, the more inclusive his definition of combat credit. Both in the survey and in the memoir literature, combat infantrymen, the most reluctant to give anybody else combat credit, generally agreed that front-line medics were combat veterans, no matter what the rear echelon might believe.[2]

*

Russell Weigley argues that what he calls 'the American way of war' originated in the Civil War. The victory of 1865 put the strategic thinking of US Grant at the centre of the army's thinking about how to win wars. Grant's conception of war was breathtakingly simple – and the product of a culture that took superiority in men and materiel for granted. 'The art of war is simple enough,' according to Grant. 'Find out where your enemy is. Get at him as soon as you can. Strike at him as hard as you can, and keep moving on.' This understanding of the North's strategy focused just where popular imagination focused – the war in Virginia. Grant was seen to have muscled his way to Richmond, destroying Lee's army in the process. All else, in this view of the war, was peripheral to Grant's decision and ability to get to grips with the CSA's leading army. Grant's army then was the point of the spear, and leaning behind it was the weight of superior Northern population, industrial base, agricultural production, wealth. This understanding of the Civil War dominated the thinking of army theorists afterwards, particularly during the wars between the First and Second World Wars. In 1922 Colonel W. K. Naylor of the Army War College summarised fundamental army strategy.

> I wish to stress this point: that warfare means fighting, and that war is never won by maneuvering, not unless that maneuvering is carried out with the idea of culminating in battle. Disabuse your mind of the idea you can place an army in a district so vital to the enemy that he will say, 'What's the use?' and sue for peace. History shows that the surest way to take the fighting spirits out of a country is to defeat its main army. All other means calculated to bring the enemy to his knees are contributory to the main proposition: which is now, as it has ever been, namely, the defeat of his main forces.[3]

This American insistence on power and mass continued into the Second World War. The British based much of their thinking on Liddell Hart, who in turned based much of his thinking on his reading of our Civil War. To his mind, the Civil War, specifically Sherman's campaigns, showed the virtue of the 'indirect approach' to warfare. I suspect Grant would have regarded the energy expended in North Africa and Italy as wasted as the time spent by McClellan seeking decisive battles. The Americans in the Second World War insisted that ultimately the only way to win the war was to concentrate the Allied forces, confront the German army, and defeat it. Or, as Colin Powell said of the Iraqi army at the beginning of the Gulf War, 'First we're going to cut it off. Then we're going to kill it.' Or, as Grant said – 'Find out where your enemy is. Get at him as soon as you can. Strike at him as hard as you can, and keep moving on.'

The American way of war has not worked in all times and all places. If the Cold War had ever led to American forces battling with the Soviets, it's hard to imagine how a strategy of direct assault would have fared against an army devoted to 'deep battle'. Nor is the American way of war very successful in non-conventional operations. Critics of the US army in the European theatre of operations, many of them British, have argued that Eisenhower's broad front strategy in 1944 slowed down Allied victory and put the Allied armies at too much risk. Here again I find Russell Weigley's analysis persuasive. In *Eisenhower's Lieutenants*, Weigley's conclusion is that the American way of war was not ill-conceived, but that the ninety-division gamble gave Eisenhower too few troops. Unlike Grant, Eisenhower did not have overwhelming manpower. 'The American army in Europe fought on too narrow a margin of physical superiority for the favoured American broad-front strategy to be anything but a risky gamble.' Thus, the German counter-offensive in the Bulge – which was successfully contained.[4]

For this paper, the critical point to make about the American way of war is that it chews up infantrymen – something that the US government has been more able to hide the longer the army's logistical tail becomes. During the Civil War 8 per cent of white adult males died – 6 per cent of northern white adult males, 18 per cent of southern white adult males. In the two-day Battle of the Wilderness, both the Union army and the Confederate lost about 17 per cent of their men. During the Second World War, casualties among American rifle companies were even higher. Omer Bradley estimated that in rifle platoons of the 30th Division at St-Lô casualties in one fifteen-day period were higher than 90 per cent. Company K of the 333rd Infantry started with 200 men. 'By the end of the war twice that number had seen actions with the company and our battle casualties had reached two hundred.'[5]

Leon C. Standifer's King Company 'paid the price' of 400 per cent casualties.[6] Eighteen infantry divisions in the European theatre of operations suffered more than 100 per cent casualties. Fortunately for the GI, army medicine had advanced considerably since the Civil War; compared to the Civil War far more Second World War casualties survived. The survival of the Second World War infantryman was not, however, because the US army had become more reluctant to put men in the killing zone. The odds were against the combat infantryman. Or, as Bill Maudlin's Willie, sitting in his foxhole under fire, expressed it, 'I feel like a fugitive from th' law of averages.'[7]

The GI could see all this. He knew there were too few infantrymen for the job; he knew that the odds were against him. He invented the concept of the 'million-dollar wound'. Knowing all this and going ahead anyway might be said to *be* the principal component of the infantryman – that and what Bill Maudlin called 'the Benevolent and Protective Brotherhood of Them What Has Been Shot At', to which the infantryman was reluctant to accept anybody other than fellow infantrymen. In *The Men of Company K*, Harold P. Leinbaugh and John

D. Campbell, two officers of the company, refer proudly to 'raggedy-assed riflemen, men who have more in common with the foot soldiers at Antietam or Chancellorsville than with anyone half a mile to their rear'. Or, as Charles B. MacDonald said in his classic *Company Commander*:

> The characters in this story are not pretty characters. They are not even heroic, if lack of fear is a requisite for heroism. They are cold, dirty, rough, frightened miserable characters: GIs. Johnny Doughboys, dogfaces, foot-sloggers, poor bloody infantry, or, as they like to call themselves, combat infantrymen. But they win wars...when you call a company a rifle company, you are speaking of the men who actually *fight* wars.[8]

There are few military traditions older or more established that the contempt of front-line soldiers for men in the rear. As the logistical tail of the American army grew, the percentage of men in the front line shrank. Leinbaugh and Campbell estimated there were twelve men behind the line for every infantryman. Another reason for the dislike of the GI for the rear echelon was the sense he had that he served under a higher command reluctant to actually come to the front lines and reckless with the lives of its rifle companies.

For a nation that prides itself on the ability to get the job done, the history of American tactics proves surprisingly dismal. The great virtue of George Washington's Continental army lay in its ability to survive, as it rarely won a battle. Americans later prided themselves on the revolutionaries' tactical innovations but the innovations were more fancy than fact. No Civil War commander solved the problem of the tactical offensive; instead, they bashed through as best they could, ordering direct assaults and bleeding their armies white. In the Plains Indians Wars, as Stephen Ambrose has argued, the cavalry deliberately allowed themselves to be ambushed, as a means of making contact with the enemy, and then relied on superior firepower to win victories. Much the same tactics were used by the US army trying to come to grips with the Viet Cong and the North Vietnamese a century later. David F. Trask has shown how poor American tactical thinking was in the First World War, and how inexperience and lack of training hurt the AEF.[9]

American tactics in the Second World War started off almost as poorly as Civil War tactics, but they improved rapidly. While inferior tanks and insufficient training certainly hurt the combat troops, the principal weakness was the army's lack of combined arms doctrine. Put simply, the military thinkers assumed that tanks and infantry worked best separately, and did not prepare for a war in which most successful attacks would be made by combined arms – tanks and infantry, principally, but with support from a range of arms including artillery, tank destroyers, armoured vehicles and airplanes.

As Michael Doubler most recently has shown, the US army combat troops improvised rapidly and successfully on the battlefield and soon mastered the craft of the combined-arms assault. Doubler praises the US army for its tactical flexibility and he is right to do so. What Doubler misses is the way the GI interpreted this process of improvisation. Where Doubler sees an army rising to meet the challenge, the GI sometimes saw an army that didn't know how to fight its war, risked its citizen-soldiers foolishly, and relied on small units to make up the deficiencies of its poor planning. It was hard for the rifle companies to give the army credit for on-the-spot improvisation.[10]

Weigley ascribes 'more ingenuity' to the soldiers and combat officers at Omaha Beach than the planners of the assault; without their initiative in getting off the beach, Bradley might have been forced to evacuate. Before the landings, nobody seems to have recognised the problems that fighting in the Normandy bocage would present; if anybody did, he failed to inform the junior officers. Doubler, author of the most recent study of American combat in the European Theatre of Operations (ETO) and a great defender of the US army, notes that the solutions to the tactical problem of fighting in Normandy came from the initiative of the small units themselves.

For example, consider the testimony of George Wilson, one of the small unit commanders who began improvising in Normandy. By his third day of combat, he realised 'while training was based on leading a full-strength unit, in actual combat I found myself at about half strength, or less, much of the time'. Wilson was successful that day, but worried. Fighting at half-strength 'makes a big difference in tactics, and I was forced to experiment as I went along – and this could have been tragic'. Robert Rasmus's first day in combat required his outfit use 'march-and-fire tactics'. 'It was a new maneuver we'd never done in training,' he remembered. 'We learned.'[11]

Consequently, after the war, American combat veterans did not usually remember the higher command with affection or respect. Early on in the ETO, infantry came to view the role of headquarters as giving them unreasonable if not impossible assignments. The infantry training exercises he'd practised in the United States amused Leon Standifer because he thought them both easy and silly. This amusement sickened when he got to France and concluded that the same idiots who planned the training exercises were planning the infantry actions he had to fight in. Charles B. MacDonald complained of majors who made him send his men on 'ridiculous missions', of colonels who hurried needlessly at the expense of the riflemen's lives.[12]

Unlike Civil War memoirs, where men write with true affection for Marse Robert or Uncle Billy, it is rare in Second World War memoirs to find much discussion of individual high-ranking officers, even Bradley, the 'soldier's general'. (Soldiers who served under Patton are the most common exceptions to this rule.) Generals were too far from the front. Even colonels and majors seemed remote. Bill Maudlin commented that, 'A lot of guys don't know the name of their regimental commander.' Unit identification usually did not rise above the company level.[13]

Because of what they often viewed as the incompetence of higher command, soldiers valued their officers by different standards than the army intended. A good company commander was one who looked after his men – who put his company first, over the requests of the army or even the battalion. Young Charles B. MacDonald overheard a private praising him, the new company commander. 'He's damn young, but he doesn't seem scared to come around and see you once and a while, no matter where the hell you are. He seems to care what happens to you.' (Some eighty years earlier, a Union lieutenant promoted from the ranks, after issuing his men their whisky ration, overhead them saying, 'By god that is the man for us; he takes some interest in his men: it is a pitty thare ant more officers like him in the army.') One man in Company K described a captain, approvingly, by saying, 'Gieszl always thought more like a sergeant than an officer.' Leon C. Standifer praised a particular lieutenant because he knew how to stall missions that were needlessly risky and how to abort them if possible.[14]

Since the GI regarded himself as only a temporary soldier, he rarely developed the respect for rank that the army thought desirable and he preferred his officers to treat him as much as possible as an equal. When Harold Leinbaugh's captain was wounded and he had to take over the company while in combat, he discussed their options with the men and took a vote on what course of action to follow. Standifer praised one platoon leader by saying, 'He respected his riflemen and always gave them a full orientation of what he knew about a mission, whether in training or in combat.' Other soldiers agreed: a good officer shared what information he had with his men.[15]

Whether the soldiers recognised it or not, this GI desire for respect and to understand, not just obey, orders had a long history behind it. Baron von Steuben, trying to explain the American soldier in the revolutionary war, said, 'The genius of this nation is not in the least to be compared with that of the Prussians, Austrians, or French. You say to your soldier, "Do this" and he doeth it: but I am obliged to say, "This is the reason why you ought to do that," and then he does it.' It was also characteristic of the Civil War soldier: Henry Pippitt admired his captain because 'he would talk with a private just as soon as he would talk with an officer', and David Nicholl wrote, 'I don't think it necessary to be so strict or exact or make so wide a difference between Officer & men in volunteers as regulars just because a man has it in his power to do so.' On this issue, the Continental, the Yankee – and the Confederate – and the GI all agreed.[16]

The substitute for hierarchy, as well as for discipline and adequate training was the old American one – inspirational leadership. In his *A Revolutionary People at War*, Charles Royster says, 'The surest way to win Continentals' respect was to be more active and able than they were – that is, to endow formal authority with the moral authority of superior service.'[17] In the Civil War American generals were so accustomed to leading from the front that their death rate was proportionally the highest in both the Union and Confederate armies. There weren't many American generals at the front in the European theatre but junior officers continued the tradition. The colonel of the 22nd Infantry gave the following speech to replacement officers shortly after D-Day: 'As officers, I expect you to lead your men. Men will follow a leader, and I expect my platoon leaders to be right up front. Losses could be very high. Use every skill you possess. If you survive your first battle, I'll promote you. Good luck.' Even more cynically, one officer observed of replacement lieutenants, 'Teach them to say "Follow me" and ship them overseas was the quickest way to replace the casualties.'[18]

This was, of course, a good way to kill lieutenants and captains. Sergeant Savage, the veteran who mentored Charles MacDonald, admonished him when he tried to lead a platoon into combat, 'Goddamit. Captain, you got to stay farther back. At least get some scouts out front.' Leo Balestri writes, 'While combat lieutenants and captains represented only 0.9 per cent of an infantry division they suffered an average of 2.7 per cent of combat casualties. A study of American combat troops in Italy showed it took a mere 88 days of combat to cause 100 per cent casualties among a division's infantry second lieutenants.'[19]

*

The foreseeable but unforeseen high casualty rate among combat infantrymen helped create another GI source of grievance: the ETO's replacement system. As Russell Weigley observes, troop replacement has not been a strong point of the US army. On the one hand, Civil War regiments fought themselves almost out of existence: on the other hand, the Vietnam era method of bringing men in and out of the war as solitary soldiers diminished unit pride and individual self-esteem. As for the policy in the ETO, Stephen E. Ambrose's assessment is blunt: the policy was 'in everyone's opinion, one of the dumbest things the Army did'. For much of the war, soldiers who had been ill or wounded but had recovered often were not returned to their old units, but were left waiting in replacement depots they regarded as little better than prisons or were sent to other units. Thus the army broke down the fundamental loyalty most soldiers felt – that to their platoon or company.[20]

An even larger problem was that there were just too few rifle companies to which to send the replacements. The 'ninety-division gamble' had left the army in Europe too few regiments. The replacement system demoralised more men than just those who passed through it. The practice of funnelling replacements into old units helped to keep those old units and the men in them on the front line. An army study of riflemen who had served in North Africa and Sicily 'showed that these veterans, while exhibiting a rather fierce pride in their outfit, were more embittered than perhaps any other soldiers who had been studied by the Research Branch'. They expressed their bitterness with wry humour: 'The Army consists of the 1st Division and eight million replacements.'[21]

Finally, insufficient attention seems to have been paid to integrating replacements into the old companies. Company K, 333rd Infantry, provides an example. Its former commanders, Leinbaugh and Campbell, were sure that the company maintained its core identity despite the number of replacements; they were also sure that the replacement soon felt at home in the company. According to Leinbaugh and Campbell, 'The veterans did not go out of their way to instruct the newcomers; they simply went about their business.

Replacements, watching and listening, learned how to go about theirs. Most of the recent arrivals began to see themselves as part of the K Company family.' One new replacement, who arrived on the eve of battle, recalled that 'not one person offered advice'. Company K did put new replacements into foxholes with the old men – apparently the most active measure the command took to welcome the newcomers to the company. For a company so rarely out of line, there was little more they could do.

*

One distinction that can be sharply drawn between the GI and the soldiers of the American Revolution and the Civil War concerns ideological motivation. The Continental, the Yankee and the Confederate were far more likely to express confidence in the justice of their cause and to respond to patriotic appeals. Patriotic appeals, particularly on the eve of battle, embarrassed and disgusted the GI.

The unideological GI has his origins partly in the culture of the Depression. This was the generation of the 'tough guy' ethos, the generation of Humphrey Bogart and John Wayne. (In the Normandy airborne assault, one commander ordered his men to 'use grenades, knives and bayonets only' until daylight, simply because 'it sounded hard-boiled'.) Orators were suspect – or as Sam Spade puts it, 'The cheaper the crook, the gaudier the patter.' *The American Soldier* identified 'reticence about emotional or idealistic matters' as one of the key criteria for American masculinity. In American literature, this reticence is most associated with Ernest Hemingway's First World War stories and novels, but it probably dates from the 'chastening' of American prose identified by Edmund Wilson as a result of the Civil War. There was also a certain revulsion from the ethos of Wilsonian rhetoric and the First World War. After the manifest failure of that war, the GI didn't intend to be suckered into another war to make the world safe for democracy. According to John Morton Blum in *V Was for Victory*, the FDR administration deliberately chose not to use propaganda to stir up a fervour comparable to that of the First World War.

It is hard to make ideology laconic. An expression of idealism, patriotism, anti-fascism put the GI at risk of being considered not only insufficiently masculine but a sap. Besides, in the end everybody knows that the Bogart character or the John Wayne cowboy will do the morally correct thing.

One alternative to emotional appeals or reasoned ideology came from a combination of advertising culture and Hollywood – simple stories intended to touch the hearts of Americans at home and the GIs at the front. The fact that these stories often work on me emotionally does not make them any less embarrassing. Ernie Pyle, whom the GIs loved, was a master of this anecdotal non-ideological ideology. For example, in *Brave Men* he tells about crew members listening to German propaganda, the radio programme of Midge. 'As usual they laughed with amusement and scorn at her childishly treasonable talk.' The point of the story? 'In a vague and indirect way, I suppose, the privilege of listening to your enemy trying to undermine you – the very night before you go out to face him – expresses what we are fighting for.' I am enough of an American sentimentalist to agree with this, but it is a remarkably inarticulate rendering of the meaning of the war against the Nazis.

Another possible alternative to ideology was simply hatred of the enemy. Another disjunction between the GI in Europe – but not in the Pacific – and his citizen-soldier predecessors was his lack of hatred for the enemy he fought. Partially, this may have been the result of the 'tough guy' ethos. The GI often regarded himself as too wise to fall for propaganda – and thus remained suspicious of stories of Nazi atrocities and the rumours of the death camps. It may also have arisen from the 'working guy' ethos of the GI – war was a job to see through, not a cause about which to feel passionate.

There is considerable discussion in the literature on the United States and the war of how the government tried hard to distinguish between the Nazis and the German people. The point usually made is that there was no such distinction between the Japanese and their government. In the ETO, Americans fought Nazis; in the Pacific they fought 'Japs'. The distinction between 'Nazis' and 'Germans' did not reach the American soldier in Europe. Invariably, they spoke of their enemy as 'the German', unless they employed more derogatory slang. Bill Maudlin dismissed the distinction between Nazis and Germans as meaningless; he wrote, 'The Nazi is simply a symbol of the German people.' Another veteran, an old sergeant, looked at his first German prisoners and announced, 'These fucking supermen are the lousiest-looking troops I've seen in all my years in the Army.' Even Bavarian-born combat doctor Klaus H. Huebner, who prayed in German at the war's conclusion because he 'never learned a prayer in English', referred to the enemy as the Krauts.[22]

David Webster, a Harvard man in the 101st Airborne, wanted Germans to feel war. He wrote his parents 'I cannot understand why you hope for a quick end of the war. Unless we take the horror of battle to Germany itself, unless we fight in their villages, blowing up their houses, smashing open their wine cellars, killing some of their livestock for food, unless we litter their streets with horribly rotten German corpses as was done in France, the German will prepare for war, unmindful of its horrors.' Historian Stephen E. Ambrose comments that most soldiers, even in the Airborne, would have preferred the war to come to an end without an invasion of Germany, and surely he is right. But W. T. Sherman would have understood Webster's harsh sentiments. Here is what Sherman told the mayor and city council of Atlanta when they protested against his orders expelling civilians from the city:

Now that war comes home to you, you feel very different. You deprecate its horrors, but did not feel them when you sent car-loads of soldiers and ammunition, and molded shells and shot, to carry war in Kentucky and Tennessee, to desolate the homes of hundreds and thousands of good people who only asked to live in peace at their old home, and under the Government of their inheritance.[23]

But relatively few GIs hated the Germans. Indeed, their attitudes toward the Germans were a source of official concern, because they seem to lack one of the most powerful motivations for military service. The GI certainly hated the Germans less than the Union soldiers hated Confederates, frontier soldiers the Indians – or soldiers in the Pacific hated the Japanese. According to the army's studies, fewer than 10 per cent of soldiers tested said they would 'really like' to kill a German soldier – fewer than 10 per cent. Close to half admitted they would 'really like' to kill a Japanese soldier. After all, the GIs were fighting in Germany because the Japanese bombed Pearl Harbor.[24]

In my experience, it is precisely those historians who are veteran officers of the Second World War who are the most reluctant to accept the idea that ideology motivates soldiers. I have heard Lawrence Stone, Peter Paret and Michael Howard all say that the men who served under them simply had very little idea what the war was about. Along these lines, it may be worth noting that in one survey discussed in *The American Soldier*, while only 5 per cent of the soldiers gave idealistic reasons as combat incentives, only 2 per cent of American officers thought idealism played a part; for the enlisted veterans' viewpoint, officers surveyed drastically overestimated the importance of revenge and vindictiveness on the one hand and leadership and discipline on the other. Furthermore, despite the tone of the bulk of literature on the American fighting man in the Second World War – and that literature that extends such analysis to Americans in other wars – loyalty to buddies or small unit cohesion, were not the reasons that veterans most often identified as their combat motivation. They most often designated their desire to end the war as the most important factor in combat.

The Second World War may present us with the odd spectacle of a democracy fighting a war with little faith in democracy as a motivational force for its citizen soldiers. The spectacle is rendered even odder by the fact that Adolf Hitler agreed. He predicated his winter offensive in 1944 on his assumptions that a coalition of democratic governments could not react quickly to an unexpected attack and that American soldiers, who would bear the brunt of the German counter-attack, were the 'Italians' of the Allies. As it was, the ideology necessary to fight the Nazis was, as George Orwell knew so well, decency – which was the ideology Ernie Pyle tried to express, as well as the cartoons of Bill Maudlin. Decency, the working-guy ethos, and something of the tough guy all played into the self-image of the GI.

*

The self-image of the American infantryman contained contradictory elements. He knew perfectly well that that rifle companies were viewed as the proper service for the less talented, less capable, less intelligent. He also believed that the infantry, in the end, was the most important element of the army. A veteran of the 333rd Infantry's K Company remembered:

When I was in training a lieutenant told us, 'You ain't nothing but gun fodder, but you are essential and nobody can take your place.' I always felt more important than gun fodder. The foot soldier was just as important as any general back in Washington or as important as Eisenhower. That foot soldier, he's the last man out there, he's got to go in and claim the territory.[25]

Another part of the soldier's self-image was that he remained a citizen-soldier. The GI distinguished himself from the old army. The American army in the Second World War resembled the Civil War armies more than might be thought. A small peacetime establishment expanded so rapidly that not only its soldiers had little military experience before the war, but its officers had just as little. Former civilians filled almost 95 per cent of all company commands in combat units. Many of these temporary soldiers would believe they won the war despite army regulations and the orders of their higher-ups, not because of them. A Bill Maudlin cartoon summarises some of this thinking: Willie says to Joe, 'You'll get over it, Joe. Oncet I wuz gonna write a book exposin' the army after th' war myself.'[26]

This vision of the citizen-soldier was bolstered by the notoriously unmilitary appearance of the GI, particularly when contrasted with the German soldier. Leon C. Standifer, always sensitive to the substance underneath the style, offers this corrective to those who might think the GI's appearance was simply slovenly. 'Image is a key to confidence in every army. Our image was as carefully cultivated and as misleading as that of the Nazi-designed army. We were sons of the pioneers, railsplitters, mountainmen, cowboys, rednecks, and lumberjacks. Our image was, and still is, pragmatism. We were massive power, competence, and ingenuity in the face of hardship.'[27]

This reluctance to look military was part of an American military tradition that began with the American Revolution. At the beginning of that war, many citizen-soldiers garbed themselves in hunting shirts, which soon turned out to be impractical. In the Civil War, Grant, who wore a private's uniform and was said to look more like a businessman than a soldier, exemplified the most successful style for command. In Vietnam – as in the Continental army – haircuts and hats were the means by which soldiers asserted their individuality within the military machine.

Perhaps Bill Maudlin's Willie and Joe cartoons did more to spread the image of the unkempt, pragmatic GI than anything else. Maudlin himself, who served as a combat infantryman in Sicily, described his work as drawing 'pictures of an army full of blunders and efficiency, irritations and comradeship'. 'But, most of all, full of men who are able to fight a ruthless war against ruthless enemies, and still grin at themselves.' Cynical, wisecracking, contemptuous of authority, spit-and-polish and chickenshit (and, therefore, particularly of George Patton, who hated them in return), Willie and Joe fought the Germans with neither enthusiasm nor hatred. They were American working men, getting the job done. The captions became catchphrases. 'Must be a tough objective. Th' old man says we're gonna have th' honor of liberatin' it.' 'Th' hell this ain't th' most important hole in th' world. I'm in it.' 'I can't git no lower, Willie. Me buttons is in th' way.' 'I need a couple guys what don't owe me no money for a little routine patrol.' 'Joe, yestiddy ya saved my life an' I swore I'd pay ya back. Here's my last pair of dry socks.' 'Just gimme a coupla aspirin. I already got a Purple Heart.' And, expressing the sentiments of every citizen-soldier who was also a combat infantryman when he encountered rear echelon authority: 'He's right, Joe. When we ain't fightin' we should ack like sojers.'

Yet while Willie and Joe may have angered George Patton, the army itself distributed Maudlin's cartoons by featuring them in *Stars and Stripes*. About 100 American newspapers carried them as well, and Maudlin's 1945 book *Up Front* was a bestseller. Maudlin's portrayal of the GI was welcomed by the fighting men, but practically it was authorised by the army itself and embraced by American commercial culture. Willie and Joe sold newspapers.

Maudlin himself identified one key reason for the popularity of his cartoons back home: their portrayal of the GI as a citizen-soldier. He observed, 'combat soldiers are made up of ordinary citizens – bricklayers, farmers, and musicians'. In general, he predicted, whatever the fears of civilians, the soldiers would reintegrate into post-war society with little problem.

This confidence in the ability of American society to reabsorb the combat soldier has also long characterised the American military tradition. At the end of most wars, the veteran has been left to shift for himself. Until the Vietnam war era, most concern about the veteran focused on his economic prospects. The veterans of the Revolution received land grants and, in their old age, pensions. The elaborate pension system which the Gilded Age built for Union veterans was in many ways the beginning of social welfare policies in America. The relatively few veterans of the First World War did not immediately receive their monetary recompense – as the famed Bonus March reminds us. The GI was rewarded with the GI Bill, a package of benefits that helped to create the modern American middle class, as well as the post-war university system. But only in the Vietnam era has much attention been directed to the psychological and social reintegration of the veteran. In general, the American veteran, like the Second World War replacement, has had to find his own way.

The healthy Second World War veteran may be in part a social construction, as is the traumatised Vietnam veteran. Wendy Bolidzar, a friend of mine who worked at a Veterans' Administration facility has told me that she's seen diagnoses of dysfunctional given to Vietnam veterans when the same test results would have been judged acceptable if a Korean War or Second World War veteran undertook the tests. Conversely, a Vietnam veteran would more likely be classified as malingering where a veteran of an earlier war would be classified as suffering from depression.

It is impossible to believe that the war experience did not mark the GI. One Second World War veteran confessed that combat was 'exhilarating'. He wasn't very successful in articulating what he meant. 'The human reaction to a situation – all those emotions combined – plus the desire to survive – and it's all so compressed in time.' This is the same sentiment that a Union soldier tried to express at the end of the Civil War; he said that in battle he felt an emotion he had no name for and that realising he would never again go into combat made him feel 'lonesome'. In *The Warriors*. J. Glenn Gray observes, 'Many men both hate and love combat. They know why they hate it: it is harder to know and to be articulate about why they love it.'[28]

*

The question how much is American about the American combat experience, as compared to how much is the product of twentieth-century technology and mass society, or how much is human, or even animal, is a troubling one. I am reminded of one of the rare observations of Jung that makes sense to me. 'The language of love is of astonishing uniformity, using the popular forms with the greatest devotion and fidelity, so that once again the two partners find themselves in a banal collective situation,' C. G. Jung wrote in 'Aion'. 'Yet they live in the illusion that they are related to one another in a most individual way.' Perhaps this is yet another way Mars and Venus are lovers – the way men at war and men in love lose their individuality.[29]

Part Five
Two Memoirs

'Puir Bluidy Swaddies Are Weary': Sicily, 1943

Hamish Henderson

In his book *Sicily* – it is not a guidebook, but a cool dispassionate account of the Sicilian campaign of July–August 1943 – Major Hugh Pond has this to say about the impression the 51st Highland Division made on other units of the Allied armies:

> The Highland Division, down from its tall gangling Commander 'Lang Tam' Wimberley down to the humblest Jock private, was very much a law unto itself. One often had a feeling that they were fighting a Holy Crusade, and at the end of the war, they would return to Scotland, take Edinburgh by storm, put a Scottish King on the throne, and form an independent country!...Wimberley was an untiring, see-for-himself commander; he did not believe in 'bumph', and the closer he was to the fighting, the better he liked it. The high morale and spirit of the division was largely due to their leader, whom the men all loved and respected.

At the first light of day, in the early morning of 10 July 1943 – D-Day! – the fisherfolk of Portopalo, the most south-easterly village in Sicily, woke up to find the normally sparsely populated searoads congested with a vast armada of landing craft and other vessels of all shapes and sizes. It must have been a stunning sight for them!... On 'Amber Beach' – Rada di Portopalo – tanks and vehicles of 154 Brigade of the Highland Division were being unloaded, and the only Sicilians to be seen were at pains to identify themselves as *borghesi* (civilians). (Some, as it turned out, were in fact members of the 'autonomous' Italian coastal battalions who had taken the first opportunity of changing into 'civvies'!)

Leading elements of the division were soon in Pachino, the first little town of any size in the area, and they were greeted with white flags and even – here and there – with what looked like a welcome on the part of the locals. The first enemy reaction was an air-raid by planes of the Luftwaffe on the beaches, but they caused comparatively little damage, and remarkably few casualties...My first job, as IO (Intelligence Officer) was to interrogate four German deserters, who were waiting to be picked up by our troops; these assured me that there were no Germans south of the plain of Catania.

I decided to see for myself, and after borrowing a motor bike from a CMP (Military Policeman) who had just disembarked, I rode north in the gathering light, with a tremendous feeling of exhilaration: we were out of Africa, and back in Europe!... Admittedly the scenery and vegetation were not particularly 'European' – white dusty roads, prickly pears, stone walls about a metre high, vineyards on either side of the road – but all the same, this was Europe, and I had a heady euphoric feeling that I could ride north for ever! There was no sign of enemy troops, or indeed of any military activity whatsoever. The *contadini* – the peasants who were working in the fields – hardly gave me a glance – although when I waved to them one or two actually waved back!... After I had ridden north for several miles I reluctantly turned the bike round and rode back the way I had come; fairly soon I encountered sections of a Black Watch company, recognisable at once by the red hackles in their bonnets, proceeding cannily northwards, strung out on both sides of the road. I was able to tell the first senior NCO I came across that they could afford to take it a bit more easy: there were no enemy troops for miles ahead.

Indeed, the first German troops whom we encountered were paratroops – three battalions of the 1st Parachute Division, sent from the Avignon area in the south of France, who landed in Sicily on 12 and 13 July, mostly on the airfields near the River Simeto. The job they had been allotted was to hold up our troops before they could reach the plain of Catania, in order to allow the defences there to be better prepared and co-ordinated. This meant that 152 Brigade of the 51st – a Camerons battalion, and two Seaforths battalions – had to deal with them, and after some very fierce fighting, mainly against No. 2 Battalion of these paratroops, the Camerons sorted them out near Francofonte (16 July), and they were pretty soon encircled and completely cut off.

The officer commanding the 2nd Battalion was one Captain Albrecht Guenther, and – against the rules of war – this officer ordered his men to discard their uniforms, and collect civilian clothes of some sort from the terrified citizens of Francofonte. This they did at gunpoint, and a number did manage to get through to the German lines, disguised – though not very convincingly – as Sicilian peasants.

On the morning of 17 July I was proceeding in a jeep up a road near Buccheri when I noticed two characters in ragged civvies standing close to the side of the road. They did not look much like Sicilians – one was a shortish tubby little man, and the other a tall lanky galoot – and I decided to have a word with them. I got my driver to stop, and walked over to this odd couple. As I approached them, the taller of the two involuntarily drew himself up, and came to attention in the German fashion, clapping his hands on his hips.

'Sind Sie deutscher?' I asked him, and he replied 'Jawohl.' I asked him his name, rank and number, which he gave – and as indeed he was bound to give under the terms of the Geneva Convention; then I asked him what his unit was – this he was not bound to give – but he at once identified it as No. 2 Battalion, 1st Fallschirmjaeger (parachute) regiment.

So far so good. Then I turned my attention to the other German; he at first refused to say anything, but after a moment or two of thought he decided he might as well divulge his own identity – and I was surprised to learn that he was the same Captain Guenther who was OC of the battalion which had held us up – though not for long – two days previously! So here I had two valuable prisoners, and was alone with the driver of my jeep, in the middle of a vast expanse of mountainous Sicily – and no other troops of either army in sight. It was obviously necessary to get these prisoners back to Corps HQ so that they could be interrogated, but there were obvious dangers involved.

I had a pistol, and the two Germans had discarded their arms when they discarded their uniforms, but there is not much room on a jeep, and my driver would be occupied driving, so I felt for a moment or two in a real quandary. If the Germans started anything when we got under way I might be able to shoot one of them – but hardly two.

I asked Guenther if he would give his parole not to try to escape on the journey back to Corps, but he not unnaturally refused. Consequently, there was nothing for it but to risk it; I put Guenther in the front seat next to the driver, and his batman/orderly – for that is what he was – in the back seat on the right; I sat, myself, as far back as possible on the left, and covered them both – but mainly the officer! – with my pistol.

Luckily other vehicles appeared, coming up the road from the direction we were travelling, and I managed to get my captives back to Corps without mishap – but I still feel a 'cauld grue' when I think what might easily have happened if Lieutenant (Acting Captain) Guenther had decided to try to disarm me and make a bolt for it.

One reason he didn't – I thought then, and still think – was that he was highly embarrassed, as a German officer, to have been captured wearing a pair of dirty ragged civvie breeks, and did not fancy roaming around any further in an area where genuine Sicilian males might easily have given him short shrift!

At Corps HQ I delivered the Jerries to the GI – remarking, incidentally, that I was pretty sure that Guenther would at some point try to escape. I was just about to get into my jeep to return to 51 Division, when who appears on the scene but a photographer from AFPU (the Army Film and Photographic Unit), who had been told by someone in the ACV (Armoured Command Vehicle) about my exploit in capturing the OC of No. 2 Battalion 1st Parachute Regiment, and wanted to take a picture of the two of us.

According to Major Pond – *Sicily*, pp. 142–3 – Albrecht Guenther continued to attract interest in various quarters as he was moved back down the line: 'Many of the captured German parachutists, taken back to rear areas for treatment and questioning, remained arrogant and hostile in the face of all threats and interrogation... One of them, 28-year-old Lieutenant Albrecht Guenther, captured behind the lines in civilian clothes, was taken to Eighth Army Headquarters, and interviewed by Major-General Francis de Guingand, where he was warned that under International Law he could be taken out and shot as a spy.

'Quite calmly the young man replied: "That is quite understood. But as a parachutist who has fought in Holland, France and Russia, it does not alarm me. I took the risk and I failed – I deserve it. Heil Hitler!"

'De Guingand ordered him to be taken to a prisoner-of-war camp, and released the story to the BBC in the hope that it might lead to better treatment for some of the Allied prisoners.'

*

When I reported back to Divisional HQ I was told to proceed immediately to the HQ of 152 Brigade – this was the brigade which had had to deal with the paratroops from Avignon – and try to placate Brigadier Gordon MacMillan of MacMillan, who was threatening to shoot another batch of paratroops in plain clothes, captured that morning. To mollify him – and I could see there was a genuine danger that he would actually carry out his threat – I explained that all captured German personnel were needed for interrogation at Army HQ and that he might well be denying valuable operational information to Monty. So, to make the point, and mollify the irate clan chief still further, I paraded the Germans in front of him, and tore strips off them – telling them that under International Law we would be quite justified in executing them summarily, and that they were darned lucky to fall in to the hands of this particular civilised Highland gentleman.

The brigadier complimented me on my command of vituperative German, and offered me a glass of vino.

*

There is a sequel to this particular story. Sixteen months later, in December 1944, I was attached as an IO to the 1st British Infantry Division on the northern slopes of the Apennines. We held Monte Grande, overlooking the Po plain, and facing us was the German 1st Parachute Division – the same outfit which had been dropped over the hills around Francofonte, and had been well and truly sorted out by the Camerons and Seaforths. (At that time we were preparing for the final offensive which – in collaboration with the great partisan insurrection of April 1945 – was going to wrap up the war in Italy.)

On 12 December, these paratroops of 1st Division launched a crazy frontal attack on Monte Cerere, with the clear intention of using it as a springboard to Monte Grande – the idea being to squeeze out our own troops at neighbouring Montecalderaro. The attack was a ghastly fiasco: 19 Brigade of 8th Indian Division, which was on our left flank, properly gave them the works; they suffered heavy losses, and their dead littered the hillside. They did manage to take Casa Nuova, but got thrown out of it again almost immediately, because our troops in Frassineto were able to shoot up their left flank. On Monte Cerere they bit the dust, and never got into our positions at all.

It was not long before prisoners arrived in our lines – some frankly deserters – and I remember reflecting grimly that at any rate they were all still wearing their uniforms!...All seemed keenly desirous to tell all they knew to the interrogator, and one and all combined to cuss the battalion commander who had sent them on this senseless enterprise. One of these prisoners was a company sergeant-major, and he bitterly accused that battalion commander of having *Halsschmerzen* (throat pains) – in other words of sacrificing his men, in order to win a *Ritterkreuz* (Knight's Cross), which was worn in front of the throat.

And what was the name of this ambitious *condottiere*? It was none other than Guenther – Major Guenther – and although that's a common enough surname in Germany, I could not help wondering if – by some extraordinary far-out chance – this bloke could possibly be the same Guenther I had captured in Sicily. I asked the sergeant-major if he knew anything about the background of this officer, and he said everyone in the battalion knew something about it, because the major could never stop boasting about his exploits. He had been in the campaign in France and Holland in 1940; had fought in Russia in 1941 and 1942; and although he had had the bad luck to be captured in Sicily, had managed to escape from a prisoner-of-war camp in southern Italy and had made his way through southern France to Germany; and that was how he had come to get his present job.

So...I was forced to face up to it – the blighter I had captured in Sicily was now facing me again on our section of the northern Apennines. I did not divulge this to the Oberfeldwebel, but I could not help reflecting – there was nothing for it now but for me to capture him again!

...However, I was not to have that pleasure. When the final offensive got under way, he was captured and bumped off by one of the Emilian partisans.

*

Back to Sicily. The campaign officially ended on 17 August 1943 when – the Italian and German High Commands having successfully completed the evacuation of their troops to the Italian mainland – a GI patrol entered Messina. Later the same day, General George S. Patton Jr accepted the formal surrender of the city in the town hall. A few moments later – according to Major Pond – a British armoured car patrol drove into Messina from the south.

In response to a GI query: 'Where youse guys been?' they greeted their Allies with a friendly, 'Hullo, you lousy bastards!' – which was misunderstood by the Americans, who were unaccustomed to such crude banter and thought they were being insulted.

*

A couple of days later I was proceeding in a jeep from Zafferana Etnea – a badly bomb-blasted village on the east side of Etna – to the little town of Linguaglossa (tongue in two lingos!), when I heard coming from a little piazza on the outskirts of the town the unmistakable sound of a massed pipe band. I hadn't heard a massed pipe band since Libya – since the big parade for Churchill in Tripoli, in fact – and although I was supposed to report to Lang Tam Wimberley at Main Divisional HQ – I decided to take five or ten minutes off, saying to myself: 'The bloody campaign's over – Lang Tam will just hae to wait!'

Getting my driver to park the jeep, and asking him to stay with it, I moved forward with some difficulty through a dense crowd of enthusiastic Sicilians, shouting things like, 'Viva la Scozia!' and 'Viva i Scozzesi', till I got to the top of the approach road leading to the piazza, and saw there a magnificent and heart-warming sight. It was the massed pipe band of 153 Brigade – two Gordons battalions, and one Black Watch battalion – and they were playing the beautiful retreat air 'Magersfontein'. Presiding over the occasion was the immense bulk of Etna, with a plume of smoke drifting lazily from its crater.

In the silence after the retreat air finished I stood wondering what the Pipe Major was going to get his boys to play for a March, Strathspey and Reel. When they struck up again, the March turned out to be one of my favourite pipe tunes – 'Farewell to the Creeks', a tune composed during the First World War by Pipe Major James Robertson of Banff. And while I listened to it, words began to form in my head – particularly one recurrent line 'Puir bluidy swaddies are weary'.

And they were too! Since Alamein some of the companies of that splendid division had been more or less totally 'made up' due to heavy losses – and I knew that shortly they were going home, presumably to take part in yet another D-Day in north-west Europe. By the time I had elbowed my way through the crowd back to the jeep, I had the beginnings of a song half completed; that night it had its first airing in a Gordons Officers' Mess, and I was soon scribbling the words out in pencil for all ranks. It took off with amazing speed – and, in the event, preceded me back to Scotland. When I was collecting songs with Alan Lomax in the northeast in 1951, we were occasionally offered it – sandwiched between classic ballads, lyric love songs and comic ditties – by ex-soldiers who had been in Sicily!

And now we backtrack again – this time to Tunisia, where – after the fall of Tunis and Bizerta on 7 May 1943, and the capitulation of the Axis armies under Von Arnim – the Highland Division had been training hard for the invasion of 'somewhere' in southern Europe. Three Greek officers were allotted to the Division, as part of an elaborate cover plan to deceive the enemy as to where we were going to invade. And when they were informed, once they had embarked, that they were heading for Sicily, they naturally became more than a little disgruntled, and I was given the task, on 9 July, of calming them down. I assured them that it would not be long before they were all back in Alexandria – the great Greek city of the ancient world!

On the evening of 6 July 1943 I had embarked at Sfax, on the south-east coast of Tunisia on LST (Landing Ship Tanks) No. 8, which was to carry us back to Europe. On deck, the following morning, I counted some sixty vessels visible to me, riding in the Sfax searoads. This was only a smallish percentage of our convoy, which was only a small percentage of the whole. The atmosphere was quite calm: rather like a Mediterranean cruise!

Our deck was crammed with vehicles; you could hardly move edgeways. We got an issue of Yank lifebelts, to go with the LST, which had only recently left the shipyards of Mr Kaiser.

About 17.20 hours on 8 July we up-anchored, and sailed north in a well-dispersed convoy of 600 vessels. Later that evening, Brigadier Davey, Chief Engineer of XXX Corps gave us a pep talk on the forthcoming battle, and read us Monty's message. Up to then the weather had been calm, but on 9 July a juicy roll developed, which got juicier through the day. I surveyed the convoy from the upper deck, and read up once again the enemy order of battle, as we knew it at that time; I idly committed to memory the units of 206 Coastal Division, which were due to enter the bag the following day.

During the night – the last night before D-Day! – the roll got worse and the whole ship shuddered. I stayed awake, and two lines of poetry – of sorts – started circulating in my head:

> To Sicily, to Sicily
> Over the dark moving waters.

Shortly I had my first stanza:

> Armour, vehicles and bodies
> Make heavy cargo that is checked and away
> To Sicily, to Sicily
> Over the dark moving waters.

This turned out to be the first stanza of the first part of what developed gradually into a quadripartite poem; I worked on it in a desultory fashion over the next months, but did not get it finished to my satisfaction until the Eighth Army was up on the Sangro Front, on the east side of Italy, and I was temporarily attached to 8th Indian Division, as an Italian and German-speaking intelligence officer. My first title for it was 'Ballad of Sicily'.

Some months later, when I was on the Anzio beach-head, I was surprised to get a letter from Nicholas Moore – the same Nicholas who had called me 'an incipient Fascist', because I had said in the CUSC clubroom in Cambridge that, irrespective of his politics, Roy Campbell was a bloody good poet – and this letter included a request for contributions to a magazine he was editing called *New Poetry*. (He had got my umquhile address – care of the Highland Division – from G. S. Fraser, who by that time was working in Asmara, in the offices of the *Eritrean Daily News*.

I decided to send 'Ballad of Sicily' to Nicholas, and posted it almost by return; later in 1944 I received a copy of *New Poetry*, No. 2 and found myself, for the first time, in the company of Conrad Aiken, Wallace Stevens, and other poets whom I had long admired from afar.

Incidentally, I have now rechristened it 'Ballad of the Simeto', to avoid confusion with the *Highland Division's Farewell to Sicily* aka *Banks of Sicily*. Primasole Bridge, over the River Simeto, was the scene of the fiercest and bloodiest fighting of the Sicilian campaign.

THE HIGHLAND DIVISION'S FAREWELL TO SICILY

The pipe is dozie, the pipie is fey,
   He winna come roon for his vino the day:
The sky ow'r Messina is unco an' grey,
   And a' the bricht chaulmers are eerie.

Then fareweel ye banks o' Sicily,
   Fare ye weel, ye valley an' shaw.
There's nae Jock will mourn the kyles o' ye.
   Puir bluidy swaddies are weary.

Then fareweel ye banks o' Sicily,
   Fare ye weel, ye valley an' shaw.
There's nae hame can smoor the wiles o' ye.
   Puir bluidy swaddies are weary.

Then doon the stair an' line the waterside,
   Wait your turn, the ferry's awa'.
Then doon the stair an' line the waterside.
   A' the bricht chaulmers are eerie.

The drummie is polisht, the drummie is braw,
   He cannae be seen for his webbin' ava.
He's beezed himsel' up for a photy an' a'
   Tae leave wi' his Lola, his dearie.

Sae fare weel, ye dives o' Sicily,
   (Fare ye weel, ye shieling an' ha').
We'll a' mind shebeens and bothies
   Whaur Jock made a date wi' his dearie.

Then fareweel, ye dives o' Sicily
   (Fare ye weel, ye shieling an' ha').
We'll a' mind shebeens and bothies
   Whaur kind signorine were cheerie.

Then tune the pipes an' drub the terror drum,
   (Leave your kit this side o' the wa'),
Then tune the pipes an' drub the tenor drum,
   A' the bricht chaulmers are eerie.

BALLAD OF THE SIMETO
(For the Highland Division)

I

Armament, vehicles and bodies
make heavy cargo that is checked and away
   to Sicily, to Sicily
   over the dark moving waters.

The battalions came back
   to Sousse and Tunis
through the barrens, and the indifferent
   squatting villages.

Red flower in the cap
   of Arab fiesta!
and five-fold domes
   on the mosque of swords!

We snuffled and coughed
   through tourbillions of dust
and were homesick and wae
   for the streams of Europe;

but the others were blind
   to our alien trouble,
they remembered the merciful,
   the Lord of daybreak.

Launches put out
   from palm-fuzzed coastline
to where landing craft lay
   in the glittering sea-roads.

The ships revolved
   through horizons of dusk –
then foregathered like revenants
   from the earth near fresh graves

Och, our playboy Jocks
   cleaned machine-guns and swore
in their pantagruelian
   language of bothies

and they sang their unkillable
   blustering songs,
ignoring the moon's
   contemptuous malice.

All the apprehensions, all the resolves
and the terrors
and all the longings are up and away
    to Sicily, to Sicily
      over the dark moving waters.

II

Take me to see the vines
    take me to see the vines of Sicily
for my eyes out of the desert
    are moths singed on a candle.

Let me watch the lighthouse
    rise out of shore mist. Let me seek
on uplands the grey-silver
    elegy of olives.

Eating ripe blue figs
    in the lee of a dyke
I'll mind the quiet of the reef
    near the lonely cape island

and by swerve of the pass
    climb the scooped beds of torrents
see grey churches like keeps
    on the terraced mountains.

The frontier of the trees
    is a pathway for goats
and convenient sanctuary
    of lascive Priapus.

Over gouged-out gorge
    are green pricks of the pines:
like strict alexandrines
    stand the verticle cypresses.

O, with prophetic grief
    mourn that village that clings
to the crags of the west
    a high gat-toothed eyrie

for the tension in the rocks
    before sunset will have formed
a landscape of unrest
    that anticipates terror.

A charge has been concealed
    in the sockets of these hills.

It explodes in the heat
of July and howitzers.

We are caught in the millennial
conflict of Sicily.
Look! Bright shards of marble
and the broken cornice.

III

On the plain of dry water-courses
on the plain of harvest and death
the reek of cordite
and the blazing stooks!

Like a lascar keeking
through the green prickly pears
the livid moon kindled
gunpowder of the dust

and whipped to white heat
the highland battalions
who stormed, savaged and died
across ditches called rivers.

Battle was joined
for crossing of the Simeto.
Pain shuddered and shrieked
through the night-long tumult.

But aloof from the rage
of projectiles and armour
our titan dreamed on
into brilliant morning,

and suspended from clouds
was asleep like a bat
in the ocean of summer
a blue leviathan.

Yes, Etna was symbol
of the fury and agony.
His heart was smouldering
with our human torment.

– Drink up your fill
parched trough of the Simeto
for blood has streamed
in too wanton libations

and tell of the hills
　　hard mastered Dittaino
where broke with our onslaught
　　the German iron.

A bonnet on two sticks
　　is tomb for the Gael.
Calum Mor, mak
　　mane for Argyll.

Infantry Combat: a GI in France, 1944

Steve Weiss

What the author, Samuel Chamberlain called, 'The poignant days of August 1944', I discovered to be the beginning of a dangerous and exciting episode during my wartime army career.

I first saw the coast of southern France, as a young combat soldier, from the deck of a landing-craft that jolted across the Gulf of Fréjus, in the tenth wave, toward the invasion beach near St Raphaël. It was 09.45, D-Day, 15 August 1944; I was eighteen years old, a first-scout in a rifle squad, part of the 36 'Texas' Infantry Division.

Off to starboard I noticed a small volcanic island, the Isle d'Or, on which was perched a fortified medieval tower; beyond, past Cape Dramont, as the coast curved away to the east and then rose steeply in height, were the sea-girded red-stained cliffs of the Esterel mountains. They were thickly covered in pine and cork forests, and as I had been told, a scenic road, the Corniche d'Or, ran through them to the seaside playground of Cannes.

Against this indivisible canvas of sea and sky, mighty warships like the *Texas*, the *Arkansas*, the *Tuscaloosa*, the *Brooklyn* and the *Emile Bertin* fired their weaponry at the enemy fortifications on the beach and deeper inland. The whooshing, rushing and rattling sounds of heavy calibre shells and rockets, mixing with the staccato hammering of machine-guns, shattered the earlier stillness of the grey morning. Allied fighters roamed overhead and bombers crossed above us in large formations to attack enemy targets beyond. I learned later that Winston Churchill, the British prime minister, observing this mighty display of firepower from the bridge of the British destroyer, *Kimberley*, puffed on his cigar and grunted in satisfaction, although he had opposed the landing in preference to action elsewhere.

Our British guided landing-craft tore through the surf and hit the beach at 09.45, just as a luminous sun dispersed the haze. The baggage that I carried ashore was typical for an infantryman at war, and most of it hung from webbing and leather straps that criss-crossed the upper part of my body like a plexus of tracks. Bandoleers of thirty-calibre ammunition, a canteen and hand grenades clinked, rocked and bumped against me, and glinted in the increasing sunlight.

I crossed the beach frightened but ready to fight. Nevertheless, I had the disconcerting thought that if an enemy bullet ignited the flame-thrower on my back, I would no longer be a part of Camel Force (composed of the 3rd, 36th and 45th Infantry Divisions), but a fragmented, unrecognisable part of the Côte d'Azur.

What followed was a series of skirmishes, but St Raphaël, its airfield and the town of Fréjus fell to our regimental assault the same day. There were few casualties. History has included the little fishing harbour of St Raphael in two great Napoleonic epics. Nepoleon arrived here triumphantly from Egypt in 1799, and embarked from here after his downfall in 1814. Both events are memorialised by a porphyry pyramid in the town square. Our arrival, 146 years later is commemorated by our own monument placed a few miles away above Green Beach.

On 17 August we captured Draguignan, a dusty town of 20,000 inhabitants rejoicing in our arrival. Draguignan, capital of the Var Department, lies about fifteen miles inland on one of the innumerable plateaux of 'inner Provence'. With enemy defences crumbling, the war became a war of movement, and early next morning, we travelled north-east in a powerful armoured convoy to gain the Castellane road above Grasse. There was an aroma of roses and lavender in the clear, bright air, as we trundled by. Napoleon had travelled this same mountain road when he returned from Elba in 1815. Compared to the coastal mountains and plateaux of inner Provence, this land is more mountainous, more rude, and more primitive. Rainfall is moderate and in August a Mediterranean drought invades these Alps almost as far inland as Grenoble.

In our headlong advance, I had no time to dwell on our recent landing and the beautiful scenery surrounding the D-Day beaches. We drove deeper into the foot-hills and then climbed higher into the bare and rugged Alps. Armed Frenchmen, dressed in baggy trousers and ill-fitting mismatched suit jackets, some with home-made cigarettes dangling from their lips, with rifles in their hands, protected our route and the heights above it from enemy surprise attack. These men and women represented the French Resistance. Thanks to them, we travelled ninety miles in fourteen hours on the Route Napoleon; in the process, we traversed breathtaking gorges, passed cascading waterfalls, and slowed through tiny hamlets. We searched for an enemy who continued to retreat.

Instead, we found villagers and townspeople who clustered beside the road, waving and throwing flowers, and shouting words of encouragement, as we sped by. If we paused for a moment's respite, amidst our own sweat and dust, they would run to greet us, arms outstretched, with tears of joy streaming down their cheeks. Five years of emotional deprivation expressed itself in a volcanic surge of feeling which engulfed us all. I was hugged and kissed until my face and ribs ached. When I shared my rations with these people or made small-talk with pretty girls of my own age, the melancholy of war seemed very far away. Amidst their wine and our cigarettes, against a backdrop of olive-green tanks and trucks, we struggled for the appropriate word in either language that somehow never expressed or matched the moment. When all towns along the highway, Castellane, Digne, Sisteron and Gap were set free in four days, the spirit of liberation was sublime.

In 1815 Napoleon returned from Elba, and at the little town of Laffrey, a few miles south of Grenoble, he was acclaimed rather than arrested by the old French soldiers chosen for this purpose. The people received him with wild enthusiasm, and their feelings of acceptance which began in these Alps, echoed throughout the rest of France.

We came as liberators not as emperors, but the acclaim we received was the same. When our motorised column rumbled into Grenoble on 22 August, thousands of people filled the streets to greet us, a huge exuberant crowd shouting its welcome, rejoicing in its freedom. French flags waved triumphantly, and the populace of this old university city, which is surrounded by snow-covered mountains and hemmed in by two bustling rivers, invited us to join them in joyous celebration. There was singing and dancing in the streets all through the night, made all the more exciting by the light of numerous bonfires. In all this jubilation, I was a young and willing participant, and the war seemed to end in momentary triumph.

I was greeted with applause wherever I went, by women queuing for bread outside their bakery, by locals sitting in their café, offering drinks, by smiling people rushing by. Having the urge to go bike riding, I found a willing proprietor who refused to take payment. He thrust the bike upon me, as a gesture of good will, insisting that since the Germans never paid, I certainly need not. 'Just return the bike when you're finished', he said. Some young French women, sitting sullenly with their heads recently shaved, retribution for fraternising with German soldiers, seemed relieved that the enemy had gone.

One attractive young woman beckoned to my comrade and me from her first-floor flat. With an eye for romance, we entered a bare room, devoid of any furniture except for a standing corner cabinet with a glass front; near the window, a few potatoes were scattered on the bare wooden floor, and a man stood silently in the half-light close by. The woman went to the cabinet and retrieved a large bottle of cognac, empty except for a tiny amount of the liquor. She found four small glasses, gave us each one, and then poured a drip into each glass until the bottle was empty. Although my French was extremely limited, I believed she said that the silent man standing next to us was her husband. Ever since the German occupation, he had guarded this dram of cognac, longing for the day when the Americans would come. He ardently believed they would. Now that his ordeal was over, his face reflected a mixture of feelings. All of us toasted the liberation, I in my schoolboy French, and he by sobbing in relief. The woman reached out to comfort him, muttering words of compassion, embracing him, the silence ended. Overcome, we left.

Thus far, it seemed as if our current army service consisted of a Mediterranean peacetime cruise and a motor tour of southern France. Two days later, we descended into the lush valley of the Rhône, replete with its famous vineyards and fruit orchards. During the late afternoon of 24 August, Company C, in which I was assigned as a first scout in a rifle squad, in units of twelve men, mounted assigned tank-destroyers to move west from Chabeuil towards Valence on Route D 68. The tank-destroyers moved cautiously down this road in single file. In the vicinity of an airfield athwart our line of advance, we ran into heavy German rifle and machine-gun fire. One of my friends was killed instantly and fell off the lead tank-destroyer in front of me.

Trying to leap out of harm's way, we sought cover in the roadside ditches; supported by the tank-destroyers, we attacked toward a number of German-held houses; in the ensuing battle we set them alight. Moments later, I manned a fifty-calibre machine-gun atop one of the tank-destroyers, but I missed hitting two Germans escaping across the flat farmland in the direction of Les Martins. The skirmish at an end, we regrouped in squad formation and moved on, covered on either side of the road by the tank-destroyers. On the night of 24 August, a night made heady by the fragrance of violets and raspberries, our elusive enemy chose to stand and fight. The fragmented whole of the German Nineteenth Army was trying to escape northwards. In open farm country with the night as black as pitch, little did I know that a violent and desperate battle was about to begin with its rearguard. As first scout, leading the rest, the enemy struck at us again, killed two men, and then quickly withdrew. Night began to fall; as I was peering into the blackness, a nearby voice shouted in German, 'Halt', immediately followed by two rifle shots. I could have reached out and touched the yellow muzzle flashes. I hit the ground immediately, still groping for the enemy in front of me. Lying there, I waited for the men behind me to fan out to form a skirmish line, but no one came. Alone and exposed, I decided to make a break for it; as I got up to run, the German tossed a grenade. Fortunately, my timing was excellent. I deduced later that the three grenades he tossed were spinning through the air, as I got up to run. Each grenade disintegrated beside me, as I hit the ground; otherwise, I would have taken the full force of any one of the explosions in an upright position. After the last grenade exploded, its fragments showering over me, I landed safely in a ditch beside my company commander, Captain Allen E. Simmons of Belfast, Maine.

He ordered my squad back into the open field I had just left to continue the attack. Moving towards our objective, a line of trees about 150 yards away, I heard the tank-destroyers moving noisily down the road. Lost was whatever surprise we had. When the Germans heard the engines, they let loose with a salvo of automatic and self-propelled cannon fire. Three of the tank-destroyers were hit immediately, burned and exploded. Tree and shell fragments caused heavy casualties. Intermittent explosions dispelled the darkness. Punctuated by gun flashes and explosions, underscored by shouts and yells, and the grinding of tank tracks, Charlie Company had been surprised and badly mauled. Out in the field, the eight of us moved slowly towards the treeline. We could see the German gun crews firing, silhouetted against their own gun flashes. A flare burst above us and all became sickly greenish daylight. We stopped, waited, and slowly sank to the ground. Caught in a withering cross-fire, enemy machine-guns blazed away at us, pinning us down, but hitting no one. We knew it was impossible to go further.

Realising that we were too exposed to stay where we were, we pulled back and found an irrigation ditch about sixty yards to our rear. Not only were we lost and possibly surrounded, but Charlie Company seemed to have disappeared. During the night, the cries of the wounded were heard above the exploding shells, most of which were our own. Ashes, smoke and the acrid odour of cordite filled the night air, making seeing and breathing difficult. The three immobilised tank-destroyers burned and popped and exploded throughout the night. 'Missing in Action Telegrams' sent to our families who knew nothing of our plight, would relay their stark message, 'We regret to inform you that your son...'

As the dawn of the 25th approached, after a night of constant shelling, with our company having pulled out, we lay bedraggled in an irrigation ditch, surrounded by an unseen enemy. Morning came, clear and bright. One German watched us from a nearby stone wall, seemingly noting our every move and gesture without sounding an alarm. We wore our knees and elbows raw crawling up and down that ditch. If we remained where we were, we would either be killed or captured. 'Where were you when we needed you, John Wayne?' We were angry at the company for abandoning us. Sergeant Scruby, our assistant squad leader, saw a farmhouse beyond an adjacent peach orchard, and thought that either might provide the shelter we needed. Exhausted, we held a desperate meeting in the ditch. It was agreed that Scruby would try to gain the refuge of the farmhouse. If he succeeded, we were to follow from the same point of the ditch, one at a time. Scruby raised up, crouched, and ran across thirty yards of field into the orchard and disappeared. Emerging from it, he disappeared again into the farmyard. I was pulling for him all the time; as if it were a horse race, I kept repeating over and over, rocking back and forth, saying, 'Come on, Scruby; come on Scruby.' Minutes that seemed like hours elapsed, but at last he signalled from the shadow of an upstairs window of a barn-like building. One by one, Reigle, Gualandi, Wohlwerth, Fawcett, Garland, Caesar and I followed his instructions and successfully crossed over.

As we gathered in the farmyard, we surprised a farmer, his wife and young daughter. Startled, they thought we came as liberators and not as desperate men. Speaking above a whisper in my broken high-school French, I explained the gravity of our situation, adding that we would not give up without a fight. When the rest of the men climbed the stairs leading to the hayloft, I remained to stand guard. Across my rifle sights soon after, I followed a cyclist in a blue uniform approaching the farm; out of sight, I considered killing him. Deciding against it for the moment, I watched him closely, as he crossed the yard to talk in low tones with the farmer. Their conversation ended quickly, he left, and I joined the rest of the men above. The farmer assured us that he would help us regardless of the risk. Even though the hay was comfortable, the air was heavy with an almost unbearable tension. I considered it more suspenseful than any of the Hitchcock films I'd seen before I enlisted, but my thoughts were interrupted by machine-gun fire close at hand that frightened the farm animals. Later we discovered that the Germans were killing some of the abandoned American wounded and that SS patrols were prowling in the vicinity.

Around mid-morning, a car approached the farmyard and stopped astride the hayloft stairs. Doors slammed, boots grated on the stairs, and voices shouted, 'Allons, allons!' We grabbed our rifles, as three excited men, followed by the farmer, burst into our preserve. Their leader, a man in civilian clothes, indicated that he and the two others, dressed in the same blue uniform worn by the cyclist, would help us escape disguised as French policemen. Since they had four uniforms between them, we would have to make two trips. To refuse was to remain and fight it out. We agreed to their plan, but if it failed, as bogus French police officers, we would be shot as spies.

Rushing, we realised that the uniforms did not fit: the 'kepi' was too small, the jacket too tight, and the trousers too short. One of my squad, squeezing the police uniform over his own, remarked, 'When we get back home, no one will believe us; it's really like the movies.' Four of us volunteered to stay behind, as the others, dressed as French 'keepers of the peace', raced down the stairs, jumped into the black Citroën police car, and drove away to an unknown destination. I was in the second group. Three of us hid all our gear and personal evidence deep in the hayloft, while the fourth kept a lookout through the cracks of the shutters, awaiting the police car's return. After the first group left the tension in the hayloft almost reached unbearable proportions. Some neighbouring farmhouses were set ablaze by marauding German patrols. Our turn would come soon, if the black Citroën failed to return. Forty-five minutes later, the car pulled up, and the original procedure was repeated. As we sped away, I realised I didn't know the farmer's name. Fifteen minutes later, an armed German patrol surrounded the farm and broke into the hayloft. The SS officers grabbed the farmer and his family and severely questioned them at gun-point. The farmer, his wife and daughter revealed nothing.

I assumed that we would return to Chabeuil, and was shocked when we entered enemy-held Valence instead. Passing slowly through the town, Germans peered into the car, close enough for me to touch them. They turned away in disinterest. Alongside a barrack-like building ranged American prisoners, standing in loose formation. Further on German officers, holding maps, surveyed their defences. In the centre of the town, the driver stopped the car at a nearby café; the civilian directing the operation was leaving. In a well-meaning but foolish gesture, I gave him a small packet of American cigarettes, its logo on view for any passer-by. Fortunately, no one saw this, but I was severely reprimanded by my three comrades for putting us all at risk.

South of the town and along the Rhône river, the adjacent land was dotted with farmhouses. Before we closed on one particular farmhouse, a shepherd guarded his sheep. The car slowed, and after an exchange of prearranged signals between him and the driver, it rolled forward. I was impressed by their cautious and deliberate behaviour, upon which our safety depended. Quickly exiting from the Citroën, we followed the policemen and ducked into the farmyard. Here, unseen by hostile eyes, and much to our delight, we were reunited with our comrades. Their journey was more harrowing because the driver, taking another route, ran into two German roadblocks; at both he was questioned intensely, but waved on. The stress was too much for him, and another officer took his place for the second journey. To reach momentary safety, we had entered a world of signs and counter-signs, of cover stories and cover names, of agents and stratagems. We discarded our police uniforms, and were given a small Italian automatic pistol each in return. Within minutes of our arrival, word was received that the Germans had discovered our escape and were after us. The eight of us were ordered to leave immediately.

We ducked out of the farmhouse, and followed our French guide to the banks of the nearby Rhône, a broad, swift-flowing river. Two small wooden boats awaited us, manned by other members of the Resistance. Each boat was heavily laden; we were defended by an old grizzled Frenchman, grasping an old, round-drum submachine-gun. The swiftness of the river carried us downstream, the pursuing Germans were at our backs; I noticed that all the major bridges were blown. Reaching the far shore, we leapt from the boats and charged up the steep bank, quickly crossed a road, and reached a two-storey turn-of-the-century house, hidden from view by its surrounding trees. For the first time in two days, protected by our anonymous friends, we were allowed to grab a few hours' sleep. Years later, I learned that for every Allied soldier helped to escape, a French, Belgian or Dutch Resistant died. There were eight of us.

As soon as darkness fell, guides shepherded us through the village of Soyons, then north for a few miles up the river to the small town of St Péray, known for its sparkling wines. In the back room of the Café du Nord, used as the local Resistance headquarters, we were introduced to some of its members. They questioned us repeatedly about our experience, and we were asked to trace our movements on a large map laying on the bar table. My limited French didn't help matters, and I had a distinct feeling that if we were in any way suspect, our lives could be jeopardised. Their preliminary doubts were not surprising, because with all bridges blown across the Rhône, with most communication facilities destroyed, in a country divided ideologically, informers and double-agents were not uncommon. To protect a Resistance network from being infiltrated or destroyed either by the Germans or French traitors, counter-measures, such as the questioning and others we had recently experienced, were taken. 1944 was the year of rough frontier justice, when old scores were settled between Frenchmen with guns and grenades, and proof of who you were and what you had done was a matter of life and death. Finally, the questioning stopped, for reasons unknown to me, but our friends seemed satisfied with our answers. They shook our hands, served up the local Ardèche red wine, and slapped us on the shoulder.

Although it was already late at night, we faced another journey. Remnants of a whole German army were approaching, in full retreat along the road we had recently followed. To avoid capture, we were taken by truck to the small roadside mountain village of Alboussière, sixteen kilometres west of St Péray. The small, three-storeys, brown-stuccoed Hôtel Serre, with its adjacent walled garden and terrace, was the central Resistance headquarters for the Ardèche Department. Here, we were introduced to Captain François Binoche, a regular French Foreign Legion officer serving General Charles de Gaulle in this area of central France. Captain Binoche, as Resistance leader, had lost an arm in action against the Germans only a few months before. For the next few weeks, under his leadership, we attacked the Germans over terrain that looked like Spain or California. Frequent British air drops brought in weapons – rifles, bazookas, submachine-guns, and one fifty-calibre machine-gun. Attempting to demonstrate the use of the bazooka to some members of the local Resistance, I fired, missed hitting the side of a farmhouse, and killed some of the farmer's sheep beyond it. I felt decidedly foolish, but we all had a good laugh, except the farmer, who was irate.

Somewhere a battle was taking place, because we could hear our own planes pounding the river valley not far away. To stem retreat, the German Nine-teenth Army sought alternate routes to reach either the Vosges hills or Germany itself; Captain Binoche denied the route through Alboussière to them. Alboussière straddles an important east–west secondary road and lies between the Rhône and the volcanoes of Le Puy. He assigned each one of us a task in the destruction of a key road bridge not far from the village. We drove east down the mountain in the darkness in two trucks to a point about midway between Alboussière and St Péray. Some of us placed the explosives, others protected the men setting the charges. As soon as the plunger was pressed, the bridge blew up and plunged into the gorge below. Mission accomplished!

There were brief moments of relaxation in the wake of the tension and stress. Sometimes over dinner, in the small rustic dining-room with its open-beamed ceiling, I would chat within my limited bounds with Captain Binoche, some of his staff, and to the hotel owners, M. and Mme Serre. From the vineyards near Avignon, I sampled glasses of Châteauneuf-du-Pape that tasted of onion to me. Captain Binoche and his men laughed; what did I know about a fine French wine? But years later I discovered that onions were planted between the grape vines, adding to its distinct flavour. I struggled with French, attempting to converse with many of the guests, who like the Haas family and Jews from Paris, were refugees. If some could no longer pay, the Serres continued to provide food and shelter; payment would come, if at all, after the long-awaited liberation. Elise and Simone, the two adolescent chamber-maids were friendly and attractive, but there was no time for romance.

On a sunny August afternoon, looking out of the garden at the Hôtel Serre, I noticed a man, accompanied by two others, being led into the village, wearing khaki shorts, a white open shirt and black shoes. His hands were tied behind his back and a halter hung from his neck. He was soon surrounded by some of the local Resistants and the residents of the hotel. From this time to the last time I saw him, he never uttered a word. We were told that he was a French collaborator, an informer under sentence of death. During my time with Charlie Company in France, we had taken several German prisoners, but he was the first collaborator that I had seen.

The firing squad was quickly chosen and the site of the execution selected. Having been asked to participate as a member of the firing squad by one of the locals, I declined because I knew nothing about the case or the role of French justice. The execution was planned for the middle of the afternoon; I noticed that some members of the firing squad with whom I sat over lunch, drank more than their usual amount of wine and their conversation seemed more animated. I stood with my comrades among the villagers and hotel residents, facing the church wall. Adjutant-Major Fernand Mathey of the local gendarmerie lined up the firing squad and called for the prisoner. Hands still tied, dressed as he had been the first time I'd seen him, the man was led to a position between the wall and the firing squad. Offered a blindfold, he shrugged the man away. I was in awe of the control he imposed upon himself. There was something noble in his bearing.

Mathey shouted the required sequence of orders, the firing squad stiffened, and at the command, 'Fire!', the bullets crashed against the man and knocked him down. He twitched, still alive. One of the firing squad ran forward, placed his rifle at the man's ear and pulled the trigger. There was a click, but no explosion. He had neglected to reload. Mathey rushed up, aimed at the man's body with his forty-five and fired. Nothing happened, the man writhed on the ground. Another rifleman fired; the man stopped moving. The last time I saw the dead man, he lay under a pile of hay on a small horse-driven cart, awaiting burial. Someone had stolen his black shoes.

One evening at dinner, Captain Binoche asked me if I'd accompany Mathey on a rabbit hunt the next morning. Of course, coming from a big city, I had never hunted, but thinking it would be fun, I agreed. When I joined Mathey the following morning, I found a contingent of twenty armed men already with him. One of the men handed me a rifle and some ammunition. I enquired of the major if it were necessary to gather so many heavily armed men to hunt a few timid rabbits. He roared with laughter, and said, 'Rabbits? We're hunting Germans!' I stood there dumbfounded, as he cleared his Sten gun. Why had Binoche chosen me rather than one of the other Americans? I thought, but it was too late to find out now. We were on our way.

Walking in loose formation to the outskirts of the village, we followed one of the tracks that overlooked the valley. I was reminded of the Spanish landscape in Hemingway's novel, *For Whom the Bell Tolls*. Half an hour later, we gathered at the edge of the road and stared at our objective down below; it was another two-storey stone farmhouse. Local intelligence reported there were Germans within. Mathey fired his Sten gun in the direction of the farm as an act of defiance and my heart sank, knowing that all surprise was lost. The rest of the men were noisy and spirited. We left the road, cautiously approaching the house with the men rushing and covering each other. The farm was well built for defence, and I anticipated one hell of a fight. Every window, every door, of which there were many, could be a gun-port for the enemy. Suddenly a door opened and three French civilians emerged with their hands up and told us that the Germans had left about a half hour ago. We found no rabbits, no Germans, and I was glad.

The group disbanded, but Mathey, two other men and I took a car and drove from village to village searching for any Germans in the vicinity. The villagers kept pointing east toward the Rhône, and we moved closer to that objective. For ten hours, we had not stopped to eat or drink. Reaching Soyons again, we parked the car; Mathey and the two other men walked into the village. I stayed behind to pump some needed air into one of the tyres. While involved, I heard a muffled explosion; hit by the blast, I was slammed against the front of the car. Confused, I rushed forward fifty or sixty yards to discover a terrible, eerie scene in the fading light of day. Men, French and German, were strewn all over the road. Mathey walked toward me dazed, wounded in the shoulder, and seven or eight men lay dead.

The retreating Germans had mined some trees along the road, and one of the trees, either by design or accident, had fallen across the road. Just before we arrived, men of the local Resistance, assisted by some German prisoners, were attempting to clear the obstacle when the mine exploded. It was dark now, and an old woman, obviously terrified, grabbed me by the hand and pulled me toward a house near the blast site. Inside, I found a young Frenchman bleeding profusely from a thigh wound. I rushed outside for help, but none was available. Using our car as an ambulance, we took the young man and some other wounded to a local cottage hospital. I stood guard over a wounded German, waiting his turn for treatment. We looked at each other, sitting in the cramped seats of the little car, both shocked at our misfortune. We spoke in halting French, looking at small wallet-size photos of our families, and despaired of war's disruption.

Weary, on the move for fourteen hours, I returned to find that my comrades had moved out of the hotel into a place that seemed crowded and stuffy in the hot summer night. In my absence, a paratrooper captain from an OSS (Office of Strategic Services) operational group operating further west had come seeking our help. I wanted to stay with Captain Binoche, regardless of the rabbit trick, but I was outvoted by the others. We left the next morning by car, driven by a former French Air Force pilot. He drove the car as if he were manœuvring a P-40 fighter plane at breakneck speed over narrow, winding mountain roads. I doubted if we would reach our destination alive.

Suddenly, he slowed down and came to a halt by three Resistance men who, as if on cue, emerged from the wood bordering the road. Enquiring if there were one of us from Brooklyn, when I said, 'Yes', much to my surprise, a woman left the wood and walked towards us. I was deeply moved when we discovered our Brooklyn homes were within walking distance of each other. Here we were meeting for the first time 3,000 miles away in a remote part of France. As an enemy alien, she had been on the run from the Germans for three years. Our chance arrival was the beginning of her personal liberation.

We shared our remaining cigarettes with her, exchanged farewells, and then watched her disappear into the wood as silently as she had emerged.

We joined the OSS Captain and his men of the 2671 Special Recon at a farmhouse near the village of Devesset. I was assigned to guard the radio operator when he transmitted to or awaited signals from Algiers. German direction-finding vans patrolled the area seeking our position. In need of more men and supplies, the captain alerted Algiers via radio. Four nights later, we positioned ourselves around a drop zone, awaiting a black four-engine Liberator bomber carrying another operational group and their supplies. In the distance, we heard the low drone of an aircraft. The Liberator circled overhead and exchanged recognition signals with us before its contents were toppled out of the sky, 500 feet above us. As the plane flew away, we gathered the new arrivals about us, and returned to the farmhouse. These reinforcements were to assist in an attack on the 11th Panzer Division, somewhere south of Lyons. For some time now I had been suffering from stomach flu so I was scratched from the operation. Once the operation was completed, my comrades would continue on to find the whereabouts of Charlie Company. War had come to an end in the Ardèche. A few weeks later, I made my way back via Grenoble and visited other OSS units gathered there. I was introduced to the Dahans, a Parisian couple in Lyons, and spent ten days with them recuperating. Vowing to meet again in Paris when the war ended, I didn't realise that the war would continue for another nine agonising months, made all the more difficult by the worst winter in fifty years; I found Charlie Company dug in along the Vosges mountains in early October, just before I reached my nineteenth birthday. Only two of the men with whom I had been surrounded were still with the company, the others had been wounded and evacuated. Scruby had lost a leg, spent a year in an army hospital in Texas, and died of his wounds in 1946. Gualandi stepped on a mine and lost a foot. He died twenty-five years ago. Wohlwerth was wounded in the chest and died of a heart attack in 1968. Reigle and Caesar are still alive; I've no idea where Fawcett and Garland are.

Epilogue

As an army photographer living in Paris during the summer of 1945, I covered news stories throughout western Europe. Happily I was reunited with Captain Binoche, the Dahans, and the Haas family in Paris. I met Captain Binoche in a revolving door at the Hotel Majestic. We looked at each other through the glass partition, momentarily transfixed, as the door slowly turned, then stopped. Seconds later, we were on the sidewalk, enveloped in the thrill of meeting again. I tracked down the Dahans through a relative, an old address, and a kind *concierge*. Thinking I'd surprise the Haas's, I arrived at their flat and rang the bell. I was greeted with shocked amazement when they let me in. It seemed that when their teenage son, Jean-Claude, had made enquiries the day before about me at army headquarters, he was told that I had been killed.

Before I returned to the States in November 1946, I spent a few tranquil days at Alboussière with the Serres, Elise, Simone and Mathey. We reminisced and toasted each other with glasses filled with wine from the nearby slopes of the Hermitage and the Côte-Rotie, overlooking my beloved Rhône.

During the summer of 1972 when my family and I travelled through France, I invited Mathey and his family to lunch with us at Alboussière. Twenty-eight years had passed since he and I had last met... Our meeting made the region's newspapers and one of the policemen who had driven the escape car telephoned the farmer about it. 'He must be one of our Americans,' the farmer replied. Our meeting, standing in the farmyard again, looking at the field, the peach orchard, the irrigation ditch and the hayloft evoked deep emotions. The farmer, Gaston Reynaud, Madeleine, his wife, and Claudia, his daughter, Marcel Volle and Richard Mathon, the two police drivers, and all their friends embraced me in thought and affection, an experience in which we all participated. I was given Scruby's helmet, the one he left behind, as a memento.

Madame Serre and Simone are companions and still live in Alboussières, across from the hotel, which is now in ruins, the garden a shambles. Elise is married to Paul Valette and they live in St Péray, their children grown. When I met her at the Mairie more than two decades later, I introduced myself, as she came down the marble staircase. 'Oh Stefan,' she said, 'I'd know you anywhere.'

Gaston told me that it was Louis Salomon, the police cyclist I had nearly shot who had come from Valence on 25 August to assess the results of the previous night's battle. Salomon's chief of police, Gérard, had asked for a volunteer to investigate. In order to pass the German roadblocks more easily, he was given a special order written in French and German addressed to Monsieur Gaston Reynaud of Les Martins. Salomon had selected Gaston, because they were good friends and knew the farm, located to allow him to see most of the surrounding land.

He passed through a number of roadblocks, not without difficulty, despite his police uniform. There were the tank-destroyers on fire, about thirty metres apart. On one side of the road, a weapons carrier was also burning; on the other, the Germans had situated their guns. Salomon was stopped three times; in front of an adjacent farm he was questioned by a German officer regarding his presence in the area. He replied that he had orders to summon Reynaud to police headquarters. After agreeing, the German officer boasted that his losses were light, with only one killed, while there were at least ten killed on the American side.

Salomon and Reynaud met at the farm. He explained his true mission and gave the farmer the false order. Quietly, he enquired, 'Maybe some of the Americans have escaped. Have you seen or observed anything?' 'Yes,' he answered, 'there are eight Americans hiding in the hayloft.' Reynaud had good reason to be worried, because a farm only a kilometre away had been set on fire and the Germans might pay him a visit at any time. If the Americans were discovered, their fate and his were already sealed. He was sure that the men in the hayloft would not give themselves up. Salomon told Reynaud that, under the circumstances, he would return to police headquarters immediately and then return to the farm. Once past the German barricades, Salomon reported to Gérard, who went to Commissaire Dubois; all three conferred with Captain Ferdinand of the 'Coty' Resistance network and explained the situation. Ferdinand devised an audacious plan: he would spirit the Americans away from the farm disguised as policemen. Ferdinand, Volle, the driver, and Salomon left immediately, taking the back roads to Reynaud's farm. They had four police uniforms between them.

Two days after the escape, Salomon met Reynaud again. Reynaud said that shortly after the second trip, the SS made several trips to his farm, holding a gun to his head and demanding to know where the Americans had gone. Despite his personal danger, he insisted that they had left in the direction of Chabeuil. The SS searched all through the farm but found nothing. In scouring the region, they had already burned the nearby farms of M. Vernet and M. Chovet. When Salomon and I met in 1984, he mused, 'What would have happened if Chief of Police Gérard had not asked for a volunteer that August morning? What if I who knew the region so well, had not volunteered? And what if I had not picked on my friend Reynaud to call upon? It is not difficult to imagine what would have happened.'

In 1984, with Salomon, Claudia, Madeleine, Volle and my daughter, Alison, and other guests, I was honoured by the mayor at the *hôtel de ville* in Valence. Sadly, Captain Ferdinand was killed the day before Valence was liberated; Reynaud had died of a heart attack in the mid-1970s, and Captain Binoche had retired as a general and was living in Nice. When the ceremony ended, the mayor said, 'You realise that you are the second person to be made a Citizen of Honour of Valence.' 'Who was the first?' I enquired. 'Napoleon,' he answered.

Part Six

Reflections

Trauma and Absence:
France and Germany,
1914–45

Omer Bartov

This chapter is concerned with the long-term effects of the First World War on French and, by way of comparison, also German conceptions of war. Most especially, I will be discussing the extent to which the traumatic experience of the trenches in 1914–18 moulded French and German consciousness in the interwar period, and influenced these two nations' very different conduct in the Second World War. The memory of the slaughter on the Western Front, and the mobilisation of that experience in the service of widely divergent political and ideological goals, ranging from pacifism to militarism, socialism to fascism, resistance to collaboration, are crucial to our understanding of the manner in which Frenchmen and Germans imagined, prepared for, and ultimately fought the next war. Moreover, the reality and memory of the First World War, that first instance of industrial killing in Europe, largely determined attitudes towards war and its representation even after 1945, and thereby had a major impact on how the Second World War too was both experienced and remembered.

*

The publication in 1990 of Jean Rouaud's novel *Les Champs d'honneur* (*Fields of Glory*) caused a major sensation on the French literary scene. Within four months the book sold over half a million copies in France and was in the process of translation into fifteen languages. Overnight the 38-year-old Rouaud was transformed from an anonymous news-stand employee in the 19th arrondissement into the winner of a prestigious literary prize, the Prix Goncourt, and was celebrated as one of the most important literary 'discoveries' of contemporary French letters. Indeed, since the publication of Michel Tournier's 1970 novel *Le Roi des aulnes* (*The Ogre*), no book had received such unanimously enthusiastic praise from French critics.[1]

The circumstances of Rouaud's rise to fame, the content and structure of his novel, and the comparison with Tournier's even more sensational 'ogre', may well be indicative of the growing importance of memory, both as mental process and as literary and scholarly trope, in current *fin-de-siècle* France. Any occasional browser in Parisian bookstores will be astonished by the transformation of titles bearing any relationship to the past. Indeed, French anxiety about time lost and past forgotten seems to compel writers and readers alike to ponder, recreate, even invent 'real', 'living' memories hitherto relegated to the category of 'mere' history. Hence books no longer claim to be about the history of the past but about its memory, even if in large part they eventually proceed to tell the same tale in more or less the same manner.

Memory is of course a vague, indeed unusable category if left to its own devices. Personal memories are often nothing more than a *mélange* of endless trivia, occasionally (if one is fortunate or unfortunate enough) brushing with events and personages of a more general interest. In order to gain a more universal relevance, memory therefore needs organisation, direction, guidance and meaning. No wonder that the first and most powerful analysis of collective memory was written during the interwar period, in a France haunted by the proximity of total war and devastation, and by the vast abyss that 1914–18 had torn between the present and the pre-war past, transformed in a series of brutal, bloody offensives into a dim, far-off, sentimental memory of a lost world. Nor is it mere coincidence that the sociologist Maurice Halbwachs, who had coined and analysed this concept, as well as the historian Marc Bloch, who had pioneered the study of collective mentalities and the role of fraud and error in history, became victims of a historical moment in which a régime determined to 'correct' the memory of the past had occupied a nation unable to be reconciled with its own memories of that same past.[2]

In what follows, I would like to raise some questions regarding the effects of the trauma of the First World War on subsequent French preoccupations with history and memory; the extent to which 1914–18 has served to block, obscure and repress more recent traumatic events, such as the débâcle of 1940, the collaboration and post-1945 colonial wars; and what can be learned about this process from a comparison between French and German reconstructions of the past. In a second part of this chapter, which due to lack of space cannot be included here, I turn to several memoirs written after 1945 by Frenchmen, Germans and Holocaust survivors, and attempt to elucidate how the trauma of war and genocide, and the loss or absence of parents and childhood memories, are reflected in these texts. For while it is difficult to establish the links between collective and individual experiences, the traumatic effects of this century cannot be understood without reference to their effects on the individual.[3]

Let me begin by suggesting that the wide-ranging preoccupation with memory in France, as well as of course in Germany, is fundamentally different from what currently appears to be a similar interest in the United States. Not only is the past remembered much more intensely in these European nations than in America; it also plays a very different role in the construction of images of the future. Present economic and social anxieties notwithstanding, American culture is deeply rooted in an optimistic vision of the future, which must by definition (although not always in fact) be better than the past. For this is a land of immigrants who came and keep coming there in vast numbers in order to improve their lot. They may have sentimental memories of their respective pasts, but their goal in coming to the United States was and is to have a better future. For many Americans, therefore, the United States is a nation with a very short past, and one that can and must be improved as it moves forward in time. As for the richer pasts brought by immigrants from their lands of origin, even if they miss them, they normally do not wish to return to the lands to which these pasts belong, and will occasionally even admit that their memories of those homelands in any case no longer reflect their present realities. Conversely, French visions of the future are filled with images of the past, some lamentably lost and gone, others cruel and destructive, making for that unbridgeable, irreparable break which inevitably diminishes one's trust in progress. After all, the greatest French novel of the century, Marcel Proust's *Remembrance of Things Past*, written before, during and after the First World War, was a series of ruminations, on a vast scale, about time lost, that is, about the inability ever to recapture and recreate the past. Significantly, Proust's relentless efforts to grasp the fleeting images of things lost were interrupted by war and illness, and finally by death, so that much of his œuvre was published posthumously, hence making the memory of the author's own existence into a final 'thing' both remembered and lost.[4] To be sure, there is no simple mechanism whereby individual memories and visions of the future come to reflect collective social or national consciousness, and in the present context I will not try to establish how such a process may take place. Yet I would argue that we have enough evidence to show the extent to which the experience of loss and trauma extends beyond personal recollection, and comes to encompass both individual and collective expectations of the future.

Can we therefore understand the enthusiastic critical reception and instant popularity of Rouaud's novel as an indication of the prevalence of thoughts of war, and especially *that* war, even in present-day France? Does it reflect a more or less widespread notion that 1914–18 was the key event of the century, and yet a riddle still to be solved in order to understand the nation's present predicament and all that has happened between then and now? Or does the preoccupation with the First World War act as a displacement for the more problematic, chronologically closer, and morally more disturbing events of the débâcle of 1940, Vichy, even Algeria? Such questions cannot be avoided considering the nature of the book and the readers' response to it. For *Les Champs d'honneur* constructs a sentimental, lightly humorous, comforting and soothingly continuous narrative of a French family of the lower Loire from 1912 to 1987, only to wreck that fabricated world in the last pages of the book, revealing the 'real' memories of a family never recovered from the trauma of the First World War, the blood-bond of death and procreation, the links between generations which still carry the memory of the wreckage of a world they had never known in the flesh. It is a novel, therefore, about the tenuous balance between memory and forgetting on which individual and collective identity is based, and about the nature of that link as it is experienced specifically by a single French family. Is this the reason that several generations of French readers shared the novel's odyssey between personal, family and historical memories, between the superficial narrative of this troubled century and the hidden diaries and skeletons which fill its closets? And is it mere coincidence that the novel was published only a few months after Bertrand Tavernier made the film *La Vie et rien d'autre* (*Life and Nothing But*), which is similarly preoccupied with personal mourning versus official commemoration, physical and psychological devastation and the unquenchable urge for life?

A comparison with Edgar Reitz's 1984 film *Heimat* on the *German* memory of this century may provide us with some clues. For while Rouaud's novel and Tavernier's film *focus* on the trauma of the First World War, Reitz merely *begins* his sixteen-hour cinematic saga at the end of the war, and represents the interwar years as still part of the sentimental memory of things now long vanished. What ruins the pastoral, idyllic existence of his imaginary village is not even the Nazi régime or the Second World War, but the modernisation, or rather Americanisation, which follows this even more disastrous defeat. And even if we dismiss Reitz's obsession with anti-Americanism and the ills of modernity, there is little doubt that it was the devastation of the Second World War, and much less overtly also the Holocaust, which displaced the trauma of the First World War for the Germans. In Germany, therefore, *Vergangenheits-bewältigung* (coming to terms with the past) must begin with Nazism, even if for a while it was sought in the Wilhelmine Reich before 1914. Conversely, in France the First World War seems to have displaced Vichy and all that it entailed for many years after 1945; coming to terms with the past had to begin with the First World War, even if it was eventually read into explanations of more recent and at least politically and personally much more troubling traumas. In yet another sense we may say that post-war German youths grew up surrounded by the debris of the Third Reich; French youths, even in the most remote villages, were confronted daily with the memorials of the First World War. The ambivalent status of the victims of 1939–45, and the blurring between them and the perpetrators, is common in both nations. Yet Germany has accepted responsibility for the atrocities committed 'in the name of the German people', and has spent much of the intervening period trying to figure out what precisely is meant by that phrase. France has commemorated the 'fallen soldiers', along with the 'executed', and 'deported' civilians of the Second World War (with the last two categories often more numerous than the first) on small plaques attached to First World War memorials or discreetly placed on street corners and bridges, parks and squares. There is no physical, plastic presence to remind one of Vichy which even remotely compares with the massive memorials and vast cemeteries of the First World War, nor, for that matter, with the concentration camp sites in Germany.

Almost a generation before *Les Champs d'honneur*, Tournier's 1970 *Les Roi des aulnes* was concerned with another war, another devastation, another (but not unrelated) perversity. The novel was part of the gradual re-remembering of defeat, collaboration and fascination with evil of the more recent past. It is neither sentimental, nor moralising, nor edifying; indeed, it is a deeply disturbing, ambivalent, unremitting work, in some ways closer to Günter Grass's 1959 novel *Die Blechtrommel* (*The Tin Drum*) than to contemporary French writers.[5] And just like Grass, Heinrich Böll, Siegfried Lenz, and other German writers,[6] Tournier is not concerned with recreating or recapturing a lost past, but rather with coming to terms, or more precisely grappling, with a personal and collective recollection, painful, troubling, unresolved, burdened with guilt and self-justification, resentment and self-contempt. The same can be said about other French works of fiction which have dealt with the roots and course of the débâcle, such as Jean-Paul Sartre's trilogy *Les Chemins de la liberté* (*The Roads to Freedom*), published between 1945 and 1949, and Julien Gracq's 1958 *Un Balcon en forêt* (*Balcony in the Forest*); memoirs of the liberation, such as Marguerite Duras's 1985 *La Douleur* (*The War*); and biting parodies of the purges, such Marcel Aymé's 1948 novel *Uranus* (interestingly also made recently into a rather unsuccessful film despite the presence of the ubiquitous Gérard Depardieu). None of them is about memory, for they constitute a reckoning with an evil not wholly spelled out, and with complicity not entirely conceded.[7] *That* memory, the memory of those who had been most directly subjected to evil, is still not, and perhaps can never be, wholly French, for the French perspective on these events is still a matter of (personal) history and (detached) documentation, or of repression and forgetting. *That* memory, the memory of evil as experienced from within, is still the domain of, or attributed to, others. It is the mythical memory of an ancient people finally devoured in Auschwitz, as portrayed in André Schwarz-Bart's 1959 novel *Le Dernier des justes* (*The Last of the Just*); the painful memory of hiding and betrayal, and the desire to make the world whole again, reconstructed with bitter-sweet irony in Romain Gary's 1979 novel, published under the pseudonym Émile Ajar, *L'Angoisse du roi Salomon* (*King Solomon*); the tormented memory of fear and rejection and the impossibility of wreaking vengeance, powerfully expressed in Boris Schreiber's semi-autobiographical 1984 book *La Descente au berceau* (*Descent to the Cradle*), and the repressed, inexpressible memory of what 'it had actually been like', delicately woven into a fragile text in Robert Bober's recent and much acclaimed 1993 novel *Quoi de neuf sur la guerre* (*What's New on the War*).[8] These are memories of the victims, or of the victims' victims, or more precisely still, of the Jews (French or otherwise). No wonder that the most important cinematic representation of memory made in recent years, the 1985 film *Shoah* by the French Jewish film-maker Claude Lanzmann, is about the *Jewish memory* of the Holocaust – with the role of the French in the event by and large left out; while the most influential cinematic representation of *France's* complicity in Nazi policies, French Jewish film-maker Marcel Ophuls's 1969 masterpiece *Le Chagrin et la pitié* (*The Sorrow and the Pity*), is a *documentary*.

By contrast, German post-war representations of Nazism are very much concerned with the *German* memory of the period, and especially of the war, and focus on German victimhood; the victims' victims, or, as some would prefer, the perpetrators' victims, are generally absent from such representations, though they feature very prominently as an absent entity, since they provide an unspoken model of victimhood for German portrayals of their own suffering, as is obvious, for instance, in Alexander Kluge's 1979 film *Die Patriotin (The Patriot)*.[9] This would suggest two different modes of remembering in France and Germany. French memories of their own suffering and victimhood in the First World War qualify or even displace the memories of those more recent victims of German-occupied and Vichy France, and along with the glorification of the Resistance have until a few years ago served to repress open and serious discussion of this second, far more disturbing and much less clear-cut traumatic event, even as it concerns its undeniably French victims. Indeed, while Vichy has by now been studied at some depth, the débâcle of 1940 is still a very much understudied event.[10] Conversely, German memories of their own suffering in the last years of the Third Reich, of the bombings and evacuations, destruction and mass rapes, have not only displaced the memory of the First World War, but have also served to repress the memory of Germany's victims in the Second World War, qualifying and relativising it by reference to what is seen as the equal victimisation of the Germans themselves.[11] Consequently, while it is formally acknowledged, the (annihilated) existence of Germany's victims in fact disappears from view into a self-contained box that can be related much more easily to other cases of atrocities and murder perpetrated by other régimes and nations than to those specifically implicated in perpetrating or being complicit in the very specific, and unprecedented genocidal policies of the Nazi régime.

If it is difficult to gauge the extent to which the trauma of the First World War has, at least indirectly, influenced the organisation of French memory and the foci of its historiography after 1945, its impact on France of the *entre-deux-guerres* was of such dimensions that it moulded not only the view of the past, but also very much the anticipation of the future, and especially of any future war. Since the next war was thought of (quite wrongly, although not untypically) in terms of the previous one, the possibility of having to endure yet another, possibly even worse slaughter, was exceedingly difficult to contemplate. Hence while the memory of the First World War came to be increasingly associated with mass destruction of human beings and property, the anticipation of another armed conflict assumed apocalyptic dimensions. This is not to say that there were not a great many Frenchmen and women willing to fight once more for the *patrie*; rather, that underlying people's perceptions of the future, no matter their political ideology and personal sympathies, was an apocalyptic vision nurtured by past destruction and trauma, and breeding profound anxiety and insecurity.

To be sure, the same traumatic event may be experienced very differently, depending, not least, on one's original expectations from it and, even more fundamentally, on individual or collective predispositions preceding its occurrence. Moreover, trauma may have very different effects even on those of substantially similar predispositions. Thus, for instance, French and German soldiers fought a very similar war on the Western Front in 1914–18 (although the German fighting on the Eastern Front and the French colonial wars greatly differed both from each other and from the Western Front); yet they arguably perceived their experiences in the trenches, and eventually reacted to their war experiences after the end of the fighting, in starkly different ways. Naturally, it would be false to assume that all French or German soldiers followed the same patterns of perception and internalisation of experience, whether during the war or in subsequent years. Indeed, in both cases one can identify a series of quite different and in some ways wholly contradictory reactions to the fighting. Nevertheless, and especially during the 1930s, both the memory of the First World War, and the anticipations of the next war, congealed into very different forms in France and Germany. At the risk of somewhat over-generalising a highly complex phenomenon, I would suggest that for large sectors of the French population, coping with the traumatic memory of the First World War was only bearable as long as it remained in the past, never to be re-enacted; while for a growing number of Germans, the unbearable memory of defeat and socio-political upheaval could be overcome only by re-enacting, and thereby 'correcting' it in a future confrontation.

Projecting memory into the future is a complex process, and in the present context I can only suggest some of its aspects. The First World War produced a series of polarities which substantially defined its perception both in France and in Germany. These included experiencing the war as an instance of defeat or victory, glory or shame, loyalty or betrayal, unity or disintegration, national endeavour or class struggle, gender equality or sexual division, hope or despair, beginning or end, accomplishment or failure, nobility or savagery, modernity or barbarism, progress or degeneration.[12] Such polarities were not neatly organised – rather, they overlapped both in the individual experience and in the national imagery – but the general and ultimate drift in France was the opposite of that in Germany. Indeed, a comparison of the interplay between these polarities in both countries may well shed light on the manner in which the experience of 1914–18 was internalised, constructed as memory, and utilised to guide (or misguide) these two nations into a radically different future.

A brief discussion of one crucial polarity, that between death and survival, may illustrate this point. What the First World War produced in unprecedented quantities was millions of corpses and further millions of physically and mentally mutilated survivors. This, the single most important outcome of the war, was of course of paramount significance for another major watershed in this century, namely the Holocaust, where the industrial killing of the First World War was perfected into a sophisticated complex of death-factories. Indeed, we can learn a great deal about both events by more rigorously studying the numerous links between them.[13] One obvious similarity, which has haunted the survivors of both slaughters, is that while they shared an intimate connection with the victims, since they had both experienced destruction at close quarters, they were also radically separated from them by the simple fact that the living cannot communicate with the dead. Hence while the victims were remembered (or forgotten), the survivors were *doing* the remembering (or forgetting). One consequence of this vast presence of death, mutilation and survival in the wake of the First World War, was the tremendous expansion of what the historian George Mosse has called 'the cult of the fallen soldier', epitomised most distinctly in the innumerable memorials and huge cemeteries which soon dotted Europe's landscape.[14] This cult of death, which associated the mass production of corpses with patriotism, heroism and religious faith, came to be related in more than one way to the legitimation of the Nazi machinery of extermination. Moreover, we might note that a comparison between First World War and Holocaust commemoration, competing claims of victimhood and political manipulation of memory, may yield disturbing conclusions regarding the potential effects of mobilising past atrocities as motivation for future actions. In some countries, and most notably in Britain and France, this 'cult' was also expressed in the invention of a wholly new institution, the tomb of the unknown soldier, no less notably absent from Germany, where Hitler's claim of being himself the unknown soldier of the *Weltkrieg* served as a remarkable indication of Germany's ultimate refusal to perceive the war as totally over, or the dead as wholly dead.[15] In Germany, we might say, the dead increasingly *scolded* the living for giving up the fight, while in France they constantly *warned* the living against thinking of ever repeating it.[16] But the cult, though in many ways similar in most European countries, indeed largely a combination of conscious emulation and adaptation to local circumstances and traditions, was understood and interpreted very differently, not least because of the differing perceptions of the relationship between the two extremes, death and survival, and of the implications of the links between them. This in turn had radically different consequences for the articulation of the memory of the First World War as well as for the constructed images of the future.

While several important studies have investigated the commemoration of the fallen and the mobilisation of the veterans in France and Germany, they have paid little attention to the fundamentally different meanings with which death and survival were endowed in these nations, all formal similarities notwithstanding. Even Mosse's pathbreaking work, though it purports to analyse the cult of the fallen soldier in Europe as a whole, is mostly about Germany (and Britain), and has little of importance to say about France.[17] Conversely, the French historian Antoine Prost's major work on the *anciens combattants* makes almost no reference to studies of German veteran associations, whose politics were crucial to the rise of fascism and differed in essential ways from those of their French counterparts.[18] Nor does the historian Annette Becker's recent important contribution to the study of commemoration in France attempt to compare the input of religion in French and German cemeteries and memorials.[19] The same can be said, of course, regarding studies of German First World War veterans and commemoration as well.[20] This is not merely a lack of a comparative perspective; rather, these works reflect a real difficulty in distinguishing between difference and similarity, experience and its constructed meaning, and, most significantly perhaps, the limitations of scholarship in dealing with the protean nature of memory (individual and collective, 'organic' and constructed, direct or mediated).

What is needed is a much closer analysis of the process whereby the same event of mass industrial killing on the Western Front, as well as its often similar formal and official representation and commemoration, nevertheless created a powerful anti-war sentiment among otherwise strongly opposed political and ideological factions in France, and a growing willingness within wide-ranging circles in Germany to go to war again, which in turn provided a convenient climate for the spread and dissemination of the Nazis' militant rhetoric even before Hitler's 'seizure of power'. To be sure, we may fall back on conventional arguments, stressing the fact that the French won the war while the Germans lost it, that France had a long (though not unbroken) republican tradition and Germany an authoritarian one, that Germany experienced a quasi-revolution in 1918 and France a largely successful one in 1789, that Germany was a frustrated, insatiate power in 1918 while France hoped to preserve the status quo, and finally, that the French débâcle of 1940 was not necessarily the product of pacifist and defeatist tendencies but merely of military incompetence and miscalculation, just as much as Germany's astonishing victory was a combination of luck, superior organisation, and brilliant tactical planning and leadership. Yet while such explanations must certainly be taken into account, I would argue that they are ultimately insufficient, both because they tend to confuse cause and effect, making the divergent memory of the First World War in both nations appear predetermined and inevitable, and because they fail to come to terms with the fact that the same traumatic event can be experienced by, and have radically different repercussions on, two substantially similar European societies and cultures. In other words, we are still groping for an explanation of why the experience of mass destruction can motivate societies (and individuals) either to reject violence or to embrace it. Nor is this question limited to the case of France and Germany in the interwar period. To cite just two examples, we may note, first, that the even greater destruction of the Second World War seemed to justify the threat of violence, indeed of total destruction by means of nuclear weapons, at least among the superpowers, while it did indeed create an abhorrence of war among many Europeans (who none the less waged several bloody colonial wars in the following two decades). And second, that even the Holocaust, which had brought the art of industrial killing to its utmost perfection, has been seen by some as a justification for violence rather than as an example of what the destructive urge of modern society can lead to if it is not strictly controlled. This is also not only a question of the victims being determined never to be victimised again, for while in the former case it was the victors who justified continued destruction and the threat of total annihilation, in the latter the justification of violence as the only means to prevent another Holocaust has

played a role in the militarisation of Israeli society even as it came to occupy
another people, and has in recent years provided an extremist, fundamentalist
Jewish fringe (whose impact on Jewish and Israeli society is far greater than its
numbers) with what it perceives as an incontrovertible legitimation for law-
lessness and violence.[21] We are therefore facing a problem whose disturbing
implications are still very much with us today. A closer analysis of the cases of
interwar France and Germany would thus shed more light also on our own
current predicament.

   After the First World War not a few anti-war activists believed that by
telling the public what the war had been all about, in graphic, uncompromising
detail, no person in his or her right mind would ever wish to go to war again.
Such arguments were voiced both in Germany and in France (and Britain).
Simultaneously, however, the terrible price of the war motivated its survivors
to endow it with meaning. Both French and German veterans could not accept
that the sacrifice had been either meaningless or in vain.[22] The story of the
German veterans' organisations such as the Stahlhelm, the paramilitary Frei-
korps and their nihilistic ethos, and the rise of the Nazi party from the
wreckage of the war, has been told often and in persuasive detail.[23] Historians
tend to accept the inner logic of this tale (without necessarily finding it
inevitable or unavoidable), while agreeing that it finally led to a terroristic
régime which eventually destroyed both Germany and much of the rest of
Europe. The story of the French veterans is perhaps more ambiguous and less
thoroughly researched and argued. Thus, for instance, Antoine Prost's interpretation of the veterans' views is not always reflected in the ample citations
from their speeches and publications included in his own study. There is more
than an occasional similarity between the French veterans' sentiments as
expressed in their own words regarding the republic, democracy, the family,
order, discipline, and authority, and those proclaimed by their German counterparts. Prost rejects the assertion that they had fascist leanings, but agrees
that at least temporarily they were wholeheartedly supportive of Pétain's
régime. This qualifies his argument that the veterans were fundamentally
moulded by the republican elementary school and the values it had inculcated
in them before 1914, and should be seen in conjunction with Annette Becker's
acute analysis of the sacralisation of death in war by the widespread involvement of established religion in the commemoration of the fallen. Nevertheless,
there is a certain inner logic to the French story as well: France was unwilling
to pay the terrible price of war once more, and therefore failed to resist the
German onslaught. Historians who accept this view (and even those who do
not are by and large of the opinion that France was reluctant to fight another
war) may find this unfortunate, but understandable. They may reject the choice
– if it was that – made by the French in 1940, but feel that they have explained
why at that point there were 'forty million Pétainists'.[24]

   While both the German and the French narratives may appear reasonable
and logical, juxtaposing them presents both a serious difficulty and a partial
resolution of the problem. Observed from the perspective I have suggested, it

may be possible to argue that since the articulation of death and survival in war went through several phases, it is only by carefully following its evolution in the interwar period that we can understand why it acquired such different meanings in France and Germany during the second half of the period. The same would hold of course for many other concepts outlined above, such as victory and defeat, loyalty and betrayal, progress and barbarism. There is little doubt that by 1933 a certain view of war, and especially that of 1914–18, was gaining the upper hand in Germany, which then achieved a position of hegemony under the Nazi régime and was widely inculcated into German youth in particular. The opposite happened in France, where by 1934 there was little agreement on the meaning and sense of the previous war as well as on what the anticipated conflict would mean and entail. To be sure, the vast majority of French men and women seem to have abhorred the idea of another war. But in this they were not different from many other Europeans, the Germans and British included. What made a difference was that the polarities could not be reconciled. What made the memory of the previous war so difficult to come to terms with, and the anticipation of another conflict almost impossible to contemplate, was not simply that 1914–18 was a horrific massacre, but that there was no generally accepted view of what it had all been about. The inability to provide the First World War with a meaning that could be mobilised to comprehend any future conflict, that could, that is, define how such a future conflict would be perceived, made conceptualising, indeed, simply imagining it, exceedingly problematic, even unbearable. Not that there had been no attempts to infuse the First World War with meaning; quite the contrary. Some viewed it as a great defensive war, others as the beginning of a proletarian revolution, yet others as the war to end all wars; there were even those who spoke of it as a grand collective and individual experience. But by the mid-1930s none of these images had managed to gain any measure of hegemony, not even the assertion that the war had been meaningless.[25]

It is doubtless difficult to determine the precise measure to which the First World War has remained a traumatic event in French collective memory, representation, and historiography. It may be possible to suggest, however, that because no agreement could be reached on its meaning, 1914–18 has maintained a complex relationship of historical and interpretative interdependence with the war that followed it. For until it actually erupted, the next war was thought of in the same terms as the last; and since it was not anything like what people had expected, but rather ended in a humiliating defeat, occupation, and more collaboration than the French cared to admit for many years after the liberation, it too became a traumatic, unbearable memory. And if that memory was also richly laced with tales of heroism and resistance, sacrifice and devotion, it was too close and too ambiguous to be confronted head-on. Hence the trauma of the First World War, which was now rapidly receding into the past, was a tempting site of remembrance, despite its own ambiguity. Of the two traumas, therefore, the more distant one was preferred. But resolving it, coming to terms with its memory, did of course lead one

eventually to the débâcle and Vichy. Consequently, while the preoccupation
with the First World War may have served to displace the more painful
confrontation with 1940–44, it eventually traced the path from one trauma
to the next, for the deep roots of Vichy are to be found in the slaughter of
1914–18, just as present-day France cannot be understood without reference
to the memory of Vichy.[26] This is not to say that a consensus has been
reached about the meaning of the First World War. Thus, for instance, while
a recent collection of articles by leading French historians was entitled *14–18:
Mourir pour la patrie*, it in fact reflects precisely those very same ambiguities
about the sense and meaning of the war already apparent during the event
itself.[27] And if the memory of the First World War is still contentious, that of
Vichy retains the qualities of a political time-bomb, as evidenced in the
revelations about the changing loyalties and convictions of François Mitter-
rand.[28] In a nation whose previous president was both a socialist and a former
Vichyite (and a youthful member of the *Croix de Feu*), and where calls for
redefining the nation and enhancing Frenchness are accompanied by growing
xenophobia, it may seem comforting to remember a distant war in which true
patriots were willing to die for the homeland in vast numbers. But whether or
not this was indeed the case, the immense sacrifice of 1914–18 can no longer
obscure the 'black years' of Vichy, just as German claims of victimhood in the
Second World War have failed to erase the memory of the régime's victims,
and Chancellor Helmut Kohl's repeated reference to his 'grace of late birth'
has not detached his nation from its history.[29]

Oral History and British Soldiers' Experience of Battle in the Second World War

Nigel de Lee

What is history? My view follows that summarised by R. G. Collingwood in his statement, 'History...is an autonomous form of thought with its own principles and its own methods.' History is not literature, for the reason given by Francis Bacon: 'Poetry...is...feigned history...submitting the shows of things to the desires of the mind.' John Clive remarked, 'Historians, unlike poets and novelists, must try to get their facts right.' Collingwood maintained that whilst the novelist and the historian must produce a coherent and internally consistent text, the historian's account must also be located in time and space, and supported by evidence. Edward Gibbon, in his *Autobiography*, insisted on 'Truth, unblushing naked truth, the first virtue of more serious history'. Hume agreed with him, that 'truth [is]...the basis of history'.[1]

Which leads to the questions: what is the nature of historical truth, and how is it established? It cannot be identical to the forms of truth claimed by the exact and natural sciences. Indeed, Descartes denied that history could be a branch of knowledge at all because its exponents could not produce the absolute and necessary proofs that his scepticism demanded. We cannot doubt that some facts about the past can be established without doubt, but as Montaigne remarked, facts are merely 'the naked and infirm matter of history'. Gibbon was grateful that he had given up the study of mathematics, the realm of absolute certainty, 'before my mind was hardened by the habit of rigid demonstration, so destructive of the finer feelings of moral evidence, which...determine the actions and opinions of our lives'. Hume wrote to Gibbon that an assumption well sustained by positive evidence must be dismissed if it was contrary to common sense.[2]

Carlyle argued that all evidence was fallible, and so its content must be suspect. This meant that the real events and conditions of history could not be measured. How could anyone measure an 'ever-working Chaos of Being'? all that could be sought was 'some more or less plausible scheme or theory...or the harmonised result of such schemes, each varying from the other and varying from truth'. He felt that in order to achieve any understanding of historical process it was necessary to abandon the sterile criteria of sceptical enlightenment, and employ subjective and intuitive modes of thought.[3]

Leopold von Ranke insisted that history must be 'true to fact...resting on thorough research', but conceded that in the face of overwhelming masses of evidence, history could hope only for an approximation to reality. Maitland remarked that once facts had been established, it was impossible to avoid guess work in interpreting them. Bury, who took the view that history is a science, made clear that it was not akin to exact or natural science by stating, 'only such things as dates, names, documents, can be considered purely objective facts. The reconstruction...the discovery of causes and motives ...depend on subjective elements'.[4]

Trevelyan took a pragmatic view. He observed, 'History...is a matter of rough guessing from all the available facts', and 'history is a matter of opinions, various and variable, playing on a body of accepted facts that is itself always expanding'. Collingwood held that some historical arguments could be proved as compelling as any in natural science, but that others could have only permissive authority. The main business of the historian was to gather evidence, then by using his judgement to select, construct and criticise it in order to produce a rational view of the past. This process must involve the use of inference from appropriate evidence, and *a priori* imagination to counter the effects of inevitable gaps in the evidence available. He claimed that history was a science in the way that it proceeded by question and answer. Marwick agreed with Collingwood that history is a systematic exploration of sources in pursuit of truth, and is as such scientific. But he differed in insisting that the 'central activity' of history is to discover new *information* from old sources, and new sources, and new means of analysing them. He does concede that difficulties inherent in imperfect sources will cause disagreement and uncertainty, but this is a much more positivist view than that of Collingwood. Recently, Jacques Le Goff has argued that history is a compound of science and art, containing elements of both narrative and explanation.[5] Perhaps we should conclude that history is an independent bastard child of three parents: literature, science and philosophy.[6]

Such a conclusion is a reminder that rather than saying what history is not (neither art nor science), we might illuminate its character by drawing an analogy with legal studies. Montaigne recommended the Roman method of making history, which, according to him, included the forensic procedures of confronting witnesses and considering objections to testimony. Trevelyan said that historians had a duty to seek 'the truth, the whole truth, and nothing but the truth'. Collingwood discussed the analogies of legal and historical thought, and indulged in a fictional example of murder to explore these.[7] In this regard, the standard of proof required by civil law – a balance of probabilities rather than the exclusion of reasonable doubt, which is used in criminal cases – is apposite. We might also argue for the utility of the Scottish not proven verdict.

Of course, the lawyer is always intent on more specific and urgent questions than the historians. But the historian would do well to approach his evidence with the suspicions and caution of the lawyer, within his own wider context of vision and interests.

But what is the historian to do with the truth once it has been discovered? He must convey it, acting as a mediator between the past and the present. Bacon wrote, 'it is the office of history to represent the events themselves together with the counsels', and Montaigne also believed in the importance of retailing motives, 'by what secret springs we move', as well as actions. Hume remarked that history must be used for purposes of instruction and amusement. Macaulay, von Ranke and Trevelyan all believed that the truth must be presented in an attractive form, in order to engage the interest of the student. Le Goff says that the historian must 'master history as experienced' in order to conceive and explain it. Carlyle had doubts if this is possible, because even well-recorded recent events were subject to incomplete understanding. Of historical events, he asked, 'Is it even possible to represent them as they were?' The realists of the last century had no doubts on the matter. They held that they had a duty to give 'a truthful objective and impartial representation of the real world, based on meticulous observations of contemporary life'. The past, beyond observation, could not be represented truthfully. The leading realist painter, Gustave Courbet, insisted, 'The history of an era is finished with that era itself and with those of its representatives who have expressed it.'[8] This theory would reduce all history to futile repetition or analysis of contemporary *reportage*, excluding long-term perspectives of change and development. History would consist of a series of self-contained and isolated occurrences, like ancient insects preserved in amber. It is more fruitful to work on the assumption that whilst we cannot aspire to convey a perfect and comprehensive representation of past reality in the present, we can convey some of it and this will be imperfect but still of interest and value.

In military history it is difficult to discover and represent reality concerning aspects of central and vital importance. Marwick remarked, 'In the past historians have been fascinated by the origins of wars; more recently they have begun to pay due regard to ... the consequences of wars'. It is significant that he does not mention the conduct of war, which must determine the outcome and consequences. But it is not easy to get a firm and clear idea of operations and fighting. Carlyle commented, 'Battles and war tumults which for the time dim every ear, and with joy or terror intoxicate very heart, pass away like tavern brawls, and ... are remembered by accident, not by desert.' Operations in modern wars are increasingly difficult to comprehend, being highly complex, fluid, incoherent and either boring or uncontrollably exciting to most of the participants. They are extremely difficult to represent; as John Clive asked, 'How is the historian able to represent experience, which is multi-dimensional, in a narrative media which is linear ... one-dimensional?'[9]

There is a saying that 'one picture is worth a thousand words', but it has proved difficult for the visual arts to portray the realities of combat in this century. The realist school of painters believed that only real and actual things could be represented by their art, and the impressionists set out to depict the effects of light on real objects dispassionately and scientifically. During the First World War, painters strove to capture and represent the 'underlying reality' of war. But, because war was such an exciting phenomenon, the artist found it difficult to observe with the detachment necessary to his vocation. The camera, whilst impartial, was so objective as to be indiscriminate, and could not convey images of real significance to the viewer.[10]

In 1939 the war artist Richard Seddon went to France, full of expectations which were disappointed. In his account of the campaign of 1939–40 he wrote:

> The war to me would not have started to be worth painting until I could paint actions and fighting…When I was in action I found it to be beyond my powers…I could paint the preparations for actions and the aftermath…but not the action itself, which is the very core and heart of war…other war artists had all found the same thing…action itself is unpaintable.[11]

It seems we must return to words, but in what form? Traditional modern academic scholars are hampered in their investigation of military operations because the forms of evidence they find most acceptable are often lacking, particularly documents. When in action, armies are mutually destructive and often careless of their records, particularly if they are being defeated or enjoying an unanticipated success. Such records as are kept are often minimal and may be unreliable. War diaries, even at Corps HQ level, are often written up several days late.[12] The use of telephone and wireless telegraphy, of personal liaison and verbal orders based on the erasable Chinagraph markings on maps, meant that many orders and decisions were not recorded in writing. So in order to obtain a comprehensive idea of the development of operations, it is necessary to use other sources and resort to the sort of *a priori* imagination described by Collingwood, and practised by Colonel A. H. Burn as 'Inherent Military Probability'.[13] The alternative course, to relate only what can be proved by written documentary evidence, would of necessity produce an extremely sterile and partial account.

The use of oral sources, eyewitness accounts, verbatim reports of proceedings and traditions has a long pedigree. Herodotus and Thucydides based their histories on the accounts of participants in and witnesses to the events they described. Macaulay believed that much of the history of common people could be learnt from their ballads. Trevelyan walked the route taken by Garibaldi and his thousand in order to interview veterans and witnesses of the Redshirts' campaign.[14]

But during the recent past historians have tended to disregard and disparage oral sources. Marwick put oral history eleventh in his list of twelve types of primary sources. He said that it was invaluable if no other form of evidence was available, but he also remarked, 'For some areas of historical study, relating to the poor and the underprivileged, this kind of source may be the main one available, the evidence it offers should, as far as is possible, always be checked against other kinds of source.' This betrays an attitude which is élitist and reactionary. To some extent, the champions of oral history agree with Marwick. Seldon and Pappworth state that oral sources can complement and supplement documentary evidence, and allow the 'non-document-producing Classes' to record their views. Paul Thompson evidently regarded oral history as inherently democratic and socialist in its character. All he said of its value for military historians is that it enables them to 'look beyond command level strategy... to the condition, recreations and morale of other ranks'.[15] These views are much too limited and dismal in regard to the potential of oral history. Other historians have taken a much broader and more optimistic view. Collingwood observed that written documents as such were not necessarily reliable or informative. He remarked that written sources were more often used because it was easier to copy them out and rely upon them than to use other forms of evidence. But to him all sources of inference were evidence, including 'The tone of voice, the involuntary gestures of a witness', and he went on, 'There is no distinction of principle between written and unwritten sources.'[16] More recently, Le Goff has claimed, 'documents include the spoken word, the image, and gestures'.[17]

My own years of experience in the fields of oral history and military history have convinced me that oral sources are of great value as evidence of the nature and events of military operations, and as a means of representing the realities of war. Some interviews contain not just testimony, but such deep and informed reflection on past experience that they amount to history as such.[18] But most are not so developed, and are simply memory. There is a current tendency to take memory as history, and 'let the age speak for itself'. Linda Nochlin observed that it is only the passage of time and reflection which can transform the raw material of memory into history.[19] Only the trained mind of the historian can give the memory a current meaning which is coherent.[20] Many a veteran has come to an interview anxious to test his memory by dialogue, and have his recollections stimulated and put into context in order to make sense of them.

Of course, there are dangers involved in the collection of oral evidence. The interviewer may impose his own preconceived ideas, or his own notion of priorities on an informant. In this way he can select or distort the evidence taken.[21] It is easy enough to cultivate and harvest the desired evidence, and some authors of popular works of oral history evidently do this. But the tape recorder is remorseless in revealing such manipulative techniques in action. Further, the interview is a dialogue, and not all informants will be suggestible. A good interviewer should be more of a hunter and gatherer. It is best to conduct an interview with topics rather than specific questions in mind,[22] and allow the informant to speak freely. In this way entirely new and unsuspected information and ideas can emerge and be pursued immediately. A written document can be examined and discussed, but it cannot be interrogated.

Anomalies can be spotted, but not challenged and explored, as they can be with a living and intelligent source.

Of course, a source is not perfect. Memory is fallible, but then so are the fruits of memory when written. An informant may seek to record false or deficient information, but it is easier to lie with calculation, or to deceive oneself undetected, when writing than when speaking. When it is necessary to evaluate and judge evidence, the spoken word can and will yield and convey more meaning than the written, especially when it carries an emotional change. R. G. Collingwood observed, 'what writing, in our clumsy notations, can represent, is only a small part of the spoken sound, whose pitch and stress, tempo and rhythm, are almost entirely ignored... The written or printed book is only a series of hints, as elliptical as the neumes of Byzantine music.'[23]

On this basis, it has been argued that evidence on film or video is even more fertile, providing visual images of body language as well as speech.[24] This is so, but personal experience has shown that the intrusive nature of video and film equipment, or the alien environment of the studio, can paralyse or inhibit an informant, at least temporarily. By comparison, sound-recording machines are unobtrusive and less unsettling.

As is the case with all forms of historical evidences, the quality and value of oral sources varies. Some authorities, such as Peter Hart, Paul Thompson and Arthur Marwick, have maintained that all oral sources are so inherently unreliable about matters of fact, such as dates and places, that they require multiple corroboration to validate them.[25] My own impression is that corroboration is reassuring, but not always necessary. Some informants have proved to be exceptionally precise in their recollections of hard facts, names, dates, times and sequences of events. These are often, but not always, people whose minds were trained to retain detailed information, such as lawyers, journalists, teachers, doctors and intelligence officers. The accuracy of their statements can be checked by reference to other oral sources, written sources, maps, photographs and actual terrain. Other informants cannot recall hard facts, but can give vivid and compelling accounts of personal experiences. The best informants, who are rare, combine both abilities, giving a stream-of-consciousness account based upon a sound framework of accurate, verifiable facts. Most oral evidence is taken from individuals, but it is possible to make the process collective, provided the participants are well disciplined. The advantage of this procedure is that the informants can stimulate each others' memories and recall areas of corroboration and dissonance, both of which are of interest. In such a process the collector of information must act as a more unobtrusive catalyst than when interviewing individuals.[26]

An old objection to oral evidence, similar to that levelled against journals, letters and memoirs, is that all accounts given by this means are personal and anecdotal. But, even statistics in official documents are merely a summary of innumerable personal replies to questions. The collection of historical evidence must be a cumulative process.[27] Carlyle held that all history was the essence of innumerable biographies.[28] It is said that oral evidence must be not only anecdotal, but also personal and subjective. However, all other forms of evidence produced by human beings must be personal and subjective. They differ from oral sources in that they cannot be interrogated in order to reveal such characteristics, their extent and their own historical significance. A deliberate lie or a falsehood based upon error, can have importance.

Social historians are inclined to the view that oral history provides a good source of social memory, and can reveal the basis of collective consciousness. They also concede that in some instances, individuals can have an 'exemplary' value.[29] Linda Nochlin argued that the only way to obtain a plausible impression of reality is to collect many examples of 'the temporant fragment, as the basic unit of perceived experience'.[30]

Even heavily contaminated or badly distorted evidence is better than none at all, so long as it is recognisable for what it is. In this spirit, von Ranke approved of Clarendon's use of personal accounts in his work. Montaigne believed that an account written by a participant in an action would be the most valuable source of insight and wisdom, but that several witnesses were better than one.[31]

When considering soldiers' descriptions of their experiences in war, the listener (or reader) usually gets an impression of fragmented perceptions. Even when comparing evidence given by members of the same sub-unit, the accounts will be found to be similar, but never exactly the same. Here, one can cite accounts from within the same tank crew, who were separated in their physical locations by mere inches for the duration of a number of major operations.[32] It is also of interest to compare the two views of a platoon in action given by its commander and by one of his NCOs.[33] In the Second World War soldiers in the fighting arms tended to have a very narrow field of interest; for the infantry private this would be centred on his section, and not extend much beyond the platoon.[34] When in action, his perceptions would be raised to abnormal levels of sensitivity, but also more closely and exclusively focused, by the physiological effects of adrenalin.[35]

However, fragmented and overlaid by excitement as it is, oral evidence is often the only source available for what happened in a fight. The story can never be complete because 'dead men tell no tales', but we must act in accordance with the military spirit of making do with what we can get, in order to acquire any knowledge at all. It can also be argued that in battle the actions of individuals can be of vital significance, and information in the possession of single persons can be of historical importance.

Oral history can address some obvious, important and obscure questions about war; questions which are not susceptible to exact measurement, calculation or cold logic. This paper will explore some of them.

1   What is it like being in battle? A commonly reported aspect is the sense of isolation, particularly for the infantry. Many soldiers were unnerved by the 'empty battlefield'. Soldiers very rarely saw their enemies; if they were well trained and fortunate, they occasionally saw light, movement and shadows.

Richard Seddon, a trained observer, remarked, 'In action in modern war, neither the enemy nor one's comrades are very visible; for any who are will soon be shot. There can be an arid emptiness in a landscape that is full of fear, death and the whine of bullets.'[36]

General Tyacke, who served in north-west Europe in 1939–40 with the Duke of Cornwall's Light Infantry, once saw a dog, presumed to be a member of an enemy patrol, whilst in no man's land near Metz. Whilst in the Dunkirk perimeter, he briefly saw a German machine-gun team move from one patch of cover to another. He saw no other enemy soldiers.[37] Brigadier Flood, then a captain with the Royal Berkshires in the same campaign, did see enemy motorcyclists while acting as rearguard east of Brussels. But he never saw the machine-gunner or sniper who wounded him whilst making an attack over open ground a few weeks later.[38] Throughout the campaign in Tunisia, as a platoon and company commander, Major Crews of the Sherwood Foresters saw enemy soldiers once, very briefly. They were advancing in company strength, and vanished as soon as the first shot was fired at them by his excited platoon.[39]

Armoured vehicles, too, hid when possible. Captain Montgomery went into action for the first time in Operation Epsom in June 1944, and was present throughout the battle, taking part in the advance to Hill 112, and subsequent withdrawal. He saw some dead enemy and friendly infantry when advancing to contact. He saw only one enemy tank, and its crew baling out after he had knocked it out. Apart from that he had no sight of the enemy.[40] His tank driver, Corporal Court, saw very little of the enemy. He said that in earlier battles in North Africa, what he could see was 'dark blobs with flashes on them'. Like other soldiers, he relied more on his ears than his eyes, listening carefully to the sounds of incoming shells.[41] Major Crews rapidly learned to distinguish between the 'crack' of the passing bullet and 'thump' of the rifle that had fired it.[42] The sense of smell could also be of importance; particularly when operating at night or in cluttered terrain.[43]

Another aspect of going into battle was the confusion caused by occasional, sudden eruptions of intense excitement when actually encountering the enemy or coming under fire.[44]

Sometimes soldiers saw large-scale actions from afar, but these panoramic views, inevitably from a safe distance, gave them a sense of unreality and detachment from the scenes witnessed.[45]

2   How does the soldier overcome apprehension and fear in order to carry on in battle? There are many answers to this question. Many veterans cite the importance of the sheer excitement and sense of adventure that they felt at the time, especially the first time.[46] The influence of training was a powerful factor, so that standard procedures would be followed as conditioned reflexes in the most difficult circumstances.[47] This was undoubtedly reinforced by the example of more experienced comrades, especially superiors.[48] Activity in itself was important, officers and NCOs in particular were often too busy to be afraid.[49] Some found that fear and anxiety were quelled by fatigue.[50] On

occasion they were driven by indignation against the enemy.[51] More often, they advanced to contact, moved by a sense of solidarity with their fellows.[52] After repeated experience of combat, some developed a comforting sense of fatalism.[53] A very few were not unduly affected, and usually felt safe, even in a pitched battle.[54]

The worst experiences were of enforced inactivity under fire,[55] or of the prelude to action.[56] Some preferred attack to defence, although offensive operations entailed a higher risk of death or injury.[57]

3   What is it like when you are wounded? Initially many casualties are unaware of their injuries, presumably anaesthetised by shock. They tend to carry on with their duties until restrained for treatment.[58]

4   What happens if soldiers cannot resist fear? And how are soldiers treated when they show signs of psychological disorder? It seems the symptoms of distress, and the response of other troops, were extremely varied.[59]

Besides these aspects of combat, oral sources can provide information about matters which were neglected because they were mundane, or suppressed because they were shameful. To a considerable extent the standard operational procedures taught in training establishments, and amply recorded in handbooks and pamphlets, were modified in the field. This was particularly so at tactical level. Many of these local procedures and modifications were recorded only in the minds and memories of those present. A few examples come to mind: the actual use of wireless in the field in 3rd RTR;[60] the division of labour amongst the officers of a field artillery battery, in particular the performance of duty as forward observation officer;[61] modifications made to armoured vehicles by their crews, contrary to official policy;[62] and the development of new tactics to suit unfamiliar ground or other material circumstances.[63] The real procedures of exerting leadership and making command decisions in the field are also illuminated by oral evidence, often upsetting 'the authorised version'.[64]

Shameful occurrences are often ignored or distorted in contemporary and historical official records. This can be done for legitimate functional reasons; armies must sustain the self-esteem and reputations of their troops in order to uphold their morale and preserve their fighting qualities. But there are other reasons as well. In Britain we are much more fortunate than the historians of the Second World War in Japan, where there has been a national effort to repress memories of defeat and bad behaviour.[65] But, there is still an authoritarian tendency amongst the senior ranks of the forces, which is inclined to conceal imperfections. The duty of the historian is clear: 'not to falsify the past by omitting ... possibly offensive aspects that were a part of that past'.[66] But it can be difficult to retrieve that past. One can encounter great reluctance to record bad behaviour, although many informants are willing, even anxious to unload such unhappy memories, if the tape recorder is switched off. Even so, oral evidence is often the only plausible form for such matters as:

1   Collective failures of morale and discipline; for example in 148 Brigade and 15 Brigade in Norway in 1940. Accounts of the progressive erosion of morale, leading to a decline of fighting spirit, decay of discipline and desertion, and rejection of officers, exist, which are of great interest in studying the inner dynamics of military failure.[67]

2   Incompetence, nervous collapse and panic on the part of officers, as reported by witnesses and, in some cases, by the sinners themselves.[68]

3   Disloyalty, defiance and murderous attacks upon superiors, usually arising from behaviour under item 2 above.[69]

4   Violations of the laws, customs and usages of war, committed both by the enemy and our own soldiers, such as killing prisoners and surrendering troops, false surrenders and attacks on wounded.[70] Conversely, there was often a considerable sense of empathy with the enemy soldier, reinforced by reciprocity of behaviour, and occasional sharing of facilities.[71]

5   Serious mistakes, such as attacks on friendly forces or civilians, or suffering from the hostile attentions of Allied troops. There is generally less inhibition in complaining about attacks on their own side by the RAF.[72]

As well as the routine and shameful, oral testimony can reveal the odd but significant aspects of war left out of the formal records. John Clive summarised these as 'the unique, the contingent, the unforeseen'.[73] Many a veteran would say that these aspects, plus long periods of unrelieved tedium, are the whole experience of war. But, the odd events and factors can be of vital importance in battle, which is the realm of uncertainty. A few examples come to mind:

1   The formation, performance and destruction of scratch units, usually created in time of catastrophe, such as the 'Champagne Battery' which flourished briefly in Petre Force in 1940. Is there another account of the use of 60-pounders in the anti tank-role?[74]

2   The use of cavalry to change field batteries, as occurred in Eritrea in 1941.[75]

3   The actual resolution of the siege of German parachutists at Dombas in Norway in 1940.[76]

4   Significant tactical details about the operations of 3 RTR in Operation Goodwood; being shelled by friendly artillery because they deployed forward of their start line before the attack began; being given timely and effective support by Typhoons against enemy anti-tank guns in a wood south of Caen–Troan railway; and being engaged by Panthers which emerged from Caen on to the Falaise road.[77]

I have not found reference to these things in other sources.

To conclude: it is a fair assumption that once all history was oral. Now we can supplement and support history with written sources and by careful study of other material evidence, such as maps, photographs, films and ground. But oral sources remain the most valuable form of evidence for historians who wish to understand and convey the reality of the most important and mysterious part of war, the nature of battle.

The Soldier's Experience in Two World Wars: Some Historiographical Comparisons

Hew Strachan

At first glance the idea of concluding a study of the soldier's experience in the Second World War with some consideration of what we know about the soldier's experience of the First World War seems eminently reasonable.

First, it makes historiographical sense. The study of morale – of why some mutinied, others deserted and most stuck it – is one of those subjects with which many historians feel uncomfortable. Men's personal convictions and inner emotions, especially when considered in the mass rather than in relation to individuals, are not as susceptible to documentary proof as many other areas of enquiry. Informed intuition plays too large a part in the process for methodological satisfaction. In such circumstances comparative approaches can prove both fertile and reassuring. For Britain and France at any rate, the examination of combat motivation is more advanced for the first of the two world wars than it is for the second. This volume, devoted as it is to the military experience of 1939–45, is important not least because it is pioneering. Therefore, the earlier conflict should be able to yield insights that will aid an understanding of the later.

Second, such a comparison makes chronological sense. The soldiers of the Second World War fought in the shadow of the First World War. When men went off to fight in 1914 their mental images of war had yet to be redefined by the grim and sustained awfulness of the Western Front. In 1939 there were none of the manifestations of popular enthusiasm which are alleged to have accompanied the advent of hostilities twenty-five years earlier. The streets of Berlin were silent on 1 September 1939; in London two days later the noise which greeted Chamberlain's announcement that Britain was at war was not that of baying crowds but of air-raid sirens. Many senior commanders had served at Verdun or on the Somme; their political coevals were persuaded that the travails of interwar politics were in part attributable to the loss of the best and brightest of their generation in those battles. And those too young to have served in the earlier war devoured the testimony of Robert Graves and Erich Maria Remarque: this was a literature of warning and they had been warned.

But on closer examination this logical and chronological progression from one war to the next is unsustainable.

First, those who were to be at the sharp end in 1939–45 did not necessarily heed the admonitions of the war literature. The 1920 edition of Wilfred Owen's poetry sold 730 copies; a second impression of 700 copies had not sold out by 1929. By the same date Rupert Brooke's collected verse had achieved total sales of 300,000.[1] Boys at school continued to read the Victorian yarns of G. A. Henty, and these military adventures were recast in twentieth-century terms by F. S. Brereton and others. Much popular fiction incorporated the First World War in the vocabulary of glorious heroism. One nineteen year old, day-dreaming on 3 September 1939, imagined the future: 'as I closed my eyes the picture which appeared on the screen of my eyelids was of myself leading a charge of cavalry'.[2]

The father of the writer of this particular passage was still living with the effects of a severe wound received at Givenchy in the winter of 1914–15, but his son's mental image owed nothing to trench warfare. The consequences of the First World War for one generation had not necessarily modified the heroic image of war for the next. Such discontinuities may illustrate nothing more profound than the belief of young men in their own immortality. All wars must trade on their recklessness. None the less, there is also a more substantive point. It required the experience of the Second World War – and perhaps even the atomic bomb – to hammer home the destructive nature of modern warfare made evident in the First World War. Benjamin Britten's *War Requiem* gave Owen's poetry a resonance it had failed to achieve hitherto, but it was not composed until 1961 and its first performance was in the rebuilt Coventry Cathedral, a victim of the bombing of 1940.

The fiftieth anniversary of the outbreak of the First World War prompted a renewed interest in the experiences of that war. Above all, the BBC's pioneering documentary series, *The Great War*, was among the first of such programmes to use 'talking heads': veterans of the front line were encouraged to reminisce in front of the cameras. But these men carried with them more than the vivid memories of the trenches: they knew that the First World War had, after all, not proved to be 'the war to end all wars'.

The result was a paradox. Much of our comprehension of the soldier's experience in the First World War was cast in the light of what subsequently happened in the Second. The demands that war made of men, of their courage and their capacity to cope, seemed unchanging across time. The principal publications on combat motivation in the First World War, which began to appear in the 1960s, and multiplied thereafter, could not rid themselves of this continuity.

One of the first was John Baynes's succinctly titled, *Morale: a Study of Men and Courage* (1967). Baynes's book was an examination of the 2nd Battalion, Scottish Rifles at Neuve Chapelle in 1915. He wanted to know how a regular army unit which went into action 900 men strong, continued to fight as a disciplined formation over six days in which it was reduced to a combat strength of 150. He dedicated his book to his father, who not only served with the Cameronians (Scottish Rifles) in the First World War, but also commanded the Second Battalion between the wars and returned to active service in the Second. The continuities went further: the younger Baynes was himself a Cameronian, and was the regiment's second-in-command at the time of *Morale*'s publication.

Paul Fussell's *The Great War and Modern Memory* (1975), an examination of the war writers, of Sassoon, Graves and the others, probably remains the most widely read text on its subject. Its merits are those of a work of literary criticism. But its treatment invited its readers to consider it also as military history, as a study of men's responses to war. In the latter capacity it was found wanting. At the time of its publication such censure seemed misplaced – a product of misunderstanding the author's intentions. But Fussell's subsequent books have made clear that the views he expressed in 1975 were in large measure the consequence of his own experience in the Second World War. *Wartime*, published in 1989 and devoted to the second war, is a sustained indictment of the military life and of military institutions.[3]

Although forerunners, neither Baynes nor Fussell launched the serious study of the soldier's experience of war. That is a claim for which the pre-eminent candidate is John Keegan's *The Face of Battle* (1976). Keegan's opening chapter was a powerful manifesto for a new form of military history – written from the perspective not of the commander but of the soldier. What, Keegan asked, *really* happens in a battle? Of his three case-studies the most up to date was the Somme. Ostensibly, therefore, Keegan had little to say about war since 1916. In reality, his agenda was set by more immediate preoccupations – the need to understand for what it was that the officer cadets, with whose education he was charged, were preparing.

Chronology was being inverted. The conceptual framework of much contained in these books and in their successors was set by studies of morale conducted during and immediately after the Second World War, not the First. Pre-eminent was the emphasis on primary-group loyalties.

It is now almost axiomatic in studies of combat motivation that the small group – the rifle section, the platoon, at most the company – is the key to morale. Men fight not for their countries or even for their regiments, but for their mates. In 1947 an American, S. L. A. Marshall, argued in *Men Against Fire: the Problem of Battle Command in Future War* that the greatest challenge on the battlefield was the sense of isolation: 'the tactical unity of men working together in combat', he averred, 'will be in the ratio of their knowledge and sympathetic understanding of each other.'[4] Marshall was a combat historian in the Pacific in the Second World War. His methods have since been called into question,[5] but at the time *Men Against Fire* seemed to represent the distillation both of extensive personal experience and of considerable research among after-action reports. The authoritativeness of its conclusions was affirmed in

1949 by an even more extensive study of the American soldier in the Second World War. This stated that 'loyalty to one's buddies' was a 'stringent group code'. But it went further: the commitment to the primary group was compatible with a simultaneous reluctance to identify with the army more generally.[6] In other words soldiers could fight for their immediate comrades while still rejecting the wider purposes of the war.

What was true for the American soldier in the Pacific war rapidly became true for the soldiers of other armies in other theatres. In 1948 Morris Janowitz and Edward Shils provided corroboration in the case of Germany. The Wehrmacht retained its combat cohesion despite defeat and allied propaganda because of 'the capacity of the primary group to resist disintegration'.[7] In 1984 John Lynn extended primary-group theory back in time, to the 1790s and the army of revolutionary France. French soldiers messed together in *ordinaires*, groups of fourteen to sixteen men under the command of a corporal.[8] Lynn's claims for primary-group theory were certainly not as extensive as those of Janowitz and Shils, but the implications were not dissimilar. Janowitz and Shils had isolated the fighting capacities of the German army from the aims and ambitions of Hitler and national socialism; Lynn's work downgraded the comparative significance of French nationalism and revolutionary ardour. In both cases political ideology was relegated as an important element in combat motivation.

Such widespread application of primary-group theory has to wrestle with two sets of difficulties. First, there is the obvious problem of time and place. Is a theory developed in the context of one army in one war applicable to the armies of other states in other epochs? The second difficulty transcends time and place and concerns the nature of combat itself. What happens to the small group when casualties are so extensive and suffered over such a brief period of time that the bonds of comradeship cannot be developed, let alone sustained? Could primary-group loyalties survive the casualty rates in the Soviet and German armies on the Eastern Front in the Second World War? Were there not also moments in the First World War when losses were destructive of comradeship?

Primary-group loyalty, although perhaps the most persuasive, was not the only explanation of combat motivation to appear in the wake of the Second World War. *The Anatomy of Courage* by Charles Wilson (or Lord Moran as he became) was first published in 1945. Wilson had been a medical officer with the Royal Fusiliers in the First World War, and his book was based on the diaries that he kept between 1914 and 1917. But *The Anatomy of Courage* was also a comparative exercise, taking its shape as lectures to the aircrews of Bomber Command in the Second World War and refined by the knowledge of the psychological stresses that they confronted. Wilson's conclusions were as relevant to airmen embarking on their second and third tours of operations as they were to soldiers about to go back up the line for the umpteenth time. Wilson believed that each man had a finite stock of courage. The battle-hardened veteran was therefore a mythical figure: sustained exposure to danger did not harden a soldier but eroded his limited resources. Therefore, what armed forces needed was a system of rotation in and out of battle, of leave and recreation, which eked out that stock of courage. For the historian of the First World War, as well as for the historian of the strategic air offensive in the Second World War, Wilson's analysis cannot fail to strike sympathetic chords. When confronted with the most dramatic of all collapses in morale on the Western Front, the mutinies of the French army in 1917, Pétain adopted just those palliatives later advocated in *The Anatomy of Courage*.[9]

The fact that these pioneering studies of the late 1940s have created such resonances for the later historians of the First World War is in part due to the fact that the fighting of the First World War has an instantly recognisable identity. This is not so true of the Second World War. The experience of the latter involves not only the strain of night-bombing, but also being on the receiving end of those bombs; it encompasses mobile operations of high tempo as well as static trench warfare; it spans a climatic range from the snow and ice of an Eastern Front winter to the desert of North Africa and the jungle of Burma. The First World War did of course embrace elements of all these. None the less, much more than the Second World War it is defined by a single sort of experience – trench warfare on the Western Front. The similarity of that experience for soldiers in different sectors and of different nationalities makes comparisons more meaningful and general conclusions more possible than is the case for the Second World War.

Within this context, the attractions of both theses – that concerning the primary group and that concerning the limited stock of each individual's courage – are that they can be related directly to the immobility and continuousness of trench war. The trench circumscribed social contact: the soldier's range of acquaintance was defined by the dug-out and the fire-step. The repetitive nature of that experience put a premium on recharging the psychological batteries when out of the line.

Their attractions go even further: they can encompass the prevalent notion that the war was futile. Persuaded that the conflict was fought without a purpose, that it was all a waste, the modern observer is potentially unable to comprehend how soldiers sustained its burden for year after year. Explanations that rely on big causes and big words are construed as big lies. Bereft of ideological explanations, the historian has to find functional reasons to explain why soldiers went on fighting. These are provided by reference to factors that operate within armies – their officers, their leave and recreational arrangements, and above all the role of comradeship.

This assumption – that most of the war's combatants for most of the time regarded their effort as futile – is a central assumption of two recent analyses of the dynamics of combat motivation in the First World War. Both appear, at least on the surface, to break the shackles of the post-1945 paradigms, but both remain strictly functional in their frameworks.

The first is Tony Ashworth's *Trench Warfare: the Live and Let Live System* (1980). Ashworth argued that soldiers learned to live with the war by managing it, by making it a routine. On quiet sectors both sides colluded in avoiding overtly aggressive behaviour; they set the war to a timetable which the other side could predict; they even negotiated tacit truces.

Their high commands did not condone this behaviour and did their best to disrupt it. But as the second of the two studies, Leonard V. Smith's *Between Mutiny and Obedience: the Case of the French Fifth Infantry Division during World War I* (1994), makes clear, the generals were not as powerful in relation to their men as their orders sometimes suggested. Troops told to hold positions at all costs did not do so: provided there was a reasonable show of resistance, their commanders colluded in such rewordings of their instructions. Command authority was therefore subject to implicit negotiation; indeed the morale of the French army was dependent on it being so.

Both books are enormously suggestive, but both come close to making a fetish of the blindingly obvious, constructing seeming singularity out of the commonplace.[10] Armies in static positions have frequently spent more time looking at each other than fighting each other. Indeed in the eighteenth century this was their preferred option, given the enormous cost in casualties of actual battle. Nor was inactivity simply mediated by the soldiers themselves; needless skirmishing was condemned by a succession of British commanders in the Napoleonic wars, including Sir Ralph Abercromby, Sir John Moore and the Duke of Wellington.[11] What was remarkable about the First World War when put in the context of previous wars was not that the armies remained comparatively inactive for such long periods, but that that inactivity still contained some elements of combat. The precise dividing line between battle and armed observation became obscure. In the Seven Years War a battle was a one-day affair and it might exhaust the fighting capacities of an army for a whole campaigning season; in the First World War the intensity of battle on any one day was not necessarily any greater than in the Seven Years War, but the battle itself could extend over months. The obvious parallel in previous wars would be not battle but siege operations. The war may at times have been fought at a less intense and a less destructive tempo than its commanders wished, but its fighting was still more continuous and more sustained than that of any war which had preceded it.

All the explanations for the motivation of soldiers so far adduced – the primary group, the limited stock of courage, the live-and-let-live system – have a common theme. They concentrate on functional aspects, on factors that are intrinsic to armies. But they gloss over one key aspect of army life – indeed the one which most obviously distinguishes it from civilian life: discipline. Furthermore, what is meant here is not discipline in some euphemistic sense – self-discipline, emulation, initiative – but discipline in a negative sense. Remarkably little is said in theories of combat motivation about punishment and deterrence.

In the 1990s the British press not infrequently dwelt on the injustices of the British military legal system in the First World War. It paraded cases of men shattered by shell-shock and yet convicted of desertion in the face of the enemy. Antony Babington's study of capital courts martial did indeed show that many of those executed were either unfit for service in the line or were soldiers with previously good records who had passed the limits of their endurance: they had expended their stock of courage. A total of 346 British soldiers received the death sentence between 1914 and 1920, 266 of them on charges of desertion.[12] But the only mutiny on any scale was that at Etaples in 1917, and the nearest the British army came to a widespread collapse in morale was the defeat of Gough's Fifth Army on 21 March 1918. The explanations are too varied for simple cause and effect, but it may be reasonable to suggest that punitive discipline was a larger factor in sustaining the British army in the First World War than the conventional wisdom allows. It is worth observing that Joffre had recourse to immediate and exemplary executions in order to sustain – successfully – the unity of the French army in the course of its massive retreat back to the Marne in 1914.

The comparison with the German army is instructive. Its response to its expansion, and to the incorporation of recruits from urban and industrial backgrounds, was to soften its disciplinary code. The German army carried out only 48 death sentences in the First World War. But by 1918 desertion and indiscipline were rampant. Richard Bessel reckons that up to 10 per cent of those moved from the Eastern Front to the Western in the winter of 1917–18 deserted.[13] Sickness rates soared. The army may have claimed that it still stood intact on enemy soil in November 1918, but many of its elements had ceased fighting long before the Armistice. In the Second World War the German army reflected the lessons of that experience by going to the opposite extreme. Manfred Messerschmidt has calculated that 15,000 German soldiers were executed.[14] The connection may not be direct, but the Wehrmacht fought on till the end in 1945. Once again, work on the Second World War may well be providing possible signposts for research on the First World War; certainly the students of both wars need to take cognisance of the possibly positive consequences of punitive procedures for combat motivation.

Nobody disputes the role of punishment in keeping order in eighteenth-century armies: the soldiers of Frederick the Great were flogged and bullied into obedience. The case for minimising this sort of discipline in the age of the mass army is that such armies have been made up of citizen soldiers, conscripts whose reasons for fighting have included their identification with the nation that gave them rights and benefits. But there is a paradox here. The functional explanations for combat motivation assume that once a man becomes a soldier he internalises military beliefs to the exclusion of others. They take at face value all those memoirs which emphasise the soldier's sense of alienation from the home front and civilian life.

The contradiction highlights the point that functional factors alone simply will not suffice to explain combat motivation in the age of the mass army. Soldiers who are also citizens take up arms for reasons that cannot be accounted for with such a narrow frame of reference. Religion may well have been significant in prompting the soldiers of the American Civil War.[15] In both world wars states fielded not regular, professional armies but nations in arms, or as near to them as Europe has got. If soldiers expressed their alienation from home, they did so with regret – they were giving voice to the hope that the gulf between them and their families would not remain forever fixed.

At least one First World War study is suggestive here. Stéphane Audoin-Rouzeau, in *Men at War 1914–18: National Sentiment and Trench Journalism During the First World War* (1992), has looked at the trench newspapers of the French army and showed, among other things, the continuing link between front and rear. The soldier may excoriate the politician and the war profiteer, but he still fights for home and hearth – for mother, wife, and children. Because historians have looked for flag-waving patriotism they have failed to detect more subtly expressed loyalties. France was fighting a defensive war: Germans stood on French soil, and to the north and east of the front line French families lived under German rule.

There is an obvious parallel here with the Soviet army in the Second World War. The Soviet Union rallied to the notion of a 'great patriotic war'. But it is an analysis which can profitably be extended to other armies in the First World War. In the case of the German army, the long line of communications, the transfer of troops between fronts through Germany, the recall and release of men from and to the army and industry – all make any fixed division between military and civil life artificial. Even for those who remained rooted either at the front or at home the maintenance of leave and the flood of regular correspondence ensured that the barrier was never impermeable. The German army collapsed in 1918 in part because the home front was collapsing; and the home front collapsed in part because the army was collapsing. The nation in arms meant that each had to validate the efforts of the other.[16]

From here it is but a short step to suggest that more attention should be given to political factors in explaining combat motivation. Historians of the First World War, for reasons which have already been discussed, have tended to dismiss any association between high morale and broader political issues, or even between high morale and a sense of victory. But, if the nation validates the sacrifices which the soldier is making, then the soldier is given the sense that the cause in which he is fighting is right.

Again the chronological sequence needs to be inverted: the historiography of the Second World War can be suggestive for that of the First. The notion that the Allied cause in 1939–45 was just has been and is widely accepted. Richard Overy has emphasised its importance to the Allied war effort in *Why the Allies Won* (1995). But the knowledge that the Allies knew the Nazi régime to be evil must not blind us to a comparably deep ideological conviction on the German side. Omer Bartov's book, *Hitler's Army: Soldiers, Nazis and War in the Third Reich* (1991) dismisses Janowitz's and Shils's emphasis on the primary group; it could not survive long in the murderous conditions of the Eastern Front. In its place he substitutes repressive discipline and terror of the enemy. But he also argues that on the Eastern Front the German army became Hitler's army; the circumstances in which the soldier found himself fitted the vocabulary of Nazism, and the ideological struggle between Teuton and

Slav, between national socialism and Bolshevism, left no possibility of quarter on either side.

Such perspectives may, for example, cause us to revisit the French mutinies of 1917. In their immediate aftermath, the French high command was quick to blame the discontent in the army on pacifism in the rear and on weak government. It argued that soldiers returning home on leave via the railway stations of Paris were badgered by potential revolutionaries. Guy Pedroncini's study of the mutinies, the first to be archivally based, rejected this interpretation as window-dressing. He saw the army's problems as self-contained: he took Pétain's measures to restore tactical and moral confidence as evidence that the high command recognised this fact. For him, the French army was not rejecting the need for a defensive war. Instead it was 'striking' because of the manner in which the war was being conducted. The root causes were professional not political. The trouble with this approach is that it assumes that the army was isolated from the prevailing currents of opinion in France, despite the fact that the army was in France and of France. The French soldier undoubtedly harboured grievances against the French military system, but that did not in itself preclude the possibility that he also picked up the views of radical socialists or that he longed for peace.[17]

Political ideology has played a part in combat motivation in the twentieth century. In the First World War, the Entente saw the war as a battle between liberalism and militarism. Of course, this was the cry of propaganda, but propaganda has to convey a viewpoint that is palatable to be effective, and big ideas do not need to find overt expression in the front lines to validate what soldiers do. For the Germans the war was one for the pre-eminence of spiritual and cultural values over materialism and capitalism. Even in 1914 they had their sights trained as much on the United States as on Britain. In the Second World War the ideologies of the fascist powers were themselves in part the fruit of the trenches. Their principal enemy was Bolshevism – whose political success owed a great deal in turn to the First World War, and which like fascism employed the vocabulary of war to project its political and economic targets.

In conclusion, we need to come back to the historian's inability to quantify morale and consequent hesitation in incorporating it in explanations for victory or defeat. It is much easier, and seemingly more objective, to resort to the determinism of advancing technology and material superiority. In such analyses, states which emphasise morale do so because they are backward and weak, because motivation must substitute for material abundance. Furthermore, the armies of these powers are inevitably attracted to the notion that the human spirit is the key to victory precisely because it is malleable; it is not, as guns and ammunitions are, capable of precise quantification, and therefore judgements concerning relative superiority or inferiority remain subjective. For the German army in both world wars the will to victory was important because it was outnumbered: quality had to substitute for quantity.

The trouble with this formulation is that it suggests a choice, a trade-off between morale and material. This will not do. What made the experience of the First World War so awful, and so radically different from every other experience in life, was that it was a war fought between industrialised powers. As another important First World War study, Eric Leed's *No Man's Land: Combat and Identity in World War I* (1979), makes clear, the individual struggled to retain his identity when caught in the maelstrom of machine war. In other words, it was the quantifiable elements of modern war, the munitions that were the consequence of economic capabilities, that put the strain on the qualitative. Men had assumed that the physical contours of war were dictated by geography and climate, by the seasons and the hours; they had reckoned that courage was allied to action and initiative. But in the First World War the timetable was set less by nature than the artillery barrage, itself the embodiment of high-level engineering and of mass production; at times, it seemed, high explosive could move mountains and turn night into day. It set the individual at nought: the test of his courage was to endure its fire with passivity, and to remain self-disciplined even when bodies round about him were smashed beyond recognisable human forms. The industrialisation of war was not a substitute for morale. Instead, it made increased demands on it: confronted by their own apparent impotence in the face of modern firepower, men had to be more brave, not less so.

Notes on Contributors

Paul Addison is Director of the Centre for Second World War Studies at the University of Edinburgh. He is the author of *The Road to 1945: British Politics and the Second World War* (1975) and *Churchill on the Home Front, 1900–55* (1992).

Martin S. Alexander is Professor of Contemporary History at the University of Salford. He is the author of *The Republic in Danger: General Maurice Gamelin and the Politics of French Defence, 1933–40* (1993).

Mark Axworthy is the author of *Third Axis, Fourth Ally: Romanian Armed Forces in the European War, 1941–45* (1995).

Omer Bartov is Professor of History at Rutgers University. He is the author of *The Eastern Front, 1941–45* (1985), *Hitler's Army* (1991) and *Murder in Our Midst: The Holocaust, Industrial Killing, and Representation* (1996).

Brian Bond is Professor of Military History at King's College, London. Among his books are *Britain, France and Belgium, 1939–40* (1975), *British Military Policy Between the Two World Wars* (1980), *War and Society in Europe, 1870–1970* (1984) and *The Pursuit of Victory: from Napoleon to Saddam Hussein* (1996).

Angus Calder is a former Reader in Cultural Studies at the Open University. He is the author of *The People's War* (1969), *Revolutionary Empire* (1981) and *The Myth of the Blitz* (1991).

Terry Copp is Professor of History and Co-Director of the Laurier Centre for Military, Strategic and Disarmament Studies, Wilfrid Laurier University, Waterloo, Ontario, He is the author of *Battle Exhaustion: Soldiers and Psychiatrists in the Canadian Army, 1939–45* and *The Brigade: The Fifth Canadian Infantry Brigade, 1939–45*.

Jeremy Crang is Lecturer in History and Assistant Director of the Centre for Second World War Studies at the University of Edinburgh. He is currently writing a book on the British army as a social institution during the Second World War.

Nigel de Lee is Senior Lecturer in the Department of War Studies at the Royal Military Academy, Sandhurst. His most recent publications are 'Post-modernism and Military History' in *War Culture and the Media* (1996) and 'Counter-Insurgency and Conscience, a British View', *Itineraris*, 1996: 2.

Gerry Douds is Head of History at Worcester College of Higher Education. He is the author of a number of articles on the political history of British India in the twentieth century.

John Ellis is the author of *Eye-Deep in Hell* (1976), *The Sharp End: the Fighting Man in World War II* (1980) and *Brute Force: Allied Strategy and Tactics in the Second World War* (1990).

John Erickson is Emeritus Professor of Defence Studies at the University of Edinburgh. He is the author of *The Soviet High Command, 1918–41* (1962), *The Road to Stalingrad* (1975) and *The Road to Berlin* (1983), *The Soviet Forces 1918–1992* (1996), and editor of *Barbarossa: The Axis and the Allies* (1994).

Jürgen Förster is Senior Research Fellow at the Militärgeschichtliches Forschungsamt, Potsdam. Among his many publications on the history of the Second World War, he is the author of *Stalingrad: Ereignis, Wirkung, Symbol* (1992), and co-author of *The Attack on the Soviet Union*, volume 4 of *Germany and the Second World War* (1996).

Diana M. Henderson is Research Director of the Scots at War Trust at the University of Edinburgh. She is the author of *Highland Soldier* (1989) and *The Scottish Regiments* (1993).

Hamish Henderson is an Honorary Fellow of the School of Scottish Studies at the University of Edinburgh. A writer and poet, he is the author of *Elegies for the Dead in Cyrenaica* (1948), *Alias MacAlias: Writings on Song, Folk and Literature* (1992) and *The Armstrong Nose: Selected Letters* (1996).

Richard Holmes is Professor and Director of the Security Studies Institute at Cranfield University. Among his publications are *Firing Line* (1985), *Riding the Retreat: Mons to the Marne 1914 Revisited* (1995) and his latest book, *War Walks* (1996), written to accompany the BBC television series which he presented.

John Keegan, formerly Senior Lecturer at the Royal Military Academy, Sandhurst, is Defence Editor of the *Daily Telegraph*. Among his other books are *The Face of Battle* (1975), *Six Armies in Normandy* (1982), *The Mask of Command* (1987), *The Second World War* (1989) and *A History of Warfare* (1993).

David Killingray is Reader in History at Goldsmiths College, University of London. He is the author of several books and articles on African, Caribbean, Imperial, and English local history, and his most recent book is *Africans in Britain* (1994).

S. P. MacKenzie is Assistant Professor of History at the University of South Carolina. He is the author of *Politics and Military Morale: Current Affairs and Citizenship Education in the British Army, 1914–50* (1992) and *The Home Guard: a Military and Political History* (1995).

Reid Mitchell is Associate Professor of History at the University of Maryland, Baltimore. He is the author of *Civil War Soldiers* (1988) and *The Vacant Chair: the Northern Soldier Leaves Home* (1993) and *All on a Mardi Gras Day* (1995).

Reina Pennington is a postgraduate student at the University of South Carolina. She is currently editing an encyclopaedia of women's military history.

Theo J. Schulte is Senior Lecturer in History at Anglia Polytechnic University (Cambridge). He is the author of *The German Army and Nazi Policies in Occupied Russia* (1989), and of a forthcoming book, *Nazi War Crimes and War Trials*, due for publication in 1997.

G. D. Sheffield is Senior Lecturer in the Department of War Studies at the Royal Military Academy, Sandhurst and Senior Research Fellow of De Montfort University. He is the joint editor of *Warfare in the Twentieth Century* (1988) and author of *The Redcaps: a History of the Royal Military Police from the Middle Ages to the Gulf War* (1994).

Hew Strachan is Professor of Modern History at the University of Glasgow. He is the author of *Wellington's Legacy: The Reform of the British Army 1830–54* (1984) and *European Armies and the Conduct of War* (1983).

Brian R. Sullivan is Senior Research Professor at the Institute for National Strategic Studies, National Defense University, Washington. He is the coauthor of *Il Duce's Other Woman* (1993), and numerous articles on the military, naval, diplomatic and colonial history of Italy.

Steve Weiss is currently Visiting Research Fellow in War Studies at King's College, London and Research Associate at the Centre for Defence Studies in London. He has written on the physical and psychological dimensions of modern warfare and his book on World War II Anglo-American negotiations, *Allies in Conflict*, was published by Macmillan in 1996.

Theodore Wilson is Professor of History at the University of Kansas. He was the editor of *D Day 1944* (1994) and his forthcoming book, *Building Warriors: the Selection and Training of US Ground Combat Forces in World War II*, is scheduled for publication in 1997.

Ian S. Wood is Senior Lecturer in History at Napier University, Edinburgh. He is the author of *John Wheatley* (1990) and the editor of *Scotland and Ulster* (1994).

Notes and References

REFLECTIONS ON THE 'SHARP END' OF WAR

John Ellis

1 Respectively, the titles are as follows (the date is that of the paperback publication): *So Few Got Through* (1956); *The Only Way Out* (1957); *From the City, from the Plough* (1953); *Journey with a Pistol* (1961); *Patrol* (1955), and *The Monastery* (1957).

THE SCOTTISH SOLDIER: REALITY AND THE ARMCHAIR EXPERIENCE

Diana M. Henderson

1 Cyril, Falls, *The Life of a Regiment*, vol. IV, *The Gordon Highlanders in the First World War, 1914–1919* (Aberdeen, 1958), pp. 6–9.
2 Brevet Colonel William Eagleson Gordon VC *v.* John Leng and Co. Ltd (the proprietors of the *People's Journal*), 1919, Scottish Court of Session Cases, p. 415.
3 MS of the 1st Battalion Gordon Highlanders in the Le Cateau area August 1914, by Second Lieutenant the Hon. S. Fraser, later Lord Saltoun. The Highlanders (Seaforth, Gordons and Camerons) Regimental Archives, Aberdeen. David Fraser (ed.), *In Good Company: the First World War Letters and Diaries of the Hon. William Fraser, Gordon Highlanders* (Salisbury, 1990).
4 See for example, Peter Cochrane, *Charlie Company: In Service with C Company 2nd Queen's Own Cameron Highlanders, 1940–44* (London, 1979).
5 Patrick Delaforce, *Marching to the Sound of Gunfire: North West Europe, 1944–45* (Stroud, 1996) and Ian MacDougall, *Voices from War* (Edinburgh, 1995).
6 The author in conversation with the late Private Reginald Lobban, 1st Battalion Queen's Own Cameron Highlanders, 1913–19, and the late Lance-Corporal Frank McFarlane, 1st Battalion Black Watch, 1912–19; Eric Lomax, *The Railway Man* (London, 1995); Paul Fussell, *The Great War and Modern Memory* (Oxford, 1975).
7 Imperial War Museum Sound Archive Collection (IWM), London.
8 Lyn Macdonald, *Somme* (London, 1983).
9 The author in conversation with the late Alexander K. Smith, formerly of 51st Highland Division, Royal Corps of Signals. See also, Seona Robertson and Les Wilson, *Scotland's War* (Edinburgh, 1995).

10 Private MS of Sir Donald Maitland GCMG.

11 Walter A. Elliott, MC, *Esprit de Corps: a Scots Guards Officer on Active Service, 1943–5* (Norwich, 1996), p. 35.

12 Lieutenant-General Sir Derek Lang, *Return to St Valery* (London, 1974), pp. 37–8.

13 Julian Thompson, *The Imperial War Museum Book of Victory in Europe: the North-West European Campaign, 1944–45* (London, 1994), p. 101.

14 The author in conversation with the late Brigadier James A. Oliver, CB, CBE, DSO, TD, DL, LL.D.

15 Shelford Bidwell, *The Women's Royal Army Corps* (London, 1977), pp. 43–4.

16 Elliott, *Esprit de Corps*, pp. 127–30.

17 Ibid., pp. 147–8.

THE SHADOW OF THE SOMME

G. D. Sheffield

I would like thank the Imperial War Museum for permission to quote from interviews held in the Department of Sound Records.

1 G. C. W. Harland, *The War History of the 1/4th Battalion the King's Own Yorkshire Light Infantry* (privately published, 1987), p. 16. I am grateful to Light Infantry Office, Yorkshire for permission to quote from this source.

2 For the importance of the issues at stake, see Brian Bond, *The Pursuit of Victory* (Oxford, 1996), pp. 104–33. For a consideration of the myth and reality of Britain's role in the First World War, see G. D. Sheffield, 'Oh! What A Futile War: Representations of the Western Front in Modern British Media and Popular Culture', in *War, Culture and the Media*, ed. Ian Stewart and Susan L. Carruthers (Trowbridge, Wilts, 1996).

3 These words appear on the cover of A. J. P. Taylor, *The First World War: an Illustrated History* (Harmondsworth, 1978), a work originally published in 1963 which powerfully restated the mythical version of the First World War.

4 George Meddemmen, Imperial War Museum Sound Archive (hereafter IWMSA) no. 004764/05, reel 1.

5 Michael Howard, *Studies in War and Peace* (New York, 1970), p. 10.

6 Sonia Orwell and Ian Angus (ed.), *The Collected Essays, Journalism and Letters of George Orwell*, vol. 1 (Harmondsworth, 1970), p. 589.

7 Ibid., pp. 589–90.

8 Patrick Hennessey, *Young Man in a Tank* (privately published, n.d.), pp. 1–2.

9 D. A. Blake, IWMSA, no. 007392/11, reel 1.

10 G. F. Andow, IWMSA, no. 007196/08, reel 1.

11 Vernon Scannell, *The Tiger and the Rose* (London, 1983), pp. 71–2. See also his poem 'The Great War'.

12 Orwell, *Collected Essays*, vol. 1, p. 589; Brian Bond, *War and Society in Europe, 1870–1970* (London, 1984), p. 138.

13 John Verney, *Going to the Wars* (London, 1955), p. 24; C. E. J. Perry, IWMSA, no. 006320/03, reel 1.

14 Andow, IWMSA; A. B. Beauman, *Then a Soldier* (London, 1960), pp. 117–18.

15 Colonel M. G. N. Stopford, 'Trench Warfare – General – Winter 1914–15', lecture (1938), p. 13, Staff College Library, Camberley; R. H. Medley, *Five Days to Live* (Abergevenny, 1990), p. 6.

16 Alexander Barron, *From the City, from the Plough* (London, 1979), p. 157; D. Appleby, IWMSA, no. 7074/03.

17 Eric Wakeling, *The Lonely War* (Worcester, 1994), p. 7.

18 I am indebted to Peter Hart of the Sound Records Department of the Imperial War Museum for allowing me to use his as yet unpublished research. See also Peter Hart, ' "When Do You Scarper? When Do You Go into Action? That's the Nightmare": The Destruction of the South Nottinghamshire Hussars at Knightsbridge, 27 May–6 June 1942', *Imperial War Museum Review*, 8 (1993), pp. 27–37.

19 See the views of Harold Macmillan – a veteran of 1914–18 – on disciplinary problems in the army: Harold Nicolson, *Diaries and Letters, 1939–45*, ed. Nigel Nicolson (London, 1970) p. 211, diary entry of 24 Feb 1942.

20 Ibid., p. 473, diary entry for 13 June 1945; John Ellis, *The Sharp End of War* (London, 1982), p. 250.

21 For the cinema as a mass medium, see Anthony Aldgate and Jeffrey Richards, *Britain Can Take It: the British Cinema in the Second World War* (Oxford, 1986), pp. 2–3.

22 Ian F. W. Beckett, *The Judgement of History* (London, 1993), pp. xx–xxi.

23 Review from *New Statesman* quoted in Sheridan Morley, *David Niven: the Other Side of The Moon* (London, 1986), p. 133.

24 Ellis, *Sharp End*, p. 214.

25 See G. D. Sheffield, 'Officer–Man Relations, Morale and Discipline in the British Army, 1902–22', PhD thesis, University of London, 1994.

26 D. Proctor, *Section Commander* (Bristol, 1990), pp. 13, 17–18, 24–5, 30, 32–3; Sydney Jary, *18 Platoon* (Bristol, 1994), pp. 4, 6, 7. I am grateful to Mr D. Proctor and Mr Sydney Jary for permission to quote from the former source.

27 Alex Bowlby, *The Recollections of Rifleman Bowlby* (London, 1989), pp. 19–20, 109–10.

28 K. W. Cooper, *The Little Men* (London, 1992), pp. 100, 107.

29 Verney, *Going to the Wars*, p. 51.

30 J. M. R. Fraser, IWMSA, no. 10259/5, 1988, reel 2.

31 John Masters, *The Road Past Mandalay* (London, 1961), p. 287.

32 Peter Cochrane, *Charlie Company* (London, 1977), p. 173.

33 Adrian Gilbert, *The Imperial War Museum Book of the Desert War* (London, 1992), p. 164.

34 John Ellis, *The World War II Databook* (London, 1993), p. 254.

35 Julian Thompson, *The Imperial War Museum Book of Victory in Europe* (London, 1994), pp. xii, 138; Jeffery Williams, *The Long Left Flank* (London, 1988), pp. 318–20; Ellis, *Sharp End*, pp. 175–89.

36 Verney, *Going to the Wars*, p. 86. See also interview, IWMSA, G. F. H. Archer.

37 Bryan Samain, *Commando Men* (London, 1988), pp. 50, 55.

38 Patrick Delaforce, *The Fighting Wessex Wyverns: from Normandy to Bremerhaven with the 43rd Wessex Division* (Stroud, 1994), pp. 21, 65.

39 Masters, *The Road Past Mandalay*, p. 236.

40 Patrick Delaforce, *Churchill's Desert Rats: from Normandy to Berlin with the 7th Armoured Division* (Stroud, 1994), p. 54; Robin Neillands, *The Conquest of the Reich: D-Day to VE Day – a Soldier's History* (London, 1995), p. 89.

41   Spike Milligan, *Mussolini: His Part in My Downfall* (Harmondsworth, 1980), p. 284.

42   For the harsh conditions at Cassino in 1944, see G. D. Sheffield, *The Redcaps* (London, 1994), pp. 123–4. See also Williams, *The Long Left Flank*, p. ix.

43   Thompson, *Imperial War Museum Book of Victory in Europe*, pp. 199, 220–1.

44   Public Records Office, WO95/1435, war diary of 1st Gordon Highlanders, Appendix to Sept. 1918.

45   See for example, *Yorkshire Post* (13 September 1940), quoted in B. N. Reckitt, *Diary of Anti-Aircraft Defence, 1938–44* (Ilfracombe, 1990), p. 24.

46   Barron, *From the City, from the Plough*, p. 136; John Brophy and Eric Partridge, *The Long Trail* (London, 1965), p. 224; George Macdonald Fraser, *Quartered Safe Out Here* (London, 1992) p. 109.

47   C. L. Potts, *Gordon and Michael: a Memoir* (privately printed, 1940), p. 96. I am indebted to Peter Caddick-Adams for this reference.

48   Masters, *The Road Past Mandalay*, p. 59; Medley, *Five Days to Live*, pp. 36–7; Proctor, *Section Commander*, p. 37.

49   Delaforce, *Desert Rats*, p. 128.

50   Bowlby, *Recollections*, p. 190; Keith Douglas, *Alamein to Zem Zem* (London, 1979), p. 16.

51   Cochrane, *Charlie Company*, pp. 16–17.

52   Harland, *War History*, p. 26.

THE BRITISH FIELD FORCE IN FRANCE AND BELGIUM, 1939–40

Brian Bond

1   Alan Bennett, *Forty Years On* (London, 1969), p. 27.

2   Field-Marshal B. L. Montgomery, *Memoirs* (London, 1961), pp. 49–50.

3   An analysis of the Territorial Army's performance in 1940 was the subject of Brigadier K. J. Drewienkwicz's unpublished thesis at the Royal College of Defence Studies (1992).

4   David Divine, *The Nine Days of Dunkirk* (London, 1959); Richard Collier, *The Sands of Dunkirk* (London, 1961); Gregory Blaxland, *Destination Dunkirk* (London, 1973); Nicholas Harman, *Dunkirk: the Necessary Myth* (London, 1980); Ronald Atkin, *Pillar of Fire: Dunkirk, 1940* (London, 1990).

5   Ewan Butler and J. Selby Bradford, *Keep the Memory Green: the Story of Dunkirk* (London, 1950); Bruce Shand, *Previous Engagements* (Salisbury, 1990); Anthony Rhodes, *Sword of Bone* (London, 1942; pbk edn, 1986); Christopher Seton-Watson, *Dunkirk – Alamein – Bologna* (London, 1993).

6   Michael Howard and John Sparrow, *The Coldstream Guards, 1920–46* (Oxford, 1951), pp. 26–7.

7   All references to Imperial War Museum (IWM) sources are to the transcripts of sound recordings (the Gort Project) unless otherwise indicated: Victor Gilbey, p. 31, H. W. Dennis, p. 16; F. P. Barclay, p. 93, G. E. Andow, pp. 87–8.

8   Alfred Baldwin, p. 14, H. W. Dennis, pp. 9, 16; D. F. Callander, p. 9, Montgomery's memorandum is cited by Atkin, *Pillar of Fire*, p. 37.

9   Barclay, p. 69.

10 Captain J. H. Patterson, Diary (IWM), entry for 19 May; Rhodes *Sword of Bone*, pp. 155–7.

11 Major I. R. English, p. 30. S. L. Rhodda, 'Recollections of the Retreat to Dunkirk in May 1940', 19 typed pages, IWM. Gilbey (p. 45) first saw Germans on the Commines Canal as late as 27 May. He mentions that twelve German prisoners were sent back to his battalion headquarters. Major E. J. Manley (IWM, 21pp. typed memoirs) who was a corporal in 92 Field Regiment Royal Artillery in 1940, displayed considerable literary skill, particularly in describing the hellish scene on the approach to Dunkirk. Callander, p. 21; Patterson, diary entries, especially for 26, 27 and 28 May.

12 J. E. H. Neville (ed.), *The Oxford and Bucks Light Infantry Chronicle*, vol. I, *September 1939–June 1940* (Aldershot, 1949). Major Elliot Viney (IWM Interview), and personal interview, 19 November 1995.

13 A. F. Johnson (IWM typed memoirs) was an RASC driver who arrived in France only on 14 April and was evacuated from La Panne on 28 May. He describes graphically the lack of control during the retreat and on the beaches (pp. 82–90). Gunner Alfred Baldwin's battery had run out of shells by 28 May. He is a severe critic of absence of control by both junior officers and NCOs (pp. 38–45), but also paid tribute to a magnificent show by Guardsmen marching on to the beach immaculately dressed and with rifles at the slope.

14 S. L. Wright, 6th Battalion Green Howards (IWM memoirs, 9 typed pp.). Major J. W. G. Cocke served with 51st Heavy Regiment Royal Artillery – see interview p. 73 for the acute lack of water. For Stan Smith's unfortunate experience, see Atkin, *Pillar of Fire*, p. 159.

15 English, pp. 39, 52. Howard and Sparrow, *The Coldstream Guards*, pp. 45–9.

16 Atkin, *Pillar of Fire*, pp. 152–3; Blaxland, *Destination Dunkirk*, pp. 286–7; Cyril Jolly, *The Vengeance of Private Pooley* (London, 1956). Another medical officer told Captain Patterson he had seen a line of eight ambulances full of wounded men shot up by German tanks, which were then themselves destroyed by British artillery.

17 Atkin, *Pillar of Fire*, p. 154; Blaxland, *Destination Dunkirk*, p. 298. The most detailed account is Leslie Aitken, *Massacre on the Road to Dunkirk* (London, 1988).

18 Harman, *Dunkirk*, pp. 98–9, 231.

19 George Self (interview on cassette only), reel 6 for the shooting of Germans in Duisans cemetery, and reel 7 for the handing over of prisoners to the French. Mr Self had read Nicholas Harman's accusation against the Durham Light Infantry before being interviewed and rebuts his charge angrily. Harold S. Sell (reel 7) confirms that he was the last person (in the DLI) to see the prisoners put in lorries to be taken to brigade headquarters.

20 Howard and Sparrow, *The Coldstream Guards*, pp. 48–58; obituary of Lieutenant-Colonel M. Ervine-Andrews, *Daily Telegraph* (1 April 1995); Callander, pp. 31–8.

21 Baldwin, p. 38, thought discipline on the beaches was good despite the absence of his battery officers. Rhodes, *Sword of Bone*, p. 227, provides a moving account of a gallant naval officer maintaining order.

22 Johnson, pp. 82–8.

23 English, p. 47; Colonel S. L. A. Carter, Sherwood Foresters (IWM), pp. 75ff.

24 J. Wheeler Bennett, *George VI* (London, 1958), p. 460.

THE REAL *DAD'S ARMY*

S. P. MacKenzie

1   Imperial War Museum Sound Archive (hereafter IWMSA), J. Perry, 1125; see J. Perry and D. Croft, *Dad's Army* (London, 1975), p. 8.

2   The sale of selected episodes from the series as BBC videos suggests that even when *Dad's Army* goes off the air for good it will retain (at the very least) a cult following.

3   IWMSA, Home Guard tapes.

4   Which is *not* to suggest that they are unworthy of serious academic study. See I. F. W. Beckett, *The Amateur Military Tradition, 1558–1945* (Manchester, 1991); I. F. W. Beckett, *Riflemen Form: a Study of the Rifle Volunteer Movement, 1859–1908* (Aldershot, 1982).

5   See Public Record Office (hereafter PRO), PREM 3/223/3.

6   S. P. MacKenzie, *The Home Guard: a Military and Political History* (Oxford, 1995), pp. 16–20.

7   PRO, INF 1/264, Public Opinion on the Present Crisis, 30 May 1940. See M. Allingham, *The Oaken Heart* (London, 1941), p. 207.

8   *The Times* 11 May 1940, p. 7.

9   See, e.g., National Library of Scotland (hereafter NLS), MS 3816, f. 167, cutting from *Daily Sketch*, 31 Oct 1944; PRO, CAB 65/7, WM 1169401; 361 H.C. (Debs) 5s col. 8; 116 H.L. (Debs) 5s, col. 329; House of Lords Record Office (hereafter HLRO), HAR 1/1, P. Harris diary, 13 May 1940; *Sunday Pictorial* (12 May 1940), p. 11; *Daily Mail* (17 May 1940), p. 4; *Sunday Express* (12 May 1940), p. 6; J. Wedgwood, *Memoirs of a Fighting Life* (London, 1941), pp. 245–6.

10 PRO, CAB 65/7, WM 141 (40) 9, 27 May 1940. See also FO 371/25189, 'Fifth Column Menace' paper by Neville Bland.

11 Churchill College Cambridge (hereafter CCC), Croft Papers, 2/4, 'An Appreciation of Present Defence Needs', 30 May 1940. See Imperial War Museum Department of Documents (hereafter IWM), H. R. V. Jordon, 'Military Security Intelligence in the Western Command, 1939–1946: a Personal Memoir', pp. 23–4; C. Graves, *The Home Guard of Britain* (London, 1943), p. 10; *The Times* (11 May 1940), p. 7; A. O. Bell (ed.), *The Diary of Virginia Woolf*, v, 1936–41 (London, 1984), p. 284; J. Langdon-Davies, *Parachutes over Britain* (London, 1940), pp. 20, 26, 30; P. Donnelly (ed.), *Mrs Millburn's Diaries: An Englishwoman's Day-to-Day Reflections 1935–45* (London, 1979), p. 37. For a more detailed discussion of the fifth column scare, see A. W. B. Simpson, *In the Highest Degree Odious: Detention Without Trial in Wartime Britain* (Oxford, 1992), pp. 105 ff.

12 Langdon Davies, *Parachutes over Britain*, pp. 35–6.

13 *The Times* (11 May 1940). See L. A. Hawes, 'The First Two Days: the Birth of the L. D. V. in Eastern Command', *Army Quarterly*, 50 (1945), p. 186; H. L. Wilson, *Four Years: the Story of the 5th Battalion (Caernarvonshire) Home Guard* (Conway, 1945), p. 19; CCC, Spears Papers, 1/278, Mullholland to Spears, 16 May 1940.

14 IWM, Hawes Papers, 'Formation of the L.D.V.', encl. Anderson to Hawes, 26 June 1941; see Graves, *The Home Guard*, pp. 15, 280; Northamptonshire Record Office (hereafter NRO), HGb/1, Shillito notes, n/d; P. Finch, *Warmen Courageous: The Story of the Essex Home Guard* (Southend-on-Sea, 1951), p. 8; E. R. Murrow,

This Is London (New York, 1941), p. 98; K. Martin, *Editor: a Second Volume of Autobiography, 1931–1945* (London, 1968), p. 280; B. G. Holloway (ed.), *The Northamptonshire Home Guard, 1940–45* (Northampton, 1945), p. 3; *'We Also Served': the Story of the Home Guard in Cambridgeshire and the Isle of Ely, 1940–43* (Cambridge, 1944), pp. 31–2; see IWMSA, W. F. Kellaway, 11283.

15 IWM, Carden Roe Papers, 77/165/1, 'Birth Pangs of the Home Guard', 1.

16 IWMSA, extract from Eden talk, I. Lavender, 11274; Liddell Hart Centre for Military Archives (hereafter LHC), Kirke Papers, 20/3, 12; PRO, WO 199/3236, 'Home Guard (Local Defence Volunteers) Origins', memo by Major J. Maxse; Hawes, 'First Two Days', *passim*; IWM, Hawes Papers, 'Formation of the L. D. V.', encl. Anderson to Hawes, 26 June 1941; PRO, HO 45/25113, sub-file 863601/1.

17 PRO, WO 199/3243, History of the Formation and Organization of the HG, 1; see also IWMSA, A. Ambler, 11229.

18 See, e.g., LHC, Liddell Hart Papers, 1/112/35, Brophy to Liddell Hart, 15 Aug 1940; NRO, HGb1, Mellows to Wake, 27 June 1940; *'We Also Served'*, p. 51; H. Gough, *Soldiering On* (London, 1954), p. 243; Wedgwood, *Memoirs*, p. 248; 361 H. C. (Deb.) 5s, cols 9, 757: 116 H.L. (Deb.) 5s, cols 467–86; 362 H.C. (Deb.) 5s, cols 62, 281–3, 458, 645, 686, 1052–7, 1105; 376 H. C. (Deb.) 5s, col. 2136; PRO, CAB 106/1202, speech by J. Wedgwood, June/July 1940; C. Mackenzie, *My Life and Times, Octave 8, 1938–46* (London, 1969), p. 95; *Picture Post* (20 July 1940), p. 34.

19 For this common rendering of LDV see, e.g., Clare Court, *A Somerset Village in Wartime: Roadwater, 1939–45* (Roadwater, 1995), pp. 16–17.

20 Central Statistical Office, *Statistical Digest of the War* (HMSO, 1951), p. 13, table 15; NLS, MS 3816, f. 400, extract from *Scotsman* (11 Oct 1940), quoting letter from Croft to Sir Samuel Chapman, MP; PRO, T 162/864/E1628/1, Humphreys-Davies to Crombie, 14 May 1940; J. Brophy, *Britain's Home Guard: A Character Study* (London, 1945), p. 17; *The Times* (18 May 1940), p. 3; ibid. (27 May 1940), p. 3.

21 The figure of 75 per cent was almost certainly too high. See MacKenzie, *The Home Guard*, p. 37. A sense of the varying number of veterans in different units can be gained from the responses to questions asked of former Home Guards by the staff of the IWMSA.

22 IWMSA, R. Freeman, 11226.

23 Greater London Record Office, HG1, LDV – London Area: Provisional List of Inner Zone Organisers.

24 HLRO, HG/1, Record of Service Book, Palace of Westminster HG; Gough, *Soldiering On*; F. Law, *A Man at Arms: Memoirs of Two World Wars* (London, 1983).

25 LHC, Pownall Diary, 20 June 1940.

26 See e.g., PRO, PREM 3/223/3; WO 199/3237, Trg Instruction No. 8; 116 H.L. (Deb.) 5s. col. 486–7; H.C. (Deb.) 5s, col. 686.

27 MacKenzie, *The Home Guard*, pp. 91–6, 120–21.

28 See PRO, CAB 65/20, 121(41)1, 28 Nov 1941; WO 162/72, ECAC/M(41)17, 27 June 1941; ibid., ECAC/M(41)30, 24 Oct 1941.

29 MacKenzie, *The Home Guard*, pp. 89–90, 118–19, 132–5, 147–9.

30 Ibid., ch. 8.

31 LHC, Pownall Diary, 12 Aug 1940.

32 J. Perry, *Literary Review* (June 1995), p. 16.

33 IWMSA, J. D. C. Graham, 008337; see, e.g., H. Holttum, 10459; G. S. Irving, 12288; W. F. Kellaway, 11283; G. Redmayne, 13850; H. A. Wright, 14928; E. R. Elliot, 10167; H. A. Wright, 14928; W. Ansel, *Hitler Confronts England* (Durham, NC, 1960), p. 227. There were, of course, exceptions: see IWMSA, H. C. Pond, 13143.

34 See, e.g., J. Lee, 'Early Days in the Home Guard', *The Back Badge*, 1 (1946), p. 61; *'We Also Served'*, pp. 63–4; J. Brady, *The History and Record of 'B' Company 10th (Torbay) Battn Devonshire Home Guard* (Torquay, 1944), p. 12; *35th City of London Battalion Home Guard* (London, 1946), pp. 1–2; J. C. Spence, *'They Also Serve': the 39th Cheshire Battalion Home Guard* (London, 1945), p. 19; L. Jackson, recollection in F. and J. Shaw (comp.), *We Remember the Home Guard* (Oxford, 1990), p. 7; P. Ziegler, *London at War, 1939–45* (New York, 1995), p. 104; N. Longmate, *How We Lived Then* (London, 1971), p. 109; D. E. Johnson, *East Anglia at War, 1939–45* (Norwich, 1978), p. 30; R. D. Brown, *East Anglia, 1940* (Lavenham, 1981), p. 85.

35 *Sunday Pictorial* (25 Aug 1940); see, e.g., *Daily Herald* (13 July 1940), p. 3; J. Jellen, 'The Home Guard', *Journal of the Royal United Services Institution* (May 1944), p. 128; *The Watch on the Braids: the Record of an Edinburgh Home Guard Company, 1940–44* (Edinburgh, 1945), p. 9; PRO, MEPO 2/7013, encl 1B; Buckinghamshire Record Office, Basden Papers, D 113/77, 'The Home Guard of Farnham Royal', p. 4; Liddle Collection, Leeds University (hereafter LC), J. Walker Papers, Walker to Ridley-Smith, 4 June 1940, YA 7517/11, SOHG (Yorks Area) to Group Cmdrs, 7 September 1940; H. Scott, *Your Obedient Servant* (London, 1959), p. 134; A. G. Street, *From Dusk Till Dawn* (London, 1947), pp. 35–8; *'We Also Served'*, pp. 78–9; *The Story of the First Berkshire (Abingdon) Battalion Home Guard* (Manchester, 1945), p. 16; P. Hallding, 'Early Days', *Choughs Annual Register*, 2 (1945).

36 Bn Order 12, quoted in A. St H. Brock (ed.), *7th Herefordshire Battalion Home Guard* (1945), 3. See NRO, HGbl, County Instruction no. 8, 11 June 1940, para 3; L. C. Jeffries, recollection in Shaw, *We Remember the Home Guard*, pp. 80–81; Birmingham Reference Library, MS 1383/2/11, 2 Platoon report, 17 Aug 1940; IWMSA, B. Brighouse, 11230.

37 *Daily Worker* (29 June 1940), p. 2, (26 June 1940), p. 2; *Daily Herald* (22 June 1940), p. 5; Johnson, *East Anglia*, pp. 28–9.

38 LC, J. Walker Papers, CRNC 2/18259/G/LDV, Northern Command circular, 14 June 1940; see IWMSA, J. Pascall, 11236.

39 MacKenzie, *The Home Guard*, pp. 77–81.

40 See, e.g., Chester City Record Office, CR 501/1, Western Cmd HG Orders, no. 2, May 1941 (ACI 544/1941); NLS, MS 3818(I), f. 338; *Illustrated London News* (28 Aug 1940), p. 250.

41 See, e.g., PRO, WO 199/363, encl. 10A; see also MacKenzie, *The Home Guard*, pp. 113 ff.; IWMSA, D. Rochford, 13202; L. Hill, 11234; J. R. Bone, 12400.

42 *Watch on the Braids*, 17; see Finch, *Warmen Courageous*, pp. 29–30; Lieutenant-Colonel C. B. Costin-Nian, 'Home Guard Notes', *Daily Express* (12 Feb 1941); see IWMSA, W. I. Watson, 10420.

43 See MacKenzie, *The Home Guard*, pp. 102–16, 122–3; IWMSA, F. W. Johnes, 9366; K. Goodchild, 13909; L. Hill, 11234; Court, *A Somerset Village in Wartime*, p. 21.

44 See, e.g., IWMSA, R. Freeman, 1126; A. Ambler, 11229; H. Weston, 11235; J. Bevis in Shaw, *We Remember*, pp. 49–50.

45 See, e.g., F. Morgan, *Peace and War: A Soldier's Life* (London, 1961), p. 141; see also Ziegler, *London at War*, p. 104; IWMSA, W. I. Watson, 10420.

46 *Daily Worker* (17 May 1940).

47 Ziegler, *London at War*, p. 105; see, e.g., W. P. McGeoch, *The Triumphs and Tragedies of a Home Guard (Factory) Company: B Company 41st Warwickshire (Birmingham) Battalion Home Guard* (Birmingham, 1946).

48 IWMSA, B. Brighouse, 11230.

49 IWMSA, W. A. Pickering, 12012; see K. A. Squires, 13484.

50 MacKenzie, *The Home Guard*, p. 38; see Ziegler, *London at War*, p. 315.

51 IWMSA, H. W. G. Gower, 10966; see J. R. Bone, 12400; D. Falgate, 11587.

52 P. Mayhew (ed.), *One Family's War* (London, 1985), p. 69; see, e.g., Chesterfield Public Library, A. E. Slack Papers, *passim*; IWMSA, W. F. Kellaway, 11283; IWM, PP/MCR/208, J. Farley memoir, 2; NRO, HGb36, H. Banks, 'Jordan's Lot: A Saga of the Home Guard', p. 72.

53 Jimmy Perry has stated that at least some of the episodes were based on actual incidents he witnessed. IWMSA, J. Perry, 11225. For negative assessments of the accuracy of the series see IWMSA, H. Holttum, 10459; F. W. Johnes, 9366. For more positive views, see, e.g., A. Bashford, 12907; J. R. Bone, 12400.

54 See, e.g., John Brophy, *The Home Guard: A Character Study* (London, 1945).

55 See Beckett, *Amateur Military Tradition*, pp. 278–9.

56 IWM, Hawes papers, 87/41/2, Short note on the HG in AA Cmd, Jan 1945, 30.

57 *Sunday Express* (7 Nov 1943); see Modern Records Centre, Warwick University, MSS. 292/881.423/5.

58 PRO, WO 199/382, encl. 11A, HG AA Progress Report, 11 June 1943.

59 F. Pile, *Ack Ack: Britain's Defence against Air Attack during the Second World War* (London, 1949), p. 258; see PRO, WO 163/74, ECAC/M(43)17, 24 Apr 1943; WO 199/402, Minutes of Army Commander's Conferences of Home Guard Commanders, 2–5 Nov 1943; J. S. Davidson recollection in Shaw, *We Remember*, pp. 154 ff.; *History of the Cheshire Home Guard: from L. D. V. Formation to Stand-Down, 1940–44* (Aldershot, 1950), p. 63; E. D. Barclay, *The History of the 45th Warwickshire (B'Ham) Battalion Home Guard* (Birmingham, 1945), p. 26.

60 IWMSA, B. Wynne, 11227; see E. Summerskill, *A Woman's World* (London, 1967), pp. 73–4.

61 MacKenzie, *The Home Guard*, pp. 127–8, 147; see also IWMSA, D. M. William, 9440; E. Selwyn, 11228.

62 IWMSA, J. D. C. Graham, 008337; see MacKenzie, *The Home Guard*, pp. 181–2. Looking back, many of the former Home Guards who later served in the British army and were interviewed by the Imperial War Museum sound archive agreed. E.g.: IWMSA, A. Ambler 11229; H. Bainbridge, 14962; A. Barclay, 13171; A. Bashford, 12907; C. J. Braithwaite, 12288; H. W. G. Gower, 10966; P. Williams, 11231; K. A. Squires, 13484G. W. Walgate, 11456. Some, however, still believed that the force would have been effective. See W. F. Kellaway, 11283; H. E. Pratt, 12320.

63 Central Statistical Office, *Statistical Digest*, 1, pp. 8–9, 13; B. R. Mitchell, *British Historical Statistics* (Cambridge, 1988), p. 9.

64 IWMSA, J. Perry, 11225. On the Home Guard as a comic subject even during the war itself, see, e.g., B. Boothroyd, *Home Guard Goings On* (London, 1943).

THE BRITISH SOLDIER ON THE HOME FRONT

J. A. Crang

1 Public Record Office, War Office papers (hereafter PRO WO), WO 163/48, War Office progress report, AE 59, Sept. 1940; PRO WO 163/50, War Office progress report, AC/P (41)28, May 1941; PRO WO 163/51, War Office progress report, AC/G(42)12, March 1942; PRO WO 163/52, War Office progress report, AC/G(43)11, March 1943; PRO WO 163/53, War Office progress report, AC/G(44)16, March 1944; PRO WO 163/54, War Office progress report, AC/G(45)6, March 1945. It is difficult to assess accurately the proportion of the army that was based in Britain during the war, since the overall size of the army and the numbers stationed in Britain did not remain constant. In February 1944, for instance, 60 per cent of troops were serving in Britain, but in September 1940 this figure was as high as 92 per cent. I am grateful to Bruce M. Gittings for his assistance with these calculations.

2 Imperial War Museum, papers of Brigadier L. R. F. Kenyon, box 84/8/3, file WO Aug–Sept 1940, Army Morale, MI(R)WO, 29 Sept 1940 (hereafter IWM, morale report, Sept 1940), p. 1.

3 PRO WO 277/16, *Morale*, compiled by Lieutenant Colonel J. H. A. Sparrow, 1949 (hereafter Sparrow, *Morale*), p. 5; PRO WO 163/86, 'The Army and the Public', memorandum by P. J. G., ECAC/P(41)96, 31 Oct. 1941; PRO WO 259/62, extract from the conclusions of a meeting of the War Cabinet, 9 March 1942.

4 PRO WO 259/62, 'Decline in Army Morale', note by P.J.G., n.d.

5 Sparrow, *Morale*, p. 5.

6 PRO WO 163/87, 'Morale in the Army', memorandum by AG, ECAC/P(42)21, 25 Feb 1942.

7 PRO WO 163/88, morale report, Jan 1942, issued with ECAC/P(42)37, 1 April 1942; Sparrow, *Morale*, p. 24; King's College, University of London, Liddell Hart Centre for Military Archives, papers of General Sir Ronald Adam (hereafter LHC Adam), viii, 'Various Administrative Aspects of the Second World War', 1960, ch. 5, pp. 2–3; PRO WO 163/87, 'Morale in the Army', memorandum by AG, 25 Feb 1942; PRO 163/87, minutes of the 48th meeting of the Executive Committee of the Army Council (ECAC), ECAC/M(42)9, 27 Feb 1942; PRO WO 163/87, 'Morale in the Army', memorandum by AG, ECAC/P(42)25, 4 March 1942; PRO WO 163/87, minutes of the 49th meeting of the ECAC, ECAC/M(42)10, 6 March 1942.

8 Sparrow, *Morale*, pp. 28–9; LHC Adam, viii, ch. 5, pp. 3–4.

9 Sparrow, *Morale*, pp. 28–30; LHC Adam, viii, ch. 5, pp. 3–5. It should be noted that the first two morale reports (Jan and Feb 1942) were produced on a monthly basis, but thereafter reports were produced every three months. See PRO WO 163/88, First Meeting of the War Office Morale Committee, note by AG, ECAC/P(42)37, 1 April 1942; PRO WO 163/88, minutes of the 53rd meeting of the ECAC, ECAC/M(42)14, 3 April 1942.

10 Sparrow, *Morale*, p. 3.

11 IWM, appendix 'A' to morale report, Sept 1940, p. 2.

12 Ibid., p. 4.

13 Ibid.

14 Ibid.

15 PRO WO 163/88, morale report, Jan 1942, p. 3; PRO WO 163/161, morale report, Nov 1942–Jan 1943, MC/P(43)2, 12 March 1943, p. 3; PRO WO 163/161, morale report, May–July 1943, MC/P(43)11, 18 Sept 1943, p. 7; PRO WO 163/162, morale report, Nov 1943–Jan 1944, MC/P(44)1, 24 March 1944, p. 4.

16 PRO WO 163/88, morale report, Jan 1942, p. 3; PRO WO 163/88, morale report, Feb 1942, issued with ECAC/P(42)37, 1 April 1942, p. 2; PRO WO 163/161, morale report, May–July 1942, MC/P(42)1, 11 Sept 1942, p. 6.

17 PRO WO 163/88, morale report, Jan 1942, p. 3; PRO WO 163/161, morale report, May–July 1942, p. 6; PRO WO 163/161, morale report, Nov 1942–Jan 1943, p. 3; PRO WO 163/162, morale report, Aug–Oct 1943, MC/P(43)13, 22 Dec 1943, p. 6.

18 PRO WO 163/162, morale report, Aug–Oct 1943, p. 6.

19 PRO 163/161, 'Army Morale and Efficiency. By a Private in the Black Watch', Dec 1942, appendix 'A' to MC/P(43)1, 1 March 1943, p. 3.

20 PRO WO 163/88, morale report, Feb 1942, p. 2; PRO WO 163/161, minutes of the 6th meeting of the Morale Committee (MC), MC/M(42)6, 21 Aug 1942; PRO WO 163/161, minutes of the 7th meeting of the MC, MC/M(42)7, 18 Sept 1942.

21 PRO WO 163/161, minutes of the 15th meeting of the MC, MC/M(43)5, 21 May 1943; PRO WO 163/162, minutes of the 22nd meeting of the MC, MC/M(43)12, 31 Dec 1943; PRO WO 163/162, minutes of the 23rd meeting of the MC, MC/M(44)1, 28 Jan 1944.

22 PRO WO 163/161, minutes of the 7th meeting of the MC, MC/M(42)7, 18 Sept 1942.

23 PRO WO 163/54, morale report, Nov 1944–Jan 1945, AC/G(45)12, 4 July 1945, p. 6; PRO WO 163/54, morale report, Feb–April 1945, AC/G(45)15, 20 Aug 1945, p. 2.

24 IWM, appendix 'A' to morale report, Sept 1940, p. 1.

25 Ibid.

26 Ibid.

27 IWM, morale report, Sept 1940, p. 6; ibid., appendix 'B', pp. 3–4.

28 PRO WO 163/88, morale report, Jan 1942, p. 3; PRO WO 163/51, morale report, Feb–May 1942, AC/G(42)20, 12 June 1942, p. 6; PRO WO 163/161, morale report, Nov 1942–Jan 1943, p. 12; PRO WO 163/161, morale report, Feb–April 1943, MC/P(43)8, 19 June 1943, p. 3; PRO WO 163/162, morale report, Nov 1943–Jan 1944, p. 4.

29 PRO WO 163/88, morale report, Jan 1942, p. 3; PRO WO 163/88, morale report, Feb 1942, p. 3; PRO WO 163/51, morale report, Feb–May 1942, p. 7; PRO WO 163/161, morale report, May–July 1943, p. 7; PRO WO 163/162, morale report, Aug–Oct 1943, p. 7; PRO WO 163/162, morale report, Nov 1943–Jan 1944, p. 4.

30 PRO WO 163/161, morale report, Nov 1942–Jan 1943, p. 12; PRO WO 163/162, morale report, Aug–Oct 1943, p. 7; PRO WO 163/53, morale report, May–July 1944, AC/G(44)39, 15 Nov 1944, p. 4; PRO WO 163/54, morale report, Aug–Oct 1944, AC/G(45)2, 17 Feb 1945, p. 3.

31 PRO WO 163/161, morale report, May–July 1942, p. 1.

32 PRO WO 163/53, morale report, Aug–Oct 1943, AC/G(44)4, 11 Jan 1944, p. 6; PRO WO 163/162, morale report, Feb–April 1944, MC/P(44)5, 24 July 1944, p.

4; PRO WO 163/162, morale report, May–July 1944, MC/P(44)10, 25 Sept 1944, p. 5; PRO WO 163/162, minutes of the 26th meeting of the MC, MC/M(44)4, 21 April 1944.

33 PRO WO 163/162, morale report, Feb–April 1944, p. 3; PRO WO 163/54, morale report, Nov 1944–Jan 1945, AC/G(45)12, 4 July 1945, p. 8; PRO WO 163/54, morale report, Feb–April 1945, p. 3.

34 PRO WO 163/162, morale report, May–July 1944, p. 4; PRO WO 163/161, minutes of the 13th meeting of the MC, MC/M(43)3, 26 March 1943.

35 PRO WO 163/54, morale report, Nov 1944–Jan 1945, pp. 7–8; PRO WO 163/54, morale report, Feb–April 1945, p. 3.

36 IWM, morale report, Sept 1940, p. 2.

37 Ibid.

38 PRO WO 32/9838, 'Simplification of Drill', Army Council Instruction 597, 9 March 1942.

39 PRO WO 163/51, morale report, Feb–May 1942, p. 3.

40 PRO WO 163/161, morale report, May–July 1942, p. 3.

41 PRO WO 163/51, morale report, Feb–May 1942, pp. 5–6; PRO WO 163/161, morale report, May–July 1942, p. 6; PRO WO 163/161, morale report, Aug–Oct 1942, p. 2.

42 PRO WO 163/161, minutes of the 14th meeting of the MC, MC/M(43)4, 22 April 1943; PRO WO 163/161, minutes of the 15th meeting of the MC, 21 May 1943; PRO WO 163/161, minutes of the 16th meeting of the MC, MC/M(43)6, 25 June 1943; PRO WO 163/161, minutes of the 17th meeting of the MC, MC/M(43)7, 23 July 1943; PRO WO 32/14687, letter from AG to GOs-C-in-C Home Commands, 31 July 1943; LHC Adam, v/iv, 'The Soldier's Well-Being', letter from AG to Corps District, Divisional, AA Group, District and Area Commanders, June 1943, p. 6; PRO WO 163/162, morale report, Aug–Oct. 1943, p. 7; PRO WO 163/162, morale report, Nov 1943–Jan. 1944, p. 6.

43 LHC Adam, v/iii, 'Administration of Discipline', letter from AG to Corps District, Divisional, AA Group, District and Area Commanders, March 1943 (hereafter LHC Adam, 'Administration of Discipline'), p. 5.

44 Penelope Summerfield, 'Education and Politics in the British Armed Forces in the Second World War', *International Review of Social History*, 26 (1981), pp. 140–46; S. P. MacKenzie, *Politics and Military Morale: Current Affairs and Citizenship Education in the British Army, 1914–1950* (Oxford, 1992), pp. 85–92, 118–20.

45 PRO WO 163/161, morale report, May–July 1943, p. 4.

46 IWM, morale report, Sept 1940, p. 2.

47 Ibid., appendix 'B', p. 2.

48 PRO WO 199/1649, minutes of the GOC-in-C's conference, 7 Aug 1940; IWM, morale report, Sept 1940, pp. 5–6.

49 PRO WO 163/161, morale report, May–July 1942, p. 2.

50 PRO WO 163/51, morale report, Feb–May 1942, p. 10.

51 Ibid., p. 3.

52 Ibid., p. 14.

53 PRO WO 163/52, morale report, Nov 1942–Jan 1943, AC/G(43)10, 7 April, p. 5.

54 Ibid.

55 Ibid.

56   PRO WO 163/162, morale report, Aug–Oct 1943, p. 3.

57   PRO WO 163/162, morale report, Nov 1943–Jan 1944, p. 2.

58   PRO WO 163/161, morale report, May–July 1942, p. 3.

59   PRO WO 163/161, morale report, Aug–Oct 1942, p. 1.

60   PRO WO 163/161, morale report, May–July 1942, p. 2.

61   Ibid., p. 3; PRO WO 163/161, morale report, Aug–Oct 1942, p. 1.

62   PRO WO 163/161, morale report, Feb–April 1943, p. 1.

63   PRO WO 163/89, 'Use of Hotels: Segregation of Officers and Other Ranks',
     memorandum by AG, ECAC/P(42)136, 13 Oct 1942; PRO WO 163/51, 'Use of
     Hotels: Segregation of Officers and Other Ranks', note by the Joint Secretaries,
     AC/P(42)11, 23 Oct 1942; PRO WO 163/51, minutes of the 16th meeting of the
     Army Council, AC/M(42)5, 28 Oct 1942; PRO WO 163/89, minutes of the 90th
     meeting of the ECAC, ECAC/M(42)51, 18 Dec 1942.

64   Parliamentary Debates (Commons), 5th ser., vol. 379 (1941–2), cols 1221–2; PRO
     WO 163/161, minutes of the 13th meeting of the MC, 26 March 1943.

65   PRO WO 163/161, morale report, Nov 1942–Jan 1943, p. 7.

66   PRO WO 163/51, morale report, Feb–May 1942, AC/G(42)24, 10 July 1942, p.
     3; Sparrow, *Morale*, p. 18; Henry Harris, *The Group Approach to Leadership-Testing*
     (London, 1949), pp. 123, 131–4.

67   PRO WO 163/51, morale report, Feb–May 1942, AC/G(42)24, p. 3.

68   LHC Adam, v/iii, 'Administration of Discipline', pp. 3–7; PRO WO 163/162,
     minutes of the 24th meeting of the MC, MC/M(44)2, 25 Feb 1944.

69   Sparrow, *Morale*, pp. 33–4.

70   PRO WO 163/161, morale report, May–July 1942, p. 6.

71   LHC Adam, v/iii, 'Administration of Discipline', p. 4.

72   Sparrow, *Morale*, pp. 31–2.

73   Ibid., p. 32.

74   Ibid.; PRO WO 163/161, morale report, Feb–April 1943, p. 9.

75   PRO WO 163/54, morale report, Nov 1944–Jan 1945, p. 6; Sparrow, *Morale*, pp.
     18–19.

76   PRO WO 163/162, morale report, Feb–April 1944, p. 1; PRO WO 163/54,
     morale report, Aug–Oct 1944, p. 2; PRO WO 163/54, morale report, Feb–April
     1945, p. 2.

77   IWM, appendix 'B' to morale report, Sept 1940, p. 10.

78   PRO WO 258/15, 'The Composition and Work of the Directorate of Public
     Relations, with Suggestions for Improvement', memorandum by Major-General,
     Director of Public Relations, 20 Aug 1940.

79   PRO WO 163/161, morale report, May–July 1942, pp. 1–2.

80   PRO WO 163/88, morale report, Jan 1942, p. 5.

81   PRO WO 163/161, 'Army Morale and Efficiency. By a Private in the Black
     Watch', Dec 1942, p. 2.

82   Captain Russell Grenfell, *Service Pay* (London, 1944), p. 52; PRO WO 163/162,
     Extract from Report from GOC, N. Ireland, appendix 'A' to morale report, Nov
     1943–Jan 1944, p. 1.

83   PRO WO 163/161, morale report, May–July 1942, p. 7.

84   PRO WO 163/51, morale report, Feb–May 1942, AC/G(42)20, p. 11; PRO WO
     163/162, morale report, Aug–Oct 1943, p. 8; PRO WO 163/162, morale report,
     Feb–April 1944, p. 1.

85 PRO WO, morale report, May–July 1942, p. 2.

86 Ibid., p. 10.

87 Ibid., p. 1.

88 PRO WO 163/88, 'Reorganisation of the Directorate of Public Relations', note by PUS, ECAC/P(42)63, 12 May 1942; WO 32/10280, 'Directorate of Public Relations', memorandum by the War Office, 8 Sept 1943; Sparrow, *Morale*, p. 31.

89 Sparrow, *Morale*, p. 33; PRO WO 163/54, morale report, Aug–Oct 1944, p. 9.

90 PRO WO 163/161, morale report, Aug–Oct 1942, pp. 2–3; PRO WO 163/162, Finance, appendix 'A' to morale report, May–July 1944, p. 1; PRO WO 163/54, morale report, Nov 1944–Jan 1945, p. 12.

91 PRO WO 163/161, morale report, Feb–April 1943, p. 1; PRO WO 163/162, morale report, Aug–Oct 1943, p. 1; PRO WO 163/162, morale report, Nov 1943–Jan 1944, p. 3; PRO WO 163/162, Finance, appendix 'A' to morale report, Feb–April 1944, p. 1; PRO WO 163/54, morale report, Nov 1944–Jan 1945, p. 1.

92 Angus Calder, *The People's War: Britain, 1939–45* (London, 1969), p. 249. This figure includes a 'compulsory allotment' deducted from a married soldier's pay.

93 PRO WO 163/161, morale report, May–July 1942, p. 7.

94 PRO WO 163/161, morale report, May–July 1943, p. 5.

95 Ibid., p. 9.

96 PRO WO 163/161, morale report, May–July 1942, p. 9.

97 PRO WO 163/161, morale report, Aug–Oct 1942, p. 3.

98 PRO WO 163/161, morale report, May–July 1943, p. 5.

99 PRO WO 163/161, morale report, Aug–Oct 1942, p. 3; PRO WO 163/162, Finance, appendix 'A' to morale report, Feb–April 1944, p. 1; PRO WO 163/161, minutes of the 17th meeting of the MC, 23 July 1943.

100 PRO WO 163/161, morale report, Nov 1942–Jan 1943, p. 3; PRO WO 163/52, morale report, Feb–April 1943, AC/G(43)17, 3 July 1943, p. 4; PRO WO 163/53, morale report, Nov 1943–Jan 1944, AC/G(44)22, 23 May 1944, p. 3; PRO WO 163/161, minutes of the 17th meeting of the MC, 23 July 1943; PRO WO 163/161, minutes of the 18th meeting of the MC, MC/M(43)8, 27 Aug 1943; PRO WO 163/162, minutes of the 27th meeting of the MC, MC/M(44)5, 26 May 1944.

101 PRO WO 163/161, minutes of the 6th meeting of the MC, 21 Aug 1942; PRO WO 163/161, minutes of the 12th meeting of the MC, MC/M(43)2, 19 Feb 1943.

102 PRO WO 277/4, *Army Welfare*, compiled by Brigadier M. C. Morgan, 1953 (hereafter Morgan, *Army Welfare*), pp. 37–8; PRO WO 163/162, morale report, Aug–Oct 1943, p. 4.

103 Morgan, *Army Welfare*, pp. 35–7; PRO WO 163/51, War Office progress report, July 1942, AC/G(42)28; PRO WO 163/161, morale report, Feb–April 1943, p. 7; PRO WO 163/161, morale report, May–July 1943, p. 4.

104 PRO WO 163/162, minutes of the 23rd meeting of the MC, MC/M(44)1, 28 Jan 1944.

105 PRO WO 163/162, Finance, appendix 'A' to morale report, May–July 1944, p. 1; PRO WO 163/162, morale report, May–July 1944, p. 3; PRO WO 163/54, morale report, Aug–Oct 1944, pp. 2–3; PRO WO 163/54, morale report, Nov 1944–Jan 1945, p. 1; PRO WO 163/54, morale report, Feb–April 1945, p. 4.

106 PRO WO 163/54, morale report, Feb–April 1945, p. 1.

107 PRO WO 163/161, morale report, May–July 1942, p. 3; PRO WO 163/161, morale report, May–July 1943, p. 2.

108 PRO WO 163/162, morale report, Nov 1943–Jan 1944, p. 1.

109 PRO WO 163/161, morale report, May–July 1943, p. 2.

110 Ibid.

111 Ibid., pp. 2–3.

112 Ibid., p. 3.

113 PRO WO 163/53, morale report, Nov 1943–Jan 1944, AC/G(44)22, 28 May 1944, pp. 6–7; PRO WO 163/54, morale report, Aug–Oct 1944, p. 1.

114 PRO WO 163/54, morale report, Aug–Oct 1944, pp. 4, 6.

115 PRO WO 163/54, morale report, Nov 1944–Jan 1945, pp. 9–10; PRO WO 163/54, morale report, Feb–April 1945, pp. 1, 3–4.

116 IWM, morale report, Sept 1940, p. 8.

117 PRO WO 163/88, morale report, Feb 1942, p. 1.

118 Ibid.

119 PRO WO 163/161, morale report, Aug–Oct 1942, p. 1.

120 PRO WO 163/54, morale report, Nov 1944–Jan 1945, p. 3.

121 Ibid., p. 1.

122 PRO WO 163/161, morale report, May–July 1943, p. 1.

123 PRO WO 163/162, morale report, Feb–April 1944, p. 1.

124 PRO WO 163/161, 'Proposed Essay Competition', memorandum by J. H. A. Sparrow, 2 Oct 1942, appendix 'A' to MC/P(42)2, 18 Nov. 1942; PRO WO 163/161, 'Assessment of Morale by Statistical Methods', report by IS2, MC/P(43)3, 23 March 1943; PRO WO 163/161, minutes of the 12th meeting of the MC, 19 Feb 1943.

125 David Englander and Tony Mason, *The British Soldier in World War II* (Warwick, 1984), p. 15.

126 Ibid., pp. 9–10.

127 Anthony Burgess, *1985* (London, 1978), p. 29.

128 PRO WO 163/161, morale report, May–July 1943, p. 2.

129 It should be recorded that several veterans have suggested to the author – for which there is some corroborative evidence in the morale reports and morale committee minutes with respect to relations between officers and men – that morale was better maintained in the fighting units than amongst those in technical or administrative units.

130 PRO WO 163/161, morale report, Aug–Oct 1942, p. 1.

''TWAS ENGLAND BADE OUR WILD GEESE GO'

Ian S. Wood

In preparing this chapter I was greatly helped by staff in Edinburgh Central Library, the National Library of Scotland, Edinburgh University Library, the libraries of the Scottish United Services Museum at Edinburgh Castle and the National Army Museum in London.

1 J. McGurk, 'Wild Geese: the Irish in European Armies', in P. O'Sullivan (ed.), *The Irish World Wide: History, Heritage, Identity*, vol. 1, *Patterns of Migration* (Leice-

ster, 1992), pp. 36–62; also H. Murtagh, 'Irish Soldiers Abroad', in *A Military History of Ireland*, ed. T. Bartlett and K. Jeffery (Cambridge, 1996), pp. 294–314.

2   *New Ulster Defender* (June/July 1994).

3   J. W. Blake, *Northern Ireland in the Second World War* (Belfast, 1956), pp. 93–4.

4   R. Fisk, *In Time of War: Ireland, Ulster and the Price of Neutrality* (London, 1983), pp. 452–3.

5   Blake, *Northern Ireland*.

6   Ibid., p. 453.

7   Fisk, *In Time of War*, pp. 390–96.

8   Blake, *Northern Ireland*, pp. 194–9.

9   Fisk, *In Time of War*, pp. 96–7, 282–3, 520–22.

10   Ibid., pp. 96–7.

11   Ibid., pp. 516–18.

12   W. Kimball, *Churchill and Roosevelt: the Complete Correspondence*, vol. 11, *The Alliance Forged* (Princeton, 1984), pp. 186–7.

13   Blake, *Northern Ireland*, pp.199–200; also D. Keogh, *Twentieth-Century Ireland: Nation and State* (Dublin, 1994), p. 123.

14   T. P. Coogan, *The IRA* (London, 1995), pp. 180–83.

15   B. Moore, *The Emperor of Ice Cream* (London, 1965), p. 15.

16   Blake, *Northern Ireland*, p. 129.

17   S. McAughtry, 'Royal Ulster Rifles Had a Part', *Irish Times* (6 June 1984).

18   S. McAughtry, *McAughtry's War* (Belfast, 1985), p. 170; also quoted in M. Goldring, *Belfast: from Loyalty to Rebellion* (London, 1991), p. 97.

19   R. Foster, *Modern Ireland, 1600–1972* (London, 1988), pp. 524–5; also E. O. Halpin, 'The Army in Independent Ireland', in Bartlett and Jeffery, *A Military History of Ireland*, pp. 409–10.

20   Foster, *Modern Ireland*, pp. 549–50; also M. Manning, *The Blueshirts* (Dublin, 1970).

21   P. Young, 'Defence and The Irish State', *The Irish Sword*, vol. xix (1993–4), pp. 1–10.

22   J. P. Duggan, *A History of the Irish Army* (Dublin, 1991), p. 180; also A. E. C. Bredin, *A History of the Irish Soldier* (Belfast, 1987), pp. 542–4; also Keogh, *Twentieth-Century Ireland*, p. 123.

23   P. Verney, *The Micks: the Story of the Irish Guards* (London, 1970), p. 133.

24   I am indebted for information about this, based upon his father's experience, to a conversation with Mr John McCabe on 27 July 1995.

25   Keogh, *Twentieth-Century Ireland*, p. 123.

26   F. H. Boland, 'A Memoir from External Affairs', *Irish Times* (8 May 1985).

27   K. Myers, 'Dublin Remembers Too', *Spectator* (18 Nov 1995).

28   F. S. L. Lyons, *Ireland Since the Famine* (London, 1971), p. 551.

29   Quoted in Fisk, *In Time of War*, p. 293.

30   Verney, *The Micks*, p. 15; also R. S. Churchill, *Young Statesman: Winston S. Churchill, 1901–14* (London, 1967), p. 505. Churchill described this incident in a letter to his wife.

31   Verney, *The Micks*, p. 85.

32   *Household Brigade Magazine* (Summer 1945), pp. 57–60.

33   Ibid.

34   Fisk, *In Time of War*, p. 526.

35   R. Doherty, *Clear the Way: a History of the 38th (Irish) Brigade, 1941–7* (Dublin, 1993), pp. 6–7.

36 S. E. Ambrose, 'Churchill and Eisenhower in the Second World War', in R. Blake and W. Roger Louis (ed.), *Churchill: a Major New Assessment of his Life in Peace and War* (London, 1993), pp. 399–400.

37 J. Ellis, *Brute Force: Allied Strategy and Tactics in the Second World War* (London, 1990), p. 293.

38 Doherty, *Clear the Way*, p. 26.

39 J. H. S. Horsefall, *The Wild Geese Are Flying* (Kineton, 1976), p. 39.

40 Doherty, *Clear the Way*, pp. 37–8.

41 Ibid., p. 57.

42 D. J. L. Fitzgerald, *A History of the Irish Guards in the Second World War* (Aldershot, 1949), p. 183.

43 Bredin, *A History of the Irish Soldier*, pp. 484–5.

44 Fitzgerald, *A History of the Irish Guards*, p. 183.

45 Doherty, *Clear the Way*, p. 42.

46 Ibid., p. 114.

47 Blake, *Northern Ireland*, pp. 462–4; also C. D'Este, *Bitter Victory: the Battle for Sicily 1943* (London, 1988), p. 404.

48 H. Henderson, 'Ballad of the Simeto (for the Highland Division)', *Cencrastus* (Winter 1995–6).

49 Doherty, *Clear the Way*, p. 73; also Bredin, *A History of the Irish Soldier*, p. 497.

50 J. Keegan, *The Second World War* (London, 1989), p. 368.

51 F. Mowat, *And No Birds Sang* (Toronto, n.d.), p. 167; also Doherty, *Clear the Way*, p. 107.

52 C. D'Este, *Fatal Decision: Anzio and the Battle for Rome* (London, 1991), p. 150; also Fitzgerald, *A History of the Irish Guards*, p. 243.

53 Fitzgerald, *A History of the Irish Guards*, p. 335.

54 Ibid.; also D'Este, *Fatal Decision*, pp. 149–50, 287–8.

55 Fitzgerald, *A History of the Irish Guards*, p. 356.

56 I am indebted to Professor Richard Holmes for this information in conversation with him on 22 September 1995.

57 Verney, *The Micks*, p. 138.

58 Doherty, *Clear the Way*, p. 116.

59 Ibid., p. 123.

60 C. Gunner, *Front of the Line* (Antrim, 1991), p. 81.

61 Ibid., p. 82.

62 Doherty, *Clear the Way*, pp. 159–60.

63 B. Harpur, *The Impossible Victory: a Personal Account of the Battle for the River Po* (London, 1980), p. 169.

64 M. Hastings, *D-Day and the Battle for Normandy* (London, 1984), pp. 144, 147–50.

65 Blake, *Northern Ireland*, pp. 495–6; also Hastings, op. cit., pp. 123–4.

66 Bredin, *A History of the Irish Soldier*, p. 510.

67 Fitzgerald, *A History of the Irish Guards*, p. 437.

68 Ibid., pp. 438–42.

69 Ibid., p. 421.

70 Ibid.

71 Fisk, *In Time of War*, pp. 522–5.

72 K. Jeffery, 'The British Army and Ireland', in *A Military History of Ireland*, ed. Bartlett and Jeffery, p. 438.

73   D. Coughlan, ' "Army Role in Ceremony Offensive", Says Fianna Fáil', *Irish Times*
     (12 Nov 1983).

74   R. Foster, 'Taking the Past Forward', *Sunday Times* (4 June 1995); also J.
     Carroll, 'Bruton's Tribute to Irish Who Fell in Last War', *Irish Times* (29 April
     1995).

'IF I FIGHT FOR THEM, MAYBE I CAN GO BACK TO THE VILLAGE'

David Killingray

1   Marika Sherwood, 'VE Day Commemorations', *West Africa*, 4038 (27 February–5
     March 1995), p. 304; 'Matchet's Diary', ibid., 4045 (17–23 April 1995), p. 578. An
     early and useful brief study of the African contribution to the war is by Rita
     Headrick, 'African Soldiers in World War II', *Armed Forces and Society*, 4 (1978), pp.
     501–26.

2   C. Mangin, *La Force noire* (Paris, 1910). There is a substantial literature on the
     French colonial military forces; e.g. Shelby Cullom Davis, *Reservoirs of Men: a
     History of the Black Troops of French West Africa* (Chambéry, 1934); Marc Michel,
     L'Appel à l'Afrique: contributions et réactions à l'effort de guerre en A. O. F., 1914–19
     (Paris, 1982); Anthony Clayton, *France, Soldiers and Africa* (London, 1988); János
     Riesz and Joachim Schultz (ed.), *Tirailleurs Sénégalais* (Frankfurt-am-Main, 1989);
     Myron Echenberg, *Colonial Conscripts: the Tirailleurs Sénégalais in French West Africa,
     1857–1960* (London, 1991); and Nancy Ellen Lawler, *Soldiers of Misfortune: Ivorien
     Tirailleurs of World War II* (Athens, Oh., 1992).

3   Brian Willan, 'The South African Native Labour Contingent, 1916–18', *Journal of
     African History*, 19 (1978), pp. 61–86.

4   David Killingray, 'The Idea of a British Imperial African Army', *Journal of African
     History*, 20 (1979), pp. 421–36.

5   Brian Sullivan, 'The Italo-Ethiopian War, October 1935–November 1941: Cause,
     Conduct and Consequence', in A. Hamish Ions and E. J. Errington (ed.), *Great
     Powers and Little Wars: the Limits of Power* (Westport, Ct, 1993), pp. 167–201.

6   Mabiala Mantuba-Ngoma, 'Les Soldats noirs de la Force Publique (1888–1945):
     contribution à l'histoire militaire du Zaire', thesis, National University of Zaire,
     Lubumbashi, 1980; Bryant Shaw, *'Force Publique, force unique*: the Military in the
     Belgian Congo, 1914–39', PhD thesis, University of Wisconsin-Madison, 1984.

7   David Killingray, 'Labour Exploitation for Military Campaigns in British Colonial
     Africa, 1870–1945', *Journal of Contemporary History*, 24 (1989), pp. 483–501; Geoff-
     rey Hodges, *The Carrier Corps: Military Labor in the East African Campaign, 1914–18*
     (Westport, Ct, 1986).

8   David Killingray, 'Labour Mobilisation in British Colonial Africa for the War
     Effort, 1939–46', in *Africa and the Second World War*, ed. David Killingray and
     Richard Rathbone (London, 1986), pp. 68–96.

9   Echenberg, *Colonial Conscripts*, p. 88.

10   Lawler, *Soldiers of Misfortune*, pp. 35, 65.

11   Much of the oral evidence collected from and about African soldiers who served
     in the Second World War includes stories which seem sensational. Ageing

memories are faulty and myths, and even the ideas of historians, have been absorbed as what actually happened all those years ago. The truth, or partial truth, of some of these stories has sometimes been verified either from other oral evidence or from archival sources. For example, there are frequent reports about the cannibalism of Congolese soldiers provided by Africans and Europeans who served in East and North Africa. The rumour mill is powerful and pervasive; caution is ever the watchword.

12 Bildad Kaggia, *Roots of Freedom, 1921–63: the Autobiography of Bildad Kaggia* (Nairobi, 1975), p. 21.

13 Public Record Office, Kew (PRO), DO35/1183/Y. 1069/1/1. Minutes of secret session of the Basutoland Council, 26 Oct 1943. Recruiting for AAPC.

14 Michael Crowder papers (MCP), 17. These papers consist of translations of soldiers' letters, and other materials, from the Botswana National Archives, collected by the late Michael Crowder for a biography of Tshekedi Khama. The papers are now deposited in the Institute of Commonwealth Studies, University of London, and subject to restricted access.

15 L. W. F. Grundlingh, 'The Participation of South African Blacks in the Second World War', unpub. D.Litt thesis, Rand Afrikaans University, 1986, p. 427. See also Ian Glesson, *The Unknown Force: Black, Indian and Coloured Soldiers Through Two World Wars* (Rivonia, 1994), chs 7 ff.

16 Deborah A. Shackleton, 'Recipe for "Failure": Integration of Botswana Soldiers into British Units during the Second World War', unpub. paper given to the 38th annual meeting of the African Studies Association of the United States, Orlando, Fl., November 1995, p. 6.

17 R. A. R. Bent, *Ten Thousand Men of Africa* (London, 1952); B. Gray, *Basuto Soldiers in Hitler's War* (Maseru, 1953); D. Kiyaga-Mulindwa, 'Bechuanaland and the Second World War', *Journal of Imperial and Commonwealth History*, 13 (1984), pp. 33–53; Brian Mokopakgosi III, 'The Impact of the Second World War: the Case of Kweneng in the Then Bechuanaland Protectorate, 1939–50', in *Africa and the Second World War*, Killingray and Rathbone (ed.), pp. 160–80; Hamilton Sipho Simelane, 'Labor Mobilization for the War Effort in Swaziland, 1940–42'. *International Journal of African Historical Studies*, 26 (1993), pp. 541–74.

18 Gerald Hanley, the Irish novelist, who served in Somalia and with East African troops in Burma, wrote a splendid account of the latter: *Monsoon Victory* (London, 1946). D. H. Barber, *Africans in Khaki* (London, 1948), is an account by an officer who served with the East African Pioneer Corps and in 1944 was appointed public relations officer for all African troops in the Middle East. See also George Youell, *Africa Marches* (London, 1949), an account by a padre who served with the West African forces.

19 There is material in Rhodes House Library, Oxford, MSS. Afr. s. 1715, for East Africa, and MSS. Afr. s. 1734, for West Africa. See further Anthony Clayton and David Killingray, *Khaki and Blue: Military and Police in British Colonial Africa* (Athens, Oh, 1989).

20 National Archives of Ghana, Accra, BF3692, SF11, contains four files of c. 600 letters from soldiers serving in India. A source of great value for this essay have been the letters from Tswana soldiers serving in the Middle East and written to Tshekedi Khama, and with the papers of the late Michael Crowder papers, MCP, 17.

21 Kaggia, *Roots of Freedom*. An earlier and much slighter autobiography is by R. H. Kakembo, *An African Soldier Speaks* (Edinburgh, 1946).

22 This material is in taped and written form and in the hands of Martin Plaut. It will be referred to here as BBC Africa Service.

23 See the discussion of figures in Echenberg, *Colonial Conscripts*, p. 88, and Lawler, *Soldiers of Misfortune*, 87.

24 Lawler, *Soldiers of Misfortune*, pp. 80–1.

25 Ibid., pp. 82, 83.

26 Ibid., pp. 85–6.

27 Ibid., pp. 143, 148.

28 Ibid., p. 161.

29 Jean Michel Domingue, 'The Experience of the Mauritian and Seychelles Pioneer Corps in, and Contribution to, the Egyptian and Western Desert campaigns, 1940–43', unpub. MA thesis, School of Oriental and African Studies, University of London, 1994, p. 18.

30 Grundlingh, 'South African Blacks in the Second World War', p. 212.

31 MCP, 17. Sgt Molefi, 1966 Coy Bechuana APC, to Tshekedi Khama, 31 Jan 1945.

32 PRO, DO35/4071, 'Swazi Regiments of the African Pioneer Corps in the war, 1941–5', Record by Major F. P. van Oudtshoorn. DO35/1183/Y. 1069/32, contains a copy of *The Swazi Pioneers, September 1941–March 1945*, a 55-page roneo'd account of the Pioneers compiled by three Swazi members of the Pioneer Corps.

33 MCP, 17, M. Thapedi, CNR no. 8161, to Tshekedi, 4 Dec 1944.

34 MCP, 16, S/133/2/1, Newsletter of Pioneers, Feb 1944.

35 MCP, 16, S/133/2/1, Chaplains (CMF) Newsletter, March 1944.

36 Lawler, *Soldiers of Misfortune*, pp. 172–3.

37 Ibid., p. 175.

38 Ibid., p. 179.

39 See ibid., ch. 5, 'The *Tirailleur* as Prisoner', and Echenberg, *Colonial Conscripts*, 96–9. A broader account of African prisoners of war during the Second World War is in David Killingray, 'Africans and African Americans in enemy hands', in *Prisoners-of-War and Their Captors in World War II*, ed. Kent Fedorowich and Robert Moore (Leamington Spa, 1996).

40 Nzamo Nogaga, 'An African Soldier's Experiences as a Prisoner-of-War', *South African Outlook* (1 Oct 1945), p. 151.

41 This is discussed in detail in a chapter on dissent, desertion and mutiny in a book that I am currently writing on 'African Soldiers in the Second World War'.

42 Kaggia, *Roots of Freedom*, pp. 24–5.

43 PRO, CO968/131, Manpower, East Africa. H. G. L. Gurney, Chief Secretary to the East African Governors' Conference to Under-Secretary of State. Confidential, 11 Dec 1942.

44 Grundlingh, 'South African Blacks in the Second World War', p. 209.

45 Domingue, 'Mauritian and Seychelles Pioneer Corps', pp. 34, 42.

46 PRO, DO35/1183/Y, 1069/1/1, 'Report of Visit to HCT Troops Serving in the Middle East, 27 May–20 July 1943'.

47 O. J. E. Shiroya, *Kenya and World War II* (Nairobi, 1985), p. 14.

48 Extract from radio talk, 'The Seychelles and the Second World War', broadcast 'L'Echo des Isles', 15 January 1969, printed in Ministry of Education and Information, Republic of Seychelles, *Histoire des Seychelles* (Paris, 1983), pp. 175–6.

49 PRO, WO 169/2440, 1504 Seychelles Pioneer Company, April–Dec 1941. Major W. F. Abercrombie, commanding Company No, 1504 P. C., to Governor, Seychelles, Nov 1941.

50 MCP, 16, S.133/2/2, Captain G. J. L. Atkinson, 1974 Coy APC, to Nettleton, 1 Aug 1945.

51 MCP, 17, N. Olebetse, of 327/106 HAA Bty RA, to Tshekedi, dd. 3 Aug 1944.

52 MCP, 17, Rahakwena, Pte no. 8373, Bechuana Coy A, Pioneer Corps, Middle East, to Tshekedi, dd. 20 Feb 1945.

53 MCP, 17, Osephile Keritetse, no. 13214, to Tshekedi, dd. 27 July 1945.

54 MCP, 17, Manaheng Lesotho, Pvt no. 9612, MEF, to Tshekesi, dd. 4 Feb 1944.

55 MCP, 17, David Kechina, 2300 Bechuana Coy, APC, MEF, to Tshekedi, dd. 22 Feb 1944.

56 Barber, *Africans in Khaki*, p. 54.

57 MCP, 16, 1981 (Bechunaland) Company, AAPC, Newsletter, May 1942.

58 Grundlingh, 'South African Blacks in the Second World War', p. 217

59 Letter from KAR soldier, quoted by Shiroya, *Kenya and World War II*, p. 51.

60 MCP, 16, Rev. A. Sandilands (LMS), *Chaplains' Magazine* (January 1943).

61 Kaggia, *Roots of Freedom*, p. 28; in chs. 3 and 4 he describes his two visits to Jerusalem.

62 Interview with Peter Ansah, by his son David Owusu-Ansah, c. 1992, in the possession of the author.

63 BBC Africa Service, Jachoniah Dlamini, Swazi Regt, 1999 Coy, 54th group, section 6.

64 Grundlingh, 'South African Blacks in the Second World War', p. 398.

65 Letter from KAR soldier, quoted by Shiroya, *Kenya and World War II*, p. 51.

66 Kenya National Archives, Nairobi, MD/4/5/116/3, 'Censorship summary on mail written by East African personnel in the MEF, 30 May–7 July 1943, quoted by Timothy Parsons, '"All *askaris* are Family Men": Military Families in the King's African Rifles, 1918–63', in *Guardians of Empire*, ed. David Killingray and David Omissi (forthcoming).

67 PRO, DO35/1184/Y 1069/1/2, F274, WO top secret, 8 April 1944.

68 BBC Africa Service.

69 Ibid.

70 Rhodes House Library, Oxford, Mss. Afr.s, 1715/24a, D. F. T. Bowie, 22 KAR, 1940–7.

71 Grundlingh, 'South African Blacks in the Second World War', p. 244.

72 H. J. Martin and N. Orpen, *South African Forces at World War II*, vol. VI, *The SAAF in Italy and the Mediterranean, 1942–5* (Cape Town, 1977), p. 99.

73 Quoted in Grundlingh, 'South African Blacks in the Second World War', p. 241.

74 David Killingray, 'Race and Rank in the British Army in the Twentieth Century', *Ethnic and Racial Studies*, 10 (1987), pp. 276–90. Since this was written I have done more research on the subject and would considerably amend these findings.

75 Roger Lambo, 'Achtung! The Black Prince: West Africans in the Royal Air Force, 1939–46', in *Africans in Britain*, ed. David Killingray (London, 1994), pp. 145–63.

76 BBC Africa Service.

77 BBC Africa Service. Former Sergeant Samson B. D. Muliango, Northern Rhodesia Regiment, no. 2054, letter to BBC 1989.

78 Kaggia, *Roots of Freedom*, p. 57.
79 BBC Africa Service. Former Regimental Sergeant-Major Joseph Chinama Mulenga, NRA 937, interviewed Lusaka, 1989.

'MATTERS OF HONOUR'

Gerard Douds

1   T. S. Eliot, opp. frontispiece in *The Tiger Triumphs*, 1946.
2   Field-Marshal Wavell, Introduction to Lieutenant-Colonel G. R. Stevens, *Fourth Indian Division* (Toronto, 1948).
3   India Office Library (hereafter IOL), L/WS/1/136.
4   Ibid.
5   Ibid.
6   F. Yeats Brown, *Martial India* (London 1945), p. 197.
7   M. Hauner, *India in Axis Strategy* (London 1981), p. 125.
8   General Auchinleck, Foreword to Lieutenant-Colonel W. G. Hingston and Lieutenant-Colonel G. R. Stevens, *The Tiger Kills* (London 1944), p. 3.
9   P. Moon (ed.), *The Viceroy's Journal* (London, 1973), p. 3.
10 J. Gallagher, *The Decline, Revival and Fall of the British Empire* (Cambridge, 1982), p. 102.
11 General J. N. Chaudhuri, 'Nehru and the Armed Forces', in *Nehru Memorial Lectures, 1966–91*, ed. J. Grigg (Delhi, 1992), p. 84.
12 Ibid., p. 78.
13 J. Nehru, *Autobiography* (London, 1936), p. 448.
14 Chaudhuri, 'Nehru and the Armed Forces', p. 82.
15 Cited in D. Omissi, *The Sepoy and the Raj* (London, 1994), p. 178.
16 Ibid., pp. 190, 191.
17 Hauner, *India in Axis Strategy*, p. 248.
18 Ibid., pp. 127, 241.
19 Ibid.
20 IOL, L/WS/1/441.
21 Hauner, *India in Axis Strategy*, p. 543.
22 IOL, L/WS/1/316.
23 N. C. Chaudhuri, *Thy Hand Great Anarch* (London, 1990), p. 550.
24 Ibid., p. 567.
25 IOL, L/WS/1/122.
26 Lieutenant-Colonel D. Holder, *Hindu Horseman* (Chippenham, 1986), p. 171.
27 Ibid., p. 173.
28 IOL, L/WS/1/122.
29 Ibid.
30 Ibid.; Hauner, *India in Axis Strategy*, p. 123.
31 IOL, L/WS/1/303; P. Mason, *A Matter of Honour* (London, 1974), pp. 153, 154.
32 Mason, *A Matter of Honour*, p. 391.
33 Hauner, *India in Axis Strategy*, p. 125.
34 Ibid., pp. 203, 211, 212.
35 Ibid., p. 215.

36 Mason, *A Matter of Honour*, p. 513.

37 IOL, L/MIL/17/5/4282, 'One More River: Story of Eighth Indian Division', p. 30.

38 IOL, L/WS/1/1172.

39 Ibid.

40 Ibid.

41 Yeats-Brown, *Martial India*, p. 183.

42 IOL, L/WS/1/441.

43 IOL, C in C I to CIGS, 22.11.42.; 16.11.43.; 15.2.44.; 9.8.44.; L/WS/1/441.

44 Testimony of Major F. W. Courtney OBE, MC, 4th Indian Division, 1939–43, interviewed Bombay, 7.2.1996 (hereafter Courtney interview); Major-General Chand Das OBE, 5th Bat. (Napiers), Rajputana Rifles, interviewed New Delhi, 13.2.96 (hereafter Chand Das interview); Brigedier Nand Lal Kapur MC, 4th Bat. (Outrams), Rajputana Rifles, interviewed New Delhi, 14.2.96 (hereafter Kapur interview); General S. C. Sinha, Mahratha Light Infantry, currently Director United Services Institute, interviewed New Delhi, 13.2.96 (hereafter Sinha interview). In gathering this field evidence, the author gratefully acknowledges the support of the Carnegie Trust.

45 'Enemy of Empire', BBC 2, 13 August 1995.

46 Hauner, *India in Axis Strategy*, pp. 547, 549.

47 IOL, L/WS/1/1536. The mutiny at Torre Muara ended Italian schemes to recruit Indian POWs. M. Bose, *The Lost Hero: a Biography of Subhas Bose* (London, 1982), p. 183.

48 Newsreel extracts reproduced in B. Lapping (ed.), *End of Empire*, 'India: Engine of War', Channel 4, 1985.

49 Hauner, *India in Axis Strategy*, p. 146.

50 Major-General I. S. O. Playfair, *The Mediterranean and Middle East*, vol. 1 (London, 1954), p. 105.

51 Field-Marshall W. J. Slim, *Unofficial History* (London, 1970), p. 147.

52 Ibid., pp. 139, 142, 174.

53 Courtney interview; Chand Das interview.

54 Slim, *Unofficial History*, pp. 120, 132.

55 Hingston and Stevens, *The Tiger Kills*, p. 14.

56 Mason, *A Matter of Honour*, p. 480.

57 Ibid., pp. 482, 483.

58 Ibid., pp. 483, 484, interview with S. Weiss, 30 July 1995, 36th Texas Division, 143 Inf, 1st Btl 'C' Co.

59 Chand Das interview; Kapur interview.

60 *The Tiger Triumphs*, pp. 16, 34, 36, 51.

61 Ibid., pp. 10, 44, 108.

62 M. Gellhorn, *The Face of War* (London, 1967), p. 100.

63 *The Tiger Triumphs*, p. 14.

64 Ibid., pp. 15, 52, 95.

65 Courtney interview.

66 *The Tiger Triumphs*, pp. 42, 88.

67 Ibid., pp. 25, 32, 103, 104.

68 Ibid., p. 39; *One More River: Story of the Eighth Indian Division* (London, 1945), pp. 34, 38.

69   Slim, *Unofficial History*, p. 217.
70   Mason, *A Matter of Honour*, p. 529.
71   Omissi, *The Sepoy and the Raj*, p. 190.
72   Mason, *A Matter of Honour*, p. 529.
73   Sinha interview.
74   Mason, *A Matter of Honour*, pp. 511, 512.
75   *The Tiger Triumphs*, pp. 2, 212.
76   A. Bowlby, *The Recollections of Rifleman Bowlby* (London, 1991), p. 230. A remark-
     ably fresh and frank account; how useful an Indian equivalent would be.
77   IOL, L/WS/1/1743.
78   Chand Das interview.
79   *Independent* (22 November 1993).
80   Y. Alibhai, 'Lest We Forget', *New Statesman and Society* (21 June 1991), p. 16.

MR WU AND THE COLONIALS

Augus Calder

1   Major-General Sir Howard Kippenberger, *Infantry Brigadier* (Oxford, 1949), p. 76.
2   Evelyn Waugh, *The Diaries of Evelyn Waugh* (Harmondsworth, 1979), pp. 509–10.
3   John Keegan, *The Second World War* (London, 1989), p. 171.
4   The first British official account was in a popular series: Christopher Buckley,
    Greece and Crete, 1941 (London, 1952; 1977 edn cited here). Then came Major-
    General I. S. O. Playfair's version in *The Mediterranean and the Middle East*, Vol. 2 of
    the UK Military Series of War Histories (London, 1956). The official Australian
    account had appeared in Gavin Long, *Australia in the War of 1939–45: Greece, Crete
    and Syria* (Canberra, 1953). Dan Davin's volume in the Official History of
    New Zealand in the Second World War... Crete (Wellington, 1953) is by far
    the fullest, representing the relative importance of this campaign for the smallest
    of the Dominions. An early unofficial account was by Alan Clark, *The Fall of Crete*
    (London, 1962) – strongly opinionated. John Hall Spencer, *The Struggle for Crete*
    (London, 1962) incorporates views supplied to the author by Colonel Laycock.
    Ian McD. G. Stewart, *The Struggle for Crete* (Oxford, 1966) has become the
    standard account, though he could not discuss ULTRA. Of the most recent
    books, Antony Beevor, *Crete: the Battle and the Resistance* (London, 1991) gives
    overdue emphasis to the role of the native population, and Callum MacDonald,
    The Lost Battle: Crete, 1941 (London, 1993) makes a very able summary of the
    accumulated knowledge of the subject.
5   See Keegan, *The Second World War*, pp. 163–4.
6   Waugh's 'Memorandum' is printed in the *Diaries*, pp. 489–517. His *Letters*, ed.
    Mark Amory (London, 1980), cast only a little further light on his experience. An
    early biography by Waugh's friend Christopher Sykes, *Evelyn Waugh*... (London,
    1975), established the view that the novelist was disqualified as field officer by the
    hatred which he roused in his men. This had been suggested by John St John's
    little memoir, *To the War with Waugh* (London, 1973). Martin Stannard's scholarly
    Evelyn Waugh, vol. II, *No Abiding City* (London, 1992) suggests a more complicated
    view. Selina Hastings's excellent *Evelyn Waugh: a Biography* (London, 1994) draws

further perceptions from 'interviews, letters and advice' from scores of people, many of them friends of Waugh. The *Sword of Honour* trilogy was published in one volume by Penguin in 1984. Though this does not in fact correspond to the conflated version devised by Waugh himself (London, 1965), differences do not affect the Crete episode.

7   Dan Davin, *The Salamander and the Fire: Collected War Stories* (Oxford, 1986), p. x.

8   Dan Davin, *Crete* (Wellington, 1953), p. 402.

9   Kippenberger, *Infantry Brigedier*, p. 72.

10 Len Deighton, *Blood, Tears and Folly: An Objective Look at World War II* (London, 1993), p. 223.

11 Keegan, *The Second World War*, p. 58.

12 W. B. Thomas, *Dare to be Free* (London, 1951), p. 19.

13 See e.g., Roy Farran, *Winged Dagger: Adventures on Special Service* (London, 1948), pp. 89–90

14 Omer Bartov, *Hitler's Army: Soldiers, Nazis and War in the Third Reich* (Oxford, 1992), passim.

15 Keegan, *The Second World War*, p. 158.

16 Buckley, *Greece and Crete*, pp. 210, 245, 184.

17 Farran, *Winged Dagger*, p. 96.

18 Hastings, *Evelyn Waugh*, pp. 427–30, gives a careful account from Waugh's point of view. Beevor, *Crete*, pp. 219–23, reckons that General Weston, left behind by Freyberg to command the rearguard, was 'collared' by Laycock after the clear order recorded by Waugh had been given, and persuaded to allow Layforce Brigade HQ to leave. As Beevor says, Laycock, undoubtedly a brave man, cannot be accused of personal cowardice. But he did want to be sure of getting off the island.

19 Kippenberger, *Infantry Brigedier*, pp. 74–5.

20 MacDonald, *The Lost Battle*, p. 289.

21 Michael Woodbine Parish, *Aegean Adventure, 1940–43* (Lewes, 1993), pp. 71–2. I owe this reference to Dr Jeremy Crang.

22 Sykes, *Evelyn Waugh*, p. 216; Waugh, *Letters*, p. 152.

23 Sykes, *Evelyn Waugh*, p. 202.

24 Waugh, *Diaries*, pp. 499–504.

25 Ibid., pp. 508–9; Beevor, *Crete*, pp. 219–22.

26 Sykes, *Evelyn Waugh*, pp. 215–16.

27 *Sword of Honour*, pp. 356–60.

28 I owe this point to an Edinburgh University undergraduate, Harry Acton.

29 Playfair, *The Mediterranean and the Middle East*, pp. 131, 144.

30 Buckley, *Greece and Crete*, p. 167; Waugh, *Diaries*, p. 507; MacDonald, *The Lost Battle*, pp. 139–43; *Freyberg: Churchill's Salamander* by Laurie Barber and John Tonkin Covell (London, 1990) is an important study.

31 *DNB*, 'Freyberg' – the biography is by Christopher Sykes; Stewart, *The Struggle for Crete*, pp. 113, 120, 123–4; MacDonald, *The Lost Battle*, pp. 157–8, 298–9.

32 Davin, *Salamander*, p. ix.

33 I. C. B. Dear (ed.), *Oxford Companion to the Second World War* (1995), pp. 796–801.

34 Davin, *Salamander*, pp. viii–xi.

35 See pp. 3–11 above, in this volume.

36 Thomas, *Dare to be Free*, pp. 8–9, 16.

37 Buckley, *Greece and Crete*, p. 220; Spencer, *The Struggle for Crete*, pp. 287, 293; Davin, *Crete*, p. 452; Stewart, *The Struggle for Crete*, pp. 467–8.

38 Farran, *Winged Dagger*, pp. 87–8, 91, 98

39 Buckley, *Greece and Crete*, p. 239.

40 Kippenberger, *Infantry Brigadier*, pp. 49–69; Farran, *Winged Dagger*, pp. 99–100.

41 Farran, *Winged Dagger*, pp. 99–101.

42 Thomas, *Dare to be Free*, pp. 21–30.

43 Davin, *Crete*, pp. 297–316.

44 Kippenberger, *Infantry Brigedier*, p. 71.

FIRST CANADIAN ARMY, FEBRUARY–MARCH 1945

Terry Copp

1 Terry Copp and Bill McAndrew, *Battle Exhaustion: Soldiers and Psychiatrists in the Canadian Army, 1939–45* (Montreal, 1990), p. 90. The verse was drawn to our attention by Colonel Strome Galloway.

2 Russell Weigley, *Eisenhower's Lieutenants* (Bloomington, 1981), p. 567.

3 The figures adjusted to include Canadian personnel are from L. F. Ellis, *Victory in the West*, vol. 1 (London, 1962), Appendix IV.

4 M. Hitsman, *Manpower Problems in the Canadian Army*, Report #63 Historical Section, Department of the National Defence, Appendix L, p. 352.

5 Ellis, *Victory in the West*, Appendix IV.

6 John Terraine, *The Smoke and Fire* (London, 1980), p. 46.

7 The figure of 70 per cent appears to be an accepted average for the proportion of infantry casualties in all Allied armies. See John Ellis, *The Sharp End of War* (London, 1990), p. 158.

8 Carlo D'Este, *Decision in Normandy* (London, 1985), p. 252.

9 C. P. Stacey, *The Victory Campaign* (Ottawa, 1960), p. 284.

10 C. P. Stacey, *Arms, Men and Government* (Ottawa, 1970), p. 435.

11 Ibid., pp. 438–9.

12 D'Este, *Decision in Normandy*, p. 262.

13 J. C. Richardson, 'Neuropsychiatry with the Canadian Army in Western Europe', National Archives of Canada Record Group (NACRG), 24, vol. 12, p. 631.

14 Copp and McAndrew, *Battle Exhaustion*, p. 110.

15 Quoted in ibid., p. 130.

16 Ibid., p. 137.

17 Ibid., p. 143.

18 The full text of Gregory's report is reproduced in Terry Copp and Robert Vogel, *Maple Leaf Route: Scheldt* (Alma, Ont., 1986), p. 102.

19 W. R. Feasby, *Official History of the Canadian Medical Services, 1939–45*, vol. II (Ottawa, 1956), p. 58.

20 On 29 August Simmonds in a letter to his divisional commander urged greater efforts to limit straggling, absenteeism and battle exhaustion and suggested that the last problem should not occur under the conditions of fighting in Normandy. War Diary, II Canadian Corps, August 1944. NAC RG 24, vol. 15, p. 143.

21   Interviews, Drs John Burch, Travis Dancey, B. H. McNeel, Clifford Richardson and John Wishart. See McNeel and Dancey, 'The Personality of the Successful Soldier', *American Journal of Psychiatry*, vol. 12, no. 3 (1945), p. 37.

22   See Terry Copp, 'Operational Research and 21 Army Group', *Canadian Military History*, vol. 3, no. 1, p. 71. See also Terry Copp, 'Scientists and the Art of War: Operational Research in 21 Army Group', *RUSI* (Winter 1991).

23   M. M. Swann (ed.), *Operational Research in North-west Europe: the Work of No.2 Operational Research Section and 21 Army Group, June 1944–July 1945*, AORG 1945, 217 pp. PRO WO 291/1331. A copy may be found in NAC RG 24, vol. 10, p. 438. There is no indication of editor or authors in the document. Swann wrote the introduction, the authors of individual reports are identified directly or indirectly in the report or other OR documents. Letter, Lord Swann to the author, 6 April 1986. See also Terry Copp, 'Counter-Mortar Operational Research in 21 Army Group', *Canadian Military History*, vol. 3, no. 2, p. 45.

24   Ibid.; War Diaries, no. 1, Canadian Radar Battery NAC RG 24, and War Diaries, no. 100, British Radar Battery WO171/5047.

25   Swann, *Operational Research*, report no. 21.

26   Ibid., report no. 16. See also M. M. Swann, 'The Effects of Close Air Support', Military Operational Research Report, no. 34, PRO WO 291/976.

27   Copp, 'Operational Research and 21 Army Group', pp. 79–80.

28   Major Tony Sargeaunt, the OR sections armour specialist, had worked closely with the British armoured units in the pre-invasion period. He continued to supervise research in the field during 1944 and 1945. Interviews and correspondence, Copp/Sargaunt, 1988–92.

29   Swann, *Operational Research*, report no. 12, 'Analysis of 75-mm Sherman Tank Casualties Suffered Between 6 June and 10 July 1944'.

30   Ibid., report no. 17, 'Analysis of German Tank Casualties in France, 6 June 1944–31 August 1944'.

31   Ibid.

32   Interview Tony Sargeant, May 1990.

33   Report no. 12.

34   There are a very large number of OR reports on artillery. A separate AORG section, no. 7, 'Lethality of Weapons', was established in 1943 to focus research on questions related to lethality and accuracy. See, for example, L. J. Huddlestone, 'The Probability of Hitting Targets with Artillery Fire', Sept 1941, PRO WO 291/1330.

35   Ibid.

36   The plan is detailed in War Diary, First Canadian Army, Feb 1945. NAC RG 24. See also C. P. Stacey, *The Victory Campaign* (Ottawa, 1966), pp. 460–69.

37   Quoted in report for quarter ending 31 March 1945 by the Adviser in Psychiatry, 21st Army Group. NAC RG 24, vol. 12631.

38   'Interrogation Report Senior Staff Officer 47 Corps', quoted in Terry Copp and Robert Vogel, *Maple Leaf Route: Victory* (Alma, Ont., 1988), p. 38.

39   Terry Copp, *A Canadian's Guide to the Battlefields of North-west Europe* (Waterloo, Ont., 1995).

40   Stacey, *The Victory Campaign*, p. 491.

41   Copp and Vogel, *Victory*, p. 40.

42   Interview Copp/Dancey, 1986.

43 B. H. McNeel, *Quarterly Report*, 1 Jan–31 March 1945. NAC RG 24, vol. 15, p. 951.

44 Ibid. I was able to discuss the decline in the percentage with both Wishart and McNeel and both stressed the high number of burnt-out cases.

45 McNeel, *Quarterly Report*.

46 Adviser in Psychiatry, 21st Army Group, 'Quarterly Report', 31 March 1945. NAC RG 24, vol. 12, p. 631.

'NO TASTE FOR THE FIGHT'

Martin S. Alexander

1   L. de Jong, *The German Fifth Column in the Second World War* (London, 1956), pp. 78–94, provides lots of examples.

2   Le général commandant en chef Gamelin, chef d'état-major général de la Défense nationale, commandant en chef les forces terrestres, à M. le ministre de la Défense nationale et de la Guerre (Cabinet militaire. Section de Défense nationale) No. 1011 Cab/F.T., 18 May 1940, reproduced in M.-G. Gamelin, *Servir*, 3 vols (Paris, 1946–7), vol. III, *La Guerre: septembre 1939–19 mai 1940*, pp. 421–7, quotation, pp. 425–6.

3   See E. du Réau, *Edouard Daladier, 1884–1970* (Paris, 1993), pp. 424–8.

4   The French historian Jean-Baptiste Duroselle, steeped in long research on this issue and after profiting from privileged access to official and private archives, concluded that 'The reproach one could make of Daladier was that, at a moment when his popularity was undeniable...he did not grasp the need for a solidly united government...There is certainly scant resemblance between the Poincaré–Clemenceau team of 1918 and the Lebrun–Daladier team of 1939.' *Politique étrangère de la France: l'abîme, 1939–45* (Paris, 1982), pp. 49, 51.

5   See the present author's *The Republic in Danger: General Maurice Gamelin and the Politics of French Defence, 1933–40* (Cambridge, 1993), ch. 12; du Réau, *Daladier*, pp. 368–418; E. M. Gates, *End of the Affair: The Collapse of the Anglo-French Alliance, 1939–40* (London, 1981); F. Bédarida (ed.), *La Stratégie secrète de la drôle de guerre: le Conseil Suprême Interallié, septembre 1939–avril 1940* (Paris, 1979); F. Bédarida, 'La Rupture franco-britannique de 1940: Le Conseil Suprême Interallié, de l'invasion à la défaite de la France', *Vingtième Siècle*, no. 25 (Jan–March 1990), pp. 37–48; P. M. H. Bell, *A Certain Eventuality: Britain and the Fall of France* (Farnborough, 1974).

6   See du Réau, *Daladier*, pp. 419–28.

7   Gamelin, *Servir*, vol. I, *Les Armées françaises de 1940*, pp. 7–8.

8   Gamelin, *Servir*, vol. III, p. 426. Elsewhere in his memoirs, Gamelin reflected further on the shortcomings of the ageing reservists of the Series B formations. These, he alleged, had less respect for authority (having been caught up, as civilian workers, in the social turbulence of 1936–9) and had been most susceptible to Phoney War defeatism and 'fifth column' propaganda. He noted that reserve officers, of whom France mobilised about 100,000 to reinforce the 30,000 active cadre officers, were usually less well-trained and had less aptitude for command. 'Certain serious and widespread lapses among the cadres must be

admitted alongside those of the rank-and-file. If the officers flee, can one expect their soldiers to stay at their posts and get themselves killed if needs be?' (Gamelin, *Servir*, vol. 1, pp. 354–7.)

9   The most celebrated reflection by a *rappelé* on the Phoney War is M. Bloch's *Strange Defeat: a Statement of Evidence Written in 1940* (New York, 1946). Cf. C. Fink, 'Marc Bloch and the *drôle de guerre*: Prelude to the "Strange Defeat"', *Historical Reflections/Réflexions Historiques*, vol. 22, no. 1 (Winter 1996), pp. 33–46. Other junior officers' or rank-and-file accounts include P. Mousset, *Quand le temps travaillait pour nous: récit de guerre* (Paris, 1941), pp. 9–166; R. Balbaud, *Cette Drôle de guerre: Alsace–Lorraine–Belgique–Dunkerque: 26 Août 1939–1er Juin 1940* (London, New York, Toronto, 1941), pp. 1–39; R. Felsenhardt, *1939–40 avec le 18e Corps d'Armée* (Paris, 1973), pp. 7–112; D. Barlone, *A French Officer's Diary: 23 August 1939 to 1 October 1940* (Cambridge, 1942), pp. 1–45; G. Sadoul, *Journal de guerre, 1939–40* (Paris, 1977), pp. 14–191; *The War Diaries of Jean-Paul Sartre, November 1939–March 1940*, trans. Q. Hoare (London, 1984).

10 P. Dunoyer de Segonzac, *Le Vieux Chef: mémoires et pages choisies* (Paris, 1971), p. 63. Lieutenant Jacques Branet was a troop commander in the 8th Dragoon Regiment, 4th Light Cavalry Division, which plunged into the Ardennes east of Dinant at the head of Corap's Ninth Army. This too, was a pre-war 'active' formation, but horsed, not mechanised. Branet's diary, *L'Escadron: carnets d'un cavalier* (Paris, 1968), pp. 9–42, describes the plight of such lightly armed troops when attacked by German armour (in this case, Rommel's 7th Panzer Division). G. de Chézal, *En Auto-mitrailleuse à travers les batailles de mai* (Paris, 1941) is a vivid memoir of action in southern Holland and Belgium with the Seventh Army in the Panhard armoured cars of the reconnaissance group of an unidentified infantry division (probably 25th, 21st or 9th Motorised).

11 J. C. Cairns, 'Along the Road Back to France 1940', *American Historical Review*, vol. 64, no. 3 (April 1959), pp. 595–6, citing J. Dutourd, *The Taxis of the Marne* (New York, 1957), p. 37.

12 In a terse exchange at Vincennes on 20 May 1940, as Gamelin transferred command to Weygand, Gamelin was treated to an unnerving foretaste of what was coming. 'All this politics [exclaimed Weygand], that's got to change. We've got to be rid of all these politicians. There's not one of them worth any more than the others.' (Gamelin, *Servir*, vol. III, p. 436.) Weygand does not mention his outburst in his own account of this meeting, noting only that 'I found General Gamelin in his office ready to leave, and looking very different from the day before. Manifestly he was relieved at being spared a heavy responsibility...This interview was briefer than that of the day before. General Gamelin bade me farewell with great dignity...He did not speak to me about the report which he had sent on May 18th to the Prime Minister, at the latter's request...Perhaps it is as well that I had no knowledge of that report until much later, when returning from the German prisons in 1945.' M. Weygand, *Recalled to Service: the Memoirs of General Maxime Weygand*, trans. E. W. Dickes (London, 1952), pp. 51–2.

13 Among the parliamentarians arrested in June and July 1940 were Reynaud, Daladier, Léon Blum (Popular Front prime minister, 1936–7 and March–April 1938), Jean Zay (the Popular Front's minister for education), Georges Mandel (minister of the interior, March–June 1940) and Pierre Mendès-France.

14 See R. O. Paxton, *Parades and Politics at Vichy: the French Officer Corps under Marshal Pétain* (Princeton, NJ, 1966).

15 Weygand was minister for national defence until 1941, and then delegate-general in French North Africa; Huntziger was minister for war in 1940–41; Admiral Darlan, chief of naval staff since 1937, became prime minister from December 1940 to April 1942. General Colson, army chief of staff, 1935–39, and General Pujo, a former chief of air staff, both became ministers, whilst General de La Porte du Theil became director of the *Chantiers de la Jeunesse*, a youth-service corps partly designed to circumvent the Franco-German Armistice's ban on military conscription. Others served as Pétain's imperial pro-consuls, including General Noguès, resident-general in Morocco, and Admiral Decoux, General Dentz and Admiral Estéva, high commissioners in Indo-China, Syria and Tunisia respectively.

16 Corap was replaced in command of Ninth Army late on 14 May by Giraud, who had begun the campaign leading Seventh Army in the dash to Breda. Gamelin placed Corap on the 'reserve commanders' list. Seeing him at Vincennes on 19 May, Gamelin reassured Corap that, the day before, 'I told prime minister Paul Reynaud and "Marshal" Pétain that your honour as a soldier was not in question.' 'Very loyal' to Gamelin in his depositions for the wartime Riom trial, Corap did not publish his own version of the battle on the Belgian Meuse in 1940. Gamelin, *Servir*, vol. III, pp. 432–3. Cf. P. Allard, *L'Enigme de la Meuse: la vérité sur l'affaire Corap* (Paris, 1941). In August 1940 Raphael Alibert, then Vichy minister of justice, announced the creation of a Supreme Court at Riom to try Gamelin, along with former prime ministers Blum and Daladier, the former air ministers Pierre Cot and Guy La Chambre, and the permanent secretary of the ministry for national defence and war from 1936–40, Robert Jacomet. Imprisoned at the Château de Chazeron, then at Bourrasol and Le Portalet in 1940–41, Gamelin prepared a voluminous documentation in his defence. But he refused to speak during the trial. The hearings lasted from February 1942 till adjourned *sine die* in April, Blum and Daladier defending themselves so skilfully as to threaten to indict Pétain himself for pre-war military incompetence. See P. Le Goyet, *Le Mystère Gamelin* (Paris, 1975), pp. 353–61; H. Michel, *Le Procès de Riom* (Paris, 1979).

17 Quoted in Felsenhardt, *1939–40*, p. 53.

18 Cairns, 'Along the Road Back', p. 596.

19 Cairns, 'Along the Road Back', p. 596 (quoting Hitler to Mussolini, 25 May 1940, in *Documents on German Foreign Policy*, Series D, vol. IX, no. 317). Cairns notes, however (citing ibid., no. 456), that on 16 June the Führer did remark to General Juan Vigon that the French and British soldiers of 1940 were worse than those of 1914.

20 See H. Dutailly, *Les Problèmes de l'armée de terre française, 1935–39* (Paris, 1980); R. A. Doughty, *The Seeds of Disaster: the Development of French Army Doctrine, 1919–39* (Hamden, Ct, 1985); E. C. Kiesling, *Arming Against Hitler: France and the Limits of Military Planning* (Lawrence, Ks, 1996).

21 See F. K. Rothbrust, *Guderian's XIXth Panzer Corps and the Battle of France: Breakthrough in the Ardennes, May 1940* (New York and London, 1990). Cf. R. A. Doughty, *The Breaking Point: Sedan and the Fall of France, 1940* (Hamden, Ct, 1990)

22 Jeffery A. Gunsburg, *Divided and Conquered: The French High Command and the Defeat of the West, 1940* (Westport, Ct, 1979), pp. 271–2.

23 B. H. Liddell Hart (ed.), *The Rommel Papers*, trans. P. Findlay (London, 1953), pp. 3–13; H. von Luck, *Panzer Commander: the Memoirs of Colonel Hans von Luck*, introduction by S. E. Ambrose (New York, 1989).

24 Luck, *Panzer Commander*, p. 38. Cf. *The Rommel Papers*, pp. 8–11.

25 Luck, *Panzer Commander*, p. 39.

26 Ibid., p. 39; *The Rommel Papers*, pp. 14–27.

27 Balbaud, *Cette Drôle de guerre*, p. 48. On the tide of refugees from Belgium and northern France in 1940 more generally, see J. Vidalenc, *L'Exode de mai–juin 1940* (Paris, 1957); V. Caron, 'The Missed Opportunity: French Refugee Policy in Wartime, 1939–40', *Historical Reflections/Réflexions Historiques*, vol. 22, no. 1 (Winter 1996), pp. 117–57, esp. pp. 146–55; M. R. Marrus, *The Unwanted: European Refugees in the Twentieth Century* (New York and Oxford, 1985), pp. 200–2.

28 See the excellent analytical account by J. A. Gunsburg, 'The Battle of the Belgian Plain, 12–14 May 1940: the First Great Tank Battle', *Journal of Military History*, vol. 56, no. 2 (April 1992), pp. 207–44. Also the first-hand recollection of the corps commander: R. Prioux, *Souvenirs de guerre, 1939–43* (Paris, 1947), pp. 55–78; and the numerous after-action reports of squadron and troop commanders from the DLM and DCR tank divisions in the Edouard Daladier Papers, 4 DA 7, Dr. 3 and Dr. 4, Archives Nationales, Paris.

29 On this fateful manœuvre, which diverted key French mechanised and motorised divisions far to the north of the main German assaults at Gembloux and across the Meuse, see D. W. Alexander, 'Repercussions of the Breda Variant', *French Historical Studies*, vol. VIII, no. 3 (Spring 1974), pp. 459–88.

30 See Field-Marshal Sir A. A. Montgomery-Massingberd to Viscount Halifax: 'Report to the Secretary of State for War on a Visit by the Chief of the Imperial General Staff to the French Army', 17 August 1935, Montgomery-Massingberd Papers, File 158/5, Liddell Hart Centre for Military Archives, King's College, London. Cf. Gamelin, *Servir*, vol. II, pp. 172, 186–8.

31 Gunsburg, *Divided and Conquered*, pp. 180, 191–2, 233–4.

32 Prioux, *Souvenirs de guerre*, p. 82.

33 Dunoyer de Segonzac, *Le Vieux Chef*, p. 69.

34 Ibid., pp. 70–71.

35 Ibid., p. 71.

36 Ibid., p. 72.

37 Detailed accounts of the deaths in action of these officers, Captain de Chatellus, Lieutenant Champsiaud and Lieutenant de Conigliano, along with vivid descriptions of the part played by other individual tank crews in the 4th Cuirassiers, are to be found in M.-A. Fabre, *Avec les Héros de '40'* (Paris, 1946), pp. 158–72.

38 Dunoyer de Segonzac, *Le Vieux Chef*, p. 72.

39 Ibid., pp. 79–162. Cf. J. Hellmann, *The Knight-Monks of Vichy France: the Ecole des Cadres at Uriage* (Montreal, 1992).

40 L. Carron, *Fantassins sur l'Aisne: Mai–Juin 1940* (Grenoble and Paris, 1943), pp. 7–26.

41 Ibid., pp. 18–19.

42 Prioux, *Souvenirs de guerre*, pp. 83–4.

43 The swift attrition of the 2nd and 3rd DLMs through wear-and-tear, breakdowns, fuel shortages and losses to enemy fire during the actions in Belgium, is apparent from ibid., pp. 78–9, 81–2, 84.

44 R. Macleod and D. Kelly (ed.), *The Ironside Diaries, 1937–40* (London, 1962), p. 321.

45 See Gunsburg, *Divided and Conquered*, pp. 193–4, 212, 273.

46 C. de Gaulle, *Lettres, notes et carnets*, vol. 11, *1919–Juin 1940* (Paris, 1980), pp. 496–7.

47 See H. Guderian, *Panzer Leader*, trans. C. Fitzgibbon (London, 1952; Futura Publications pbk edn, 1974, p. 111); also P. Huard, *Le Colonel de Gaulle et ses blindés: Laon, 15–20 mai 1940* (Paris, 1980).

48 See A. Clayton, *Three Marshals of France: Leadership after Trauma* (London, 1992), pp. 94–5.

49 This emerges clearly in Prioux, *Souvenirs de guerre*, pp. 78–9, 81–3.

50 F. W. Von Mellenthin, *Panzer Battles: a Study of the Employment of Armor in the Second World War* (Norman, Ok, 1956; Ballantine Books, 12th imprint, 1990, p. 24).

51 Luck, *Panzer Commander*, p. 43.

52 Ibid., pp. 44–5.

53 Ibid., pp. 45–9.

54 S. David, *Churchill's Sacrifice of the Highland Division: France 1940* (London, 1994).

55 Mellenthin, *Panzer Battles*, pp. 24–5.

56 Ibid., p. 25.

57 Ibid.

58 Ibid., pp. 26–7.

59 Gunsburg, *Divided and Conquered*, pp. 274–5.

60 Sadoul, *Journal de guerre*, pp. 353–4.

61 P. M. H. Bell, *France and Britain, 1900–40: Entente and Estrangement* (Harlow, 1996), pp. 240–41.

62 Statistics in Gunsburg, *Divided and Conquered*, p. 275.

63 Luck, *Panzer Commander*, p. 50.

64 Gunsburg, *Divided and Conquered*, p. 275.

65 For a first-hand account, see R. Guerlain, *Prisonnier de guerre* (London, 1944). Uncertainty surrounds the total number of 1940 prisoners-of war, whilst officers incarcerated in Germany displayed widely divergent views on what had caused the defeat, and on the subsequent politics of Vichy. See J. M. d'Hoop, 'Propagande et attitudes politiques dans les camps de prisonniers: le Cas de Oflags, *Revue d'Histoire de la Deuxième Guerre Mondiale*, no. 122 (April 1981); Yves Durand, *La Captivité: histoire des prisonniers de guerre français, 1939–45* (Paris, 1980); Christophe Lewin, *Le Retour des prisonniers de guerre français* (Paris, 1986). I am indebted to Dr Martin Thomas of the Faculty of Humanities, University of the West of England, Bristol, for these three references.

66 Gunsburg, *Divided and Conquered*, pp. 275–6.

67 D. Porch, 'Arms and Alliances: French Grand Strategy and Policy in 1914 and 1940', in P. M. Kennedy, *Grand Strategies in War and Peace* (New Haven and London, 1991), pp. 125–43.

68 See P. Rocolle, *L'Hecatombe des généraux* (Paris, 1985).

69 M.-G. Gamelin, *Manœuvre et victoire de la Marne* (Paris, 1954).

70 Carron, *Fantassins sur l'Aisne*, p. 8.

71 Cairns, 'Along the Road Back', p. 597.

72 Typical of the genre is Gamelin's comment (*Servir*, vol. 111, p. 454), that 'Despite the flagrant inferiority of our aviation, no one will ever get me to believe that the

Germans could have forced the Meuse... if there hadn't been mistakes or failures of nerve, at least locally.'

73 General L. Menu, *Lumière sur les ruines: les combattants de 1940 réhabilités* (Paris, 1953). To this should be added M. Lérécouvreux, *L'Armée Giraud en Hollande, 1939–40* (Paris, 1951) and Pierre Porthault, *L'Armée du sacrifice, 1939–1940* (Paris, 1965). For more recent attention to 1940 (albeit with no special focus on the reputation of the ordinary French soldier) see A. C. Pugh, 'Defeat, May 1940: Claude Simon, Marc Bloch and the writing of disaster', in I. Higgins (ed.), *The Second World War in Literature: Eight Essays* (Edinburgh, 1989), pp. 59–70; N. Jordan, 'Strategy and Scapegoatism: Reflections on the French National Catastrophe, 1940' and S. Hoffmann, 'The Trauma of 1940: A Disaster and its Traces', in *Historical Reflections/Reflexions Historiques*, vol. 22, no. 1 (Winter 1996), pp. 11–32, 287–301, respectively.

THE ITALIAN SOLDIER IN COMBAT, JUNE 1940–SEPTEMBER 1943

Brian R. Sullivan

1 A single American infantry regiment served in Italy in the First World War. In the Second World War, the US army fought Italian troops in Tunisia from January to May 1943 and on Sicily in July–August 1943.

Beginning in April 1917, the British army deployed thirteen heavy artillery batteries to provide fire support on the Italian Front. Britain sent an entire army to Italy in November–December 1917, following the Italian defeat at Caporetto. Three British divisions remained in Italy until the end of the war and played a prominent part in the Vittorio Veneto campaign that brought the war with Austria–Hungary to a victorious conclusion on 4 November 1918.

British and Commonwealth experience with the Italian army was even more widespread in the Second World War. British and Commonwealth troops engaged Italian ground forces almost continuously from June 1940 until September 1943. These encounters included the East African campaign from June 1940 to November 1941, the campaigns in Egypt, Libya and Tunisia from June 1940 until May 1943, the invasion of Sicily in July–August 1943 and some skirmishing in Calabria in early September 1943, prior to the Armistice on 8 September. Finally, the British government equipped six Italian light divisions in the summer of 1944. Four of these took part in the British Eighth Army's final offensive in Italy in the spring of 1945.

2 Brian R. Sullivan, 'The Italian Armed Forces, 1918–40', in *Military Efficiency*, vol. II, *The Interwar Period*, ed. Allan R. Millett and Williamson Murray (Boston, 1988), pp. 173–4.

3 Sullivan, 'The Italian Armed Forces,' pp. 171–4, 183; Lucio Ceva and Andrea Curami, *Industria bellica anni trenta: Commesse militari, l'Ansaldo ed altri* (Milan, 1992), pp. 30–33, 103–5; Lucio Ceva and Andrea Curami, *La meccanizzazione dell'esercito fino al 1943*, vol. 1, *Narrazione* (Rome, 1989), pp. 138–43, 147–50, 157–75, 257–75; Andrea Curami and Fulvio Miglia, 'L'Ansaldo e la produzione bellica', in *L'Italia nella seconda guerra mondiale e nella Resistenza*, ed. Francesca Ferratini Tosi, Gaetano Grassi and Massimo Legnani (Milan, 1988), pp. 266–74.

4   Brian R. Sullivan, 'The Impatient Cat', in *Calculations: Net Assessment and the Coming of World War II*, ed. Williamson Murray and Allan R. Millett (New York, 1992), pp. 112–35; MacGregor Knox, *Mussolini Unleashed, 1939–41: Politics and Strategy in Fascist Italy's Last War* (New York, 1982), pp. 87–125.

5   Enzo Forcella and Alberto Monticone, *Plotone d'esecuzione: I processi della prima guerra mondiale* (Bari, 1968), pp. 468–512; Giorgio Rochat, 'La giustizia militare nella guerra italiana, 1940–43: Primi dati e spunti di analisi', *Rivista di storia contemporanea*, 4 (Oct 1991), pp. 536–7.

6   Knox, *Mussolini Unleashed*, pp. 108–12; Brian R. Sullivan, 'Caporetto: Causes, Recovery, and Consequences' in *The Aftermath of Defeat: Societies, Armed Forces and the Challenge of Recovery*, ed. George J. Andreopoulos and Harold E. Selesky (New Haven, Ct, 1994), pp. 64–7; Lucio Ceva, *Le forze armate* (Turin, 1981), pp. 356–64.

7   Luigi Emilio Longo, *Francesco Saverio Grazioli* (Rome, 1989), pp. 426, 429, 431–5; Filippo Stefani, *La storia della dottrina e degli ordinamenti dell'esercito italiano*, vol. 11, tomo 1, *Da Vittorio Veneto alla 2a guerra mondiale* (Rome, 1985), pp. 78–9, 206–7.

8   Unlike British army terminology, 'regiment' in Italian military usage refers to a unit composed of several battalions, almost always three.

9   Sullivan, 'The Italian Armed Forces', p. 174; W. Victor Madej (ed.), *Italian Army Handbook, 1940–43* (Allentown, Pa, 1984: reprint of US Military Intelligence Service, *Handbook on the Italian Military Forces*, Washington, 1943), p. 16; Ugo De Lorenzis, *Dal primo all'ultimo giorno: Ricordi di guerra (1939–1945)* (Milan, 1971), p. 347; Mario Montanari, *L'esercito italiano alla vigilia della 2a guerra mondiale* (Rome, 1982), p. 224.

10 Antonio Gandin, 'Relazione sul soggiorno in Germania presso il 109 reggimento fanteria', *Memorie storiche militari, 1982* (Rome, 1983), pp. 328–33.

11 B. H. Liddell Hart, *The Liddell Hart Memoirs*, vol. 1 (New York, 1965), pp. 104–5; Enrico Serra, *Tempi duri: guerra e resistenza* (Bologna, 1996), p. 8; Alberto Rovighi and Filippo Stefani, *La partecipazione italiana alla guerra civile spagnola (1936–9)*, vol. 1, *Dal luglio 1936 alla fine del 1937: Testo* (Rome, 1992), pp. 354–7; Stefani, *Da Vittorio Veneto alla 2a guerra mondiale*, pp. 556–7; Giulio De Giorgi, *Con la Divisione Ravenna: Tutte le sue vicende sino al rientro dall Russia, 1939–43* (Milan, 1973), p. 28; Rex Trye, *Mussolini's Soldiers* (Osceola, Wi, 1995), pp. 25–9.

12 Stefani, *Da Vittorio Veneto alla 2a guerra mondiale*, pp. 486–7, 651–4; De Giorgi, *Con la Divisione Ravenna*, p. 24.

13 Longo, *Francesco Saverio Grazioli*, pp. 429, 432; Stefani, *Da Vittorio Veneto alla 2a guerra mondiale*, pp. 78–9.

14 Serra, *Tempi duri pp.* 37, 42; Sullivan 'The Italian Armed Forces,' p. 200; Mac-Gregor Knox, 'The Italian Armed Forces, 1940–3', in *Military Efficiency*, ed. Millet and Murray, vol. 3, *The Second World War*, p. 166; Trye, *Mussolini's Soldiers*, pp. 35, 59–84; letters from Lucio Ceva to author, 1 and 4 Sept 1995.

15 Montanari, *L'esercito italiano alla vigilia*, pp. 225–9; Irish National Archives, Department of Foreign Affairs, Confidential Reports, 19/1 Rome, MacWhite to Secretary, Department of External Affairs, 4 April 1939; Giovanni Pirelli, *Un mondo che crolla: lettere, 1938–43* (Milan, 1990), pp. 70–71; Giovanni Bladier, *Per il duce, per il re . . .* (Milan, 1972), pp. 29–30; Sullivan, 'The Italian Armed Forces', p. 200; Nuto Revelli, *La strada del Davai* (Turin, 1966), p. 326; Enrico Martini-Mauri, 'Gli italiani nel deserto', in *Trent'anni di storia italiana*, ed. Franco Antonicelli (Turin, 1961), pp. 273–4; Oderisio Piscicelli Taeggi, *Diario di un combat-*

tente nell'Africa Settentrionale (Milan, 1971), pp. 17–24; Serra, *Tempi duri*, pp. 10–11, 50.

16 Rovighi and Stefani, *La partecipazione italiana*, pp. 354, 357; Revelli, *La strada del Davai*, pp. 325–7; National Archives, Microform Series T821, roll 384, frames 377–418; De Lorenzis, *Dal primo all'ultimo giorno*, pp. 20–23, 39; Bladier, *Per il duce*, pp. 15–16; Serra, *Tempi duri*, pp. 9–10, 37–8.

17 *Italian Army Handbook, 1940–43*, p. 20; Serra, *Tempi duri*, p. 8; Lucio Ceva, 'Riflessioni e notizie sui sottufficiali', *Nuova Antologia*, 2182 (Apr–June 1992), pp. 345–9; John Ellis, *Brute Force: Allied Strategy and Tactics in the Second World War* (New York, 1990), p. 234.

18 Montanari, *L'esercito italiano alla vigilia*, pp. 221–3; Serra, *Tempi duri*, pp. 8–11, 35–9; De Lorenzis, *Dal primo all'ultimo giorno*, p. 24; Trye, *Mussolini's Soldiers*, pp. 30–33; Ugo Marchini, *La battaglia delle alpi occidentali* (Rome, 1947), pp. 113–15.

19 Edward Clinton Ezzell, *Small Arms of the World: a Basic Manual of Small Arms* (Harrisburg, 1983), pp. 575–9; Ian Hogg and John Weeks, *Military Small Arms of the Twentieth Century* (Northfield, Il, 1973), 2.24–.28, 3.25–.26 [*sic*]; Gerald Posner, *Case Closed: Lee Harvey Oswald and the Assassination of JFK* (New York, 1993), p. 104; Nicola Pignato, *Armi della fanteria italiana nella secondo guerra mondiale* (Parma, 1971), pp. 15–17, 22–8; *Italian Army Handbook*, pp. 52–4; Trye, *Mussolini's Soldiers*, pp. 121, 124–6.

20 Pignato, *Armi della fanteria italiana*, pp. 29–43, 54–60; *Italian Army Handbook*, pp. 55–7, 61–4; De Giorgi, *Con la Divisione Ravenna*, p. 24; Ezell, *Small Arms of the World*, pp. 584–6; Hogg and Weeks, *Military Small Arms*, 5.31–.33; Trye, *Mussolini's Soldiers*, pp. 112–13, 118–23.

21 Pignato, *Armi della fanteria italiana*, pp. 43–53, 71–4; Trye, *Mussolini's Soldiers*, pp. 110–11; Knox, *Mussolini Unleashed*, pp. 72–3; Stefani, *Da Vittorio Veneto alla 2a guerra mondiale*, p. 487; National Archives, Record Group 226, Entry 51, Box 3, Folder 6 Jan–15 March 1945, #32224, Conley (Rome) to Magruder and Langer, 13 Feb 1945; Trye, *Mussolini's Soldiers*, pp. 110–11.

22 Lucio Ceva, 'Un intervento di Badoglio e il mancato rinnovemento delle artiglierie italiane', *Il Risorgimento*, 28 (June 1976), pp. 124–35; Curami and Miglia, 'L'Ansaldo e la produzione bellica', pp. 275–81; Carlo Montù, *Storia dell'artiglieria italiana*, vol. 16 (Rome, 1955), p. 39; Tito Montefinale, 'L'artiglieria italiana durante e dopo la guerra europea', part 3, *Rivista di artiglieria e genio*, 52 (Nov 1933), pp. 1,613–14; I. S. O. Playfair et al., *The Mediterranean and Middle East*, vol. 3, *September 1941 to September 1942: British Fortunes Reach their Lowest Ebb* (London, 1960), pp. 427–32; Montanari, *L'esercito italiano alla vigilia*, 239–47.

23 Ceva and Curami, *La meccanizzazione*, 1, pp. 229–31, 243–4, 283–4; ibid. vol. 11, 470, 502, 538; Montanari, *L'esercito italiano alla vigilia*, pp. 237–8; Giulio Benussi, *Carri armati e autoblindate del Regio Esercito italiano*, 1918–43 (Milan, n.d. [but 1974]), pp. 20–23, 34–41, 49–56; Ralph Riccio, *Italian Tanks and Fighting Vehicles of World War 2* (Henley-on-Thames, 1975), pp. 18–20; David Hunt, *A Don at War* (London, 1966), pp. 51–2; Ronald R. Greenman, 'Beda Fomm', *Army Quarterly and Defence Journal*, 111 (Oct 1981), p. 446, n. 21.

24 Ceva, *Le forze armate*, pp. 349–50; Montanari, *L'esercito italiano alla vigilia*, pp. 250–51; Ceva and Curami, *La meccanizzazione*, 1, pp. 234–46.

25   Montanari, *L'esercito italiano alla vigilia*, pp. 9–20; Stefani, *La storia della dottrina*, p. 487; Dorello Ferrari, 'Dalla divisione ternaria all binaria: una pagina di storia dell'esercito italiano' in *Memorie storiche militari, 1982* (Rome, 1983), pp. 69–77; Uffico Storico, Stato Maggiore dell'Esercito, (hereafter, use) *L'esercito italiano tra la la e la 2a guerra mondiale (novembre 1918–giugno 1940)* (Rome, 1954), pp. 240–59, 307.

26   Mario Cervi, *The Hollow Legions: Mussolini's Blunder in Greece, 1940–41* (London, 1972), pp. 126–37, 151–2, 211, 224; Angelo Del Boca, *Gli Italiani in Africa Orientale*, vol. III, *La caduta dell'impero* (Bari, 1982), pp. 411–33; Salvatore Loi, *Le operazioni delle unità italiane in Jugoslavia (1941–43)* (Rome, 1978), pp. 77–87.

27   Ceva, *Le forze armate*, p. 284.

28   I. S. O. Playfair et al., *The Mediterranean and Middle East*, vol. 1, *The Early Successes against Italy (to May 1941)* (London, 1954), pp. 267–8, 291, 359–61; Ceva, *Le forze armate*, p. 288; Mario Montanari, *Le operazioni in Africa Settentrionale*, vol. 1, *Sidi el Barrani (Giugno 1940–Febbraio 1941)* (Rome, 1990), pp. 423–42.

29   Cervi, *The Hollow Legions*, p. 308.

30   Ibid., pp. 29–204; Knox, *Mussolini Unleashed*, pp. 189–238, 249–51; Gian Paolo Melzi d'Eril, *Inverno al caposaldo (Albania 1941)* (Milan, 1970); *La campagna di Grecia* (Rome, 1980), pp. 417–32, 930 ff.

31   Anthony Mockler, *Haile Selassie's War: The Italian Ethiopian Campaign, 1935–41* (New York, 1984), pp. 320–83; *L'esercito italiano tra la la e la 2a guerra mondiale*, p. 332; Galeazzo Ciano, *Diario, 1937–43* (Milan, 1980), p. 564; Del Boca, *La caduta dell'impero*, pp. 529–30; Antonio Giacchi, *Truppe coloniali italiane: Tradizioni, colori, medaglie* (Florence, 1977), pp. 123–4, 134, 139. The Italian official history of the campaign is contained in Alberto Rovighi, *Le operazioni in Africa Orientale*, 2 vols (Rome, 1988).

32   Emilio Canevari, *Graziani mi ha detto* (Rome, 1947), p. 43; Serra, *Tempi duri*, pp. 35–49; Lucio Ceva, *La condotta italiana della guerra Cavallero e il Comando Supremo, 1941–42* (Milan, 1975), pp. 72–3, 161–4; Ceva, *Le forze armate*, pp. 307–8.

33   Ceva, *La condotta italiana della guerra*, pp. 73–5; Virgilio Ilari, 'Il partito armato del fascismo: la milizia dallo squadrismo alla RSI', in Virgilio Ilari and Antonia Sema, *Marte in orbace: Guerra, esercito e milizia nella concezione fascista della nazione* (Ancona, 1988), pp. 329–33, 375–6, 405–8; Giorgio De Vecchi and Ettore Lucas, *Storia delle unità combattenti della milizia volontaria per la sicurezza nazionale, 1923–43* (Rome, 1976), passim.

34   Gerhard L. Weinberg, *A World in Arms: a Global History of World War II* (New York, 1994), pp. 523–6; Loi, *Le operazioni delle unità italiane in Jugoslavia*, pp. 139–263; Knox, *Mussolini Unleashed*, p. 4; Gian Paolo Melzi d'Eril, *Un'estate nella Slovenia in fiamme (1943)* (Milan, 1990), pp. 9–62; Virgilio Ilari, *Storia del servizio militare in Italia*, vol. 4, *Soldati e partigiani* (Rome, 1991), pp. 28–9; Ceva, *Le forze armate*, pp. 329–31.

35   *Le operazioni delle unità italiane in Jugoslavia*, pp. 32, 54; Montanari, *Sidi el Barrani*, pp. 346, 396–8, 442; Montanari, *Le operazioni in Africa Settentrionale*, vol. II, *Tobruk (Marzo 1941–Gennaio 1942)* (Rome, 1993), pp. 12–42; Jack Greene and Alessandro Massignani, *Rommel's North African Campaign, September 1940–November 1942* (Conshohocken, Pa, 1994), pp. 75–6.

36   Lucio Ceva, *Africa Settentrionale 1940–43 negli studi e negli letteratura* (Rome, 1982), p. 37; Ceva and Curami, *La meccanizzazione*, vol. I, pp. 336–76, vol. II, pp. 494–531; Riccio, *Italian Tanks and Fighting Vehicles*, pp. 22–3, 26–33, 38–46, 60–66; Montanari, *Tobruk*, pp. 75, n. 35, 79–175, 269, 299, 353–4, 443–742, 761–2, 789–92; Piscicelli Taeggi, *Diario di un combattente*, pp. 40–46; B. H. Liddell Hart (ed.),

The Rommel Papers (New York, 1953), pp. 109–78; Greene and Massignani, *Rommel's North Africa Campaign*, pp. 51–2, 54–6, 63–4, 70–5; Serra, *Tempi duri*, pp. 56–119.

37 Alpheo Pagin, *Mussolini's Boys: La battaglia di Bir el Gobi* (Milan, 1976); Greene and Massignani, *Rommel's North African Campaign*, p. 74, 88–9.

38 Montanari, *Tobruk*, pp. 16, 25, 322, 739–41, 777–83; Liddell Hart (ed.), *The Rommel Papers*, pp. 91, 156, 177–8.

39 Montanari, *Tobruk*, pp. 216–21, 267, 276–7; Ceva, *La condotta italiana della guerra*, pp. 50–51, 143, 148.

40 Ugo Cavallero, *Diario, 1940–43*, ed. Giuseppe Bucciante (Villa S. Lucia, 1984), p. 188; Montanari, *Tobruk*, p. 221; Ceva, *La condotta italiana della guerra*, pp. 57–61; Weinberg, *A World at Arms*, pp. 179–81, 188, 205.

41 Ceva, *La condotta italiana della guerra*, pp. 76–7, 169–70; Ceva, *Le forze armate*, p. 306.

42 Ceva, *Le forze armate*, p. 306; *L'esercito italiano tra la la e la 2a guerra mondiale*, pp. 309–10.

43 Giovanni Messe, *La guerra al fronte russo* (Milan, 1964), pp. 50–223; Brian R. Sullivan, 'Russia, Italian Forces in' in *Historical Dictionary of Fascist Italy*, ed. Philip V. Cannistraro (Westport Ct, 1982); *I servizi logistici delle unità italiane al fronte russo, 1941–3* (Rome, 1976), pp. 69–70; Aldo Gianbartolomei, 'La campagna in Russia del CSIR e dei suoi veterani nell'ARMIR', in *L'Italia in guerra: il terzo anno*, ed. R. H. Rainero and A. Biagini (Gaeta, 1993), pp. 278–84.

44 Ceva, *Le forze armate*, pp. 306–7, 591–4, 598; Ceva, 'Rapporti fra industria bellica ed esercito' in *L'Italia in Guerra: il secondo anno – 1941*, ed. R. H. Rainero and A. Biagini (Rome, 1992), pp. 219, 237–8.

45 Serra, *Tempi duri*, pp. 121–45; Mario Montanari, *Le operazioni in Africa Settentrionale*, vol. III, *El Alamein (Gennaio–Novembre 1942)* (Rome, 1989), pp. 7–156 (division dispositions and readiness levels appear on p. 96.); Ceva and Curami, *La meccanizzazione*, I, pp. 347, 352–3, 361, 426–9.

46 Ceva, *Le forze armate*, pp. 313–15, 322–3; Ceva, *La condotta italiana della guerra*, pp. 91–108; Sullivan, 'Russia, Italian Forces in', p. 477; Messe, *La guerra al fronte russo*, pp. 227–315; *Le operazioni delle unità italiane al fronte russo (1941–1943)* (Rome, 1977), p. 188.

47 Weinberg, *A World at Arms*, pp. 348–52; Ceva, *La condotta italiana della guerra*, pp. 108–18; Montanari, *El Alamein*, pp. 157–626; Serra, *Tempi duri*, pp. 147–73; Marco Di Giovanni, *I paracadutisti italiani* (Gorizia, 1991), pp. 130–72; Gaetano Pinna, *In battaglia nel deserto: diario di guerra di un artigliere paracadutista* (Milan, 1985), pp. 41–153; Renato Migliavacca, *La Folgore nella battaglia di El Alamein* (Milan, 1983), pp. 15–19, 101–5; Paolo Caccia Dominioni and Giuseppe Izzo, *Takfir* (Milan, 1967).

48 Montanari, *El Alamein*, pp. 719–847; Montanari, *Le operazioni in Africa Settentrionale*, vol. IV, *Enfidaville (novembre 1942–maggio 1943)* (Rome, 1993), pp. 5–203; Liddell Hart (ed.), *The Rommel Papers*, pp. 337–58; Gerolamo Pedoja, *La disfatta nel deserto* (Rome, 1946), pp. 169–92: Ceva, *Le forze armate*, pp. 320–22.

49 *Le operazioni delle unità italiane al fronto russo*, pp. 435, 464–5, 47–71; David M. Glantz and Jonathan House, *When Titans Clashed: How the Red Army Stopped Hitler* (Lawrence, 1995), pp. 139–41; Messe, *La guerra al fronte russo*, pp. 316–87; Ceva, *Le forze armate*, pp. 323–5; Alessandro Massignani, *Alpini e Tedeschi sul Don* (Vicenza, 1991); Egisto Corradi, *La ritirata di Russia* (Milan, 1965); Ilari, *Soldati e partigiani*,

pp. 28–30. A total of 257,000 Italian troops had served on the Eastern Front at one time or another.

50 Montanari, *Enfidaville*, pp. 206–551; Ceva, *Le forze armate*, pp. 325–9; Ceva, *Africa Settentrionale*, pp. 132–4; Ilari, *Soldati e partigiani*, p. 34.

51 Alberto Santoni, *Le operazioni in Sicilia e in Calabria (luglio–settembre* 1943) (Rome, 1989), pp. 137–401; Ceva, *Le forze armate*, p. 332; Ilari, *Soldati e partigiani*, pp. 31–2.

52 Santoni, *Le operazioni in Sicilia e in Calabria*, pp. 413–35; Elena Aga Rossi, *Una nazione allo sbando: l'armistizio italiano del settembre 1943* (Bologna, 1993), pp. 61–111.

53 Aga Rossi, *Una nazione allo sbando*, pp. 113–46; Claudio Dellavalle (ed.), *8 settembre 1943: storia e memoria* (Milan, 1989); Marziano Brignoli, *Raffaele Cadorna, 1889–1973* (Rome, 1982), pp. 73–107; Filippo Stefani, 'L'8 settembre e le forze armate italiane', in *L'Italia in guerra: il quarto anno*, ed. Romain H. Rainero (Gaeta, 1994); Ceva, *Le forze armate*, pp. 334–5.

54 Ilari, *Soldati e partigiani*, pp. 28–9; Lucio Ceva, 'The North African Campaign, 1940–43: A Reconsideration', in *Decisive Campaigns of the Second World War*, ed. John Gooch (London, 1990), pp. 100–1.

THE ITALIAN JOB

Richard Holmes

1   Fridolin von Senger und Etterlin, *Neither Fear nor Hope* (London, 1963), p. 215.

2   John Ellis, *Cassino: The Hollow Victory* (London, 1984), p. xiv.

3   Christopher Buckley, *The Road to Rome* (London, 1945), p. 97.

4   Quoted in Carlo D'Este, *Fatal Decision: Anzio and the Battle for Rome* (London, 1991), p. 107.

5   Quoted ibid., p. 277.

6   Quoted in Ellis, *Hollow Victory*, p. 199.

7   Dominick Graham and Shelford Bidwell, *Tug of War: the Battle for Italy, 1943–5* (London, 1986), p. 19.

8   Ibid., p. 403.

9   *The US Army Campaigns of World War II: Rome–Arno* (Washington, n.d.), p. 30.

10 Quoted in D'Este, *Fatal Decision*, p. 316.

11 Quoted in Ellis, *Hollow Victory*, p. 183.

12 Quoted in D'Este, *Fatal Decision*, p. 324.

13 Personal account by C. Richard Eke, Department of Documents, Imperial War Museum, (hereafter IWM), p. 83.

14 Ibid., p. 117.

15 Ibid., p. 142.

16 Personal account by J. B. Tomlinson, Department of Documents, IWM, p. 405.

17 Samuel A. Stouffer *et al., The American Soldier: Adjustment During Army Life* (New York 1965), pp. 70, 75.

18 Ibid., p. 74

19 Sir David Cole, *Rough Road to Rome* (London, 1983), p. 206.

20 Quoted in D'Este, *Fatal Decision*, p. 322.

21 Samuel A. Stouffer *et al.*, *The American Soldier: Combat and its Aftermath* (Princeton, 1965), p. 169.

22 Ibid., p. 124.

23 Ibid., p. 143.

24 Eke, personal account, IWM, p. 93.

25 Alex Bowlby, *The Recollections of Rifleman Bowlby* (London, 1969), p. 115.

26 Geoffrey Curtis, *Salerno Remembered* (Canterbury, 1988), p. 81.

27 Bowlby, *Recollections*, p. 114.

28 David Williams, *The Black Cats at War: the Story of the 56th (London) Division TA, 1939–45* (London, 1995), p. 73.

29 C. E. Montague, *Disenchantment* (London, 1922), p. 56.

30 Bowlby, *Recollections*, p. 134.

31 A. G. Oakley, diary, Department of Documents, IWM.

32 Von Senger, *Neither Fear nor Hope*, p. 178; Oakley, diary, 8 October 1944; Philip Brutton, *Ensign in Italy* (London, 1992), p. 50.

33 E. D. Smith, *Even the Brave Falter* (London, 1992), p. 131.

34 Curtis, *Salerno*, p. 118.

35 T. de F. Jago, diary, Department of Documents, IWM, 25 Jan. 1944.

36 Eke, personal account, IWM, p. 114.

37 Quoted in Saul David, *Mutiny at Salerno: An Injustice Exposed* (London, 1995), p. 48.

38 Jago, diary, 2 Feb 1944.

39 Edward A. Shils and Morris Janowitz, 'Cohesion and Disintegration in the Wehrmacht in World War II', *Public Opinion Quarterly*, vol. XII, no. 2 (Summer 1948).

40 Iris Origo, *War in the Val d'Orcia* (London, 1947), pp. 61, 216.

41 Von Senger, *Neither Fear nor Hope*, pp. 200–1.

42 Martin van Creveld, *Fighting Power: German and US Army Performance, 1939–45* (Westport, Conn., 1982), p. 79.

43 Von Senger, *Neither Fear nor Hope*, p. 206.

44 Quoted in Ellis, *Hollow Victory*, p. 140.

45 W. Anders, *An Army in Exile* (London, 1949), p. 174.

46 Fred Majdalany, *The Monastery* (London, 1950), p. 68.

47 John Horsfall, *Fling Our Banner to the Wind* (Kineton, 1978), p. 33.

48 Tomlinson, personal account, IWM, p. 136.

49 Eke, personal account, IWM, p. 94.

50 Tomlinson, personal account, IWM, p. 157.

51 Eke, personal account, IWM, p. 96.

52 Cole, *Rough Road*, p. 119.

53 Smith, *Even the Brave Falter*, p. 80.

54 Quoted in Arthur Bryant, *Triumph in the West* (London, 1959), p. 141.

55 Cole, *Rough Road*, p. 207.

56 Smith, *Even the Brave Falter*, p. 87.

57 Major Jago served in HQ VIth US Corps at Anzio, and considered American senior officers to be 'too old, [with] no idea of the technique of modern war...' He thought that Major General Lucas, the corps commander, 'was far too old a man, who seemed dazed by what was happening'. Jago, diary, passim. For the issue in general see Graham and Bidwell, *Tug of War*, pp. 156–8 and D'Este, *Fatal Decision*, pp. 220–21, 253–6, 264–79, 340–42, 418–23.

58   Bill Mauldin, *Up Front* (New York, 1945), p. 138, 135.

59   H. Morus Jones, personal account, Department of Documents, IWM.

60   *Fifth Army at the Winter Line* (Washington, 1950), p. 23.

61   D'Este, *Fatal Decision*, p. 243. See also John Ellis, *The Sharp End of War* (Newton Abbot, 1980), p. 119.

62   Quoted in Ellis, *Hollow Victory*, p. 86.

63   Smith, *Even the Brave Falter*, p. 11.

64   Cole, *Rough Road*, p. 153.

65   Majdalany, *Monastery*, p. 57.

66   Cole, *Rough Road*, p. 153.

67   Morus Jones, personal account, IWM.

68   Brutton, *Ensign in Italy*, p. 117.

69   Bowlby, *Recollections*, pp. 16, 198.

70   J. D. A. Stainton, *Memoirs* (privately published, Wilton 1988), p. 141.

71   D'Este *Fatal Decision* p. 151, Brutton *Ensign in Italy*, p. 67.

72   D'Este, *Fatal Decision*, pp. 250, 179.

73   Witold Madeja (ed.), *The Polish IInd Corps and the Italian Campaign* (Allentown, Pa, 1984), p. 18. See also Ellis, *Hollow Victory*, p. 468, which suggests a loss rate of at least 42.5 per cent in Polish rifle companies between 11 May and 5 June 1944.

74   Graham and Bidwell, *Tug of War*, p. 365.

75   Raleigh Trevelyan *The Fortress* (London, 1956), p. 200.

76   Smith, *Even the Brave Falter*, p. 137.

77   Von Senger, *Neither Fear nor Hope*, p. 229.

78   Smith, *Even the Brave Falter*, p. 28.

79   John Terraine, *The Western Front, 1914–18* (London, 1964), p. 15.

80   Tomlinson, personal account, IWM, p. 236.

81   Bowlby, *Recollections*, p. 50, Eke, personal account, IWM, p. 82.

PEASANT SCAPEGOAT TO INDUSTRIAL SLAUGHTER

Mark Axworthy

1   A crude measure of the relative value placed by the Germans on the armies of their two allies was the award of Ritterkreuzen. The Romanian army got sixteen, the Italian army got six. For the overall history of the Romanian army in the Second World War see Mark Axworthy, *Third Axis, Fourth Ally: Romanian Armed Forces in the European War, 1941–5* (London, 1995).

2   On 19 November 1942 the Romanian army had 361,000 men in 26 divisions east of the Bug. On 23 October 1942 Italy had 360,000 men in 19 divisions in Africa and Russia.

3   John Manolescu, *Permitted to Land* (London, 1949), p. 19.

4   Ibid., p. 17.

5   Ministry of Economic Warfare, *Romania Handbook* (London, 1944).

6   Compare this with the casualties of three German infantry divisions at Kursk in 1943: 106th Division had lost 3,224, 320th Division 2,839 and 168th Division 2,671. They were considered to have been 'smashed up'.

7 German losses between 3 August and 30 September 1941 were 309,000 men. Romanian losses between 3 August and 16 October 1941 were 98,000, relatively few of which were suffered after 23 September.

8 Italian combat losses in their six-month war with Greece were 13,755 dead, 50,874 wounded, and 25,067 missing. Romanian losses in two months at Odessa were 18,730 dead, 67,955 wounded and 11,471 missing. However, Italian losses were compounded by 12,368 maimed by frostbite and 52,108 hospitalised for other reasons.

9 This section is primarily based on an untitled, after-action report produced by a Colonel Damaceanu of the Romanian general staff in late 1941. It was itself compiled from reports commissioned from the divisions which served at Odessa.

10 *Revista Infanteriei* (February 1944).

A note on sources

Very few personal memoirs of Romania's war against the Soviet Union have been published in Romania because of the Communist domination of the years between 1947 and 1990. A few memoirs have been published abroad, all apparently by officers. Even were this not the case, the illiteracy of the average infantryman of the times would make it difficult to recover his personal experience. Regrettably, there is at present no oral history programme to fill the breach.

Since 1990 two particularly important books have been published which provide a Romanian overview of the campaigns in the Crimea and at Stalingrad. This paper goes a small way to complementing them. They are:

Adrian Pandea, I. Pavelescu and Eftimiu Ardeleanu, *Romanii la Stalingrad* (Bucharest, 1992).
Adrian Pandea and Eftimiu Ardeleanu, *Romanii in Crimeea* (Bucharest, 1995).

RED ARMY BATTLEFIELD PERFORMANCE, 1941–45

John Erickson

1 Quoted in L. G. Beskrovnyi, *Russkaya armiya i flot v XVIII veke.* (Moscow, 1958), p. 307. On the Russian army and the Russian soldier in the eighteenth century, see A. A. Kersnovskii, *Istoriya Russkoi Armii* (Moscow, 1992), vol. 1, *Ot Narvy do Parizha, 1700–1814* gg. (a reprint of a famous work published in four volumes in Belgrade in the 1930s); also Major-General V. A. Zolotarev, *Apostoly armii rossiiskoi* (Moscow, 1993) which reproduces many contemporary documents and prints, and Christopher Duffy, *Russia's Military Way to the West: Origins and Nature of Russian Military Power, 1700–1800* (London, 1981).

2 Interview with Lieutenant-General N. G. Pavlenko, 'Obrechennye triumfatory', *Rodina* (Moscow, 1991), nos 6–7, pp. 87–90, expanded in General Pavlenko's monograph, *Byla voina...Razmyshleniya voennovo istorika* (Moscow, 1994), 416 pp.

3   General N. N. Golovin, *Voennye usiliya Rossii v mirovoi voine*, vol. 1 (Paris, 1939), pp. 167–8. In common with many other pre-1917 and émigré works this is now being republished in *Voenno istoricheskii zhurnal* (cited as *ViZh*). Also Nicholas N. Golovine, *The Russian Army in the World War* (New Haven, Conn., 1931; facsimile reprint, Archon Books, 1969).

4   N. Galay, *The Relationship Between the Structure of Society and the Armed Forces, as Illustrated by the Soviet Union* (Institute for the Study of the USSR, Munich, June 1966), Paper for the Sixth World Congress of Sociology, Tables, pp. 26–9.

5   Soviet wartime literature is a vast and specialised subject, but wartime writing presented vivid front-line portraits. For a collection see, *Literaturnoe nasledstvo: Sovetskie pisateli na frontakh Velikoi Otechestvennoi voiny* (Moscow, 1966), vol. 78 (in two volumes); for wartime reporting see, *Reportazh s frontov voiny, 1941–5* (Moscow, 1970) and the two volumes *Ot Sovetskovo informbyuro... 1941–5 Publitsistika i ocherki voennykh let* (Moscow, 1982), with a list of war correspondents and writers on the back. The post-war war novel, even if suffused with 'ideological conformity' roused fierce controversy, Grossman being a case in point. See Frank Ellis, 'Army and Party in Conflict: Soldiers and Commissars in the Prose of Vasily Grossman', in J. and C. Garrard (ed.), *World War 2 and the Soviet People*, ed. J. and C. Garrard (New York, 1993), pp. 180–201. We now have Grossman's *Life and Fate* (brought to the West in microfilm) completed in 1960, which was intended originally as a sequel to *For a Just Cause* (*Za pravoe delo*) the epic novel of Stalingrad.

6   E. S. Senyavskaya, 'Chelovek na voine: opyt istoriko psikhologicheskoi kharakteristiki rossiiskovo kombatanta', *Otechestvennaya istoriya* (1993), no. 3, pp. 7–16; see also Baurdzhan Momysh-uly, *Psikhologiya voiny* (Alma Ata, 1990).

7   There are multi-volume collections of wartime letters, here *Zhivye stroki voiny...* (Sverdlovsk, Sredne Ural. knizh. izd, 1984), 352 pp. with photographs and reproductions, also *Pis'ma slavy i bessmertiya* (Baku, 1987), 220 pp., soldiers from Azerbaidzhan.

8   Major General F. W. Mellenthin, *Panzer Battles: a Study of the Employment of Armour in the Second World War* (London, 1955; reprinted 1979, 1984), here Part Three, 'Russia, the Red Army', pp. 356–62; see also Colonel N. K. Shishkin, 'General Mellentin: "Tankisty Krasnoi Armii zakalilis v gornile voiny..."', *ViZh* (1995), no. 4, pp. 4–9. In February 1942 Red Army Intelligence forwarded to Stalin General Guderian's draft report on Red Army performance, presumably a captured document; see *Istoricheskii arkhiv* (Moscow, 1991), no. 1, pp. 57–61.

9   Mellenthin, *Panzer Battles* (1984 edn), p. 361.

10 The most up-to-date study using recently declassified sources is David M. Glantz and Jonathan M. House, *When Titans Clashed: How the Red Army Stopped Hitler* (Kansas, 1995), 414 pp., a comprehensive operational narrative with a documentary supplement. An extremely important and highly illuminating addition to both Russian and non-Russian studies is presented by Colonel David Glantz in 'The Great Patriotic War Revisited: the Failures of Historiography: Forgotten Battles of the German Soviet War (1941–5)', *Journal of Slavic Military Studies* (Dec 1995), pp. 768–808. The first volume of a new Russian military history of the Great Patriotic War, projected in four volumes, has now appeared, *Velikaya Otechestvennaya voina 1941–1945 gg. Voennoistoricheskii ocherk* (Moscow, 1995), vol. 1, *Surovye ispitaniya*, 455 pp. with a documentary supplement.

11 P. N. Knishevskii *et al.*, *Skrytaya prvada voiny: 1941 god. Neizvestny dokumenty* (Moscow, 1992), pp. 164–6, document, 25 Dec 1941 signed by Mekhlis to Sixteenth Army, criticising the 'improper attitude and indifference' of commanders and commissars to 'their prime duty', the 'timely and with appropriate honours burial of those fallen in battle for the Motherland'. In many cases the dead had been left lying for days 'with no one bothering'. Order No. 138 1941 required registration of the details of those killed in action (burial place, home address, party status).

12 House, *When Titans Clashed*, presents an indispensable review of recently declassified Soviet military records under 'Soviet Archival Materials', pp. 311–19. Colonel David Glantz and Dr Harold Orenstein were also responsible for publishing key declassified documents in the *Journal of Soviet Military Studies* (now designated *Journal of Slavic Military Studies*) published by Frank Cass, London. Two detailed wartime Soviet general staff evaluations of Soviet performance are reproduced in *The Battle for Moscow* and *The Battle for Stalingrad*, ed. Michael Parrish and Louis Rotundo respectively (London, Washington, 1989). *ViZh* also increasingly publishes military archive material. The voluminous German files of *Fremde Heere Ost* (*GenStH/Abt.FHO*), German military intelligence on the Eastern Front, were previously the main source of 'raw intelligence' on the wartime Red Army and still retain their utility.

13 *Boevoi sostav Sovetskoi armii* (Moscow), *Voenno nauchnoe upravlenie General'novo Shtaba: VAGSh* (1963) and *Tsentral'nyi arkhiv Ministerstvo oborony SSSR* (1990), in all five volumes.

14 Colonel General G. F. Krivosheyev (ed.), *Grif sekretnosti snyat: Statisticheskoe issledovanie* (Moscow, 1993), 416 pp.; also General Krivosheyev on losses, 'Eshche raz o tsene Pobedy', interview, *Krasnaya Zvezda* (18 June 1991); General M. A. Moiseyev, interview, 'Tsena pobedy', *ViZh* (1990), no. 3, pp. 14–16; on losses in offensive operations 1941–5, Colonel S. N. Mikhalev, 'Boevye poteri storon v strategicheskikh nastupatel'nykh operatsiyakh Sovetskoi armiii. 1941–1945 gg.', *ViZh* (1991), no. 11, pp. 11–18 (the structure of losses); on overall losses Colonel V. T. Eliseev, 'Tak skol'ko zhe lyudei my poteryali v voine?' *ViZh* (1992), nos 6–7, pp. 31–4. A further addition to his protracted and furious controversy appeared in *Moscow News* (28 April–4 May 1995), Boris Sokolov's 'New Estimates of World War II Losses'. See also Edwin Bacon, 'Great Patriotic War: Soviet Military Losses in World War II', *Journal of Slavic Military Studies* (Dec 1993), pp. 613–33. The German military records also contain a mass of data on Soviet losses, *Verluste, Gefangene u. Beute* files.

15 Detailed expositions in House, *When Titans Clashed*, and Walter S. Dunn Jr, *Hitler's Nemesis: the Red Army, 1930–45* (New York, 1994), 231 pp.

16 Tabulation in *Krasnaya Zvezda* (22 June 1993), 'Balans spisochnoi chislennosti lichnovo sostava VS SSSR...' The figure of 29 million mobilised is generally accepted, including 800,000 women, so that 28,200,000 men represented 59.8 per cent of the age contingents liable for mobilisation.

17 *Grif sekretnosti snyat* on officer losses, pp. 314–21, including 421 generals and admirals but lieutenants incurred the heaviest casualties, 353,040 with junior lieutenants close behind, 279, 967.

18 No work on Soviet prisoners-of-war has surpassed Christian Streit's *Keine Kameraden: Die Wehrmacht und die sowjetischen Kriegsgefangen, 1941–5* (Bonn, 1991 edn), 448

pp. Excerpts from *Keine Kameraden* were recently published in *ViZh. Grif sekretnosti snyat*, pp. 330–40, gives the figure of 4,059,000 for Soviet prisoners-of-war but with qualification and reservation about 'non-returning' former Red Army soldiers. German figures are used for prisoner total in 1941, 2,561,000.

19 Colonel Fritz Stoeckli (Swiss army) presented a reasoned analysis of this furiously debated topic in 'Wartime Casualty Rates: Soviet and German Loss Rates during the Second World War: the Price of Victory', *Journal of Soviet Military Studies* (Dec 1990), pp. 645–51.

20 Figures and tabulations in Major General V. V. Gurkin and Major General A. I. Kruglov, 'General fel'dmarshal E. von Manshtein: "Oshibka, v kotoruyu vpal Gitler"', *ViZh* (1995), no. 3, pp. 48–59.

21 Figures in *Grif sekretnosti snyat*, 'Vooruzhennie i boevaya tekhnika: proizvodstvo i poteri', pp. 340–84.

22 The long-standing controversy about Red Army preparedness in 1941, the debate over a possible Soviet 'pre-emptive' strike has not abated, indeed was intensified with the publication here and in Russia of Viktor Suvorov (pseud.), *Icebreaker: Who Started the Second World War?* (London, 1990). A 'round table' of opinion, little of it sympathetic to 'Suvorov', was presented in G. A. Bordyugov (ed.), *Gotovil li Stalin nastupatel'nuyu voiny protiv Gitlera?* (Moscow, 1995). Colonel General Yu. Gor'kov in a recent study *Kreml', Stavka, Genshtab* (Tver, 1995), examines the famous 15 May 1941 document outlining a limited pre-emptive strike, but makes the point that while this was advanced by the General Staff the operational documents of the Western frontier military districts contained no orders for offensive operations. To support that statement Colonel-General Gor'kov and Colonel Yu. Semin published the operational orders for the Baltic Special Military District in *ViZh* (1996), no. 2, pp. 2–15, with more promised from declassified General Staff May–June 1941 directives. For an example of opposing points of view see Yu. A. Gor'kov, 'Gotovil li Stalin uprezhdayushchii udar protiv Gitlera v 1941 g.' *Novaya i noveishaya istoriya* (1993), no. 3, pp. 29–39, and I. Khoffman (Joachim Hoffmann), 'Pogotovka Sovetskovo Soyuza k nastupatel'noi voine 1941 god', *Otechestvennaya istoriya* (1993), no. 4, pp. 19–31.

23 Three studies in the *Journal of Soviet Military Studies* (Sept 1992) are indispensable for investigating Soviet mobilisation, force structure and order of battle: David M. Glantz, 'Soviet Mobilization in Peace and War, 1924–42: a Survey', pp. 323–62; James Goff, the *doyen* of force structure analysis, 'Evolving Force Structure, 1941–5: Process and Impact', pp. 363–404; and Walter S. Dunn Jr, 'Deciphering Soviet Wartime Order of Battle: in Search of a New Methodology', pp. 405–25; see also Robert G. Poirier and Albert Z. Conner, *The Red Army Order of Battle in the Great Patriotic War* (Presidio Press, 1985) 408 pp.

24 See A. D. Kolesnik, *Narodnoe opolchenie gorodov geroev* (Moscow, 1974), tables and text pp. 45–8 for examples of shortages: in one Leningrad militia division, light and heavy machine-guns (10 instead of 175), automatic weapons (none), machine pistols (none instead of 1,135), no mortars, no anti-tank or anti-aircraft artillery.

25 On strategic reserves, see text and tables, Army-General M. Kazakov, 'Sozdanie i ispol'zovanie strategicheskikh rezervov', *ViZH* (1972), no. 12, pp. 45–53: n. 5, p. 46 puts Soviet front-line strength on 10 July 1941 at 201 divisions, 90 almost up to full strength, the remainder having lost half or more of their manpower, Stavka reserve 31 divisions.

26 Zhukov to Stalin 15 July 1941: 'Our army has considerably underrated the significance of cavalry', recommending cavalry for raids in the German rear: *Istoricheskii arkhiv* (1992), no. 1, p. 56. On Red Army cavalry see A. Ya. Soshnikov, *Sovetskaya kavaleriya Voenno istoricheskii ocherk* (Moscow, 1984), p. 161, Red Army cavalry strength June, 1941, 3 cavalry corps (II, V, VI) 13 cavalry divisions (4 mountain cavalry), divisional establishment 9,240 men (average strength 6,000), 34 light tanks, artillery and mortars, 7 cavalry divisions deployed in the western frontier military districts, on operations as front mobile groups, Moscow 1941, and order of battle (8 cavalry corps), pp. 183–98.

27 Marshal F. I. Golikov, *V Moskovskoi bitve. Zapiski komandarm* (Moscow, 1967), assumed command of Tenth Army shortages, pp. 25–9 and pp. 46–51.

28 The German study, *Die Kriegswehrmacht der UdSSR* (Stand, December 1941) envisaged no existing Soviet offensive capability. The *Feindlage* attached to the reports on the tasks of the *Ostheer*, 1941–2, *Weisung für die Aufgabe des Ostheeres im Winter 1941/42* (General Staff, Operations Section No 1693/41) concluded that 'no large [Soviet] reserve formations on any significant scale' existed at this time.

29 Soviet rifle division establishments (rifle divisions, rifle regiments, Guards rifle divisions and regiments) for April 1941–December 1942 tabulated in *Fremde Heer Ost* compilations *Soll-Kopfstarken*, but actual strength fell far below establishments; also Fremde Heer Ost / (II) file *Kriegsgliederungen der Roten Armee (Ablage)* data on rifle divisions, brigades, cavalry divisions armoured units, for period 10 December 1941–10 September 1943, examined in great detail in Dunn, *Hitler's Nemesis*, ch. 5, 'The Rifle Division', pp. 77–87.

30 The most reliable work on Red Army defensive and offensive operations amidst a plethora of publication is Marshal V. D. Sokolovskii (ed.), *Razgrom nemetsko fashistskikh voisk pod Moskvoi* (Moscow, 1964), 444 pp. and map supplement. The classified version is Marshal B. M. Shaposhnikov (ed.), *Razgrom nemetskikh voisk pod Moskvoi (moskovskaya operatsiya Zapadnovo fronta 16 noyabrya 1941 g. – 31 yanvarya 1942 g.)*, 3 parts (Moscow, 1943; declassified 1965). See also the critique of Soviet performance in Parrish (ed.), *Battle for Moscow*, 1942, General Staff Study. A full documentary collection on the defensive and offensive operations (to April 1942) using Ministry of Defence archives has now been published *Bitva za Stolitsu. Sbornik dokumentov* (Moscow, 1994), vols 1–2, 260 pp., 242 pp., respectively.

31 On the 1942 tank armies Colonel General I. M. Anan'ev, *Tankovye armii v nastupleniii. Po opytu Velikoi Otechestvennoi voine 1941–1945 gg.* (Moscow, 1988), ch. 2, 'Voina potrebovala tankovye armii', pp. 36–70.

32 Background in 'The Reluctant Warriors: The Non-Russian Nationalities in Service of the Red Army During the Great Patriotic War, 1941–5', *Journal of Slavic Military Studies* (Sept 1993), pp. 426–45.

33 Details in Dunn, *Hitler's Nemesis*, ch. 6, 'The Replacement System', pp. 89–105.

34 Fremde Heer Ost (IIc), *Übersicht über Ersatzeinheiten und Offizierschulen der Roten Armee, Stand Dezember 1943*, lists replacement regiments and officer training schools; also the 19 March 1945 study *Feld Ersatzeinheiten der Roten Armee*, identifies numerical designations of field replacement units and their locations with fronts and armies.

35 Anan'ev, *Tankovye armii v nastuplenii*, pp. 64–5, State Defence Committee (GKO) instruction no. 2791, 28 January 1943 on establishment of 'homogeneous tank armies' and their establishments, First Tank Army formed by 23 February 1943;

details of tank army organisation and operations are in Marshal of Armoured Troops O. A. Losik (ed.), *Stroitel'stvo i boevoe primenenie sovetskikh tankovykh voisk v gody Velikoi Otechestvennoi voiny* (Moscow, 1979), 414 pp.

36   On artillery organisation, armament and deployment see Marshall of Artillery G. E. Peredel'skii *et al.*, *Artilleriya v boyu i operatsii* (Moscow, 1980); also 'Artillery Doctrine and Organization' in Dunn, *Hitler's Nemesis*, ch. 11, pp. 169–88 and Chris Bellamy, *Red God of War: Soviet Artillery and Rocket Forces* (London, Washington, 1986), 247 pp.

37   On evolving force structures see Part 4, *Boevoi sostav Sovetskoi armii* (Moscow, 1988) for January–December, 1944 and Part 5 (1990) for January–September 1945; see also *Fremde Heere Ost, Truppen Übersicht und Kriegsgliederungen Rote Armee. Stand: August* 1944 for details of Red Army deployments, structure and organisation; *Unterlagen für grosse Kräftegegenüberstellung*, for comparison of Soviet and German manpower, tank strength for the period July 1943 to October 1944.

38   Tables for June 1944 and January 1945 are in vol. 4, p. 125 and vol. 5, p. 27 respectively of the six-volume 'official history', *Istoriya Velikoi Otechestvennoi voiny Sovetskovo Soyuza, 1941–5* (Moscow, 1962 and 1963).

39   Colonel M. Glantz, *From the Don to the Dnepr: Soviet Offensive Operations, December 1942–August 1943* (London, 1991) 430 pp., is an indispensable study of Soviet operations in 1942–3 with much detail on changing Soviet organisation and structures.

40   Details and analysis are in James M. Goff, 'Evolving Soviet Force Structure, 1941–5', *Journal of Soviet Military Studies* (Sept 1992), which covers all types of Red Army units for each period of the war.

41   On the pre-June 1941 situation see James Barros and Richard Gregor, *Double Deception: Stalin, Hitler and the Invasion of Russia* (Illinois, 1995), 307 pp; Soviet intelligence reports March–June 1941 (central archives, Russian Federal Security Service), *Sekrety Gitlera na stole u Stalina* (Moscow, 1995), 253 pp.

42   *Journal of Soviet Military Studies*, starting with March 1991 issue, published classified Soviet documents from *Sbornik dokumentov Velikoi Otechestvennoi voiny* (43 volumes 1947–60), vols 34–43 for the period 22 June–5 November, 1941 – invaluable material; also Soviet reaction in German operational analysis, see David M. Glantz (ed.), *The Initial Period of the War on the Eastern Front, 22 June–August 1941* (London, 1993), 511 pp. This is merely the tip of the iceberg of a mass of material, see, for example, revealing documentation in *Leto 1941 Ukraina* (Kiev, 1991), 337 documents.

43   Army General D. Pavlov, interrogation and verdict, Case No. R 2, 4000, *Voennye znaniya* (1991), no. 11, pp. 9–11, 25; list of senior commanders 'repressed', missing, taken prisoner 1941, *Skrytaya pravda voiny: 1941*, pp. 343–8.

44   Text in *Skrytaya pravda voiny: 1941*, pp. 254–5.

45   General Kazakov, 'Sozdanie...rezervov' *ViZh* (1972), no. 12, p. 48 reported sixty militia divisions, 'hundreds of independent regiments and battalions' involving two million men. Forty divisions transferred to the Red Army, twenty-six divisions fought throughout the entire war. For detailed listing of divisions, brigades, regiments in a definitive study of the organisation and deployment of the DNO, see A. D. Kolesnik, *Opolchenskie formirovaniya Rossiiskoi Federatsii v gody Velikoi Otechestvennoi voiny* (Moscow, 1988), lists of DNO divisions, brigades, regiments pp. 281–7.

46 For a rare discussion of panic in the Red Army see Major General I. Kolesni-
chenko, 'O prichinakh, porozhdayushchikh paniku v voiskakh, i merakh ee
predotvrashcheniya', *ViZh* (1963), no. 1, pp. 46–59; see also Stanislaw Konieczny,
Panika wojenna (Warsaw, 1969), 282 pp., for a Polish military study of panic.

47 Party documents, *Skrytaya pravda voiny: 1941*, 'Vsegda li kommunisty byli vpredi?'
('Were Communists always in the vanguard?'), pp. 276–86.

48 Report, 325th Rifle Division, 10th Reserve Army, 12 November 1941, *Skrytaya
pravda voiny: 1941*, pp. 220–21.

49 The full list is in *Skrytaya pravda voiny: 1941*, pp. 261–2.

50 *Grif sekretnosti snyat*, pp. 171, 174–5.

51 Marshal A. I. Yeremenko, *V nachale voiny* (Moscow, 1964), p. 406. Front staff not
only refused help but 'liberated' Yeremenko's reserve stocks, which were rapidly
exhausted; the war diary of Four Shock Army 8 January 1942 reported 360th and
332nd Rifle Divisions without supplies, due to the 'criminal negligence' of the
North-Western Front rear services. The chief and commissar of rear services
were put in front of a military tribunal.

52 376th Pskov Red Banner Rifle Division, River Volkhov operations, December,
1941–January 1942, *Skrytaya pravda voiny: 1941*, pp. 224–5.

53 General Vlasov, commander Twentieth Army, instructions for improved tactical
handling, Army General Zhukov on 'criminally negligent attitude' to losses,
Order No. 3750, Western Front, 30 March 1942 in *Skrytaya pravda voiny: 1941*,
pp. 225–6, 228–9.

54 Text of Order No. 227 in *ViZh* (1988), no. 8 'Dokumenty i materialy', pp. 73–5.
Stalin found the first draft of this order prepared by the Main Political Administration 'toothless and liberal' and rewrote it himself, Pavlenko, *Byla voina...*, p.
323. On the organisation of penal companies and conditions of *shtrafniki*, see
Regulation 26, Sept 1942, issued by General Zhukov, Deputy Defence Commissar, text in *Skrytaya pravda voiny:* 1941, pp. 363–5; on *strafbats* in action, 'V proryv
idut shtrafnye batal'ony...', *Krasnaya Zvezda* (25 December 1991).

55 Stalin's hand-written correction to the draft Stavka directive 4 June 1942, lessons
from the Kerch disaster. Personal copy.

56 Marshal V. I. Chuikov's own account of the defensive fighting in Stalingrad,
Nachalo puti (Moscow, 1959, 1962) stands supreme; see also operational narrative
and analysis in A. M. Samsonov, *Stalingradskaya bitva* (Moscow, 1989) 4th edn, 630
pp. and Marshal K. K. Rokossovskii (ed.), *Velikaya pobeda na Volge* (Moscow,
1965), 527 pp., map supplement; see also Louis Rotundo (ed.), *Battle for Stalingrad*
(1943), General Staff study, detailed observations on performance (lack of artillery
infantry co-ordination – 'artillery commanders cannot conduct observation,
correct fire directions and fight as riflemen all at the same time'). Casualty figures,
Grif sekretnosti snyat, pp. 178–9, 181–2.

57 On medical services at Stalingrad and in the wartime, see Red Army Chief of
Medical Services, E. I. Smirnov, *Voina i voennaya meditsina, 1939–45* (Moscow,
1979), on Stalingrad pp. 281–9. A very detailed and virtually unique study of
medical services is A. I. Burnazyan, *Bor'ba za zhizn ranenykh i bol'nykh na Kalinins-
kom-i-m Pribaltiiskom fronte, 1941–5* (Moscow, 1982), 304 pp. See also Amnon Sella,
The Value of Human Life in Soviet Warfare (London, 1992), 237 pp. and *Voenno
meditsinskii Zhurnal*.

58 John Erickson, 'Soviet Women at War', *World War 2 and the Soviet People*, p. 62.

59   On Soviet women at the front, V. Ya. Galagan, *Ratnyi podvig zhenshchin v gody Velikoi Otechestvennoi voiny* (Kiev, 1986), pp. 271–5, lists rank and assignment of women awarded Hero of the Soviet Union decoration.

60   On anti-epidemic work, see B. L. Ugryumov, *Zapiski infektsionista* (Moscow, 1973), 167 pp., bibliography. The diseases and illnesses treated by this specialised hospital included dysentery, typhoid fever, tularaemia, influenza, infectious hepatitis, typhus, malaria, angina, pneumonia and gastro-enteritis.

61   On Vlasov and the ROA, Russian Liberation Army, see W. Strik Strikfeldt, *Gegen Stalin und Hitler: General Wlassow und die russische Freiheitsbewegung* (Mainz, 1979), trans. David Footman, *Against Hitler and Stalin* (London, 1970). *Grif sekretnosti snyat* records that 150,000 Soviet citizens also served with the SS. Vlasov claimed that he was fighting against Stalin, but for many of the captured Red Army soldiers the stark choice was between collaboration, 'a mug of soup and a warm greatcoat', or agonising death in circumstances which beggar description.

62   The wartime Red Army was predominantly 'Russo Ukrainian' (58 per cent and 22 per cent respectively in January 1944, 2.6 per cent Belorussian, 1.3 per cent Armenian). There were large-scale defections by non-Russian nationalities, only to be exploited in turn by the Germans, policies analysed in Alex Alexiev, *Soviet Nationalities in German Wartime Strategy, 1941–1945* (RAND Corporation R-2772 NA, 1982), 39 pp., bibliography, also Joachim Hoffmann, *Kaukasien 1942/43 Das deutsche Heer und die Orientvölker der Sowjetunion* (Freiburg, 1991), 534 pp. The largest contingents serving with the Wehrmacht or SS were central Asians 180,000, Caucasians 110,000, Ukrainians 220,000, Volga and Crimean Tartars 60,000. On the other hand non-Russian divisions fought well, 'national' units such as 77th, 416th, 402nd, 396th, 227th, 271st, 151st Rifle Divisions recruited in Azerbaidzhan, north Caucasus, Georgia and Armenia, duly recorded in Z. Buniyatov and P. Zeinalov, *Ot Kavkaza do Berlina* (Baku, 1990), 221 pp.

63   For Soviet planning see G. A. Koltunov and B. G. Solov'ev, *Kurskaya bitva* (Moscow, 1970), pp. 27–40. Special training for Soviet troops involved dealing with 'Tiger fear', countering the new German tanks. On the defensive battle and the role of infantry, see Documents (from *Sbornik materialov po izucheniya opyta voiny*, vol. 11 1944), *Journal of Slavic Military Studies* (Dec 1993), pp. 656–700.

64   This reinforcement was emphasised by Stalin in May 1944 as the Red Army crossed the Soviet frontiers – 'the Soviet Union is waging a *class, just war (klassovaya, spravedlivaya voina)* against Hitlerism, each commander and political worker must become an ardent propagandist of the socialist idea'. For this and details of party political work, see Colonel-General G.V. Sredin, *Ideologicheskaya rabota KPSS v deistvuyushchei armii* 1941–5 gg. (Moscow, 1985), p. 111. Yu. P. Petrov, *Partiinoe stroitel'stvo v Sovetskoi armii i flote (1918–1961 gg.)* (Moscow, 1961); ch. III, still remains a useful and even a frank account, for example, General Gordov in 1943 proposing the abolition of the entire 'political apparatus', p. 337.

65   On the wartime role of the General Staff and Frunze academies, see *Akademiya General'novo shtaba* (Moscow, 1987) 2nd edn, ch. 3 and *Voennaya akdemiya imeni M.V. Frunze* (Moscow, 1988) 3rd edn, ch. 3.

66   Recorded interviews with women veterans in S. Alexiyevich, *War's Unwomanly Face* (Moscow, 1988), translation of *U voiny nie zhenskoe litso* . . . .

67   On supply see Army-General S. K. Kurochkin (ed.), *Tyl Sovetskikh vooruzhennykh sil v Velikoi Otechestvennoi voine 1941–5 gg.* (Moscow, 1977), pp. 158–225. Front and

army supply dumps carried only imperishables (flour, canned foods, millet, sugar), field rations supplemented locally included 1 lb 15 oz bread in winter, 5 oz meat, 3 oz fish, tea, salt, tomato paste, tobacco, additional ration of cigarette paper (or rolled-up *Pravda*).

68 Documents and individual testimony in Harold J. Berman and M. Kerner, *Soviet Military Law and Administration* (Cambridge, Mass., 1955).

69 A generalised description is Colonel Louis B. Ely, *The Red Army Today* (Harrisburg, 1953), 3rd edn., in 'Infantry – The Massed Majority'. The series *Taktika v boevykh primerakh, Polk* (Moscow, 1974), *Batal'on* (Moscow, 1974), *Rota* (Moscow, 1974), 'Regiment', 'Battalion', 'Company', respectively, presents 'instructional models' of infantry actions. Among many thousands of publications I. N. Pavlov, *Legendarnaya Zheleznaya* (Moscow, 1987), a history of the Iron Division illustrates infantry actions throughout the war.

70 Alexander Werth, *Russia at War 1941–5* (London, 1964), 'Some Characteristics of 1944', pt 7, ch. 1, pp. 759–70.

OFFENSIVE WOMEN

Reina Pennington

1 John Keegan, *A History of Warfare* (New York, 1993), p. 76.

2 John Keegan, *The Face of Battle* (New York, 1978), p. 30.

3 D'Ann Campbell, 'Women in Combat: the World War II Experience in the United States, Great Britain, Germany, and the Soviet Union', *Journal of Military History*, 57.2 (1993), p. 301.

4 Only rough estimates of the number of women involved are possible. Available figures indicate that 5–6,000 women fought in the First World War; while around 80,000 women fought in the civil war (including medical personnel). See Anne Eliot Griesse and Richard Stites, 'Russia: Revolution and War', *Female Soldiers – Combatants or Noncombatants?: Historical and Contemporary Perspectives*, ed. Nancy Loring Goldman (Westport, Ct, 1982), p. 67.

5 Nancy Loring Goldman, introduction to *Female Soldiers*, p. 7.

6 Vera Semenova Murmantseva, *Zhenshchiny v soldatskikh shineliakh* (Women in Soldiers' Overcoats) (Moscow, 1971), p. 9. See also Griesse and Stites, 'Russia', p. 73. The Soviets have given these figures consistently for the past fifty years; however, precise statistical breakdowns are not available. Some corroboration is possible through examination of Komsomol histories (the Komsomol claims to have mobilised 500,000 women in five major mobilisation drives) and through extensive reading of memoir literature. The opening of Soviet archives on this period should finally permit confirmation of the numbers. Even if the actual number of women who fought was far lower than that claimed (and there is no evidence to suggest that the Soviets have exaggerated in any significant way), it would still be an impressive figure, far outweighing the participation of women in any other country.

7 George H. Quester, 'Women in Combat', *International Security* (Spring 1977), p. 81.

8 Jeff M. Tuten, 'The Argument Against Female Combatants', in *Female Soldiers*, p. 243. Tuten makes the interesting claim that 'early in the war, the Germans had captured well over 100,000 Russian female soldiers who held full combatant status'

and implies that this is the basis for the German's scorn for women soldiers (p. 55); this assertion seems implausible, since Soviet women were not mobilised until 1942, and then only for air defence and support services. This myth of the mass capture of women soldiers has unfortunately been picked up in other works; see Shelley Saywell, *Women in War* (Markham, Ont. 1985), p. 149. It must be remembered that the Germans were often contemptuous of the performance of *all* Soviet troops, not just the women. Many of these criticisms lack substance.

9   Many people remain unaware that Germans frequently praised Soviet skill. See for example Generalleutnant D. W. Schwabedissen, *The Russian Air Force in the Eyes of German Commanders*, prepared by the USAF Historical Division (New York, 1968), p. 48.

10   Gordon Williamson, *Loyalty Is My Honour: Personal Accounts from the Waffen SS* (London, 1995), p. 167.

11   Albert Seaton and Joan Seaton, *The Soviet Army: 1918 to the Present* (New York, 1986), p. 292.

12   George Minde, letter to author, 5 Mar 1995. Minde is a PhD candidate at Indiana University studying the experience of Stalingrad during and after the war.

13   For example, Inna Semenovna Mudretsova was commander of a rifle/sniper company of the 702nd Rifle Regiment of the 213th Novoukrainsk Rifle Division; see Yu. E. Dmitrienko, 'Tip velichavoi slavianki...' ('Character of a Great Slav...') *Voenno istoricheskii zhurnal*, 3 (1995), pp. 89–93. Other examples of women who served as platoon and company commanders can be found in Svetlana Alexiyevich, *War's Unwomanly Face* (U voiny – ne zhenskoe litso...) trans. Keith Hammond and Lyudmila Lezhneva (Moscow, 1988), pp. 166–76; V. V. Chudakova, *Chizhik – ptichka s kharakterom* (The Siskin Is a Strong-Willed Bird) (Leningrad, 1965), and Vera Semenova Murmantseva, *Sovetskie zhenshchiny v Velikoi Otechestvennoi voine 1941–5* (Soviet Women in the Great Patriotic War, 1941–5) (Moscow, 1974), p. 124. For an example of a medic at company level, see N. S. Modorov, *V trude i v boiu: Ocherki o zhenshchinakh Gornogo Altaia* (In Labor and in Battle) (Gorno-Altaisk, 1990), pp. 109–15. These are just a few of the numerous examples of women at the company and platoon level.

14   Griesse and Stites, 'Russia', p. 79.

15   Sharon Macdonald, 'Drawing the Lines – Gender, Peace and War: an Introduction,' *Images of Women in Peace and War*, ed. Sharon Macdonald et al. (London, 1987), p. 6.

16   Anne Elliott Griesse, 'Soviet Women and World War II: Mobilization and Combat Policies' (M.A., Georgetown, 1980); Griesse and Stites, 'Russia', pp. 61–84; Richard Stites, *The Women's Liberation Movement in Russia; Feminism, Nihilism and Bolshevism, 1860–1930* (Princeton, NJ; 1978); K. Jean Cottam (ed.), *The Golden-Tressed Soldier* (Manhattan, 1984); K. Jean Cottam, *In the Sky Above the Front: a Collection of Memoirs of Soviet Air Women Participants in the Great Patriotic War* (Manhattan, Ks; 1984); *My Fire-Scorched Youth: the Story of a Woman Machine-Gunner*, translation of *My Fire-Scorched Youth* by Zoya Smirnova Medvedeva (unpublished manuscript, n.d.); K. Jean Cottam, *Soviet Airwomen in Combat in World War II* (Manhattan, Ks, 1983); K. Jean Cottam, 'Soviet Women in Combat in World War II: the Ground Forces and the Navy', *International Journal of Women's Studies*, 3.4 (1980), pp. 345–357; K. Jean Cottam, 'Soviet Women in Combat in World War II:

the Rear Services, Resistance Behind Enemy Lines, and Military Political Workers', *International Journal of Women's Studies*, 5.4 (1982), pp. 363–78; 'Soviet Women in Combat in World War II: The Ground/Air Defense Forces', *Women in Eastern Europe and the Soviet Union*, ed. Tova Yedlin (New York: Praeger, 1980), pp. 115–27; John Erickson, 'Night Witches, Snipers and Laundresses', *History Today*, 40 (1990), pp. 29–35; John Erickson, 'Soviet Women at War', *World War II and the Soviet People*, ed. John Garrard et al. (New York, 1993); Valentina Yakovlevna Galagan, *Ratny podvig zhenshchiny v gody Velikoi Otechestvennoi voiny* (Military Feats of Women during the Great Patriotic War) (Kiev, 1986); Yu. N. Ivanova, 'Vypusknitsy voennykh akademii' ('Women Graduates of Military Academies'), *Voennoistoricheskii zhurnal*, 8 (1993); 'Prekrasneishie iz khrabrykh' ('The fairest of the brave'), *Voenno istoricheskii zhurnal* 3 (1994), pp. 93–6; 'Problem khvatalo i bez nikh, no...' *Voennoistoricheskii zhurnal*, 6 (1994), pp. 75–7; 'Zhenshchiny v Istorii Rossiiskoi Armii' ('Women in the History of the Russian Army') *Voenno istoricheskii zhurnal*, 3 (1992), pp. 86–9; Vera Semenova Murmantseva, *Zhenshchiny v soldatskikh shineliakh* (Women in Soldiers' Overcoats) (Moscow, 1971); *Sovetskie zhenshchiny v Velikoi Otechestvennoi voine, 1941–5* (Soviet Women in the Great Patriotic War, 1941–5) (Moscow, 1974); 'Ratny i trudovoi podvig sovetskikh zhenshchin' ('Military and Labour Feats of Soviet Women'), *Voenno istoricheskii zhurnal*, 5 (1985), pp. 73–81. Two recent collections of interviews are also important: Svetlana Alexiyevich, *War's Unwomanly Face* (U voiny – ne zhenskoe litso...), trans. Keith Hammond and Lyudmila Lezhneva (Moscow, 1988), and Anne Noggle, *A Dance with Death: Soviet Airwomen in World War II* (College Station, Texas, 1994).

17 Erickson, 'Night Witches, Snipers and Laundresses', p. 29.

18 Cottam, 'The Rear Services' p. 363. According to Cottam, 'no separate army women's organizations or policies on conscription and promotion appear to have existed'.

19 Griesse and Stites, 'Russia', pp. 68–9; Erickson, 'Soviet Women at War', p. 51.

20 Raisa E. Aronova, *Nochnye ved'my* ('Night Witches'), revised and expanded 2nd edn (Moscow, 1980), p. 21.

21 Erickson, 'Soviet Women at War', p. 70; Alexiyevich, *War's Unwomanly Face*, pp. 52, 128–33.

22 Alexiyevich, *War's Unwomanly Face*, p. 154.

23 For further discussion of the physical strength issue, see Reina Pennington, 'Wings, Women and War: Soviet Women's Military Aviation Regiments in the Great Patriotic War', Master's thesis (University of South Carolina, 1993), p. 41, and Kenneth D. Slepyan, '"The People's Avengers": Soviet Partisans, Stalinist Society and the Politics of Resistance, 1941–4', PhD dissertation (University of Michigan, 1994), p. 260.

24 Aronova, *Nochnye ved'my*, 2nd edn, p. 24; Alexiyevich, *War's Unwomanly Face*, p. 53.

25 A. M. Bereznitskaia, 'Snaiperskii ekipazh', *V nebe frontovom*, ed. M. A. Kazarinova and A. A. Poliantseva. 1st edn (Moscow, 1962), p. 23. A former aircraft mechanic commented that when the women finally received uniforms there were still problems. 'In 1943 we got skirts for parades and for attending dances. But we had no stockings. Another problem was how to hold up the stockings. Should we try to pin them or use rubber bands?' Irina Favorskaia Luneva, interview with author, 10 May 1993.

26 Alexiyevich, *War's Unwomanly Face*, pp. 30, 82–4.

27 Ibid., p. 156.

28 Erickson, 'Soviet Women at War', p. 75, n. 36; Cottam, 'Soviet Women in Combat in World War II: The Ground Forces and the Navy', p. 353.

29 See Pennington, 'Wings, Women and War'; Reine Pennington 'Wings, Women and War,' *Air & Space Smithsonian* (Dec/Jan 1993–4), pp. 74–85; and Reina Pennington, '"Do Not Speak of the Services You Rendered": Women Veterans of Aviation in the Soviet Union', *Journal of Slavic Military Studies*, 9.1 (1996), pp. 120–51, as well as works by K. Jean Cottam and Anne Noggle.

30 Griesse and Stites, 'Russia', p. 70.

31 Alexiyevich, *War's Unwomanly Face*, p. 47.

32 Cottam (ed.), *My Fire-Scorched Youth*, pp. 14–15.

33 Griesse and Stites, 'Russia', pp. 70–71; Erickson, 'Soviet Women at War', p. 61.

34 Alexiyevich, *War's Unwomanly Face*, p. 97.

35 Ibid., p. 64.

36 Erickson, 'Soviet Women at War', p. 62; Alexiyevich, *War's Unwomanly Face*, pp. 96–7.

37 Alexiyevich, *War's Unwomanly Face*, 53.

38 Ibid., pp. 55–6.

39 *Geroi i Podvigi: Sovetskie listovki Velikoi Otechestvennoi voiny 1941–5 gg.* ('Heroes and Victories: Soviet Leaflets of the Great Patriotic War, 1941–5') (Moscow, 1958), pp. 71–2. Leaflet from the political directorate of the Black Sea Front, summer 1942.

40 Alexiyevich, *War's Unwomanly Face*, pp. 108–9.

41 See A. Barmin, 'Kavaler ordena Slavy', *Docheri Rossii*, ed. I. Cherniaeva (Moscow, 1975), pp. 65–70, translated as A. Barmin, 'A Bearer of the Order of Glory', *The Golden-Tressed Soldier*, ed. Cottam, pp. 99–107; and V. Kaverin, 'Partizanskii vrach', *Docheri Rossii*, ed. I. Cherniaeva (Moscow, 1975), pp. 93–6, translated as Venyamin Kaverin, 'The Story of a Partisan Doctor', *The Golden-Tressed Soldier*, ed. Cottam, pp. 85–91.

42 Shelley Saywell, *Women in War* (Markham, Ont., 1985), pp. 147–8; Cottam (ed.), *The Golden-Tressed Soldier*, p. 243.

43 See Barmin, 'Kavaler ordena Slavy'.

44 Cottam, *My Fire-Scorched Youth*, 2, citing T. K. Kolomiets, 'Chapaevtsy stoiali na smert,' *U chernomorskikh tverdyn* (Moscow, 1967), pp. 216–17.

45 See for example Mariya Grudistova, 'Na zashchite neba stolitsy', *Ikh slavit Rodina*, ed. T. Lil'in (Moscow: Politizdat, 1959), pp. 129–38, translated as Mariya Grudistova, 'Defending the Skies of Moscow', *The Golden-Tressed Soldier*, ed. Cottam, pp. 125–35, and Slepyan, 'The People's Avengers', pp. 259–60.

46 Cottam, 'Soviet Women in Combat in World War II: the Ground Forces and the Navy', p. 352.

47 Alexiyevich, *War's Unwomanly Face*, p. 10.

48 Erickson, 'Soviet Women at War', 61; Cottam, 'Soviet Women in Combat in World War II: the Ground Forces and the Navy', p. 352.

49 Alexiyevich, *War's Unwomanly Face*, p. 30, report of Senior Sergeant Mariia Kaliberda, signaller with 129th Independent Communications Regiment of the 65th Army. See also Cottam, 'Soviet Women in Combat in World War II: The Ground/Air Defense Forces', p. 115.

50 Erickson, 'Soviet Women at War', p. 52.

51 Cottam, 'Soviet Women in Combat in World War II: the Ground Forces and the Navy', p. 25; Murmantseva, *Zhenshchiny v soldatskikh shineliakh*, 68–73.

52 Cottam, 'Soviet Women in Combat in World War II: the Ground/Air Defense Forces,' p. 117, citing T. Klopina, 'Khronika, fakty, nakhodki', *Voenno istoricheskii zhurnal*, 3 (1976), p. 126, and N. A. Kirsanov and V. F. Cheremisov, 'Zhenshchiny v voishakh protivo vozdushnoi oborony v gody Velikoi Otechestvennoi voiny', *Istoriia SSSR*, 3 (1975), p. 66.

53 Erickson, 'Soviet Women at War', p. 62.

54 Erickson, 'Soviet Women at War', p. 52; Alexiyevich, *War's Unwomanly Face*, pp. 161–2; Cottam, *The Golden-Tressed Soldier*, p. 125.

55 Erickson, 'Soviet Women at War', p. 52; Murmantseva, *Zhenshchiny v soldatskikh shineliakh*, p. 217; Griesse and Stites, 'Russia: Revolution and War', p. 71.

56 Cottam, 'Soviet Women in Combat in World War II: the Rear Services', p. 368.

57 Slepyan, 'The People's Avengers', pp. 250–65.

58 Cottam, 'Soviet Women in Combat in World War II: the Rear Services', p. 368.

59 Griesse and Stites, 'Russia: Revolution and War', p. 711.

60 Kenneth D. Slepyan, 'Partisans, Soviet women as', in *Military Women Worldwide: a Biographical Dictionary* ed. Reina Pennington (Westport, Ct, in progress).

61 Griesse and Stites, 'Russia: Revolution and War', p. 72.

62 Alexiyevich, *War's Unwomanly Face*, pp. 193–5.

63 Ibid., pp. 208–9.

64 Mary Penich Motley (ed.), *The Invisible Soldier: The Experience of the Black Soldier, World War II* (Detroit, 1975), p. 234.

65 Erickson, 'Night Witches, Snipers and Laundresses', p. 34; Alexiyevich, *War's Unwomanly Face*, pp. 149–150, 166.

66 Alexiyevich, *War's Unwomanly Face*, pp. 169–70.

67 Ibid., p. 176.

68 See N. Borchenko, 'Ee boevoy shchet', *Voenniia znaniia*, 3 (1966), pp. 16–17 and V. Sokolov, 'V zhelzhnykh nochakh Leningrada', *Devushki v pogonach: sbornik*, ed. A. Nesterskii (Mosow, 1964), pp. 159–66, translated as 'The Score of Yadviga Urbanovich', *The Golden-Tressed Soldier*, ed. Cottam, pp. 153–63; and Andrei Viatskii, 'Kommandir s pushistymi resnitsami', *Sovetskii voin*, 19 (1976), pp. 30–31, translated as Andrei Viatskii, 'The Commander with Thick Eyelashes', *The Golden-Tressed Soldier*, ed. Cottam, pp. 164–72.

69 Erickson, 'Soviet Women at War', pp. 65–6.

70 Cottam, 'Soviet Women in Combat in World War II: the Ground Forces and the Navy', p. 350; Murmantseva, *Zhenshchiny v soldatskikh shineliakh*, pp. 120–21, 130.

71 Alexiyevich, *War's Unwomanly Face*, pp. 67–72.

72 Cottam (ed.), *The Golden-Tressed Soldier*, pp. 273–4; Erickson, 'Soviet Women at War', p. 66.

73 Erickson, 'Soviet Women at War', p. 66.

74 Griesse and Stites, 'Russia: Revolution and War', p. 70; Murmantseva, *Zhenshchiny v soldatskikh shineliakh*, pp. 126–7; 'Tank "boevaia podruga"', *Voenno istoricheskii zhurnal*, 6 (1972), pp. 118–19; I. F. Verevkin and V. Ya. Karlin (ed.), *Byla ty otvazhnym boitsom: ocherki o podvigakh sibiriachek na frontax Velikoi Otechestvennoi voiny* 'You were a Brave Soldier' (Novosibirsk, 1982), pp. 17–31.

75 Erickson, 'Night Witches, Snipers and Laundresses', p. 33; Erickson, 'Soviet Women at War', p. 65; Alexiyevich, *War's Unwomanly Face*, pp. 92–3.

76 Murmantseva, *Zhenshchiny v soldatskikh shineliakh*, pp. 128–9.

77 Griesse and Stites, 'Russia: Revolution and War', p. 69; Erickson, 'Soviet Women at War', p. 65; V. S. Murmantseva, 'Ratny i trudovoi podvig sovetskikh zhenshchin', 76. The graduates include 1,061 snipers and 407 instructors.

78 Alexiyevich, *War's Unwomanly Face*, pp. 14–17.

79 Erickson, 'Soviet Women at War', p. 65; Murmantseva, 'Ratny i trudovoi podvig sovetskikh zhenshchin', p. 76, Murmantseva, *Zhenshchiny v soldatskikh shineliakh*, pp. 106–7, 147–8; Cottam, 'Soviet Women in Combat in World War II: the Ground Forces and the Navy', pp. 347–348.

80 Murmantseva, *Sovetskie zhenshchiny v Velikoi Otechestvennoi voine, 1941–5*, pp. 128–9.

81 Cottam, *My Fire-Scorched Youth*, pp. 14–15.

82 Ibid.

83 Cottam, 'Soviet Women in Combat in World War II: The Ground/Air Defense Forces', p. 348; Murmantseva, *Zhenshchiny v soldatskikh shineliakh*, p. 110.

84 Murmantseva, *Sovetskie zhenshchiny v Velikoi Otechestvennoi voine, 1941–5*, pp. 128–9.

85 Erickson, 'Soviet Women at War', p. 63; Murmantseva, 'Ratny i trudovoi podvig sovetskikh zhenshchin', p. 76. L. F. Toropov (ed.), *Geroini: Ocherki o zhenshchinakh Geroiakh Sovetskogo Soiuza* ('Heroines: Biographical Sketches of Women Heroes of the Soviet Union') (Moscow, 1969), says 5,200 women soldiers were produced (p. 69). Murmantseva and Erickson cite 3,892 privates, 986 NCOs and 2,987 officers, plus a further 514 officers and 1,504 NCOs in 1943; 500 of the 1943 group became front-line veterans.

86 Erickson, 'Soviet Women at War', p. 63; Murmantseva, 'Ratny i trudovoi podvig sovetskikh zhenshchin', p. 76.

87 Saywell, *Women in War*, p. 149.

88 Alexiyevich, *War's Unwomanly Face*, p. 8.

89 Ibid., p. 194.

90 Saywell, *Women in War*, p. 149.

91 Alexiyevich, *War's Unwomanly Face*, p. 103.

92 Ibid., pp. 102–3.

93 Ibid., pp. 223–6.

94 Evodokia Bershanskaia, et al. (ed.), *46 Gvardeiskii Tamanskii zhenskii aviatsionyi polk* ('The 46th Guards Taman Women's Aviation Regiment') (n.p., Tsentral'nyi Dom Sovetskoi Armii imeni M. v. Frunze, n.d.); V. F. Kravchenko (ed.), *125 Gvardeiskii Bombardirovochnyi Aviatsionnyi Borisovskii ordenov Suvoroba i Kutuzova polk imeni Geroia Sovetskogo Soiuza Mariny Raskovoi* (Moscow, privately published, 1976).

95 Erickson, 'Night Witches, Snipers and Laundresses', p. 30.

96 Griesse and Stites, 'Russia: Revolution and War', p. 78.

97 These issues are discussed in detail in Reina Pennington, ' "Do Not Speak of the Services You Rendered": Women Veterans of Aviation in the Soviet Union', *Journal of Slavic Military Studies*, 9.1 (1996), pp. 120–51.

98 Olga Mishakova, 'Sovetskaia zhenshchina velikaia sila,' *Pravda* (8 March 1945), p. 3. This was a distinct change in tone from some of her earlier works; see *Sovetskaia zhenshchina v Velikoi Otechestnnoi voine* (Soviet Woman in the Great Patriotic War) (Moscow, 1943), where she extolls only the bravery and combat

skill of the women; or *Sovetskie devushki v Otechestnnoi voine* ('Soviet Girls in the Great Patriotic War') (Moscow, 1944).

99 M. I. Kalinin, *On Communist Education: Selected Speeches and Articles* (Moscow, 1953), p. 428.

100 Barton Hacker, 'Women and Military Institutions in Early Modern Europe: a Reconnaissance', *Signs: Journal of Women in Culture and Society*, 6.4 (1981), pp. 643–71; see also Hacker's 'Where Have All the Women Gone? The Pre-twentieth-Century Sexual Division of Labor in Armies', *Minerva: Quarterly Report on Women and the Military*, 3.1 (1985), pp. 107–48.

MOTIVATION AND INDOCTRINATION IN THE WEHR-MACHT, 1933–45

Jürgen Förster

1 Max Hastings, *Overlord: D-Day and the Battle for Normandy* (London, 1984), p. 319.

2 Ibid., p. 315.

3 Martin van Creveld, *Fighting Power: German and U.S. Army Performance, 1939–45* (Westport, Ct, 1982), p. 4.

4 Ibid., p. 163.

5 J. P. Stern in his article about German literature and war in *Times Literary Supplement* (17 May 1985), p. 547.

6 *Wofür kämpfen wir?* (Jan. 1944), pp. 80–81.

7 Printed in Klaus Jürgen Müller (ed.), *Armee und Drittes Reich, 1933–9* (Paderborn, 1987), p. 161.

8 Ibid., pp. 164–5 (24 May 1934).

9 Ibid., p. 151 (17 March 1933).

10 Printed in Michael Salewski, 'Von Raeder zu Dönitz. Der Wechsel im Oberbefehl der Kriegsmarine 1943', *Militärgeschichtliche Mitteilungen*, 14 (1973), p. 144.

11 See Blomberg's address to the commanders in Müller (ed.), *Armee und Drittes Reich*, p. 235 (12 Jan 1935).

12 Ibid., p. 171.

13 Weichs' order (2 March 1937) is printed in Müller (ed.), *Armee und Drittes Reich*, pp. 174–6. Schörner's communication of 2 July 1938 can be found in *Bundesarchiv Militärchiv*, Freiburg (hereafter *BA-MA*), RH 46/309.

14 Oberkommando der Wehrmacht (ed.), *Nationalpolitischer Lehrgang der Wehrmacht vom 15.–23. Januar 1937* (Berlin, 1937). The course was repeated in December 1938; this time the participants were commandants and teachers at officer and weapons schools.

15 Adresses of 10 November 1938 and 18 January 1939 in *Bundesarchiv*, Koblenz (hereafter *BA*), NS 11/28. See Gerhard L. Weinberg, 'Propaganda for Peace and Preparation for War', *Germany, Hitler and World War II*, ed. Gerhard L. Weinberg (Cambridge, 1995), p. 73.

16 See Colonel Hermann Reinecke's lecture notes of December 1938 in *BA-MA*, RW 6/ v. 156. Cf. Hermann Foertsch, 'Wer soll Offizier werden?', *Deutsche Infanterie*, 9 (1938), p. 4.

17 Müller (ed.), *Armee und Drittes Reich*, pp. 180–1.

18 See *Heeresdienstvorschrift*, g(eheim) 82 (1 August 1939), p. 91.

19 Organizational order of OKW (25 March 1939) in *BA-MA. RW* 4/v, 143.

20 Cited from Robert J. O'Neill, *The German Army and the Nazi Party. 1933–1939* (London, 1966), p. 83

21 Helmuth Groscurth, *Tagebücher eines Abwehroffiziers 1938–1940*, ed H. Krausnick and H. C. Deutsch, (Stuttgart, 1970), p. 216, in (10 October 1939).

22 See Brauchitsch's order of 7 October 1940, in *BA-MA*, RH 19 III / p. 157.

23 Brauchitsch's order of 21 February 1941, op. cit., p. 152.

24 Order of the commander of the 12th Infantry Division (29 October 1940). Cited from Omer Bartov, *The Eastern Front 1941–1945, German Troops and the Barbarisation of Warfare* (London, 1985) p. 151.

25 OKW, *Mitteilungen für die Truppe*, June 1941. Cited from Bartov *Eastern Front*, 83. See the intelligence report of Army Group South, Rear Area, from 19 July 1941, in *BA-MA.*, RH 22/170. Cf. Jürgen Förster, 'Zum Rußlandbild der Militärs', *Das Rußlandbild im Dritten Reich* (ed.) Hans-Erich Volkmann (Köln, 1994), pp. 141–63.

26 See Jürgen Förster 'New Wine in Old Skins? The Wehrmacht and the War of "Weltanschauungen", 1941' in *The German Military in the Age of Total war (ed.)*, Wilhelm Deist (Leamington Spa, 1985), pp. 304–22.

27 OKII, PA (2) la, communications of 22 May, 6 and 11 November 1942, in *BA-MA*, RH 15/v, p. 15.

28 Communication to all commanders of 13 April 1944, in *BA-MA, RW* 4/v. p. 878.

29 Further order of 22 December 1943. See Arne W.G. Zocpf, *Wehrmacht zwischen Tradition und Ideologie. Der NS-Führungsoffizier im Zweiten Weltkrieg* (Frankfurt/M. 1988) and Jürgen Förster, 'Vom Führerheer der Republik zur national sozialist-ischen Volksarmee. Zum Strukturwandel der Wehrmacht; *Deutschland in Europa. Kontinuität und Bruch*, (ed.) Jost Dolffer. Bernd Martin, Günther Wollstein (Berlin. 1990). pp. 321–3. For the organization of ideological indoctrination in the Waffen SS. see Bernd Wegner, *The Waffen-SS. Organization. Ideology and Function* (Oxford, 1990), pp. 198, 220.

30 Party statistics of 20 December 1944 in BA–MA, NS 6/782.

31 *Hitlers Weisungen für die Kriegführung 1939–1945* (ed). Walther Hubatsch, 2nd edn (Frankfurt/M, 1983), pp. 310–11.

32 Cited from Klaus Hildebrand *The Third Reich* (London, 1964), p. 83.

33 *The Testament of Adolf Hitler : The Hitler–Bormann Documents, February–April 1945* (London, 1961), pp. 58–9.

THE GERMAN SOLDIER IN OCCUPIED RUSSIA

Theo J. Schulte

1 For a more detailed discussion of the creation of this myth, as well as some of the material in this article presented in a somewhat different manner, see: Theo J. Schulte, 'German Army Occupation Policy in Eastern Europe: the State of the Debate Half a Century After the End of the Second World War, Half a Decade After German Reunification', in *The German Lands and Eastern Europe*, ed. Roger Black and Karen Schönwälder (London, 1996) (hereafter, *German Lands*).

2   On the precise numbers of German prisoners-of-war who died in Soviet hands, see: Christian Streit, *Keine Kameraden. Die Wehrmacht und die sowjetischen Kriegsgefangenen*, 1941–5, 2nd edn (Bonn, 1991), p. 301, n. 2.

3   Theo J. Schulte, *The German Army and Nazi Policies in Occupied Russia* (Oxford, 1989) (hereafter *German Army*), pp. 1–27.

4   Hannes Heer, 'Killing Fields: Die Wehrmacht und der Holocaust', in Hannes Heer (ed.), *Vernichtungskrieg. Vebrechen der Wehrmacht 1941–4* (Hamburg, 1996) (hereafter *Vernichtungskrieg*), pp. 74–5. One might also add *'violent anti Semitism deeply embedded in German culture* (Daniel Jonah Goldhagen, *Hitler's Willing Executioners: Ordinary Germans and the Holocaust* (London, 1996)) (hereafter *Willing Executioners*).

5   Christopher Browning, *Ordinary Men: Reserve Police Battalion 101 and the Final Solution in Poland* (New York, 1992) (hereafter *Ordinary Men*); Omer Bartov, *Hitler's Army: Soldiers, Nazis and War in the Third Reich* (Oxford, 1991) (hereafter *Hitler's Army*); Goldhagen, *Willing Executioners*: Heer, 'Killing Fields'.

6   Christopher Browning, 'Hitler and the Euphoria of Victory', in David Cesarani (ed.), *The Final Solution: Origins and Implementation* (London, 1994) (hereafter *Final Solution*), pp. 137–58.

7   An up-to-date exhibition itinerary is available on + 49 221 31 76 68. The number of reviews, particularly in German, is far too extensive to cite, but the Hamburg Institute itself has produced a special 'review': Klaus Naumann, 'Wenn ein Tabu bricht. Die Wehrmachtausstellung in der Bundesrepublik', *Mittelweg* 36 (February / March 1996), pp. 11–24; see also, Walter Manoschek, 'Die Wehrmachtsaustellung in Österreich. Ein Bericht', ibid., pp. 25–32. In English, see: Ian Traynor, 'Hitler's Army Shares SS Guilt', *Guardian* (6 April, 1995); Robin Gedye, 'Wehrmacht Atrocities Exposed at Last', *Daily Telegraph* (23 March 1995).

8   Hannes Heer (ed.), *Vernichtungskrieg. Verbrechen der Wehrmacht, 1941–4. Ausstellungskatalog* (Hamburg, 1996).

9   Ernst Klee et al., *The Good Old Days: the Holocaust as Seen by Its Perpetrators and Bystanders* (New York, 1991); Klaus Latzel, 'Tourismus und Gewalt. Kriegswahrnehmungen in Feldpostbriefen', *Vernichtungskrieg*, pp. 447–59; Dieter Reifahrth / Viktoria Schmidt Linsenhoff, 'Die Kamera der Täter', ibid., pp. 475–503; Bernd Hüppauf, 'Der entleerte Blick hinter der Kamera', ibid., pp. 504–27.

10 *Vernichtungskrieg*, op. cit.

11 Horst Boog et al., *Das Deutsche Reich und der Zweite Weltkrieg*, vol. 6 (Stuttgart, 1990), pp. 778–85.

12 Bartov, *Hitler's Army*, pp. 106–78; Heer, 'Killing Fields', in *Vernichtungskrieg*, pp. 74–5.

13 *Zeit – Punkte Nr.* 3/1995. 'Gehorsam bis zum Mord? Der verschwiegene Krieg der deutschen Wehrmacht – Fakten, Analysen, Debatte', ed. Theo Sommer (Hamburg, 1995) (hereafter *Zeit Punkte*).

14 Ibid., pp. 91–7; *Mittelweg*, 36 (February/ March 1996), pp. 11–24.

15 Emnid Umfrage, 18–23 April, in *Der Spiegel*, 19 (8 May 1995) p. 77.

16 As well as the extensive 'case studies' in *Vernichtungskrieg*, see also Walter Manoschek (ed.), *Die Wehrmacht im Rassenkrieg* (Vienna, 1996); Hannes Heer, *Dienststelle Minsk: Deutsche Vernichtungspolitik in Weißrußland 1941–4* (Hamburg, 1996); Hannes Heer (ed.), *Der Minsker Prozeß. Protokoll der Gerichtsverhandlung gegen 18 deutsche Kriegsverbrecher 1946* (Hamburg, 1996); G. R. Ueberschär (ed.), *Militärs und ihre*

Verstrickung in die NS Politik (Darmstadt, forthcoming, 1997); Paul Kohl, *Der Krieg der deutschen Wehrmacht und der Polizei, 1941–4. Sowjetische Überlebende Berichten* (Frankfurt, 1995) (originally published in 1990 as: *Ich wundere mich, daßich noch lebe*); Siegfried Frech, 'Gehst mit Juden erschießen?', *Die Unterrichtspraxis*, Heft 6, 23.9 (1995). See also W. Grussman and I. Ehrenburg (ed.), *Das Schwarzbuch. Der Genozid an den sowjetischen Juden* (Stuttgart, 1994); Roland Headland, *Messages of Murder: a Study of the Einsatzgruppen of the Security Police and the Security Service, 1941–3* (Ontario, 1992); Hans Heinrich Wilhelm, *Rassenpolitik und Kriegführung. Sicherheitspolizei und Wehrmacht in Polen und der Sowjetunion, 1939–2* (Passau, 1991).

17   *Willing Executioners*. An extensive collection of review articles is to found in nos 16–25 of *Die Zeit*, covering the weeks 12 April – 14 June 1996. See also *Der Spiegel*, no. 21 (20 May 1996), pp. 48–77; Jeremy D. Noakes, 'No Ordinary People', in *Times Literary Supplement* (7 June 1996), pp. 9–10.

18   Wolfram Wette, 'Erobern, zerstören, auslöschen. Die verdrangte Last von 1941. Die Rußlandfeldzug war ein Raub und Vernichtungskrieg von Anfang an', in *Die Zeit*, no. 48 (28 Nov 1987); Andreas Hillgruber, *Zweierlei Untergang. Die Zerschlagung des Deutschen Reiches und das Ende des europaischen Judentums* (Berlin, 1986); Hans Ulrich Wehler, *Entsorgung der deutschen Geschichte?* (Munich, 1988), p. 46; Omer Bartov, 'Historians on the Eastern Front', *Tel Aviver Jahrbuch für deutsche Geschichte*, 16 (1987), pp. 325–45.

19   For the original remark, the emphasis of which I have inverted, see Förster, 'Relation', in Cesarani (ed.), *Final Solution*, p. 97.

20   Jürgen Förster, Das Sicherung des Lebensraumes, in MGFA (ed.), *Das Deutsche Reich und der Zweite Weltkrieg*, vol. 4 (Stuttgart, 1983), p. 1054.

21   Schulte, *German Army*.

22   Christian Streit, 'Wehrmacht, Einsatzgruppen, Soviet POWs and anti-Bolshevism', in Cesarani (ed.), *Final Solution*, pp. 123–4; Bartov, *Hitler's Army*, pp. 12–28; Schulte, *German Army*, pp. 42–52.

23   *Zeit Punkte*, pp. 78–9.

24   Bernd Boll and Hans Safrian, 'Auf dem Weg nach Stalingrad, Die 6. Armee 1941/42', Boll, 'Auf dem Weg nach Stalingrad', in *Vernichtungskrieg Ausstellungskatalog*, pp. 62–101. See also Schulte, *German Army*, pp. 81–5, 287–8; *Hitler's Army*.

25   Schulte, *German Army*, documents a–d, pp. 317–23.

26   R. C. Fattig, 'Reprisal: the German Army and the Execution of Hostages During the Second World War', Ph.D., Michigan Microfilm No. JWK81 07460 (San Diego, Cal, 1980).

27   Christian Streit, 'Wehrmacht', in Cesarani (ed.), *Final Solution*, pp. 121–2; Schulte, *German Army*, pp. 135–7; Gustav Strübel, 'Anatomie eines Massakers' in *Zeit Punkte*, pp. 58–63. See also Mark Mazower, *Inside Hitler's Greece: the Experience of Occupation, 1941–4* (Yale, 1993), pp. 199 ff; Goldhagen, *Willing Executioners*, pp. 401 ff.

28   Hans Mommsen, 'The Realization of the Unthinkable: the Final Solution of the Jewish Question in the Third Reich', in *The Policies of Genocide*, ed. Gerhard Hirschfeld (London, 1986), pp. 128–9.

29   cf. Detlev J. Peukert, 'Four Political Generations', *The Weimar Republic* (Harmondsworth, 1991), pp. 14–18.

30   Bernhard R. Kroener, 'Die Personellen Resourcen des Dritten Reiches im Spannungsfeld zwischen Wehrmacht, Bürokratie und Kriegswirtschaft, 1939–1942', in

MGFA (ed.), *Das Deutsche Reich und der Zweite Weltkrieg*, vol. 5.1 (Stuttgart, 1988), pp. 693 ff.

31 Omer Bartov, 'The Missing Years: German Workers, German Soldiers', in *Nazism and German Society, 1933–45*, ed. David F. Crew (London, 1994), pp. 41–66.

32 H. E. Volkman (ed.), *Das Rußlandbild im Dritten Reich* (Cologne, 1994). Alf Lüdtke, 'The Appeal of Exterminating "Others": German Workers and the Limits of Resistance', in *Resistance Against the Third Reich, 1933–90*, ed. Michael Geyer and John W. Boyer (Chicago, 1990).

33 R. W. Davies, *Soviet History in the Gorbachev Revolution* (London, 1988), p. 102 ff.

34 Jonalthan Steinberg, 'The Third Reich Reflected German Civil Administration in the Occupied Soviet Union', *English Historical Review*, vol. cx (June 1995), pp. 649–50; Michael Geyer, 'The Stigma of Violence: Nationalism and War in Twentieth Century Germany', in *German Studies Review*, Special Issue on German Identity (Winter, 1992); Frank Bajohr et al., *Zivilisation und Barbarei. Die widersprüchlichen Potentiale der Moderne. Detlev Peukert zum Gedanken* (Hamburg, 1991).

35 W. Manoschek (ed.), *Es gibt nur eines für das Judentum: Vernichtung. Das Judenbild in deutschen Soldatenbriefen, 1939–44* (Hamburg, 1995).

36 Schulte, *German Army*, pp. 131–2, n. 54.

37 Jonathan Steinberg, *All or Nothing: the Axis and the Holocaust* (London, 1990), pp. 206–42; W. Manoschek, *Serbien ist judenfrei. Militärische Besatzungspolitik und Judenpolitik in Serbien 1941/2* (Munich, 1994).

38 Schulte, *German Army*, p. 198; Goldhagen *Willing Executioners*, pp. 268–9.

39 Jens Elert, *Stalingrad – eine deutsche Legende* (Hamburg, 1992).

40 F. Andrae, *Auch gegen Frauen und Kinder. Der Krieg der deutschen Wehrmacht gegen die Zivilbevölkerung in Italian, 1943–5* (Munich, 1995); Menachem Shelah, 'Die Ermordung italienischer Kriegsgefangener, September bis November 1943', in *Vernichtungskrieg*, pp. 191–207; Michael Geyer, ' "Es muß mit schnellen und drakonischen Maßnahmen durchgreifen werden". Civitella in Val di Chiana am 29 Juni 1944', ibid., pp. 208–38; Mazower, *Inside Hitler's Greece*.

41 Konrad Kwiet, 'From the Diary of a Killing Unit', in *Why Germany? National Socialist Anti-Semitism and the European Context*, ed. John Millfull (Oxford, 1993), pp. 75–90; Browning; *Ordinary Men*; Goldhagen, *Willing Executioners*; Bernhard Chiari, 'Deutsche Zivilverwaltung in Weißrußland, 1941–44. Die lokale Perspektive der Besatzungsgeschichte', in *MGM*, vol. 2, (1993), pp. 67–88; Steinberg, 'Third Reich Reflected', pp. 620–51.

42 Goldhagen, *Willing Executioners*, pp. 239–62, 375–415.

43 Schulte, 'Korück 582', *Vernichtungskrieg*.

44 Streit, *Keine Kameraden*, pp. 17–19.

45 Schulte, *German Army*, p. 334.

46 Mark Mazower, 'Military Violence and National Socialist Values: the Wehrmacht in Greece, 1941–4', in *Past and Present*, 134 (1992); p. 135.

47 David Kitterman, *Refusing Nazi Orders to Kill: Germans in Uniform who Resisted the Holocaust* (Oxford 1996); Browning, *Ordinary Men*, pp. 159 ff.; Goldhagen, *Willing Executioners*, pp. 263–80; Schulte, *German Army*, pp. 136–7.

48 I understand that Monika Kaiser is working on this topic. See *Journal of Contemporary History*, vol. 30 (1995). See also Hans Jürgen Brandt, *Priestersoldaten in der Wehrmacht 1939 bis 1945*, in Hans Jürgen Brandt (ed.), *Priester in Uniform: Seelsorger, Ordensleute und Theologen als Soldaten im Zweiten Weltkrieg* (Augsburg, 1994), pp. 7–24.

49 Ortwin Buchbender and R. Sterz (ed.), *Das andere Gesicht des Krieges. Deutsche Feldpostbriefe, 1939–45* (Munich, 1982); V. Osipov (ed.), *Ich will raus aus diesem Wahnsinn. Deutsche Briefe von der Ostfront 1941–45. Aus sowjetischen Archiven* (Wuppertal, 1991); H. J. Schröder, *Die gestohlenen Jahre: Erzählgeschichten und Geschichtserzählung im Interview. Der Zweite Weltkrieg aus der Sicht ehemaliger Mannschafts-soldaten* (Tübingen, 1992); Manfred Messerschmidt, 'June 1941: Seen through German Memoirs and Diaries', in *Operation Barbarossa*, ed. Department of Humanities (Utah, 1992); H. Edler and T. Kufus (dirs), *Mein Krieg. Ein Dokumentarfilm* (Germany, 1990).

50 Schulte, *German Army*, pp. 30, 253–76; Mazower, 'Military Violence', p. 135; Goldhagen, *Willing Executioners*, pp. 266–8.

51 Mazower, 'Military Violence', p. 148; Ian Kershaw, *The Nazi Dictatorship: Problems and Perspectives of Interpretation* (London, 1993), pp. 150–79.

52 Schulte, *German Army*, pp. 145–9, 260–76; cf. Gabriele Salvatores (dir.), *Mediterraneo* (Italy, 1991).

53 Norbert Hasse and Gerhard Paul (ed.), *Die anderen Soldaten. Wehrkraftzersetzung, Gehorsamsverweigerung und Fahnenflucht im Zweiten Weltkrieg* (Frankfurt, 1995); F. Ausländer, *Verräter oder Vorbilder? Deserteure und ungehorsame Soldaten im Nationalsozialismus* (Bremen, 1995); F. W. Seidler, *Fahnenflucht. Der Soldat zwischen Eid und Gewissen* (Munich, 1993); Karsten Bredemeier, *Kriegsdienstverweigerung im Dritten Reich. Ausgewählte Beispiele* (Baden-Baden, 1991).

54 See my forthcoming book, *Nazi War Crimes and War Crimes Trials: 1945–95* (Oxford, 1997).

WHO FOUGHT AND WHY?

Theodore A. Wilson

1 The best recent assessment of the Victory Program's origins is Charles Kirkpatrick, *An Unknown Future and a Doubtful Present: Writing the Victory Plan of 1941* (Washington, DC, 1991). Also see Mark S. Watson, *Chief of Staff: Pre war Plans and Preparations* (Washington, DC, 1950, pp. 337–40.

2 A recent summary of these issues is Theodore A. Wilson, 'Leviathan: the American Economy in World War II', in *Allies at War: The Soviet, American, and British Experience of World War II*, ed. David Reynolds, Warren Kimball and A. O. Chubarian (New York, 1994).

3 Kent R. Greenfield, Robert R. Palmer and Bell I. Wiley, *The Organization of Ground Combat Troops* (Washington, D.C., 1947); Robert R. Palmer, Bell I. Wiley and William R. Keast, *The Procurement and Training of Ground Combat Troops* (Washington, DC, 1948).

4 These issues have been dealt with in such recent studies as Lee Kennett, *GI: the American Soldier in World War II* (New York, 1987), John Sloan Brown, *Draftee Division: The 88th Infantry Division in World War II* (Lexington, Ky, 1986), and Theodore A. Wilson, 'Deposited on Fortune's Far Shore: the 2nd Battalion, 8th Infantry', in *D-Day 1944*, ed. Theodore A. Wilson (Lawrence, Ks, 1994). Also see Theodore A. Wilson, *Building Warriors: the Selection and Training of US Ground Combat Forces in World War II* (Lawrence, Ks, forthcoming).

5   See the discussion in Theodore A. Wilson, 'Leviathan: the American Economy in World War II', in *Allies at War* op. cit., pp. 182–5.

6   For full consideration of these issues, see my forthcoming study, *Building Warriors*, esp. chs 3, 6 and 17.

7   The reasons for rejection broke down roughly as follows: physical defects (66.2 per cent), emotional disorders (18.5 per cent), mental or educational deficiency (13.6 per cent) and administrative (1.7 per cent), Eli Ginzberg et al., *The Ineffective Soldier*, 3 vols (New York; 1959), vol. 1, pp. 30–36.

8   Ginzberg et al., *The Ineffective Soldier*, vol. 1, pp. 12–15. Also see Maphaeus Smith, 'Population Characteristics of American Servicemen During World War II', *Scientific Monthly* (Sept 1947), pp. 246–52.

9   Kennett, *GI*, pp. 26–7.

10 Ginzberg et al., *The Ineffective Soldier*, pp. 27–8.

11 Kennett, *GI*, pp. 17–18; George Q. Flynn, *The Draft*, 1940–73 (Lawrence, Ks, 1993), pp. vii–ix.

12 See Palmer, et al., *Procurement and Training of Ground Combat Troops*, pp. 3–11, and also Samuel A. Stouffer et al., *The American Soldier*, 4 vols (Princeton, NJ; 1949), vol. 11, p. 234.

13 Interview with Brigadier General James G. Christiansen, 12 May 1944, Army Ground Forces Historical Files, Box 26, RG 319, NARA.

14 General Willard Paul stated, 'I don't believe that anyone can honestly dispute this statement', Folder: 'Personnel and Administrative: 1946–48', Box 2, Arthur Trudeau Papers, Military History Institute, Carlisle Barracks, Pa.

15 US War Department, Office of the Adjutant General, US army, *Training Circular No. 25*, 'Reading Courses for Officers' (5 April 1941); this release replaced War Department, *Bulletin No. 46* (1931), as amended by *Bulletin No. 6* (1938) and *Bulletin No. 4* (1939).

16 Daniel Katz and Richard L. Schank, *Social Psychology* (New York, 1935); Leonard W. Doob, *Propaganda: Its Psychology and Technique* (New York, 1935); Ordway Tead, *The Art of Leadership* (New York, 1927); Walter Van Dyke Bingham, *Aptitude and Aptitude Testing* (New York, 1937); Morris S. Viteles, *Industrial Psychology* (New York, 1932).

17 Aside from Freeman's biography of Robert E. Lee, not one of the above works showed up in non-fiction bestseller tabulations for the years 1933–40. Middle-class Americans were relaxing with such Depression era escapism as *Life Begins at Forty*, *Life with Father*, Marjorie Hillis's *Live Alone and Like It* (with sales of 100,000 in 1936), and her frothy sequel, *Orchids on Your Budget*; penetrating the mysteries of three no trumps with the help of Ely Culbertson's *Contract Bridge Blue Book*; acquiring religious homilies to live by from Father Vincent Sheean and dollops of self-confidence from Dale Carnegie's *How to Win Friends and Influence People* (a fixture on bestseller lists for 1937 and 1938 when it sold 729,000 copies); and enjoying biographies of Nijinsky, Marie Antoinette, the also beheaded Mary Queen of Scots, and – yet another coquette – Fanny Kemble.

18 See Loren Baritz, *The Servants of Power: a History of the Use of Social Science in American Industry* (Middletown, Ct, 1960) for a helpful discussion of the relationship between 'approving intellectuals' and institutions such as the US army.

19 Max Lerner, *America as a Civilization: Life and Thought in the United States Today* (New York, 1957), p. 227.

20 JoAnne Brown, *The Definition of a Profession: the Authority of Metaphor in the History of Intelligence Testing, 1890–1930* (Princeton, NJ, 1992), p. 97.

21 The number of professionally educated engineers multiplied tenfold between 1890 and 1900, and the linear language of engineering pervaded American society over the next several decades. For example, Thorndike wrote in 1922: 'Education is one form of human engineering and will profit by measurements of human nature and achievement as mechanical and electrical engineering have profited by using the foot pound, calorie, volt and ampère.' Edward L. Thorndike, 'Measurement in Education', in *Twenty-First Yearbook of the National Society for the Study of Education*, ed. J. K. Whipple (1922), Part 1, p. 1.

22 Edward Bellamy, *Looking Backward, 2000–1887* (New York 1962), p. 67.

23 Ronald Schaffer, *America in the Great War: the Rise of the War Welfare State* (New York, 1991), pp. 135–6.

24 'Committee on Classification of Personnel in the Army', *History of the Personnel System* (Washington DC, 1919), p. 12.

25 Ibid., p. 8.

26 Research by Dwight Boring, of the Yerkes group, revealed that of the 18 per cent who were immigrants, those from England, Holland, Denmark, Scotland, and Germany did quite well compared with immigrants from Turkey, Greece, Russia, Italy and Poland. Paul D. Chapman, *Schools as Sorters: Lewis M. Terman, Applied Psychology and the Intelligence Testing Movement, 1890–1930* (New York, 1988), p. 71.

27 Edward W. Coffman, *The War to End All Wars: the American Military Experience in World War I* (New York, 1968), p. 61.

28 Penn Borden, *Civilian Indoctrination of the Military* (Westport, Ct, 1986), p. 41.

29 James Reed, 'Robert M. Yerkes and the Mental Testing Movement', *Psychological Testing and American Society*, ed. Michael M. Sokal, (Rutgers University Press 1987), p. 77.

30 Chapman, *Schools as Sorters*, p. 71.

31 Ronald Schaffer has written: 'Yerkes himself knew that sex, race, language, and social and economic factors affected performance on intelligence tests...But he administered a program in which the examining instruments were not only shaped by the requirements of mass testing but were culturally biased.' Schaffer, *America in the Great War*, p. 137.

32 Hamilton Cravens, *The Triumph of Evolution: American Scientists and the Heredity–Environment Controversy, 1900–41* (Philadelphia, Pa, 1978), p. 252.

33 Ibid.

34 For attitudes toward class in this context, see Kenneth W. Eells, et al., *Intelligence and Cultural Differences: a Study of Cultural Learning and Problem Solving* (Chicago, Il., 1951), pp. 20–25.

35 Major General Henry Jervey, 'Mobilization of the Emergency Army', Lecture for the General Staff College, 3 January 1920, quoted in Kreidberg and Henry, *History of Military Mobilization in the United States Army*, (Washington DC, 1955) p. 308.

36 Reed, 'Robert M. Yerkes and the Mental Testing Movement', pp. 84–5.

37 Directive No. 5: 'Methods to be prescribed to insure the readiness in First Phase [of mobilisation] of all Regular Army inactive units', Army War College Syllabus, War College Files, Military History Institute, Carlisle Barracks, Pa.

38 Ibid.

39 For a full discussion, see Cravens, *Triumph of Evolution*, pp. 228–52, and Carl Degler, *In Search of Human Nature* (New York, 1991).

40 Robert M. Yerkes (ed.), 'Psychological Examining in the United States Army' XV, *Memoirs of the National Academy of Sciences* (Washington DC; GPO, 1921), pp. 104–14.

41 Quoted in Degler, *In Search of Human Nature*, p. 168.

42 Ralph E. Bowers to Adjutant-General, Subject: 'Report of Army Intelligence Studies', 15 June 1928, Box 838 (AG 201.6), RG 94, NARA.

43 Attachment, 'Personnel Research Agencies: War Department General Staff', Ethelbert Stewart, Commissioner of Labor Statistics, to A. Gerhard, Chief Clerk, WDGS, 30 Nov 1929, Testing folder, Box 837 (AG 201.6), RG 94, NARA.

44 Donald S. Napoli, 'The Mobilization of American Psychologists, 1938–41', *Military Affairs* (Feb 1978), p. 32.

45 Major Charles A. Drake, 'Psychology and Modern War', *Infantry Journal*, 47 (July–Aug 1940), p. 379.

46 Ibid., p. 380.

47 Minutes of ASF Personnel Conference, 'Physical Profile Serial or Classification and Assignment of Enlisted Men Based Upon Physical Capacity or Stamina', [1943], ASF Personnel Conferences folder, Box 664, RG 160, NARA.

48 Addendum for Signal Corps Officers (AR 140–38) to 'Training Circular No. 25', 5 April 1941, War Department Publication: TC 25.

49 Memorandum for the Director, Training Requirements Division, Army Service Forces, Re: 'Identification and Preservation of Historical Materials on Psychology and Psychological Techniques During the War', 28 September 1944, (AG 040: National Research Council File), RG 319, NARA.

50 The careers of the two senior officers in the Adjutant-General's Office during these years serve as cases in point. Both, notably, came from the ranks. Emory S. Adams was born in Manhattan, Kansas in 1881. Educated at hometown Kansas State Agricultural College, he enlisted in the 20th Kansas Infantry in June 1898, shortly after the United States initiated hostilities with Spain, shifted to the regular army in 1900 and two years later was commissioned a second lieutenant of infantry. Lieutenant Adams fought in 'numerous engagements' during the Philippine Insurrection and during 1904–5 as Quartermaster and Commissary on the army transports, *USS Ingalls* and *USS Seward*. Following a second tour with the 14th Infantry stationed at Vancouver Barracks, Washington, detached service with the Military Intelligence Division, Manila, another jaunt as QM/Commissary aboard the *USS Warren*, he joined the 15th Infantry and served at Fort Douglas, Utah, Fort Sam Houston, San Antonio, Texas, and in Tientsin, China. Returning from China in 1913 Adams spent five years at the Recruiting Depot, Columbus Barracks, Ohio. That assignment was to prove fateful, for Adams never again commanded troops in the field. Brought to Camp Devens in July 1918 as Assistant Chief of Staff of the 12th Division by former Adjutant-General McCain and promoted to the temporary rank of lieutenant-colonel, he went to France in October 1918, serving on the AEF Headquarters staff and at Base Section 5, Brest. Adams returned in December 1919 to become Assistant to the Depot Quartermaster at Jeffersonville Arsenal and, then, Assistant Personnel Adjutant and Sixth Corps Area Headquarters Adjutant, Chicago, Illinois. In May 1922, accepting his fate, Adams transferred from infantry to the Adjutant General's department. He spent the next four years in the Office of the Adjutant General,

then alternated between Washington and AG postings in various Corps Areas. Promoted to Major-General, Adams was appointed Adjutant-General in May 1938 and served in that capacity until February 1942.

His successor, Major-General James A. Ulio, followed a remarkably similar path. Born in Walla Walla, Washington in June 1882, Ulio enlisted in the regular army in September 1900 and four years later was commissioned a Second Lieutenant of Infantry. He also served in the Philippine Islands and spent a four-year tour at Vancouver Barracks. Having been promoted to First Lieutenant in 1911, in May 1912, Ulio joined the 1st Infantry, then stationed at Schofield Barracks. He returned in 1917 to a posting with the 32nd Infantry and was subsequently transferred to Camp McClellan, Alabama, serving as Adjutant, 29th Division. Ulio went to France in March 1918 and served as Assistant Chief of Staff for Personnel, 35th Division, then did the same job at IV Corps, as Chief, Statistical Division, and Chief, Central Records Office, AGO, AEF General Headquarters. After a brief stint with the US relief mission to Armenia, Ulio was ordered to the Adjutant-General's Office for a three-year tour. He bounced from the Second Corps Area to Washington, then to the Command and Staff School, then back to the AGO with an interlude as White House Aide. Ulio graduated from the Army War College in 1934 and returned to TAGO. Following an interlude as General Hugh A. Drum's aide in Hawaii, he went to the Second Corps Area, thence back to TAGO and an appointment as Executive Officer and then Assistant Adjutant-General, with the rank of Brigadier-General, .201 files, Center of Military History Archives, Washington, DC.

51 Reflecting both those assumptions and the widespread confusion about who was in charge, a civilian consultant to the army enthused in 1944:

> Modern warfare has put a premium on labor skills and technical proficiencies and made the proper assignment of manpower an absolute necessity in the conservation of human resources. It is fortunate that a few farsighted men in the War Department foresaw the implications of this type of warfare. These men were responsible for the philosophy of change that has enveloped the Army personnel program. Unlike many oldline Army officers, they perceived that this would not be a war of numbers in which men would be assigned to a military unit based upon a table of organization – a war in which full responsibility would be placed upon the individual organization commander to make the most of personnel literally 'dumped into his lap'. They foresaw in this war of machines a body of Army manpower carefully selected, technically trained, and properly assigned, and a personnel system that would follow through, procedure by procedure, the metamorphosis of each soldier in the development of his capacities to meet Army needs.

Ernest Engelbert, 'The Army Personnel Process: Trends and Contributions', *Public Administration Review*, 4 (Winter 1944), pp. 51–2.

52 Brown, *The Definition of a Profession*, 97.

53 Notes of conference on classification and assignment, National Research Center, n.d., Folder: 'Personnel and Classification'; Box 2, Trudeau Papers, MHI.

54 Walter V. Bingham, 'How the Army Sorts Its Man Power,' *Harper's Magazine* (March, 1942), p. 434.

55 Palmer et al., *Procurement and Training of Ground Combat Troops*, pp. 5–9.

56 Marvin A. Kreidberg and Merton C. Henry, *History of Military Mobilization in the United States, 1775–1945*, DA Pamphlet, 20–212 (Washington, DC, 1955), pp. 635–6;

Ginzberg et al., *Ineffective Soldier*, vol. 1, p. 56; Louis E. Keefer, 'Birth and Death of the Army Specialized Training Program', *Army History*, 33 (Winter 1995), pp. 1–7.

57 'Illiteracy Among Replacements', 6 Sept 1943, and other materials in Francis Mallon Papers, MHI; Ginzberg, *Ineffective Soldier*, vol. 1, pp. 151–62.

58 The US army's policies toward African-Americans in combat are documented in Ulysses Lee, *The Employment of Negro Troops* (Washington, DC, 1970), and in such other sources as Stouffer et al., *The American Soldier*, vol. 1, pp. 486–96; and Thomas St John Arnold, *Buffalo Soldiers: the 92nd Infantry Division and Reinforcements in World War II, 1942–5* (Manhattan, Ks, 1990).

59 Palmer et al., *Procurement and Training of Ground Combat Troops*, p. 18; Headquarters Army Ground Forces, Ground Statistics Section, 'Army Ground Forces Statistical Data', 31 Aug 1945, HRC 321, Center of Military History Archives, Washington, DC.

60 L. J. McNair to George C. Marshall, re Distribution of Manpower, 17 Dec 1943, AGF D. F. 327.3, Box 60, RG 337, NARA.

61 Ibid.

62 Palmer et al., *Procurement and Training of Ground Combat Troops*, pp. 66–72.

63 Ginzberg et al., *Ineffective Soldier*, vol. 1, pp. 44–5.

64 Engelbert, 'Army Personnel Process', p. 94.

65 Palmer et al., *Procurement and Training of Ground Combat Troops*, p. 49; AGF, 'Statistical Data', R6337, NARA.

66 A. T. Trudeau Memorandum, 13 April 1943, Correspondence folder, 1943–4, Box 2, Arthur T. Trudeau Papers, MHI.

67 According to the AGF, the casualty rate for infantry was 582 per thousand and that for air combat crews was 475.6 per thousand, AGF, 'Statistical Data', p. 35.

68 Samuel A. Stouffer et al., *The American Soldier* (Princeton, 1949), vol. 1, pp 192–4; the 'Doolittle Board' conducted highly publicised hearings in 1946, 'Report', HRC 321, CMH Archives.

THE GI IN EUROPE AND THE AMERICAN MILITARY TRADITION

Reid Mitchell

1 Ernie Pyle, *Brave Men* (New York, 1944), p. 380.

2 Samuel A. Stouffer et al., *The American Soldier* (Princeton, 1949), vol. 11, pp. 62–5.

3 Russell F. Weigley, *The American Way of War* (Bloomington, Ind, 1977).

4 Russell F. Weigley, *Eisenhower's Lieutenants: the Campaign of France and Germany, 1944–5* (Bloomington, Ind, 1981), p. 464.

5 Harold P. Leinbaugh and John D. Campbell, *The Men of Company K: the Autobiography of a World War II Rifle Company* (New York, 1987), p. xiv.

6 Leon C. Standifer, *Not in Vain: a Rifleman Remembers World War II* (Baton Rouge, 1992), p. 212.

7 Bill Maudlin, *Up Front* (New York, 1945), p. 39.

8 Stouffer, *American Soldier*, vol. 11, p. 69; Leinbaugh and Campbell, *The Men of Company K*, p. xvi; Charles B. MacDonald. *Company Commander* (New York, 1978), Preface.

9 David F. Trask. *The AEF and Coalition Warmaking, 1917–18* (Lawrence, Ka, 1993).

10   Michael D. Doubler, *Closing with the Enemy: How the GIs Fought the War in Europe, 1944–1945* (Lawrence, Kansas, 1994).

11   Weigley, *Eisenhower's Lieutenants*, p. 89; George Wilson, *If You Survive* (New York, 1987), p. 41; Studs Terkel, *'The Good War': an Oral History of World War Two* (New York, 1985), p. 41.

12   Standifer, *Not in Vain*, p. 148; MacDonald, *Company Commander*, pp. 99, 246, 301–2, 362.

13   Maudlin, *Up Front*, p. 60.

14   MacDonald, *Company Commander*, p. 96.

15   Leinbaugh and Campbell, *The Men of Company K*, p. 56; Standifer, *Not in Vain*, p. 79.

16   Charles Royster, *A Revolutionary People at War: the Continental Army and American Character, 1775–83* (Chapel Hill, 1979), p. 219.

17   Ibid., p. 216.

18   Wilson, *If You Survive*, p. 10; Stephen E. Ambrose, *Band of Brothers* (New York, 1993), p. 235. Doubler also tells the story from Wilson as well as the Charles MacDonald story which follows. Compared to the Civil War, there seems to be a paucity of good ETO memoirs – MacDonald's and Leinbaugh and Campbell's are the two most often cited. Standifer's is in many ways both the most reflective and practical of the ones I've read – the one where the reader comes away knowing best what a rifleman did. The most interesting experiment in creating a memoir is Alice M. and Howard S. Hoffman. *Archives of Memory; a Soldier Recalls World War II* (Kentucky, 1990), in which the author – a historian and an ETO veteran who is also an experimental psychologist – argue for the existence of what they call 'archival memory'. This book should be better known.

19   Leo Balestri, 'Combat Command: U.S. Frontline Officers in Europe: 1942–5', a paper written for History 411: War and Society, at Princeton University, May 1992.

20   Ambrose, *Band of Brothers*, p. 109.

21   Stouffer, *The American Soldier*, vol. II, pp. 3–4.

22   Mandlin, *Up Front*, p. 11; Leinbaugh and Campbell, *The Men of Company K*, p. 19; Klaus H. Huebner, *Long Walk Through War: a Combat Doctor's Diary* (College Station, Texas, 1987), p. 196.

23   Ambrose, *Band of Brothers*, pp. 117–18; W. T. Sherman, *Memoirs of General William T. Sherman by Himself* (Bloomington, Ind, 1957), vol. 2, 127.

24   Doubler, *Closing with the Enemy*, p. 258; Stouffer, *The American Soldier*, vol. II, pp. 32–5.

25   Leinbaugh and Campbell, *The Men of Company K*, p. 305.

26   John Brown, *Draftee Division* (Lexington, Ky, 1986), p. 26, quoted in Balestri, 'Combat Command'.

27   Standifer, *Not in Vain*, p. 280.

28   Leinbaugh and Campbell, *The Men of Company K*, p. 303; Reid Mitchell, *Civil War Soldiers* (New York, 1988), p. 208; J. Glenn Gray, *The Warriors: Reflections on Men in Battle* (New York, 1967), p. 28.

29   Violet S. de Laszlo, *Psyche and Symbol: a Selection from the Writings of C. G. Jung* (New York, 1958), p. 14.

INFANTRY COMBAT

Steve Weiss

Further Reading

A Pictorial History of the 36th 'Texas' Infantry Division (Austin, 1946).
R. H. Adelman and G. Walton, *The Champagne Campaign* (Boston, 1969).
J. J. Clarke and R. R. Smith, *Riviera to the Rhine* (Washington, D.C., 1993).
L. F. Ducros, *Montagnes Ardèchoises dans la guerre III* (Valence, 1981).
A. L. Funk, *Hidden Ally* (New York, 1992).
J. P. de Lassus Saint Genies, *Combat pour le Vercors et pour la liberté* (Valence, 1984).
P. Leslie, *The Liberation of the Riviera* (New York, 1980).
V. M. Lockhart, *T-Patch to Victory* (Canyon, Texas, 1981).
J. Robichon, *The Second D-Day* (New York, 1969).
A. F. Wilt, *The French Riviera Campaign* (Carbondale, Il, 1981).

TRAUMA AND ABSENCE

Omer Bartov

This is the first part of a longer paper, versions of which were delivered in 1995–6 at the Society for French Historical Studies Annual Meeting, Emory University; the colloquium 'Traumatic Events: Historical Constructions of Violence in Modern France', Rutgers University; the conference 'The Soldier's Experience of War, 1939–1945', Edinburgh University; the European History Colloquium, Cornell University; and the History and German Departments, University of California, Berkeley. The second part of this chapter will be published in Helmut Peitsch (ed.), *European Memories of World War II* (Bergman Books, forthcoming). I would like to thank the participants in these meetings for their comments and criticism. Thanks are also due to Paul Addison, James Amelang, Yehudit and Hanoch Bartov, Angus Calder, John Erickson, Sarah Farmer, Carole Fink, Paul Holdengräber, Steven Kaplan, Jacob Meskin, Barry Strauss and most especially to Wai yee Li.

1   Jean Rouaud, *Les Champs d'honneur* (Paris, 1990); Michel Tournier, *Le Roi des Aulnes* (Paris, 1970), translated as *The Ogre* (New York, 1972).
2   Maurice Halbwachs, *On Collective Memory* (Chicago, 1992), translated from *Les Cadres sociaux de la mémoire* (Paris, 1952), and *La Topographie légendaire des évangiles en terre sainte: Etudes de mémoire collective* (Paris, 1941); see also the posthumous work *The Collective Memory* (New York, 1950). Marc Bloch, *L'Étrange Défaite: Témoignage écrit en 1940* (Paris, 1957), translated as *Strange Defeat: a Statement of Evidence Written in 1940* (New York, 1968 [1948]); Marc Bloch, *Apologie pour l'histoire ou Métier d'historien* (Paris, 1949), translated as *The Historian's Craft* (New York, 1953); and Marc Bloch, *Souvenirs de guerre, 1914–15* (1969), translated as *Memoirs of War, 1914–15*, 2d edn (Cambridge, 1988). See ibid., Introduction by Carole Fink, esp. pp. 64–7.
3   In his introduction to Halbwachs, *Collective Memory*, p. 22, Lewis A. Coser cites Halbwach's assertion that 'While the collective memory endures and draws

strength from its base in a coherent body of people, it is individuals as group members who remember.' Citing also Barry Schwartz's statement that 'the collective memory comes into view as both a cumulative and an episodic construction of the past' (p. 30), Coser concludes that Halbwachs's single most important contribution was in his understanding 'that our conceptions of the past are affected by the mental images we employ to solve present problems, so that collective memory is essentially a reconstruction of the past in the light of the present... Memory needs continuous feeding from collective sources and is sustained by social and moral props' (p. 34).

4   Even those parts of the novel published in his lifetime retained the quality of fragile, incomplete, uncertain (drafts of) memories. In December 1919 Proust wrote to Paul Souday regarding the publication of *Swann* and *Jeunes Filles*: 'La guerre m'a empêché d'avoir des épreuves; la maladie m'empêche, maintainant, de les corriger.' Marcel Proust, *A la recherche du temps perdu*, 3 vols (Paris, 1954), vol. 1, p. xxiii. Translated as *Remembrance of Things Past*, 3 vols (Harmondsworth, 1983). German novels which come to mind here include Thomas Mann's *Buddenbrooks*, first published in 1901, which laments the decline and fall of Hanseatic *Bürgertum* in anticipation of a future cataclysm, and *The Magic Mountain*, begun before the First World War but published in 1924, which paints a more disturbing, but not unsentimental picture of a lost past, and still reflects the exhilaration of 1914 and the prospects of a post-war 'brave new world'. Perhaps even more appropriate is Robert Musil's *The Man Without Qualities*, whose writing covered the whole interwar period (the first volume was published in 1930, the second in 1932, and the third only after Musil's death in 1942). A critical portrayal of the vanished civilisation of Habsburg from the perspective of an increasingly ominous interwar Europe, this vast, unfinished work has been rightly compared to Proust's similarly unfinished search for things (time) lost, lost again with the demise of the rememberer. See the following English translations: Thomas Mann, *Buddenbrooks* (New York, 1952 [1901]); Thomas Mann, *The Magic Mountain* (Harmondsworth, 1960 [1924]); Robert Musil, *The Man Without Qualities*, 3 vols (London, 1979 [1930, 1932, 1951]).

5   Günter Grass, *Die Blechtrommel* (Darmstadt, 1974 [1959]). Translated as *The Tin Drum* (Harmondsworth, 1978 [1961]).

6   I have in mind here, for instance, Siegfried Lenz's *Deutschstunde* (Hamburg, 1973), translated as *The German Lesson* (New York, 1972), and Heinrich Böll's *Billiard um halbzehn* (Munich, 1959), translated as *Billiards at Half-Past Nine* (New York, 1962). See also Judith Ryan, *The Uncompleted Past: Post-war German Novels and the Third Reich* (Detroit, 1983), chs 5 and 7; and Lawrence L. Langer, *The Holocaust and the Literary Imagination* (New Haven and London, 1975), pp. 265–84.

7   Marcel Aymé, *Uranus* (Paris, 1980 [1948]). Jean-Paul Sartre, *Les Chemins de la Liberté*, 3 vols: *L'Âge de raison* (Paris, 1945); *Le Sursis* (Paris, 1945); *La Mort dans l'âme* (Paris, 1949); translated as *The Age of Reason* (Harmondsworth, 1961); *The Reprieve* (New York, 1947); *Iron in the Soul* (Harmondsworth, 1963). Julien Gracq, *Un Balcon en forêt* (Paris, 1958), translated as *Balcony in the Forest* (New York, 1959). Marguerite Duras, *La Douleur* (Paris, 1985), translated as *The War: a Memoir* (New York, 1986). It should be pointed out, however, that Claude Simon's *La Route de Flandres* (Paris, 1960), *is* about memory and disintegration, both of physical reality and of its recollection. Yet this unique work, which can probably be rated as the

greatest French novel of the débâcle, and one of the most important modernist works of post-war European literature, is also intensely concerned with individual memory of a kind not at all unlike Tournier's. It remains to be explained why Simon's novel has received so little recognition in France. One might also mention here Jean-Paul Sartre's *Les Carnets de la drôle de guerre: Novembre 1939–Mars 1940* (Paris, 1983), which, among other things, seems to be constructing the memory of the débâcle even before it happened (a tendency characteristic of several other contemporary texts).

8 André Schwarz-Bart, *Le Dernier des justes* (Paris, 1959), translated as *The Last of the Just* (New York, 1961); Romain Gary (Émile Ajar), *L'Angoisse du roi Salomon* (Paris, 1979), translated as *King Solomon* (New York, 1983); Boris Schreiber, *La Descente au berceau* (Paris, 1984); Robert Bober, *Quoi de neuf sur la guerre* (Paris, 1993).

9 On this see in greater detail Omer Bartov, 'War, Memory, and Repression: Alexander Kluge and the Politics of Representation in Postwar Germany', *Tel Aviver Jahrbuch für deutsche Geschichte*, 23 (1994), pp. 413–32; and Omar Bartov, '"...seit die Juden weg sind": Germany, History, and Representations of Absence', in *A User's Guide to German Cultural Studies* (ed.) S. Denham, I. Kacandes and J. Petropoulos (forthcoming).

10 A good example of the resistance of French scholarship to the sensitive issues of Vichy and the post-liberation purge is the fact that these topics were first seriously treated by American historians, whose books had to wait several years before finally being translated into French. See Peter Novick, *The Resistance Versus Vichy: the Purge of Collaborators in Liberated France* (London, 1968), and Robert O. Paxton, *Vichy France: Old Guard and New Order, 1940–44*, 2nd edn (New York, 1975 [1972]).

11 For two recent works on the topic, see Atina Grossmann, 'A Question of Silence: the Rape of German Women by Occupation Soldiers', *October 72* (Spring 1995), pp. 43–63; Heide Fehrenbach, *Cinema in Democratizing Germany: Reconstructing National Identity after Hitler* (Chapel Hill and London, 1995).

12 For some interesting recent works on these issue, see, e.g., Gerhard Hirschfeld et al. (ed.), *Keiner fühlt sich hier mehr als Mensch... Erlebnis und Wirkung des Ersten Weltkriegs* (Essen, 1993); Stéphane Audoin Rouzeau, *Men at War, 1914–18: National Sentiment and Trench Journalism in France during the First World War* (Providence and Oxford, 1992); Leonard V. Smith, *Between Mutiny and Obedience: the Case of the French Fifth Infantry Division during World War I* (Princeton, 1994); Michael Jeismann, *Das Vaterland der Feinde: Studien zum nationalen Feindbegriff und Selbstverständnis in Deutschland und Frankreich, 1792–1918* (Stuttgart, 1992); Daniel Pick, *War Machine: the Rationalisation of Slaughter in the Modern Age* (New Haven and London, 1993); Margaret R. Higonnet et al. (eds.), *Behind the Lines: Gender and the Two World Wars* (New Haven and London, 1987); Mary L. Roberts, *Civilization Without Sexes: Reconstructing Gender in Postwar France, 1917–27* (Chicago and London, 1994); and the still indispensable volume, Renate Bridenthal et al. (ed.), *When Biology Became Destiny: Women in Weimar and Nazi Germany* (New York, 1984).

13 See further in Omer Bartov, *Murder in Our Midst: the Holocaust, Industrial Killing, and Representation* (New York and Oxford, 1996).

14 George Mosse, *Fallen Soldiers: Reshaping the Memory of the World Wars* (New York and Oxford, 1990).

15 K. S. Inglis, 'Entombing Unknown Soldiers: from London and Paris to Baghdad', *History & Memory*, 5/2 (1993), pp. 7–31.

16 Thus, for instance, while Georg Wilhelm Pabst's 1930 film *Westfront 1918* was a stark depiction of the First World War, Karl Ritter's 1938 *Pour le mérite* showed the betrayed heroes of the last war preparing eagerly for the next. Conversely, Abel Gance's 1919 *J'accuse*, filmed during the war, in which dead soldiers warn the living from continuing the war (some of the soldiers acting the dead were actually killed later on in battle), began a tradition of anti-war French cinema that continued all the way to Jean Renoir's 1937 *La Grande Illusion*. See, e.g., Jay W. Baird, *To Die for Germany: Heroes in the Nazi Pantheon* (Bloomington and Indiana-polis, 1992), ch. 8; David Welch, *Propaganda and the German Cinema, 1933–45*, 2nd edn (Oxford, 1990), ch. 6; Alan Williams, *Republic of Images: a History of French Film-making* (Cambridge, Mass., 1992), pp. 86–8; Alexander Sesonske, *Jean Renoir: the French Films, 1924–39* (Cambridge, Mass., 1980), pp. 282–322. Most recently, see Jay Winter, *Sites of Memory, Sites of Mourning: The Great War in European Cultural History* (Cambridge, 1995), especially pp. 15–28.

17 See in this context Jay M. Winter, 'Catastrophe and Culture: Recent Trends in the Historiography of the First World War', *Journal of Modern History*, 64 (1992), pp. 525–32.

18 Antoine Prost, *In The Wake of War: 'Les Anciens Combattants' and French Society, 1914–39* (Providence, 1992).

19 Annette Becker, *La Guerre et la foi: de la mort à la mémoire, 1914–30* (Paris, 1994); Annette Becker, 'From Death to Memory: the National Ossuaries in France after the Great War', *History & Memory*, 5/2 (1993), pp. 32–49.

20 See note 24.

21 On militarism in Israel, see, most recently, Uri Ben Eliezer, *Through the Sights: the Creation of Israeli Militarism, 1936–56* (Tel Aviv, 1995 [in Hebrew]), and his 'A Nation in Arms: State, Nation, and Militarism in Israel's First Years', *Comparative Studies in Society and History*, 37/2 (1995), pp. 264–85. The literature on religious fanaticism in Israel will have to be revised following the assassination of Prime Minister Yitzhak Rabin; currently the best sources are the media reports in Israel and elsewhere. But see, e.g., Ehud Sprinzak, *The Ascendance of Israel's Radical Right* (New York and Oxford, 1991). One can of course add here the case of the former Yugoslavia, where the memory of past atrocities (dating all the way back to the wars with the Turks) has fed the destructive momentum of the conflict.

22 In this context, see the preface written by Roland Dorgelès to the 1964 edition of his Great War novel, *Les Croix de bois*, originally published in 1919. Reminded of a visit to a meeting of German First World War veterans in Berlin shortly before the Second World War, he writes: 'Eux aussi, cinquante ans plus tard, doivent s'inter-roger sur leurs raisons de se battre et mesurer l'inutile horreur de ces égorgements. Eux aussi se demandent à cette heure, si nos deux pays sont condamnés à se haïr perpétuellement et à payer leurs ruines par la ruine du voisin. Alors je fais un effort, j'essaie de l'offrir, cette main qui résiste. Je m'oblige à croire que les hommes de mon âge ont connu les dernières guerres, que les nations de demain, enfin réconciliées, ne formeront plus qu'un grand bloc uni pour des tâches pacifiques. Je me le répète: "Tous frères!" Je l'écris pour m'en persuader.' But then he reaffirms his belief in the meaning of that war in which he fought: 'De même je reste fidèle à l'Histoire dont

les faits glorieux nous émerveillaient, et suis fier que les hommes de mon âge y aient ajouté un chapitre digne du passé', p. xii.

23 See, e.g., Volker R. Berghahn, *Der Stahlhelm, Bund der Frontsoldaten, 1918–35* (Düsseldorf, 1966); Robert G. L. Waite, *Vanguard of Nazism: the Free Corps Movement in Postwar Germany, 1918–23* (New York, 1969); James M. Diehl, *Paramilitary Politics in Weimar Germany* (Bloomington, 1977); Klaus Theweleit, *Male Fantasies*, 2 vols (Minneapolis, 1987, 1989).

24 Henri Amouroux, *Quarante millions de pétainistes, Juin 1940–Juin 1941* (Paris, 1977).

25 See further in Omer Bartov, 'Martyrs' Vengeance: Memory, Trauma, and Fear of War in France, 1918–40', *Historical Reflections / Réflexions Historiques*, 22/1 (1996), pp. 1–21.

26 See Henri Rousso, *The Vichy Syndrome: History and Memory in France since* 1944 (Cambridge, Mass., 1991).

27 *14–18: Mourir pour la patrie* (Paris, 1992). Nevertheless, note the last paragraph of the Introduction by Antoine Prost: 'Plus que les réalités, ce sont les discours de légitimation qui ont changé. L'enflure héroïque, le ton moralisateur sonnent faux désormais, et nous supportons mal les cérémonies collectives. Cela ne signifie pas que l'identité nationale soit moins forte, ni les Français moins capables de la défendre, le cas échéant. Ne regrettons pas qu'ils n'en aient pas l'occasion: le vice du patriotisme, comme du courage, c'est qu'il faille des guerres pour en révéler la profondeur.'

28 See, e.g., the review article by Sarah Fishman, 'The Power of Myth: Five Recent Works on Vichy France', *Journal of Modern History*, 67 (1995), pp. 666–73.

29 On the normalisation of the memory of the war and the attempt to associate the victims of the bombing with those of the Holocaust by way of official commemoration, see, most recently, Klaus Naumann, 'Dresdener Pietà. Eine Fallstudie zum "Gedenkjahr 1995"', *Mittelweg 36*, 4 (1995), pp. 67–81. On the penetration of Nazi terminology into the language of both left and right-wing activists, see Uli Linke, 'Murderous Fantasies: Violence, Memory, and Selfhood in Germany', *The New German Critique*, 64 (Winter 1995), pp. 37–59; and Uli Linke, 'The Politics of Memory and Denial: Holocaust Language and Street Violence in Germany', in *Responses to Violence*, ed. Allen Feldman (Chicago, forthcoming).

ORAL HISTORY AND BRITISH SOLDIERS' EXPERIENCE OF BATTLE IN THE SECOND WORLD WAR

Nigel de Lee

1 R. G. Collingwood, *The Idea of History*, ed. J. van der Dussen (Oxford, 1994), pp. 140, 245–6; Bacon, *The Works* (London, 1826), vol. 1, pp. 88–9; J. Clive, *Not By Fact Alone* (London, 1990), p xiv; E. Gibbon, *Autobiography*, ed. Lord Sheffield (London, 1907), p 1; D. Hume, *Essays, Moral, Political and Literary* (London, 1903), p. 559.

2 Collingwood, *The Idea of History*, pp. 59–60; Montaigne, *The Essays* (London, 1905), vol. 2, p. 96; E. Gibbon, *Autobiography*, pp. 74, 181–2.

3 T. Carlyle, *Critical and Miscellaneous Essays: Collected and Republished* (London, 1872), vol. 11, p. 257; Clive, *Not By Fact Alone*, p. 89.

4 L. von Ranke, *A History of England, Principally in the Seventeenth Century* (Oxford, 1875), pp v–vi; E. L. G. Stones (ed.), *F. W. Maitland, Letters to George Neilson* (Glasgow, 1976), p. 9; Gibbon, *Autobiography*, p. xiv.

5 G. M. Trevelyan, *An Autobiography, and Other Essays* (London, 1949), pp. 56, 72; Collingwood, *The Idea of History*, pp. 266, 209–39, 176, 241, 9–10; A. Marwick, *The Nature of History* (London, 1989), pp. 291, 198; J. le Goff, *History and Memory* (New York, 1992), p. xix.

6 A view taken by John Buchan, in Trevelyan, *An Autobiography*, p. 81.

7 Montaigne, *The Essays*, vol. II, p. 98; Trevelyan, *An Autobiography*, p. 81; Collingwood, *The Idea of History*, pp. 266–8.

8 Bacon, *The Works*, p. 85; Montaigne, *The Essays*, vol. I, p. 162; Hume, *Essays*, p. 558; Lord Macaulay, *The History of England* (London, 1865), vol. I, p. xxii; L von Ranke, *A History of England*, p. v; Trevelyan, *An Autobiography*, p. 65; Le Goff, *History and Memory*, p. xi; Carlyle, *Essays*, p. 256; L. Nochlin, *Realism* (London, 1990), pp. 13, 28.

9 Marwick, *The Nature of History*, p. 169; Carlyle, *Essays*, p. 256; Clive, *Not By Fact Alone*, p. 104.

10 G. Mack, *Gustave Courbet* (New York, 1984), pp. 94, 101, 162; J. E. Crawford-Flitch, in C. R. W. Nevinson, *The Great War, Fourth Year* (London, 1918), pp. 25, 8, 14.

11 R. Seddon, *A Hand Uplifted* (London, 1963), p. 71.

12 Interview in progress with Major-General M. Forrester. General Forrester served on the HQ Staffs of Western Desert Force and XIII Corps in North Africa; interview with Major J. Jephson, Imperial War Museum (hereafter IWM), 8941.

13 Colonel A. H. Burn, *The Battlefields of England* (London, 1951), pp. xi, xii.

14 Collingwood, *The Idea of History*, pp. 18–20, 25; Macaulay, *The History of England*, p. xvi; Trevelyan, *An Autobiography*, pp. 31–2.

15 Marwick, *The Nature of History*, pp. 208–10, 215; A. Seldon and J. Pappworth, *By Word of Mouth: Elite Oral History* (London, 1983), pp. 37, 42, 43, 50, 216; P. Thompson, *The Voice of the Past: Oral History* (Oxford, 1988), pp. viii, 6.

16 Collingwood, *The Idea of History*, pp. 247, 372, 489.

17 Le Goff, *History and Memory*, p. xvii.

18 Interview with Sir David Hunt. Sir David was an archaeologist, then a soldier, then a diplomat, and a historian. One is rarely so fortunate in the qualities of informants. IWM, 14232/20.

19 Nochlin, *Realism*, p. 52.

20 Le Goff, *History and Memory*, p. xi.

21 Seldon and Pappworth, *By Word of Mouth*, pp. 27–8.

22 Margaret Brooks, in ibid., p. 91.

23 R. G. Collingwood, *The Principles of Art* (Oxford, 1958), p. 243.

24 I. Evans, 'French Soldiers' Memories of the Algerian War', paper at conference, 'War and Memory in the Twentieth Century', Portsmouth University, May 1994.

25 P. Hart, 'The South Notts Hussars', paper at conference, 'War and Memory in the Twentieth Century'; Thompson, *The Voice of the Past*, p. 137; Marwick, *The Nature of History*, p. 215.

26 Interviews with: His Honour Judge Jalland, IWM, 11944/14; Major Dowson, IWM 8911/3; Major P. Lewis, IWM 10657/42; Sergeant K. Lovell, IWM 13251/

28; Dr R. Olver, IWM, 16577/4; Sir David Hunt, IWM 14232/20; General D. Tyacke, IWM, 16053/46; General M. Forrester, Brigadier R. Flood and Colonel R. Dawnay IWM, 13645/3.

27 Clive, *Not By Fact Alone*, pp. 15, 305.

28 Carlyle, *Essays*, p. 255.

29 S. Sokoloff, 'Birmingham Ex-Servicemen's Memories of World War II', paper at conference, 'War and Memory in the Twentieth Century'; C. Ginzburg, *Clues, Myths and the Historical Method* (Baltimore, 1992), p. 16.

30 Nochlin, *Realism*, p. 31.

31 von Ranke, *A History of England*, vol. vi, pp. 5, 15–17, 28–9; Montaigne, *Essays*, vol. ii, p. 97.

32 Interviews with S. Montgomery, IWM, 16426/15, and J. Court, IWM, 16059/151. Both served in 3 RTR in Normandy.

33 Sydney Jary, *18 Platoon* (Carshalton Beeches, 1987) and D. Proctor, *Section Commander* (RMAS, 1989). Both served in the Somerset Light Infantry in north-west Europe, 1944–5.

34 Interview with Sergeant K. Lovell, IWM, 13251/28.

35 E. Dinter, *Hero or Coward?* (London, 1985), p. 31.

36 Seddon, *A Hand Uplifted*, p. 71.

37 Interview with General Tyacke, IWM, 16053/47.

38 Interview in progress with Brigadier Flood.

39 Interview in progress with Major D. Crews.

40 Interview with Captain S. Montgomery, IWM, 16426/15.

41 Interview with J. Court, IWM, 16059/151.

42 Interview in progress with Major Crews.

43 Interviews with Judge Jalland, IWM, 11944/14 and Major Patrick, Dr R. Olver, IWM, 16577/4.

44 Interview with Judge Jalland, IWM, 11944/14; interview in progress with Major Crews.

45 Interviews with Sergeant-Major Sinclair, IWM 11954/23, Major P. Lewis IWM, 10657/42, Major Blake IWM, 7392/42.

46 Information from Sydney Jary, Group Captain P. Hennessy (in the ranks in the 13/18 Hussars in 1944–5), and Major J. Jephson (served in North Africa in Royal Artillery).

47 Interview with Captain Montgomery, IWM, 16426/15.

48 Interview with Corporal. J. Court, IWM, 16059/15.

49 Interview with Captain Montgomery, IWM, 16426.

50 Interview with Lord Campbell of Croy, IWM, 11955/24.

51 Interviews with Sergeant Lovell, IWM, 13251/28 and Lord Campbell of Croy, IWM, 11955/24.

52 Interviews with Sergeant Lovell, IWM, 13251/28, Judge Jalland, IWM, 11944/14 Captain Montgomery, IWM, 16426/15.

53 Interview with Sergeant Lovell, IWM, 13251/28.

54 Interview with Corporal Court, IWM, 16059/15.

55 Interview with Captain Bickersteth, IWM, 9225/3; interview in progress with Major Crews.

56 Interview with Captain Montgomery, IWM, 16426/16 and Lieutenant-Colonel Cowling, IWM, 9272/4.

57   Interview with Sergeant Lovell, IWM, 13251/28.

58   Interviews with Major Lewis, IWM, 10657/42, Lord Bramall, IWM, 12502/18, Captain Montgomery, IWM, 16426/15.

59   Interviews with Judge Jalland, IWM, 11944/14, Corporal Court, IWM, 16059/15, Dr Olver, IWM, 16577/4, Mr Wildman, IWM, 9139/3, interview in progress with Major Crews.

60   Interview with Captain Montgomery, IWM, 16426/15.

61   Interview with Lord Campbell of Croy, IWM, 11955/24.

62   Interview with Corporal Court, IWM, 16059/15.

63   Interviews with Captain Montgomery, IWM, 16426/15, Brigadier Sir Alexander Stanier, IWM, 7175/7, interview in progress with Major Crews.

64   Interview with Captain Montgomery, IWM, 16426/15, Judge Jalland, IWM, 11944/14, Brigadier Sir A. Stanier, IWM, 7175/7, General D. Tyacke IWM, 16053/42.

65   H. T. Cook and T. F. Cook, *Japan at War: an Oral History* (New York, 1992), pp. 8–13.

66   Clive, *Not By Fact Alone*, p. 76.

67   Interviews with Brigadier Vickers, IWM, 9271/5, Private A. Carter, IWM, 8910/2, Captain H. Dolphin, IWM, 8912/5, Private M. Barthorpe, IWM, 9081/4, Major G. Shields, IWM, 8333/12, Colonel Mackenzie, IWM, 8882/4, Colonel Stansfeld, IWM, 9081/3.

68   Interviews with Judge Jalland, IWM, 11944/14, Sergeant Lovell, IWM, 13251/28, Major General C. Lloyd, IWM, 7266/9, Brigadier Sir C. Dunphie, IWM, 6633/4, Brigadier F. X. Barclay, IWM, 8912/7, Colonel S. Carter, IWM, 7039/5, Captain P. Haynes, IWM, 7918/8, Major R. Linton, IWM, 8849/7, Sergeant Cossens, IWM, 14898/16, Major-General W. Fox Pitt, IWM, 7038/4, Major T. Ward, IWM, 16427/16.

69   Interviews with Captain I. Hammerton, IWM, 8939/16; Sergeant K. Lovell, IWM, 13251/28; Judge Jalland, IWM, 11944/14, Captain Bickersteth, IWM, 9225/3.

70   Interviews with Judge Jalland, IWM, 11944/14, Major P. Lewis, IWM, 10657/42 Sergeant Savin, IWM, 10424/17, Mr Carter, IWM, 8910/2, Mr Barthorpe, IWM, 9081/4.

71   Interviews with Sergeant Lovell, IWM, 13251/28, Sergeant Cossens, IWM, 14898/10.

72   Interviews with Mr Carter, IWM, 8910/2, Mr Barthorpe, IWM, 9081/4, Brigadier Scott, IWM, 8235, Major T. Ward, IWM, 16427/16; Judge Jalland, IWM, 11944/14, Captain I. Hammerton, IWM, 8939/16, Captain Montgomery, IWM, 16426/15.

73   Clive, *Not By Fact Alone*, p. 11.

74   Interview with Major-General Lloyd, IWM, 7266/9.

75   Interviews with Mr Mitchell, IWM, 9207, Mr Nash, IWM, 8656, Colonel Wyld, IWM, 7366/4.

76   Interview with Lindvig and Sveen, IWM, 16309.

77   Interview with Captain Montgomery IWM, 16426/15.

THE SOLDIER'S EXPERIENCE IN TWO WORLD WARS

Hew Strachan

1 Samuel Hynes, *A War Imagined: the First World War and English Culture* (London, 1990), pp. 300–1.

2 M. F. S. [Strachan], 'Experiences of an Interrogation Officer', *Blackwood's Magazine*, 236 (Feb 1948), p. 113.

3 For a summary of these criticisms, see Robin Prior and Trevor Wilson, 'Paul Fussell at War', *War in History*, 1 (March 1994), pp. 63–80.

4 S. L. A. Marshall, *Men Against Fire: the Problem of Battle Command in Future War* (New York, 1966), p. 150.

5 Roger J. Spiller, 'S. L. A. Marshall and the Ratio of Fire', *RUSI Journal*, 133 (Winter 1988), pp. 63–71.

6 Samuel A. Stouffer et al., *The American Soldier: Combat and Its Aftermath* (New York, 1965), p. 135.

7 Morris Janowitz and Edward A. Shils, 'Cohesion and Disintegration in the Wehrmacht in World War II', *Public Opinion Quarterly*, 12 (Summer 1948), pp. 280–315; reprinted in Morris Janowitz, *Military Conflict: Essays in the Institutional Analysis of War and Peace* (Beverly Hills, 1975).

8 John A. Lynn, *The Bayonets of the Republic: Motivation and Tactics in the Army of Revolutionary France, 1791–4* (Urbana, 1984), pp. 30–37, 163–4.

9 Guy Pedroncini, *Les Mutineries de 1917* (Paris, 1967), pp. 234–9, 304–5.

10 See M. L. Buck's review of *Between Mutiny and Obedience* in *War in History*, 3 (Jan 1996), pp. 107–19.

11 Piers Mackesy, *British Victory in Egypt, 1801: the End of Napoleon's Conquest* (London, 1995), p. 105.

12 Antony Babington, *For the Sake of Example: Capital Courts-martial, 1914–20* (London, 1983), pp. 189–90.

13 Richard Bessel, 'The Great War in German Memory: the Soldiers of the First World War, Demobilisation and Weimar Political Culture', *German History*, 6 (April 1988), p. 24; see also Wilhelm Deist, 'The Military Collapse of the German Empire: the Reality Behind the Stab-in-the-Back Myth', *War in History*, 3 (June 1996); Hew Strachan, 'The Morale of the German Army in 1917–18', in *Facing Armageddon: the First World War Experienced*, ed. Hugh Cecil and Peter Liddle (Barnsley, 1996).

14 Omer Bartov, *Hitler's Army: Soldiers, Nazis and War in the Third Reich* (Oxford, 1992), pp. 95–6; I. C. Dear (ed.), *The Oxford Companion to the Second World War* (Oxford, 1995), pp. 297, 464.

15 Samuel J. Watson, 'Religion and Combat Motive in the Confederate Armies', *Journal of Military History*, 58 (January 1994), pp. 29–55.

16 I have developed these points in 'The Morale of the German Army in 1917–18', in *Facing Armageddon*, ed. Cecil and Liddle.

17 In English Richard M. Watt, *Dare Call It Treason* (London, 1964) is the fullest version to embody this perspective. It is reworked to good effect in Leonard V. Smith, 'War and "Politics": the French Army Mutinies of 1917', *War in History*, 2 (July, 1995), pp. 180–201.

Index